Lecture Notes in Computer Science 1026

Edited by G. Goos, J. Hartmanis and J. van Leeuwen

Advisory Board: W. Brauer   D. Gries   J. Stoer

Springer
Berlin
Heidelberg
New York
Barcelona
Budapest
Hong Kong
London
Milan
Paris
Santa Clara
Singapore
Tokyo

P.S. Thiagarajan (Ed.)

Foundations of Software Technology and Theoretical Computer Science

15th Conference
Bangalore, India, December 18-20, 1995
Proceedings

 Springer

Series Editors

Gerhard Goos
Universität Karlsruhe
Vincenz-Priessnitz-Straße 3, D-76128 Karlsruhe, Germany

Juris Hartmanis
Department of Computer Science, Cornell University
4130 Upson Hall, Ithaca, NY 14853, USA

Jan van Leeuwen
Department of Computer Science, Utrecht University
Padualaan 14, 3584 CH Utrecht, The Netherlands

Volume Editor

P.S. Thiagarajan
SPIC Science Foundation, School of Mathematics
92, G.N. Chetty Road, 600 017 T. Nagar, Madras, India

Cataloging-in-Publication data applied for

Die Deutsche Bibliothek - CIP-Einheitsaufnahme

**Foundations of software technology and theoretical computer
science** : 15th conference, Bangalore, India, December 18 - 20,
1995 ; proceedings / P. S. Thiagarajan (ed.). - Berlin ;
Heidelberg ; New York ; Barcelona ; Budapest ; Hong Kong ;
London ; Milan ; Paris ; Santa Clara ; Singapore ; Tokyo :
Springer, 1995
   (Lecture notes in computer science ; Vol. 1026)
   ISBN 3-540-60692-0
NE: Thiagarajan, Pazhamaneri S. [Hrsg.]; GT

CR Subject Classification (1991): F.1-4, G.2.2, I.2.3, I.3.5

ISBN 3-540-60692-0 Springer-Verlag Berlin Heidelberg New York

Typesetting: Camera-ready by author
SPIN 10512368   06/3142 – 5 4 3 2 1 0   Printed on acid-free paper

Preface

The Foundations of Software Technology and Theoretical Computer Science (FST & TCS) conference has become a well established event in the international computer science calendar. It serves as a forum for the presentation of new results as well as an important occasion for the theory community in India to meet each other and to interact with their counterparts from abroad.

This year's conference, the 15th in the series, attracted 106 submissions from 19 countries. For the first time the programme committee included members residing outside India. I thank the programme committee members for ensuring that each paper received at least three reviews. I also thank the subreferees for contributing their time and effort to provide the reviews. My warm thanks also go to John Barretto, Madhavan Mukund and Jaikumar Radhakrishnan who put in a great deal of effort towards organizing the review process. The PC meeting took place during August 4th and 5th at the Tata Institute of Fundamental Research, Bombay.

I am grateful to the invited speakers Herbert Edelsbrunner, Mogens Nielsen, Prakash Panangaden and Mike Paterson for providing written submissions to be included in the proceedings.

This year's conference is being organized at the Indian Institute of Science, Bangalore. The conference has received valuable financial support from the following institutions: Institute for Robotics and Intelligent Systems, Bangalore, SPIC Science Foundation, Madras, SUN Microsystems, Bangalore, Tata Institute of Fundamental Research, Bombay, and Texas Instruments, Bangalore.

Finally, I am happy to thank Alfred Hofmann and the staff at Springer-Verlag for the excellent support in producing the proceedings.

Madras, October 1995   P.S. Thiagarajan.

Programme Committee:

E. Allender (Rutgers Univ., New Brunswick)
V. Arvind (IMSc., Madras)
I. Castellani (INRIA, Sophia Antipolis)
V. Chandru (IISc., Bangalore)
T.K. Dey (IIT, Kharagpur)
G. Huet (INRIA,Rocquencourt)
R. Kannan (CMU, Pittsburgh)
D. Kapur (SUNY, Albany)
H. Karnick (IIT, Kanpur)
H. Narayanan (IIT, Bombay)
C. Pandu Rangan (IIT, Madras)
A. Pnueli (Weizmann Inst., Rehovot)
S. Prasad (IIT, Delhi)
J. Radhakrishnan (TIFR, Bombay)
R. Ramanujam (IMSc., Madras)
R.K. Shyamasundar (TIFR, Bombay)
P.S. Thiagarajan **(Chair)** (SPIC Sci. Found., Madras)
V. Vazirani (IIT, Delhi)

Organizing Committee:

V. Chandru **(Chair)** (IISc., Bangalore)
P. Shankar (IISc., Bangalore)
Y.N. Srikant (IISc., Bangalore)
A. Subramanian (IISc., Bangalore)
C.E. Veni Madhavan (IISc., Bangalore)
H. Venkateswaran (CAIR, Bangalore)

List of Referees

L. Aceto
M. Agrawal
R. Alur
R. Amadio
N. Amato
P. Audebaud
L. Bachmair
V. Bafna
R. Balasubramanian
S. Bensalem
P. Bhaduri
P. Bhattacharyya
S. Biswas
D. Boneh
M. Le Borgne
G. Boudol
F. Boussinot
C. Brown
P. Caspi
G. Castagna
G.-L. Cattani
S. Chandran
R. Cleaveland
P. Codognet
H. Comon
T. Coquand
J. Courant
J.-M. Couveignes
G. Das
A. A. Diwan
U. H. Engberg
P. Enjalbert
J. Esparza
M. Farach
M. Fernandez
R. Fleischer
R. Fraer
T. Genet
K. Georgatos
R. K. Ghosh
S. K. Ghosh
J. L. Gross

R. Grosu
D. Guaspari
M. Halldorsson
S. Hawkin
M. Hennessy
J. G. Henriksen
F. Honsell
O. I. Hougaard
M. Huhn
M. Jacob
J.-P. Jouannaud
S. Kapoor
S. Khuller
A. Kiehn
M.R.K. Krishna Rao
K.N. Kumar
S. Kumar
T.-M. Kuo
Y. Laknech
C. Laneve
K. Larsen
K.S. Larsen
K. Lodaya
M. Mahajan
A. Maheshwari
O. Maler
M. Mauny
T. McCarty
S. Merz
P.D. Mosses
V. Mueller
S. Mukhopadhyay
M. Mukund
Y. Narahari
T.K. Nayak
P. Niebert
W. Niehaus
M. Nielsen
J. van Oosten
C. Palamidessi
P. Panangaden
S. Patkar

C. Paulin
B.C. Pierce
R. Platek
D. Plump
K.V.S. Prasad
C. Prehofer
N. Raja
K.V.S. Rajeev
Y.S. Ramakrishna
C.R. Ramakrishnan
I.V. Ramakrishnan
V. Raman
S. Ramesh
E. Ramos
S.S. Ravi
L. Regnier
D. Remy
S. Rengarajan
A.W. Richa
R. Robbana
D.J. Rosenkrantz
S. Saluja
D. Sangiorgi
A. Sanyal
H. Saran
V. Sassone
P. Sen
S. Sen
A. Seth
P. Sewell
N. Shah
P. Shankar
J.F. Sibeyn
R. de Simone
G. Singh
G. Sivakumar
M. Sohoni
Y.N. Srikant
E.W. Stark
K.V. Subrahmanyam
A. Subramanian
M. Subramanian

Table of Contents

Algorithms

Invited Talk

Computational Geometry

Temporal Logics and Verification Theory

XII

Looking for MUM and DAD:
Text-Text Comparisons Do Help*

Mike Paterson,[1] Shlomit Tassa,[2] Uri Zwick[2]

[1] Department of Computer Science, University of Warwick, UK,
[2] Department of Computer Science, Tel Aviv University, Israel

Abstract. It is known that about $\frac{4n}{3}$ comparisons are needed, in the worst case, to find all the occurrences of the string **aba** in a text of length n if only pattern–text comparisons are allowed. We show that if text–text comparisons can be used then only about $\frac{5n}{4}$ comparisons are required. This gives the first example in which text–text comparisons provably help.

1 Introduction

We consider one of the simplest possible string matching problems: what is the number of comparisons needed, in the worst case, to find all the occurrences of the pattern **aba** in a text of length n. Surprisingly, the exact answer to this question is not yet known.

There has been much recent research analysing the exact complexity of algorithms on strings. Particular attention has been given to comparison-based string matching algorithms. The only operations such algorithms are allowed to perform on the pattern and text characters are pairwise comparisons. To each such comparison the algorithm gets a 'yes' (the characters compared are equal) or a 'no' (the characters compared are not equal) answer.

The well-known Knuth–Morris–Pratt (KMP) algorithm [13] is a comparison-based algorithm. It uses only pattern–text comparisons and performs at most $2n$ comparisons in order to find all the occurrences of a pattern of length m in a text of length n. In contrast, the also well-known Boyer–Moore (BM) algorithm [3] is not comparison-based as it uses the pattern and text characters as indices to tables. A comparison-based variant of the BM algorithm was shown by Cole [4] to perform at most $3n + o(n)$ comparisons.

Apostolico and Crochemore [1] and Colussi [5] designed variants of the KMP algorithm requiring at most $\frac{3n}{2}$ comparisons. Galil and Giancarlo [11] improved their results and obtained an algorithm that performs at most $\frac{4n}{3}$ comparisons. All these algorithm use only pattern–text comparisons. Furthermore, they are *on-line* algorithms, i.e., they only access the text through a sliding window whose size is equal to the length of the pattern.

---

* Supported in part by the ESPRIT Basic Research Action Programme of the EC under contract No. 7141 (project ALCOM II).

Breslauer and Galil [2] and Cole and Hariharan [6] showed that the string matching problem actually becomes easier as the length of the pattern increases. Breslauer and Galil [2] developed an algorithm that performs at most $(1+O(\frac{\log m}{m}))n$ character comparisons for texts of length n and patterns of length m. Cole and Hariharan [6] improved this by obtaining an algorithm that performs at most $(1+\frac{8}{3(m+1)})n$ comparisons for patterns of length m. It is therefore interesting to study the complexity of finding the occurrences of particular short patterns. In particular, **aba** (or MUM or DAD) is probably the hardest pattern to find.

Galil and Giancarlo [10] and Cole et al. [7] proved lower bounds on the number of comparisons that comparison-based string matching algorithms must perform in the worst case. In particular, Galil and Giancarlo showed that any on-line algorithm must perform, in the worst case, at least $\frac{4n}{3}$ comparisons to find all occurrences of **aba** in a text of length n. Cole et al. showed that $\frac{4n}{3}$ comparisons are also necessary for *off-line* algorithms, i.e., algorithms that have general random access to the text and pattern, provided that only pattern–text comparisons are allowed. These two lower bounds are easily seen to be tight.

In this work we show that the $\frac{4n}{3}$ lower bound fails to hold if text–text comparisons are allowed. We describe an off-line algorithm that finds all occurrences of **aba** in a text of length n using at most $\frac{5n}{4}$ comparisons. This exhibits the first example in which text–text comparisons provably help. However, the best lower bound known is that of $\frac{6n}{5}$ obtained by Cole et al. [7], so the exact complexity of the problem is not yet resolved.

2 Notation and Preliminaries

We consider algorithms which access the text t and the pattern w only through queries of the form "$w[i] = t[j]$?" or "$t[i] = t[j]$?", and measure the complexity in terms of the number of such comparisons required in the worst case.

It can be shown that for any string w there is a constant $c(w)$ such that the worst-case comparison complexity of finding all occurrences of the pattern w in a text of length n is $c(w)n + o(n)$. We define also the constant $c^*(w)$ corresponding to algorithms which can make only pattern–text comparisons, i.e., those of the form "$w[i] = t[j]$?".

These constants are for unrestricted *off-line* algorithms which have random access to all the characters of the text. We will also consider *on-line* algorithms, where access to the text is only through a sliding window of size $|w|$. The window is moved from left to right and the algorithm is only allowed to slide the window past a text position when it has already reported whether or not an occurrence of the pattern starts at that text position. The corresponding constants for on-line algorithms are denoted by $c_0(w)$ and $c_0^*(w)$ respectively. Algorithms using larger windows are also of interest. We call these *window* or *finite look-ahead* algorithms. If the window is of size $|w| + k$, the corresponding two constants will be denoted by $c_k(w)$ and $c_k^*(w)$.

Suppose $|w| = m$. We say that z $(1 \leq z \leq m)$ is *a period* of w if and only if $w[i] = w[i + z]$ for every $1 \leq i \leq m - z$. Let z_1 be the minimal period of w. (A minimal period exists since m is always a period of w.) Let z_2 be the minimal period of w which is not divisible by z_1. If such a second period does not exist we set $z_2 = \infty$. We call z_1 *the period* of w and z_2 *the second period* of w.

We consider text strings over the (finite or infinite) alphabet Σ which contains the symbols **a**, **b**, and at least one other symbol. In this paper we make extensive use of bounds proved in [7], but focus our attention on patterns of the form $\mathbf{a}^k\mathbf{ba}^\ell$, and even more on the single pattern **aba** in particular (as suggested by the title of the paper). Determining the complexity of looking for **aba** already presents difficult combinatorial problems. It also seems likely that **aba** is the hardest pattern to find!

3 Lower Bounds

The following lower bounds are proved in [7].

Proposition 1. *If w is a string and z_1, z_2 are its first and second periods then:*

1. $c(w) \geq 1 + \frac{1}{z_1 + z_2}$;
2. *if $2z_2 - z_1 \leq |w|$ then $c(w) \geq 1 + \frac{1}{z_2}$;*
3. $c_0(w) \geq 1 + \frac{1}{z_2}$.

Fact. For $w = \mathbf{a}^k\mathbf{ba}^\ell$, where $k, \ell \geq 1$, we have $z_1 = \max\{k, \ell\} + 1$ and $z_2 = \max\{k, \ell\} + 2$.

Corollary 2. *If $p = \max\{k, \ell\}$ and $q = \min\{k, \ell\} \geq 1$ then:*

1. $c(\mathbf{a}^k\mathbf{ba}^\ell) \geq 1 + \frac{1}{2p+3}$;
2. *if $q \geq 2$ then $c(\mathbf{a}^k\mathbf{ba}^\ell) \geq 1 + \frac{1}{p+2}$;*
3. $c_0(\mathbf{a}^k\mathbf{ba}^\ell) \geq 1 + \frac{1}{p+2}$.

For small look-ahead we have a new generalization of Proposition 1.

Theorem 3. *If z_1 and z_2 are the first and second periods of w, and $r = \min\{z_1, z_2 - z_1\}$, then $c_r(w) \geq 1 + \frac{1}{z_1 + z_2 - r} = 1 + \frac{1}{\max\{2z_1, z_2\}}$.*

Proof. The Adversary will construct stage-by-stage a text string which is covered by (usually overlapping) occurrences of w. In such cases the algorithm requires at least n 'yes' responses for a text of length n (see Lemma 3.1 [7]). We will show that, even if an algorithm can take advantage of the restricted strategy employed by the Adversary, it will still be forced to get a 'no' response for each shift of the window by at most $\max\{2z_1, z_2\}$ steps.

Suppose that the algorithm has just found an occurrence of w and has then been able to move its window z_1 places right. The Adversary will now guarantee that the next part of the text consists of either (a) two overlapping occurrences of w displaced by z_1 and $2z_1$ to the right relative to the found occurrence, or else (b) an occurrence of w displaced by distance z_2 to the right. Since the window size is only $|w| + \min\{z_1, z_2 - z_1\}$, the algorithm can access nothing to the right of these alternative continuations until it has decided whether or not case (a) holds.

As long as the algorithm asks questions whose answer is the same for each of these two continuations, this common answer will be given. At some stage the algorithm must ask a question that differentiates between the two continuations. The answer to this question will be 'no'. The algorithm may now take advantage of the guarantee from the Adversary and verify the continuation (a) or (b). As a result, the algorithm finds two or one occurrences of w respectively and achieves a shift of at most $\max\{2z_1, z_2\}$ for the price of one 'no' response. ☐

Corollary 4. *If $w = \mathsf{a}^k \mathsf{ba}^\ell$, $p = \max\{k, \ell\}$ and $q = \min\{k, \ell\} \geq 1$ then $c_1(w) \geq 1 + \frac{1}{2p+2}$.*

4 Previous Upper Bounds for $\mathsf{a}^k \mathsf{ba}^\ell$

To represent the state of knowledge about a text string during the course of a window algorithm we shall use the *information string* notation introduced in [7]. The text in the window is represented by a string over the alphabet $\{0, \mathsf{a}, \mathsf{b}, \mathsf{A}, \mathsf{B}\}$, where a 0 indicates that no information is recorded about that position, a or b indicates that the character is known to be a or b respectively, while A or B means that the character is known *not* to be an a or b respectively. Further information may be added in other ways. A pattern–text query will be called an a-*query* or a b-*query* according to whether the $w[i]$ used is a or b. We sometimes point out the next position to be queried in an algorithm by underlining that position in the information string: a single underline for an a-query, a double underline for a b-query.

A very simple on-line algorithm for finding occurrences of $\mathsf{a}^k \mathsf{ba}^\ell$ was given in [7]. It makes a sequence of 'a?' and 'b?' queries. In the case of a mismatch or if an occurrence of the pattern has been verified, the window is shifted to the next position at which a pattern occurrence is possible. The information string representing the state before each step has the form uxv, where $u \in \{0, \mathsf{a}\}^k$, $v \in \{0, \mathsf{a}\}^\ell$, and $x \in \{0, \mathsf{A}, \mathsf{b}\}$. The state can be written in the specified form because, after any necessary window shift, the information string must be consistent with the pattern, and we *choose to forget* any negative information represented by 'B'. The $(k+1)$-st position in the window is called the b-*position* and all the others a-*positions*. An a-position is always queried for an a. A b-position is always queried for a b. The algorithm is described below.

```
REPEAT
      IF there is some 0 in the information string
          THEN query the rightmost 0
      ELSE {x = A} query the b-position
UNTIL text string exhausted.
```

In [7] we analyse the worst-case complexity of this on-line algorithm and derive the following result.

Proposition 5. *For every* $k, \ell \geq 1$ *we have* $c_0^*(\mathbf{a}^k \mathbf{b} \mathbf{a}^\ell) \leq 1 + \frac{1}{\max\{k,\ell\}+2}$.

With Corollary 2(2), this yields tight bounds, provided $\min\{k, \ell\} \geq 2$.

Corollary 6. *If* $w = \mathbf{a}^k \mathbf{b} \mathbf{a}^\ell$, $p = \max\{k, \ell\}$, *and* $\min\{k, \ell\} \geq 2$ *then* $c(w) = c^*(w) = c_r(w) = c_r^*(w) = 1 + \frac{1}{p+2}$ *for every* $r \geq 0$.

Thus, when $\min\{k, \ell\} \geq 2$, an on-line algorithm using only pattern–text comparisons performs as well as any arbitrary off-line algorithm. However, when $\min\{k, \ell\} = 1$, a potential gap emerges. From the above proposition and Corollary 2, we have the following corollary.

Corollary 7. *If* $w = \mathbf{a}^k \mathbf{b} \mathbf{a}^\ell$, $p = \max\{k, \ell\}$, *and* $\min\{k, \ell\} = 1$ *then* $c_0(w) = c_0^*(w) = 1 + \frac{1}{p+2} \geq c(w) \geq 1 + \frac{1}{2p+3}$.

For the pattern **abaa**, we were able to improve on the above algorithm by using a window of size 8. This shows that look-ahead is indeed helpful.

Proposition 8 [7]. $c_0^*(\mathbf{abaa}) = c_0(\mathbf{abaa}) = \frac{5}{4} > c_4^*(\mathbf{abaa}) = c_4(\mathbf{abaa}) = c(\mathbf{abaa}) = \frac{8}{7}$.

5 Looking for aba

In the remainder of the paper we focus entirely on the pattern string **aba**. We shall see that certain problems or sets of problems are *equivalent*, in the sense that any optimal off-line algorithm for the one can be trivially transformed into an algorithm for the other, which has an isomorphic decision tree and so uses the same numbers of comparisons in corresponding cases. We note without proof some obvious equivalences arising from the special nature of the pattern **aba**.

Lemma 9. *Let* α *and* β *be arbitrary strings over the information alphabet* $\{0, \mathbf{a}, \mathbf{b}, \mathbf{A}, \mathbf{B}\}$.

(i) *The problem* $\mathbf{A}\alpha$ *is equivalent to the problem* α.
(ii) *The problem* $\mathbf{B}\alpha$ *is equivalent to the problem* 0α.
(iii) *The problem* α *is equivalent to the problem* α^R, *where* α^R *denotes the reversal of the string* α.
(iv) *The problem* $\alpha \mathbf{a} \beta$ *is equivalent to the pair of problems* $\alpha \mathbf{a}$ *and* $\mathbf{a}\beta$.
(v) *The problem* $\alpha \mathbf{A} \mathbf{A} \beta$ *is equivalent to the pair of problems* α *and* β.
(vi) *The problem* $\alpha \mathbf{B} \mathbf{B} \beta$ *is equivalent to the pair of problems* $\alpha 0$ *and* 0β.

Thus for example the problem $\mathbf{a00AA00a000a}$ is equivalent to the set of three problems $\{\mathbf{a000a}, \mathbf{a00}, \mathbf{a00}\}$.

Using such decompositions, we derived a lower bound of 4/3 for $c^*(\mathbf{aba})$.

Proposition 10 [7]. $c_0(\mathbf{aba}) = c_0^*(\mathbf{aba}) = c^*(\mathbf{aba}) = \frac{4}{3} \geq c(\mathbf{aba}) \geq \frac{6}{5}$.

Our main results in the next two sections will show that, for the pattern **aba**, text–text comparisons do indeed help.

6   A Finite Configuration

We show in Figure 1 the number of comparisons which are sufficient and necessary in the worst case to solve some small problems when *only pattern-text comparisons* are used. These numbers can be determined by careful analysis considering at each stage the set of possible remaining places for occurrences of **aba**. In some cases we have indicated one possible next comparison in an optimal algorithm.

configurationt	comparisons		configuration	comparisons
a00	2		a0Ba	2
a000	3		a0Aa	2
a0000	4		a00Aa	3
a0a	1		a00A00a	5
a00a	3		a00B00a	6
a000a	4		a00000a	7

Fig. 1. Small configurations and their individual costs using pattern-text comparisons.

In addition, it is not difficult to show by induction the following result for a simple family of texts.

Theorem 11. *Let $T(n)$ be the cost of finding* **aba** *'s in the configuration* **a0^na** *using pattern-text comparisons only. Then*

$$T(n) = \left\lfloor \frac{4n+1}{3} \right\rfloor .$$

When text-text comparisons are allowed the difficulty of analysing small configurations increases. We used a computer to extend our search and soon discovered, to our surprise, that only six comparisons are now needed for the text string **a00000a**; an algorithm is shown in Figure 2. The first step is a comparison between the second and fourth positions of the text string. The left and right branches of the decision tree show the continuation according to whether the result is positive or negative respectively. On the left the next step is an **a**-query on the second text position; on the right, a **b**-query on the same position. The leaf configurations show the remaining queries to be made. For the leftmost leaf, we see from Figure 1 that three comparisons suffice for **a00a**. At the third leaf, the notation **C** means "not **a** and not **b**", so the leaf configuration is equivalent to **00a** (or **a00**).

From Figure 1 it can be seen that seven comparisons are needed for the text **a00000a** with the pattern **aba** in the worst case when text-text comparisons are forbidden. This gives the smallest concrete situation that we know in which text-text comparisons do help.

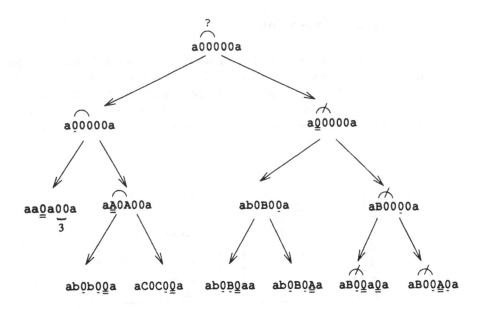

Fig. 2. Algorithm for the text string a00000a.

7 Showing That $c(\text{aba}) < c^*(\text{aba})$

Having seen the computer-generated algorithm for **a00000a**, we were able to take advantage of the ingenious first (text–text) comparison used there to design a more efficient general **aba**-finding algorithm.

Theorem 12. $c(\text{aba}) \leq c_1(\text{aba}) = \frac{5}{4} < c^*(\text{aba})$.

Proof. The lower bound, $c_1(\text{aba}) \geq \frac{5}{4}$, is given by Corollary 4. The upper bound is established by the algorithm presented in Figure 3 in the form of a decision tree. The configurations are information strings for an algorithm with a window of size four. Each leaf of the tree is one of the configurations **0000** or **a000**, and represents a return to the root of the tree or its left successor. After the result of each query, the window is shifted to the right where possible. A label $x{:}y$ on an arc indicates that x (obvious) queries are made and the window is shifted right by y places. (An unlabelled arc corresponds to the default 1:0.) For example, consider the root query **0000**. If the result is positive the information string is **00a0**, and after the two queries **0 0a0** the window can be shifted two places to the right.

In the diagram the arcs corresponding to the worst-case executions of the algorithm are shown as heavy lines. For executions following these arcs the average number of window shifts per query is 4/5, which establishes the claimed upper bound: $c_1(\text{aba}) \leq 5/4$. □

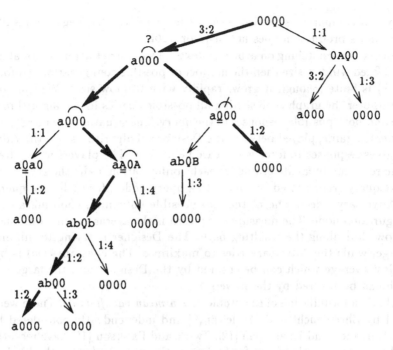

Fig. 3. Algorithm showing $c(\text{aba}) \leq 5/4$.

8   Further Results

As we have seen in the two previous sections, computer assistance has been valuable in searching for efficient pattern-matching algorithms. We have investigated the complexity of some short text strings as well as window algorithms with small window size. In this section we report on some of these results. Work is still continuing.

For finding **aba**'s in fixed text strings, we can try to establish the complexity by exploring the full tree of possible algorithms, taking advantage of reductions, decompositions and equivalences where possible.

We would like to know whether **a00000a** is the smallest problem for which text–text comparisons are helpful. An interesting extension of this kind of question is to consider *sets of* configurations. When text–text comparisons are permitted, it may no longer be true that the query complexity of a set of configurations is equal to the sum of the complexities of the individual configurations. There could be the possibility of economies in *mass production*, i.e., solving ℓ disjoint pattern matching problems might sometimes require fewer than ℓ times the number of comparisons needed to solve each of these problems separately. (A similar phenomenon has been usefully employed in the context of selection algorithms over ordered sets. Mass production of partial orders was used by Schönhage, Paterson and Pippenger [14], and recently by Dor and Zwick [8], to reduce the number of comparisons needed to find the median.) One negative result so far is that mass-production does not help for **a00a**.

For pattern-matching on arbitrary texts, if we restrict attention to algorithms with a fixed window size then the number of possible configurations (information strings) is finite (though it grows rapidly with window size). We can construct by computer the graph representing all possible queries for an algorithm at each stage and both possible responses to the query. The situation can be regarded as a two-person game, played between the Designer of algorithms and the Adversary, who gives responses to force worst-case behaviour. The players move alternately on the resulting bipartite graph: for each configuration node, the Designer selects a next query (represented by an arc to a query node); for each such query node, the Adversary selects one of the two possible outcomes, choosing an arc to a configuration node. The measure of concern is the average number of queries per window shift along the resulting path. The Designer is trying to minimise this average, while the Adversary tries to maximise. The constant $c(w)$ is both the smallest average which can be ensured by the Designer and the largest average which can be ensured by the Adversary.

This is a formulation of the problem as a *mean payoff game*. These were introduced by Ehrenfeucht and Mycielski [9] and independently considered by Gurvich, Karzanov and Khachiyan [12]. Zwick and Paterson [15] have recently given a polynomial-time algorithm for finding optimal strategies for the kind of mean payoff games arising from pattern-matching (though only pseudo-polynomial for the most general formulation). Using this kind of algorithm we have obtained the following lower bound on algorithms for **aba**-finding with a window of size 5.

Theorem 13. $c_2(\mathbf{aba}) \geq \frac{5}{4}$.

Corollary 14. $c_0(\mathbf{aba}) = \frac{4}{3} > c_1(\mathbf{aba}) = c_2(\mathbf{aba}) = \frac{5}{4} \geq c(\mathbf{aba}) \geq \frac{6}{5}$.

9 Conclusion and Open Problems

We have reviewed some recent results on comparison-based algorithms for pattern matching. In this paper we focussed particularly on the single pattern string **aba**, which already offers challenging problems. This string also seems to the hardest pattern to find, though this is not yet proven. An unexpected outcome of our research is that comparisons between pairs of positions in the text (rather than just those between pattern and text) are required for optimal algorithms.

We know that text–text comparisons help for finding **aba** in the text string **a00000a**, but do not know if this is the simplest such example. A more fundamental open problem is whether text–text comparisons give rise to economies using mass-production (see Section 8).

An interesting structural question concerning window algorithms is whether there is an infinite hierarchy with window size for any fixed pattern. Is there some pattern w, perhaps even **aba**, such that for every k there exists a k' for which $c_k(w) < c_{k'}(w) < c(w)$?

For any w, $c_k(w)$ and $c_k^*(w)$ must be rational numbers for all k, but we do not know whether this is also true for $c(w)$ and $c^*(w)$.

Is it always the case that $c_0(w) = c_0(w^R)$ and $c_0^*(w) = c_0^*(w^R)$, where w^R is the reverse of w? Clearly this holds for c and c^*.

The most immediate open problems are to determine the exact value of $c(\text{aba})$ and to see whether **aba** is indeed the hardest pattern to find.

References

1. A. Apostolico and M. Crochemore, *Optimal canonization of all substrings of a string*, Information and Computation, 95 (1991), pp. 76–95.
2. D. Breslauer and Z. Galil, *Efficient comparison based string matching*, Journal of Complexity, 9 (1993), pp. 339–365.
3. R. Boyer and S. Moore, *A fast string matching algorithm*, CACM, 20 (1977), pp. 762–772.
4. R. Cole, *Tight bounds on the complexity of the Boyer-Moore algorithm*, SIAM J. on Computing, 23 (1994), pp. 1075–1091.
5. L. Colussi, *Correctness and efficiency of pattern matching algorithms*, Information and Computation, 95 (1991), pp. 225–251.
6. R. Cole and R. Hariharan, *Tighter upper bounds on the comparison complexity of string matching*, in preparation. A preliminary version appeared in Proceedings of the 33rd Annual IEEE Symposium on the Foundations of Computer Science, (1992), pp. 600–609.
7. R. Cole, R. Hariharan, M. Paterson, U. Zwick, *Tighter lower bounds on the exact complexity of string matching*, SIAM J. on Computing, 24 (1995), pp. 30–45.
8. D. Dor and U. Zwick, *Selecting the median*, Proceedings of the 6th Annual ACM-SIAM Symposium on Discrete Algorithms, (1995), pp. 28–37.
9. A. Ehrenfeucht and J. Mycielski, *Positional strategies for mean payoff games*, International Journal of Game Theory, 8 (1979), pp. 109–1139.
10. Z. Galil and R. Giancarlo, *On the Exact Complexity of String Matching: Lower Bounds*, SIAM J. on Computing, 6 (1991), pp. 1008–1020.
11. Z. Galil and R. Giancarlo, *On the Exact Complexity of String Matching: Upper Bounds*, SIAM J. on Computing, 3 (1992), pp. 407–437.
12. V. Gurvich, A. Karzanov, L. Khachiyan, *Cyclic games and an algorithm to find minimax cycle means in directed graphs*, USSR Computational Mathematics and Mathematical Physics, 28 (1988), pp. 85–91.
13. D.E. Knuth, J. Morris, V. Pratt, *Fast pattern matching in strings*, SIAM J. on Computing, 6 (1977), pp. 323–350.
14. A. Schönhage, M. Paterson, N. Pippenger, *Finding the median*, J. of Computer and System Sciences, 13 (1976), pp. 184–199.
15. U. Zwick and M. Paterson, *The complexity of mean payoff games*, Computing and Combinatorics (COCOON '95), LNCS 959 (Springer-Verlag 1995), pp. 1–10.

Branch and Bound on the Network Model

Sanjay Jain[1]

Department of Information Systems and Computer Science
Lower Kent Ridge Road
National University of Singapore
Singapore 0511, Republic of Singapore
Email: sanjay@iscs.nus.sg

Abstract. Karp and Zhang developed a general randomized parallel algorithm for solving branch and bound problems. They showed that with high probability their algorithm attained optimal speedup within a constant factor (for $p \leq n/(\log n)^c$, where p is the number of processors, n is the "size" of the problem, and c is a constant). Ranade later simplified the analysis and obtained a better processor bound. Karp and Zhang's algorithm works on models of computation where communication cost is constant. The present paper considers the Branch and Bound problem on networks where the communication cost is high.

Suppose communication in a p processor network takes $G = O(\log p)$ time and node expansion (defined below) takes unit time (other operations being free). Then a simple randomized algorithm is presented which is, asymptotically, nearly optimal for $p = O(2^{\log^c n})$, where c is any constant $< 1/3$ and n is the number of nodes in the input tree with cost no greater than the cost of the optimal leaf in the tree.

1 Introduction

Branch and Bound algorithms are frequently used for solving optimization problems. Because of the importance of several such optimization problems, and the fact that most of these jobs are computation intensive, it is useful to look at parallelizing such algorithms.

Karp and Zhang [KZ88, KZ93] presented a simple randomized optimal algorithm for such problems. They showed that the algorithm was optimally fast (upto a constant multiplicative factor) on every problem instance with high probability. Their analysis was later simplified and improved by Ranade [Ran90]. Karp and Zhang (Ranade) assumed a constant time communication between the processors (with local memory only). However, in reality networks are not fully connected and communication cost could go up depending on the number of processors. We consider Branch and Bound on a model in which communication cost depends on the number of processors. In the next section we present the generic Branch and Bound problem and our model. In section 3 we present our algorithm. Sections 4 to 6 contain a sketch of the analysis of our algorithm.

2 Model

Karp and Zhang model a generic Branch and Bound problem as follows. A tree H with costs associated with each node is given. The cost function has the property that if v is the parent of w then $\text{Cost}(v) < \text{Cost}(w)$. The goal is to find a least cost leaf. For simplicity we assume that all costs are different. The input to the algorithm is the root of H, and the other nodes are generated during execution using a procedure called node expansion. When this procedure is applied to a node v, it either determines that v is a leaf, or it generates the children of v along with the associated costs. A node can be expanded only if it is a root or it has been generated by expanding its parent earlier. We assume that the degree of nodes of H are bounded by a constant. Specifically, we assume that H is a binary tree, even though our analysis works for any b-ary tree. In most applications the degree of the branch and bound tree can be made constant by increasing the height by at most a logarithmic factor. Let $\hat{H}$ denote the subtree of H formed using nodes with cost no greater than the cost of the minimal cost leaf. Let n denote the number of nodes in $\hat{H}$ and h denote the height of $\hat{H}$. It is easy to see that any sequential algorithm for branch and bound must do at least n node expansions, and that any parallel algorithm must take at least $\Omega(n/p+h)$ steps, where p is the number of processors used. In a parallel computation model with communication cost being constant, Karp and Zhang [KZ88, KZ93] gave an algorithm which achieves the above lower bound for $p < n/(\log n)^k$, for some constant k (Ranade [Ran90] simplified the analysis and improved the processor bound to $n/\log n$).

We use the following model of parallel computation (see [BNK92, CKP+93, KSSS93] for similar models). There are p processors numbered from 1 to p. Each processor has local memory and there is no global memory. There is no global control but the processors operate synchronously. We assume that the processors have capability to do independent random coin tosses. Expanding a node takes one unit of time, sending a message takes G units of time, and receiving a message is free. Thus, a message sent by processor P_i to processor P_j at time t will be received by processor P_j at time $t + G$ and processor P_i is unavailable for doing any work between time t and $t + G$. In any message, a processor can transmit at most one node of the tree H.

In our model, as in the model of Karp and Zhang and that of Ranade, we charge a processor just for node expansions (1 unit) and communication. Everything else is free. This is based on the assumption that in real applications cost of communication and node expansion would dominate.

Karp and Zhang's algorithm (as modified by Ranade) is essentially the algorithm given below for the case when $G = 1$. Our main result is that for communication time logarithmic in the number of processors, branch and bound can be implemented in nearly optimal randomized time, for processor bound of $O(2^{(\log n)^c})$ (for $c < 1/3$).

3 Branch and Bound Procedure

The following algorithm is a local best first algorithm with a bias towards children remaining in the processor where their parents are expanded. Initially root is present at processor 1. Each processor maintains two data structures: (1) a local priority queue of unexpanded nodes, and (2) a bound LCLD on the least cost leaf discovered.

Let LCLD = ∞.
Loop
1. Node Expansion Phase
Repeatedly execute the following for a total time usage of $G \cdot \log p$ steps.
begin
    Expand the least cost unexpanded node, v, in the local queue.
    If v is a leaf then let LCLD = $\min(\{\text{LCLD}, \text{Cost}(v)\})$.
    Otherwise, for each child w of v:
        With probability $1/G$, transmit w to a random processor.
        With probability $1 - 1/G$, add w to the local queue.
    Update the local queue with any arrivals from outside.
end
2. Termination Detection Phase
    Find the minimum, m, of LCLD on all the processors.
    Broadcast m to all the processors.
    Halt if no processor has any unexpanded node with weight less than m.
forever

Note that the termination detection phase can be executed in $O(G \log p)$ steps. Thus, from now on we assume that the processors do not execute the Termination Detection Phase but automatically know when to stop. (This would only affect the run time by a constant multiplicative factor and an additive factor of $O(G \log p)$). Let n denote the number of nodes in $\hat{H}$. Note that no processor expands any node in $H - \hat{H}$, if it has some node in $\hat{H}$ left to expand. Clearly, only the expansion of nodes in $\hat{H}$ can affect the performance of the algorithm. Thus, for our analysis we assume that $H = \hat{H}$.

We now show that for an arbitrary tree H, for certain bounds on p and G ($p = 2^{\log^{1/4} n}$ and $G = \log p$ satisfy our constraints), the time taken before the least cost leaf is expanded is bounded by $O(n/p + h \cdot \alpha)$, where α is determined below. It should be noted that for the analysis below we assume that n is large enough.

4 Analysis of the Algorithm

We first prove the following lemma which is useful in bounding the number of nodes that could potentially get stuck in the same processor due to a single node arrival. Size of a tree denotes the number of nodes in the tree.

Lemma 1. *Suppose a tree T is given with $l \leq n$ nodes. Let F_T denote a forest obtained when each edge of T is retained with probability $1 - 1/G$. Let $P_T(w)$ denote the probability that the tree in F_T containing the root of T has size at least w. Suppose r is given such that $G \log^2 n \leq r \leq l$. Let $B(l, r) = O(\frac{l \log l}{r} + l \cdot e^{-\log l/(8G \log r)})$ (where the multiplicative constant is determined from the analysis below and is independent of r, l and G). If T is a binary tree, then $P_T(B(l, r)) \leq n^{-c \log n}$ for some constant c.*

Proof. For the sake of the following argument, we will first transform T into a tree T' (which may have a large degree) such that, for all w, $P_T(w) \leq P_{T'}(w)$. Then we will prove the bound on $P_{T'}(B(l, r))$. It is easy to observe that if v_1 is an ancestor of v_2 in T_1 then, if T_1' is obtained from T_1 by deleting the edge from parent(v_2) to v_2 and adding v_2 as a child of v_1, then $P_{T_1}(w) \leq P_{T_1'}(w)$. Now consider the following procedure, Balance(T_1, v).

Balance(T_1, v)
(Here v is a node of tree T_1.)
    Transform the tree T_1 according to the following operations.
    Let s be the size of the subtree rooted at v.
    while there exists a node v', such that
        (i) v' is a proper descendant, but not a child of v,
        (ii) subtree rooted at v' is of size at least s/r, and
        (iii) none of the subtrees rooted at a proper descendant of v' has size
            at least s/r.
    Delete the edge from parent(v') to v' and make v' a child of v.
    end
end

Note that the above construction makes the node v nearly "balanced." If the subtree rooted at v was originally a binary tree with s nodes, then after the execution of the above procedure, v has a degree between $\min(\{r/3, (s - 1)\})$ and $r + 2$. Moreover, subtrees rooted at each child of v have size between s/r and $\max(\{2s/r, 1\})$, except possibly, for 2 children (which have subtrees of size less than s/r). Note that the above construction does not increase the number of children for any of the proper descendants of v.

We convert the tree T to T' by repeatedly using the above procedure for balancing the nodes (in a top down fashion, starting from the root). In the tree T' so formed, mark every node v such that the subtree rooted at v is of size less than $1/r$ of the size of the subtree rooted at parent(v).

Thus, at the end of the process we will have a tree T' that has the following four properties.

(1) Suppose the size of the subtree rooted at v is s. Then v has between $\min(\{r/3, s - 1\})$ and $r + 2$ children.
(2) Each node has at most two marked children. Suppose v is a root of a subtree with size s. Then the size of the subtree rooted at each unmarked child of v

is between s/r and $\max(\{2s/r, 1\})$. On the other hand the size of the subtree rooted at each marked child of v is $< s/r$.

(3) For any node with size s, the sum of sizes of its marked children, if any, is bounded by $2s/r$. Thus at any level of the tree T', the sum of sizes of the marked nodes is bounded by $2l/r$. Also, since the size of any node is at most half of its parent (we assume n is large enough so that $r > 4$), we can conclude that the depth of the tree T' is $\leq \log l$. Thus the total number of nodes which have a marked ancestor is at most $\frac{2l(\log l)}{r}$.

(4) For each w, $P_{T'}(w) \geq P_T(w)$.

We now bound $P_{T'}(B(l, r))$. Consider a forest $F_{T'}$ formed by keeping each edge of T' with probability $1 - 1/G$.

Since the size of any unmarked child is at least $\frac{1}{r}$ fraction of its parent's size, we have that all unmarked leaves of T' are at depth at least $\frac{\log l}{\log r}$.

For any node in T', which is a root of a subtree of size at least r, the probability that greater than $1 - 1/(2G)$ fraction of its unmarked children are still its children in $F_{T'}$ is bounded by $2^{-c_1 \cdot r/G}$ (for some constant c_1, by Chernoff bounds). Thus, with probability at least $1 - l \cdot n^{-c_1 \log n}$, every node in T', which is a root of a subtree of size at least r in T', has at most $(1 - \frac{1}{2G})$ fraction of its unmarked children in T' as its children in $F_{T'}$ (and thus the sum of sizes of subtrees rooted at unmarked children of the node is at most $(1 - \frac{1}{4G})$ fraction of the size of the node).

Thus, with probability at least $1 - n^{-c \log n}$, the size of the tree in $F_{T'}$ containing the root of T' is at most $O(\frac{l \log l}{r} + [l(1 - \frac{1}{4G})^{\frac{\log l/2}{\log r}}])$, which is $O(\frac{l \log l}{r} + l \cdot e^{-(\log l)/(8G \log r)}) = B(l, r)$. ∎

5 Modified Node Expansion Phase

We now consider the following modification to the Node Expansion Phase.

First, each node in H is ticked with probability $1/G$.

Node expansion Phase

Loop

Expand the least cost node, v, in the local queue.

If any of v's children is ticked, then transmit it to a random processor. Unticked children are retained in the local queue.

Update the local queue with any arrivals.

end

(We dropped the reference to LCLD since it is used only for detection of termination.)

It is easy to see that the running time of the above procedure has same probabilistic time bounds as the procedure given in Section 3.

Let $H_0, H_1, H_2, \ldots$ denote a disjoint partition of H into subtrees, such that the subtrees are of size between q and $2q$, except possibly for one subtree which is of size $< q$ (we will choose an appropriate q later; for now we assume that $q > G \log^2 n$). Let r be given (we will pick r later satisfying $G \log^2 n \leq r \leq q$). Consider a forest F formed using the following probabilistic game on H. Each edge $(\text{parent}(v), v)$ in H is deleted with probability $1/G$. If $(\text{parent}(v), v)$ gets deleted, then we place G marks on $\text{parent}(v)$. The relationship of F with the above node expansion phase is obvious (note that there is a one-one correspondence between the ticked nodes v in the node expansion phase and the roots of trees in F). In the following lemma we claim certain properties that F satisfies with high probability. After that we will fix F, having these properties, and prove the bound on runtime of the node expansion phases, for such F. $B(l, r)$ is as defined in Lemma 1.

Lemma 2. *With probability at least* $1 - n^{-c_2 \log n}$ *(for some constant c_2), the following hold:*

(a) For any i, the intersection of any tree in F with H_i has at most $B(2q, r)$ nodes.

(b) The number of marks on nodes in any subtree H_i is at most $8q$.

Note that any tree in F which is not a subtree of any H_i must contain an edge connecting two distinct H_i's in H. Now since there are are at most $n/q + 1$ such edges, we have that whenever (a) holds, the sum of sizes of trees in F which are not subtrees of some H_i is at most $B(2q, r)[n/q + 1]$

Proof. (a) follows using Lemma 1. (b) follows using Chernoff bounds. ∎

We assume from now on that F satisfies (a) and (b) above.

For S, a tree in F, let weight of S, $WT(S) =$ number of nodes in $S+$ the number of marks on nodes in S (note that weight is different from size of the tree).

If a tree S in F is a subtree of some H_j, then we say that S is of type A; otherwise we say that S is of type B. Note that each tree of type A has weight at most $10q$. Also, the total number of nodes in F which belong to a tree of type B is bounded by $[n/q + 1](B(2q, r))$. Let n_i denote the number of trees S in F of type A with weight i. Using Lemma 2 (b), we have that $\sum_i [i \cdot n_i] \leq (2q + 8q)(n/q) = 10n$.

We now consider any node v in H (F). We will show that with high enough probability (of form $n^{-c \log n}$) v gets expanded (by the algorithm above) in time $O(n/p + n^{1/4}q^2 + hq^2)$ for p, G satisfying certain bounds. It immediately follows that all nodes in H would get expanded with high enough probability (of form $n^{-c' \log n}$) in time $O(n/p + n^{1/4}q^2 + hq^2)$.

We call ancestors of v (in H) special nodes and the trees in F which contain any ancestor of v as special trees. Clearly, at any particular time instant, at most one processor has a special node in its local queue. We refer to such a processor as the special processor (for that time instant).

Note that the total time taken before v is expanded consists of (a) time taken to expand its ancestors and possibly transmit its children and (b) delay encountered by the ancestors of v due to special processor expanding a non-special node (of lower cost) (and subsequent possible transmission of its children). Time taken due to (a) is at most $O(hG)$. We consider (b) below. Time delays in (b) can be caused by one of the following reasons:

1. Special processor expands a node (or transmits a child of a marked node) from a tree of type B.
2. Special processor expands a node (or transmits a child of a marked node) in a special tree of F.
3. (post-delay) Special processor expands a node in a type A tree S of F (or transmits a child of a marked node in S), such that the root of S arrived at the special processor after the root of the tree (in F) containing the special node.
4. (pre-delay) Special processor expands a node in a type A tree S of F (or transmits a child of a marked node in S), such that the root of S arrived at the special processor before the root of the tree (in F) containing the special node.

The above division of delays in groups is similar in spirit to the division in [KZ88]. Note that the total delay due to delays of type 1, can be at most $O(G \cdot (n/q + 1) \cdot B(2q, r))$. Delay due to delays of type 2, which do not fall in type 1, are bounded by $h * B(2q, r)$.

Now let us consider post-delays. For any tree of type A to cause a post-delay, the root of the tree must be sent to the special processor. Probability of this happening is $1/p$. Thus if $n_i \geq n^{1/4}$, then the probability that the total amount of post-delay due to nodes in trees of type A and weight i is greater than $(2n_i/p)i$, is bounded by $c_3^{-n_i/p} \leq c_3^{-n^{1/4}/p}$ (for some constant $c_3 > 1$, using Chernoff bounds). On the other hand, if $n_i < n^{1/4}$, then the total amount of post-delays due to nodes from trees of type A and weight i is clearly $\leq n^{1/4} \cdot i$.

Thus, if $p < n^{1/5}$, then with probability at least $1 - n^{-c_4 \log n}$ (for some constant c_4), total amount of post-delay is bounded by $O(n^{1/4}q^2 + n/p)$ (since $\sum_i n_i \cdot i = O(n)$).

We now consider pre-delays. We will prove that, as long as G and p satisfy certain bounds, the probability that, the pre-delays is more than $O(n/p + n^{1/4}q^2 + hq^2)$ is small.

5.1 A queuing model

To model the pre-delays we consider the following queuing problem. In this problem the goal of the adversary is to get a high payoff with significant probability.

There are k groups of customers, $X_1, X_2, \ldots, X_k$. Some groups of customers are of type A and some are of type B. Number of customers in any group of type A is bounded by $10q$. Number of groups of customers of type A having i customers is n_i. Total number of customers in groups of type B is bounded by $O(G[n/q + 1]B(2q, r))$. Also $\sum_i n_i \cdot i \leq 10n$. Some groups contain special customers (such groups are called special). Total number of special customers in groups is bounded by h. X_1 is a special group.

At the start of the queuing process, group X_1 is assigned to processor 1. The queuing process alternates between sequences of arrival and service phases. At each service phase, one of the customers (if present) in each queue is serviced. In an arrival phase, a group of customers arrives. Each such group is assigned to a random processor (by assigning a group to a processor, we mean assigning all members of the group to the processor). A special group can arrive only if no special customers are present at any processor. The choice of arrival/service phase and the group at any arrival phase depends only on the the random choices (of processor assignment) made earlier in the queuing process. Note that at any step at most one processor has a special customer. Let the processor having a special customer be called special processor.

Payoff: The payoff (to adversary) for service phases is computed as follows. If the special processor (if any) in the service phase serves a customer from a group, X_i, such that

X_i is of type A,
X_i is not a special group, and
X_i arrived before the special customers,

then the payoff is 1 unit. Otherwise there is no payoff.

It is easy to see that the following holds.

Proposition 3. *There exists an adversarial strategy, so that the probability of the adversary getting a payoff of $\geq x$, in the above model, is at least as much as the probability of pre-delays (as defined in the previous subsection) being $\geq x$.*

Further, we may assume without loss of generality that any group contains at most 1 special customer (we can consider the last special customer served in any special group to be the special customer of that group). Also, without loss of generality we can assume that the special groups consist only of one customer which is a special customer.

The goal of the adversary is to maximize the chances that the payoff in the above process exceeds $O(n/p + n^{1/4}q^2 + hq^2)$ (where the constant in big O comes from the analysis below).

A destination sequence, $d_1, d_2, \ldots, d_k$, is a sequence of numbers, each from the set $\{1, \ldots, p\}$. We interpret d_i as the processor to which the ith arrival in the above process goes. Clearly, each sequence of random choice in the above process has an associated unique destination sequence.

The following proposition can be proved essentially along the lines of Proposition 7 in Chapter 5 of [Zha89].

Proposition 4. *The following strategy maximizes the payoff for the adversary for any fixed destination sequence.*
  1. Schedule no arrivals while a special customer is present.
  2. Always serve other customers before a special customer.
  3. Schedule no service phases, if there is no special customer present.

We omit the proof of above proposition which is essentially the same as given in [Zha89].

We can thus assume without loss of generality that the adversary follows the strategy as in Proposition 4. Before proceeding with the proof, we first state a special case of the above queuing process which was analyzed by Karp and Zhang.

Lemma 5. *[Zha89] Suppose there are $\leq h$ groups containing one special customer each and l groups (of type A) containing one non-special customer each. Then the probability that the payoff to the adversary is greater than $c(l/p + h)$ is bounded by $2^{-l^{c'}}$, for some constants c and c', if $p < l^{1/3}$.*

We now modify the process in favour of adversary as follows.

1. If at the special processor a customer from a group of type B or a special customer is being served, then no service takes place at the other processors.
2. If, at the special processor, a non-special customer from a group of type A and size i, is being served then only a customer from a group of type A and of size i, if present, is served at any other processor.

Lemma 6 below shows that the payoff to the adversary, due to service phases in which the special processor serves a customer from a non-special group of type A and of size i, is bounded by $O(\max(i \cdot n^{1/4}, i(n_i/p + h)))$, with probability at least $1 - n^{-c_5 \log n}$ for some constant c_5. Thus the probability that the adversary gets a payoff of more than $O(\sum_i [\max(i \cdot n^{1/4}, i(n_i/p + h))])$ is bounded by $n^{-c_6 \log n}$ for some constant c_6.

Lemma 6. *Fix i. Probability that the adversary gets a payoff of at least $c(\max(i \cdot n^{1/4}, i(n_i/p + h)))$, (for c determined from the analysis below) due to service phases in which the special processor serves a customer from a non-special group of type A and of size i, is bounded by $n^{-c_5 \log n}$, for some constant c_5.*

Proof. If $n_i < n^{1/4}$ then clearly the bound holds. So assume $n_i \geq n^{1/4}$.

To prove the bound as above, we assume (as an advantage to the adversary) that the goal of the adversary is to maximize the probability that the payoff due to service phases in which the special processor serves a customer from a group of type A and of size i, is at least $c(\max(i \cdot n^{1/4}, i(n_i/p + h)))$. As an added advantage to the adversary (just for proving the bound for this fixed i) we allow the adversary to choose, the destination for every group except the special groups and the groups of type A with i customers. Clearly, in this situation, all groups of customers, except the special groups and the groups of type A with i customers, can be ignored.

We are thus left with the following groups of customers:

(a) n_i groups of i customers each.
(b) $\leq h$ special customers.

Now it follows from Lemma 5 that the payoff to the adversary due to service phases in which a special customer from a group of type A and size i is served is bounded by $c(\max(i \cdot n^{1/4}, i(n_i/p + h)))$, with probability at least $1 - n^{-c_8 \log n}$, for some constant c_5, if $p < n^{1/12} \leq n_i^{1/3}$. ∎

6 Back to Branch and Bound Problem

From the analysis in the previous section we have that the total amount of pre-delay is bounded by $c \sum_i [\max(i \cdot n^{1/4}, i(n_i/p + h))]$ with probability at least $1 - n^{-c_8 \log n}$. Now $\sum_i [\max(i \cdot n^{1/4}, i(n_i/p + h))] \leq O(n^{1/4} \cdot q^2 + n/p + h \cdot q^2)$.

Thus we have that, with probability at least $1 - n^{c_7 \log n}$, for some constant c_7, the total delay is bounded by $D(n, q, p, G, r) = O((n/q)B(2q, r)G + hq^2 + n^{1/4}q^2 + n/p)$.

Assuming $q, p < n^{1/12}$, $r > Gp \log q$, and $\log q > c_q G \log r \log(Gp)$ (for some constant c_q), we have delay $D(n, q, p, G, r) = O(hq^2 + n/p)$. Thus

Theorem 7. *Suppose H is such that $|\hat{H}| = n$ and $height(\hat{H}) = h$. Let p ($<$ $n^{1/12}$) and G be such that there exist r, q satisfying:*
(a) $q < n^{1/12}$,
(b) $r > \max(Gp \log q, G \log^2 n)$, and
(c) $\log q > c_q G \log r \log(Gp)$.
Then the algorithm given in Section 3 completes execution in time $O(hq^2 + n/p)$ with probability $\geq 1 - n^{-c \log n}$, for some constant c. This is optimal as long as h is not too large.

Now suppose $G = O(\log p)$. Then the requirements on G, p, r, q are satisfied by choosing $r = O(Gp \log^2 n)$ and $q = 2^{O(\log^2 p(\log p + \log \log n))}$, as long as $q < n^{1/12}$. The above constraints can be satisfied for $p = 2^{\log^c n}$, $c < 1/3$. Thus for communication time logarithmic in the number of processors, branch and bound can be implemented in nearly optimal randomized time, for $p < 2^{\log^c n}$, $c < 1/3$.

7 Lower Bound on the Runtime of Our Branch and Bound Procedure

Consider a tree such that $\hat{H}$ is a complete balanced binary tree. Then the expected number of nodes which are expanded by processor 1 (which starts with the root) is at least $n(1 - 1/G)^{\log n}$. Thus with significant probability ($> \frac{1}{Gn}$), processor 1 does at least $\frac{n}{2}(1 - \frac{1}{G})^{\log n})$ node expansions (since the maximum work for any processor is bounded by nG). Thus for $G = \log p$, the algorithm can be optimal (with probability $> 1 - \frac{1}{Gn}$) only for $p < 2^{O(\log^{1/2} n)}$.

8 Conclusion

In this paper we gave a simple parallel algorithm (which is a modified version of an algorithm due to Karp and Zhang) for Branch and Bound problems. We showed that this algorithm performs nearly optimally for a modest number of processors, in a model where communication costs are high. It is easy to formulate several variants of the algorithm by allowing the probability bias to be dependent on the size of the local queue. However all such variants, though seemingly better, are very hard to analyze.

Our analysis is built upon the methods developed by Karp and Zhang. We could not use the simplification of Ranade, since Ranade's analysis crucially depended on symmetry of distribution of different nodes to different processors. It will be interesting to see if the techniques used here can be combined with Ranade's methods to simplify the analysis.

9 Acknowledgements

I am specially grateful to Raghu Raghavan for introducing the problem to me and for constantly urging me to solve the problem. I would like to thank Mohan Kankanhalli, Timothy Poston, Raghu Raghavan and Weiguo Wang for several discussions we had on different ways of analyze the algorithm given in this paper and its variants.

References

[BNK92] A. Bar-Noy and S. Kipnis. Designing broadcasting algorithms in the postal model for message-passing systems. In *Proceedings of the fourth Annual ACM Symposium on Parallel Algorithms and Architectures*, pages 11–22. ACM Press, 1992.

[CKP+93] D. Culler, R. Karp, D. Patterson, E. Santos, A. Sahay, K. Schauser, R. Subramonian, and T. von Eicken. Towards a realistic model of parallel computation. In *Proceedings of the Principles and Practices of Parallel Programming*, 1993.

[KSSS93] R. Karp, A. Sahay, E. Santos, and K. Schauser. Optimal broadcast and summation in the logp model. In *Proceedings of the 5th Symposium on Parallel Algorithms and Architectures*, July 1993.

[KZ88] R. Karp and Y. Zhang. A randomized parallel branch and bound procedure. In *Proceedings of the ACM Annual Symposium on Theory of Computing*, pages 290–300. ACM Press, 1988.

[KZ93] R. Karp and Y. Zhang. Randomized parallel algorithms for backtrack search and branch and bound computation. *Journal of the ACM*, 40(3):765–789, 1993.

[Ran90] A. Ranade. A simpler analysis of the Karp-Zhang parallel branch-and-bound method. Technical Report UCB/CSD 90/586, University of California, Berkeley, 1990.

[Zha89] Y. Zhang. *Parallel Algorithms for Combinatorial Search Problems*. PhD thesis, University of California, Berkeley, 1989.

A Near Optimal Algorithm
for the Extended Cow-Path Problem
in the Presence of Relative Errors

Pallab Dasgupta, P.P.Chakrabarti and S.C.DeSarkar

Dept. of Computer Sc. & Engg.,
Indian Institute of Technology, Kharagpur,
INDIA 721302.
{pallab,ppchak,scd}@cse.iitkgp.ernet.in

Abstract. In classical path finding problems, the cost of a search function is simply the number of queries made to an oracle that knows the position of the goal. In many problems, we want to charge a cost proportional to the distance between queries (e.g., the time required to travel between two query points). With this cost function in mind, the original w-lane cow-path problem [8] was modeled as a navigation problem in a terrain which consists of w-concurrent avenues. In this paper we study a variant of this problem where the terrain is an uniform b-ary tree, and there is a lower-bound estimate of the cost function. We present a strategy $CowP$ for this class of problems where the relative error is bounded by a known constant and show that its worst case complexity is less than or equal to $\lceil 4b/(b-1) \rceil$ times optimal.

1 Introduction

The cow-path problem [8] is defined as follows. Consider a cow standing at an intersection of w concurrent rays. The goal lies at an unknown distance along one of these rays and cannot be identified until the cow reaches the goal. The objective is to devise a strategy for the cow so that it reaches the goal while traveling the least distance possible.

Abstractions of this problem has been addressed by researchers from the theoretical computer science community (Papadimitriou and Yannakakis [14, 13], Blum, Raghavan and Schieber [2]). A previous algorithm for this problem was used by Fiat, Rabani and Ravid [6] in presenting the first competitive algorithm for the online k-server problem. The task of navigating in unknown terrains has also been of much interest to researchers on computational geometry [7, 10, 11]. A similar problem known as layered graph traversal, has been studied by Papadimitriou and Yannakakis [14] and Fiat, Foster, Karloff, Rabani, Ravid and Vishwanathan [5].

The idea of using a wandering agent (such as the cow) to model the search mechanism has been used for analyzing the task of searching different terrains. The problem of searching rectangular grid terrains (under the agent model) have been studied by Baeza-Yates, Culberson and Rawlins [1]. The task of searching a tree is an important problem since many terrains can be conveniently modeled as trees. One variant of the agent searching problem have been studied by Karp, Saks and Widgerson [9] to analyze the essential elements common to all branch-and-bound procedures. In their work the agent (or cow) had been referred to as a *wandering ram*, and the terrain considered was a binary tree. In one of our earlier works [3] we had studied the task of agent searching in an uniform *b-ary* tree.

An important feature of the cow-path problem is that it characterizes the costs of backtracking from a path and retracing a previously visited path in terms of the distance traversed by the cow. This motivates the use of this model as a means for accurately analyzing the complexities of backtracking strategies where the act of backtracking and retracing play a significant role. Interestingly, the overall cost of backtracking and retracing in most cow-path problems is not reflected by the asymptotic complexity of the strategy, but appears as a coefficient of the highest order complexity term. This necessitates the use of the following strong optimality measure for analyzing the optimality of cow-path strategies. We shall call a strategy optimal only if it is strongly optimal.

Definition 1. Strong Optimality: A strategy is optimal in the strong sense if the function measuring its complexity is optimal up to lower order terms, that is, the coefficient of the highest order term is also optimal. □

In this paper we study the extended cow-path problem (where the terrain is an uniform *b-ary* tree) in the presence of a lower-bound estimate of the cost function. In terms of robot navigation problems the lower-bound estimate could be computed from sensory information gathered by the robot. In recursive backtracking strategies, it simply represents a heuristic evaluation function. The objective is to determine how to use the lower-bound estimate so that the worst case complexity of the cow-path strategy is optimal or near optimal. The worst case complexity of a cow-path strategy is defined as follows.

Definition 2. Worst Case Complexity: The worst case complexity of a cow-path strategy for trees is a function $C(d)$, where d denotes the depth of the goal, such that:

1. There exists a constant ψ such that for all d, $d > \psi$, the distance traversed by the cow is upperbounded by $C(d)$, and
2. For every constant ϕ, there exists a problem instance where the goal is at a depth d, $d > \phi$, such that the distance traversed by the cow is equal to $C(d)$.

□

If the depth of the goal is bounded by some known constant then it is likely that for this specific class of problems some specific strategy will be optimal. Thus for different values of the constant bound on the depth of the goal different optimal strategies may exist. In this paper our objective is not to study such specific cases, but to address the entire class of problems where there is no known bound on the depth of the goal. This is the reason for incorporating the first clause in the definition of worst case complexity. The second clause simply ensures that the bound is tight.

It is often natural to assume that there is some known upperbound on the error of the estimating function [1]. Typically it is difficult to obtain a tight upperbound on the absolute error of an estimating function, rather it is often more natural to obtain an upperbound on the relative error [15]. In this paper we consider such situations, that is, when there is a known upperbound on the *relative error*. Since this paper considers only terrains which are uniform *b-ary* trees having a single goal, throughout this paper the phrase *"the class of problems"* refers to *"the class of cow-path problems where the terrain is an uniform* b-ary *tree with a single goal"*. The major result presented in this paper is as follows.

- For the class of problems where there is a known upperbound on the *relative* error of the lower-bound estimating function, we present a strategy called *CowP*, and show that its worst case complexity is upperbounded by $[4b/(b-1)]$ times the optimal worst case complexity.

The paper is organized as follows. Section 2 contains some basic definitions and preliminary results. The proposed cow-path strategy and the analysis of its complexity is presented in 3. Part of the complexity analysis is given in an appendix.

2 Preliminaries

In the model of the cow-path problem considered in this paper, the cow can use a lower-bound estimating function h on each node n it visits to determine a real valued underestimate $h(n)$ of the distance of the goal from that node. In order to determine the estimate at each of the successor nodes of the present node, the cow has to actually visit each of them. When the cow has visited all the successors of a node and evaluated their estimates we say that the node has been *expanded*. The notions of node expansion and node cost follow directly from heuristic search terminology (see [15]), and may be defined as follows.

Definition 3. Node Expansion: A node n is said to be "expanded" when the cow has visited all the successors of n. □

Definition 4. Node Cost: The cost of a node n, denoted by $f(n)$ is the sum of the distance of n from the origin (denoted by $g(n)$) and the estimate $h(n)$ at node n. □

It is known that for tree search problems any admissible algorithm must expand all nodes in the tree whose cost is less than the cost of the optimal solution path [4]. The following lemma (whose proof follows from adversary argument) shows that this result will extend to the cow-path framework as well (for uniform *b-ary* trees).

Lemma 5. *If N denotes the depth containing the goal node, then in the worst case, every node up to depth $(N-1)$ whose cost is less or equal to N will have to be expanded by the cow.* □

---

[1] In case no such upperbound is known, we have already established in a previous work that the iterative deepening strategy is optimal [3]

3   Relative Error Bounded by a Constant

The relative error bound of a lower-bound estimating function is defined as follows.

Definition 6. Relative error bound: ϵ
Let $h^*(n)$ denote the distance of a node n from the goal. Then the relative error of the estimate $h(n)$ is $[h^*(n) - h(n)]/h^*(n)$. The constant ϵ is a relative error bound if for every node n,

$$\frac{h^*(n) - h(n)}{h^*(n)} \leq \epsilon$$

□

In this section we consider the problem of developing a cow-path strategy for the entire class of problems for which the relative error is bounded by a given constant ϵ. Our approach to this task is as follows.

1. We first identify the set of nodes that will have to be visited by any strategy in the worst case over all problem instances where the relative error is bounded by ϵ.
2. Having identified the set of nodes that will be *surely visited* we evaluate the distance traversed for an optimal tour of these nodes. This gives us a lowerbound L on the worst case complexity of all possible strategies.
3. We try to develop a strategy whose worst case complexity is close to the lowerbound L.

Let us first analyze the set of nodes that will have to be visited in the worst case for the class of problem where the absolute error is bounded by ϵ. A typical search tree T is depicted in Fig 1 where (without loss of generality) the solution path $(s, n_{N-1}^s, \ldots, n_j^s, \ldots, n_1^s, \gamma)$ is represented on the extreme right. The trees $T_1, \ldots, T_j, \ldots, T_N$ are subtrees of T, stemming from the solution path; each *off-course* subtree T_j is identified by its root node n_j^s on the solution path, (n_j^s is the only node in T_j having $(b-1)$ sons).

Let S_N^i denote the set of nodes at depth i in T whose cost can be less than or equal to N. Thus, in the worst case (by lemma 5), the set of nodes expanded is as follows:

$$S_N = \bigcup_{i=1}^{N-1} S_N^i$$

Note that by definition 3, the set of nodes visited comprises of those that are expanded and their children. Let:

$$\theta_i = \left\lfloor \frac{i\epsilon}{2-\epsilon} \right\rfloor \qquad \text{and} \qquad \theta = \left\lfloor \frac{N\epsilon}{2-\epsilon} \right\rfloor$$

The following lemmas help in identifying the set of nodes that will have to be expanded by every strategy in the worst case.

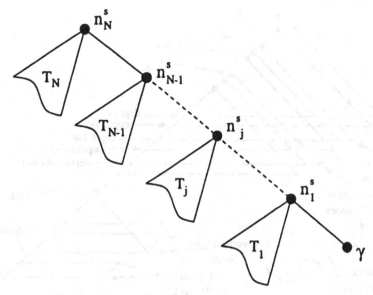

Fig. 1. A typical tree containing a single goal

Lemma 7. *In each subtree T_j of T (as shown in Fig 1) all nodes up to depth θ_j belong to S_N. No node at depth greater than θ_j can belong to S_N.*
Proof: Let us consider a node n at depth k in T_j.

$$g(n) = (N - j) + k \qquad \text{and} \qquad h^*(n) = k + j$$

Since the relative error is bounded by ϵ we have $h(n) \geq (1 - \epsilon)(k + j)$. Therefore:

$$f(n) \geq (N - j) + k + (1 - \epsilon)(k + j)$$

Simplifying the expression for $f(n)$ we have $f(n) \geq N - j\epsilon + (2 - \epsilon)k$. Thus $f(n)$ is less than or equal to N provided the following condition holds:

$$N - j\epsilon + (2 - \epsilon)k \leq N$$
$$\Rightarrow k \leq \frac{j\epsilon}{2 - \epsilon}$$
$$\Rightarrow k \leq \theta_j \qquad (\text{since } k \text{ is an integer})$$

Thus all nodes up to depth θ_j in T_j belong to S_N. If $k \geq \theta_j + 1$ then:

$$f(n) \geq N - j\epsilon + (2 - \epsilon)\left(\left\lfloor \frac{j\epsilon}{2 - \epsilon} \right\rfloor + 1\right) > N - j\epsilon + (2 - \epsilon)\left(\frac{j\epsilon}{2 - \epsilon}\right)$$

$$\Rightarrow f(n) > N$$

Thus no node at depth greater than θ_j can belong to S_N. □

Lemma 7 identifies the members of S_N that belong to each subtree T_j in Fig 1. The following lemma (whose proof is straight-forward) uses this result to identify the members of S_N in the entire tree T.

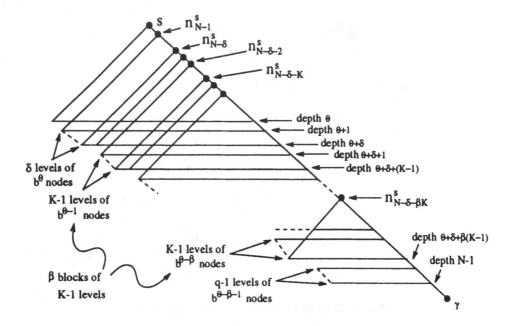

Fig. 2. Set of nodes belonging to S_N^i when the relative error is bounded by a constant

Lemma 8. *In the subtree rooted at n_j^s, all nodes up to depth θ_j belong to S_N.* □

We initially consider only those cases where $[(2 - \epsilon)/\epsilon]$ is a finite integer. Let:

$$K = \frac{2 - \epsilon}{\epsilon}$$

Since K is an integer, it is easy to see that the if we consider the value of θ_i at every depth i starting from zero, then there will be an increase after every K depths. This property helps in keeping our analysis simple. Later, we shall also extend the analysis to those situations where $[(2 - \epsilon)/\epsilon]$ is not an integer.

When the estimate at every node is zero (that is, $\epsilon = 1$) we have $\theta = N$. In this special case it is already known that the iterative deepening algorithm [12] is an optimal cow-path strategy [3]. In this study, we are interested in the cases where $\epsilon < 1$. In such cases, the goal occurs at depths greater than θ.

Let N denote the depth of the goal and $N > \theta$. Let $\delta = N - \theta K$. Then without loss of generality, we may rewrite N as follows:

$$N = \theta + \delta + \beta(K - 1) + q \qquad 0 \le \delta \le (K - 1) \qquad \beta \ge 0 \qquad 0 \le q \le (K - 1)$$

Throughout this analysis we will refer to this alternate expression for N. The reason for rewriting N in this way will become apparent through the following analysis.

The set of nodes in S_N: Using lemma 8 we find that the following nodes belong to S_N. The set of nodes are also shown in Fig 2.

1. Up to depth θ all nodes belong to S_N.
2. At depth $\theta + j$, where $0 < j \leq \delta$, S_N contains b^θ nodes which belong to the subtree rooted at n^*_{N-j}. This may be shown as follows:

$$\theta_{N-j} = \left\lfloor \frac{N-j}{K} \right\rfloor = \left\lfloor \frac{N-\delta}{K} + \frac{\delta-j}{K} \right\rfloor = \left\lfloor \theta + \frac{\delta-j}{K} \right\rfloor = \theta$$

Therefore by lemma 8, in the subtree rooted at n^*_{N-j}, all nodes at the depth θ belong to S_N. In T, these b^θ nodes are at the depth of $\theta + j$.

3. At a depth $\theta + \delta + i(K-1) + j$, where $0 \leq i \leq \beta$ and $0 < j \leq q$, S_N contains $b^{\theta-i-1}$ nodes which belong to the subtree rooted at $n^*_{N-\delta-iK-j-1}$. This may be shown as follows.

$$\theta_{N-\delta-iK-j-1} = \left\lfloor \frac{N-\delta-iK-j-1}{K} \right\rfloor = \left\lfloor \frac{N-\delta}{K} - i - \frac{j+1}{K} \right\rfloor = \theta - i - 1$$

Therefore by lemma 8, in the subtree rooted at $n^*_{N-\delta-iK-j-1}$, all nodes at the depth $\theta - i - 1$ belong to S_N. In T, these $b^{\theta-i-1}$ nodes are at a depth of $\theta + \delta + i(K-1) + j$.

Thus, after the depth θ (up to which all nodes are in S_N), we have δ levels each having b^θ members of S_N. Then we have β number of blocks of $(K-1)$ levels each (which we shall refer to as the β-blocks) and the remaining $(q-1)$ levels before the level N containing the goal node. Each level in i^{th} β-block $(1 \leq i \leq \beta)$ from the top contains $b^{\theta-i}$ members of S_N. Each of the last $(q-1)$ levels before the level N contains $b^{\theta-\beta-1}$ members of S_N.

In the worst case (by lemma 5) all the nodes belonging to S_N will have to be expanded (and therefore all nodes in S_N and their children will have to be visited). The distance traversed for an optimal tour of these nodes can be determined by computing the distance traversed in a depth-first traversal of the tree shown in Fig 2. This distance is as follows:

$$
\begin{aligned}
D_{min} &= 2 \sum_{i=1}^{\theta+1} b^i + 2\delta b^{\theta+1} + 2 \sum_{i=0}^{\beta-1} (K-1) b^{\theta-i} + 2(q-1) b^{\theta-\beta} \\
&= \frac{2b}{b-1} b^\theta \left[b + \delta(b-1) + (K-1)\left(1 - \frac{1}{b^\beta}\right) + \frac{(q-1)(b-1)}{b^{\beta+1}} \right] \quad (1)
\end{aligned}
$$

D_{min} is a lowerbound for the worst case distance that has to be traversed by any strategy when the relative error is bounded by ϵ.

3.1 The strategy $CowP$

The analysis made so far shows that if we can identify the nodes that belong to S_N a priori, then for every problem instance where the relative error is bounded by ϵ, the cow can find the goal by traversing no more distance than D_{min}. Since the cow cannot identify the nodes belonging to S_N without actually visiting them, our objective is to devise some strategy for the cow so that it can identify (and expand) the nodes in S_N in an efficient way.

One way to guarantee that only nodes belonging to S_N are expanded, is to expand nodes in a best-first manner. The drawback of this approach is that the cow may frequently traverse to and fro between alternate paths in the tree and spend most of the time in backtracking and retracing. One way to restrict the amount of backtracking is to plan the set of nodes to be expanded next. Once a plan is generated, the cow expands the nodes in the current plan by tracing an optimal tour of those nodes.

The strategy $CowP$ that we are about to present uses the idea of planning the set of nodes to be expanded next. To describe the set of nodes that the strategy incorporates in a plan we define a *slice* as follows.

Definition 9. Slice: A set of b^i nodes at depth d is called a "slice of size b^i" if these b^i nodes have a common ancestor n at depth $d - i$. In other words, the b^i nodes can be called a slice, if there exists a node n in the tree such that the b^i nodes are the leaf nodes of a subtree of depth i rooted at n. $\Box$

Properties of a slice: We enumerate three properties of a *slice* that will be useful in the subsequent analysis.
1. Given a node n at depth d $(d \geq i)$, there exists an unique slice of size b^i that contains n. This is because of the fact that in a tree the node n has an unique i^{th} ancestor.
2. Let A be a slice of size b^i and B be a slice of size b^j such that $j < i$. If there exists any node n that belongs to both A and B, then B is a subset of A.
3. The nodes of a given *slice* of size b^i can be expanded by a depth-first traversal during which no more distance than $[2b/(b-1)]b^{i+1}$ is covered.

It should be noted (see Fig 2) that due to lemma 8, each S_N^i is a slice.

In the proposed strategy $CowP$, instead of selecting a single node for expansion (as is the policy of strictly best-first strategies), the cow will select a *slice* for expansion. The slice selected by $CowP$ is the one containing the minimum cost frontier node. If the best cost node is at a depth i, then the size of the slice selected by $CowP$ is b^{θ_i}, where:

$$\theta_i = \left\lfloor \frac{i\epsilon}{2 - \epsilon} \right\rfloor$$

After expanding the set of nodes belonging to the slice at depth i, the cow returns to the source node s. Therefore, (by the third property of a slice) the distance traversed by the cow to expand all nodes of a slice at depth i is as follows:

$$DS_i = \frac{2b}{b - 1} b^{\theta_i + 1} + 2i$$

where the distance $2i$ is traversed to reach the slice from the source node s and to return to s after expanding all nodes in the slice.

The outline of the strategy $CowP$ is given below. At any instant, the *frontier* is the set of nodes that have been visited but not expanded (that is, their children have not been visited). Initially, the frontier consists of the source node s. By the term *"expansion of a slice"* we mean *"the expansion of all nodes belonging to the slice"*.

> **Algorithm** *CowP*
> 1. Select the minimum cost frontier node n. Let n be at depth i.
> 2. Expand the slice of size b^{θ_i} containing n as follows:
>   2.1 Move to the θ_i^{th} ancestor of n. Let it be node m.
>   2.2 Perform a depth-first traversal with m as root visiting
>   all nodes up to depth $\theta_i + 1$.
> 3. If the goal is found, then terminate, else return to s.
> 4. Goto step 1.

3.2 The complexity of *CowP*

When the cow selects the minimum cost frontier node, it will definitely select a node belonging to S_N (because until the goal is found there will always be one node whose cost is less than or equal to N). Let $CARD(S_N^i)$ denote the number of nodes in the set S_N^i. If the selected minimum cost node is at depth i, then we can have either of the following two situations.

$\mathbf{CARD}(S_N^i) \geq \mathbf{b}^{\theta_i}$: Since the minimum cost node belongs to both the chosen slice and S_N^i, therefore by the second property of a slice, we can infer that all the nodes belonging to the chosen slice also belong to S_N^i.

$\mathbf{CARD}(S_N^i) < \mathbf{b}^{\theta_i}$: In this case, all nodes belonging to S_N^i also belong to the chosen slice. In addition, the cow will have to visit some nodes outside S_N^i. However, no other slice at depth i will be chosen in future because depth i will not have any remaining frontier node of cost less than or equal to N.

Therefore, the number of slices at depth i that have to be expanded by the cow in the worst case is

$$NS_i = \left\lceil \frac{CARD(S_N^i)}{b^{\theta_i}} \right\rceil$$

Note that since $CARD(S_N^i)$ is a power of b, the ceiling is only to ensure that when $CARD(S_N^i)$ is less than b^{θ_i}, at least one slice is expanded. The total distance traversed by the cow to expand these slices is independent of when and in what sequence these slices are expanded because the cow returns to the source node s after expanding each slice. The total distance traversed by the cow to expand the NS_i slices at depth i is therefore:

$$DD_i = NS_i * DS_i = \left(\frac{2b}{b-1} b^{\theta_i+1} + 2i \right) \left\lceil \frac{CARD(S_N^i)}{b^{\theta_i}} \right\rceil$$

Since the sequence in which *CowP* expands the slices at different depths is independent of the distance traversed to expand a slice, we can analyze the complexity of *CowP* simply by adding up the distance it traverses to expand the NS_i slices at each depth i, $(i < N)$. The worst case distance traversed before locating the goal at depth N is therefore given by $D_N = \sum_{i=1}^{N-1} DD_i$. Since $N = \theta + \delta + \beta(K-1) + q$ we can rewrite D_N as follows:

$$D_N = \sum_{i=1}^{\theta} DD_i + \sum_{i=\theta+1}^{\theta+\delta} DD_i + \sum_{i=\theta+\delta+1}^{\theta+\delta+(K-1)\beta} DD_i + \sum_{i=\theta+\delta+(K-1)\beta+1}^{N-1} DD_i$$

$$= D_\theta + D_\delta + D_\beta + D_q$$

where D_θ, D_δ, D_β and D_q (above) are respectively the first, second, third and fourth terms (of the previous expression). To compute the upperbound on D_N, we will separately compute upperbounds on the values of D_θ, D_δ, D_β and D_q. The detailed analyses and derivations are given in appendix A. The following upperbounds are established.

$$D_\theta \leq \frac{2b^2}{(b-1)^2} b^{\theta+1} \tag{2}$$

$$D_\delta \leq \frac{2b^2}{(b-1)^2} b^\theta \delta(b-1) + \sum_{i=\theta+1}^{\theta+\delta} \frac{2b}{b-1} b^{\theta_i+1} \tag{3}$$

$$D_\beta \leq \frac{2b^2}{(b-1)^2}(K-1)\left(1 - \frac{1}{b^\beta}\right) b^\theta + \sum_{i=\theta+\delta+1}^{\theta+\delta+(K-1)\beta} \frac{2b}{b-1} b^{\theta_i+1} \tag{4}$$

$$D_q \leq \frac{2b^2}{(b-1)^2} b^\theta \left(\frac{(q-1)(b-1)}{b^{\beta+1}}\right) + \sum_{i=\theta+\delta+(K-1)\beta+1}^{N-1} \frac{2b}{b-1} b^{\theta_i+1} \tag{5}$$

Using these upperbounds we now prove the following result.

Theorem 10. *An upperbound on the worst case distance traversed by a cow using CowP to locate a goal at depth N (on problem instances where $[(2-\epsilon)/\epsilon]$ is an integer) is as follows.*

$$D_N \leq \frac{4b^2}{(b-1)^2} b^\theta \left[b + \delta(b-1) + (K-1)\left(1 - \frac{1}{b^\beta}\right) + \frac{(q-1)(b-1)}{b^{\beta+1}}\right]$$

Proof: Using the upperbounds on D_θ, D_δ, D_β and D_q, we have:

$$D_N = D_\theta + D_\delta + D_\beta + D_q$$

$$\leq \frac{2b^2}{(b-1)^2} b^\theta \left[b + \delta(b-1) + (K-1)\left(1 - \frac{1}{b^\beta}\right) + \frac{(q-1)(b-1)}{b^{\beta+1}}\right] + \sum_{i=\theta+1}^{N-1} \frac{2b}{b-1} b^{\theta_i+1}$$

Let Z denote the second term, that is, $Z = \sum_{i=\theta+1}^{N-1}[2b/(b-1)]b^{\theta_i+1}$. The term denotes the distance traversed to expand one slice each at each depth $i > \theta$. Since $N = \theta + \delta + (K-1)\beta + q$, therefore:

$$\theta + 1 = N - \delta - (K-1)\beta - q + 1 \geq N - \delta - \beta K - q + 1$$

Thus:

$$Z \leq \frac{2b}{b-1}\left[\sum_{i=N-\delta-\beta K-q+1}^{N-1} b^{\theta_i+1}\right]$$

At depth $i = N - \delta - \beta K - 1$, we have:

$$\theta_i = \left\lfloor \frac{N - \delta - \beta K - 1}{K} \right\rfloor = \left\lfloor \frac{(\theta - \beta)K - 1}{K} \right\rfloor = \theta - \beta - 1$$

Thus for the $q - 1$ depths between $N - \delta - \beta K - q + 1$ and $N - \delta - \beta K - 1$, we have $\theta_i \leq \theta - \beta - 1$. At depth $i = N - \delta - \beta K = (\theta - \beta)K$, we have $\theta_i = \theta - \beta$. Also for $N - \delta \leq i \leq N$, it is easy to see that $\theta_i = \theta$. Combining these results, we have:

$$Z \leq \frac{2b}{b-1}\left[(q-1)b^{\theta-\beta} + \sum_{i=\theta-\beta}^{\theta-1} Kb^{i+1} + \delta b^{\theta+1}\right]$$

$$\leq \frac{2b}{b-1}\left[(q-1)b^{\theta-\beta} + (K-1)b^{\theta-\beta+1}\frac{b^\beta - 1}{b-1} + \frac{b^{\theta+1}}{b-1} + \delta b^{\theta+1}\right]$$

$$\leq \frac{2b^2}{(b-1)^2}b^\theta\left[1 + \delta(b-1) + (K-1)\left(1 - \frac{1}{b^\beta}\right) + \frac{(q-1)(b-1)}{b^{\beta+1}}\right]$$

Substituting the above expression for Z in the expression for D_N, we have:

$$D_N \leq \frac{4b^2}{(b-1)^2}b^\theta\left[b + \delta(b-1) + (K-1)\left(1 - \frac{1}{b^\beta}\right) + \frac{(q-1)(b-1)}{b^{\beta+1}}\right]$$

□

Corollary 11. *The worst case distance traversed by a cow using CowP (on problem instances where $[(2 - \epsilon)/\epsilon]$ is an integer) is less than or equal to $[2b/(b-1)]$ times optimal.*
Proof: Follows from theorem 10 and expression 1. □

We have so far considered only those situations where $[(2-\epsilon)/\epsilon]$ is an integer. It is our hunch that for other situations too corollary 11 holds, though we have not been able to prove it. However, we can prove the following weaker result.

Theorem 12. *The worst case distance traversed by a cow using CowP is less than or equal to $[4b/(b-1)]$ times optimal.*
Proof: Note that the value of θ is not affected by the choice of K. When $[(2 - \epsilon)/\epsilon]$ is an integer, we had β equal blocks of $(K - 1)$ levels each (see Fig 2), such that in a block, each level had similar number of nodes belonging to S_N. If $[(2 - \epsilon)/\epsilon]$ is not an integer, then the number of levels in a block is at least $\lfloor(2 - \epsilon)/\epsilon\rfloor - 1$ and at most $\lceil(2 - \epsilon)/\epsilon\rceil - 1$. Let us define K as follows:

$$K = \left\lceil \frac{2 - \epsilon}{\epsilon} \right\rceil$$

Surely, $N \leq \theta + \delta + (K - 1)\beta + q$. Therefore, for this value of K, the distance traversed by the cow using *CowP* is:

$$D_N \leq \frac{4b^2}{(b-1)^2}b^\theta\left[b + \delta(b-1) + (K-1)\left(1 - \frac{1}{b^\beta}\right) + \frac{(q-1)(b-1)}{b^{\beta+1}}\right]$$

Now:

$$(K - 1)\left(1 - \frac{1}{b^\beta}\right) \leq (K - 2)\left(1 - \frac{1}{b^\beta}\right) + b$$

Therefore:

$$D_N \leq \frac{4b^2}{(b-1)^2}b^\theta \left[2b + \delta(b-1) + (K-2)\left(1 - \frac{1}{b^\beta}\right) + \frac{(q-1)(b-1)}{b^{\beta+1}}\right]$$

$$\leq \frac{8b^2}{(b-1)^2}b^\theta \left[b + \delta(b-1) + (K-2)\left(1 - \frac{1}{b^\beta}\right) + \frac{(q-1)(b-1)}{b^{\beta+1}}\right]$$

Let us now analyze the value of D_{min} under this situation. The size of a block (as described above) is at least $\lfloor (2-\epsilon)/\epsilon \rfloor - 1$ which is obviously greater than $K-2$. Therefore:

$$D_{min} \geq \frac{2b}{b-1}b^\theta \left[b + \delta(b-1) + (K-2)\left(1 - \frac{1}{b^\beta}\right) + \frac{(q-1)(b-1)}{b^{\beta+1}}\right]$$

By comparing D_N and D_{min}, we have $D_N \leq [4b/(b-1)]D_{min}$. $\Box$

4 Conclusion

It may be noted that to evaluate the lowerbound on the optimal complexity for a cow-path problem with a known bound on the error we had computed the cost of the optimal tour of those nodes that have to be visited by every strategy in the worst case. Achieving this lowerbound is clearly impossible since in that case the cow must be able to identify those nodes a priori. Computing the actual optimal complexity remains an open problem. Consequently it remains to be investigated whether the strategy $CowP$ is actually optimal.

Appendix

A Upperbounds for D_θ, D_β and D_q

In this section we compute the upperbounds on the values of D_θ, D_β and D_q. We first make an observation that will be used from time to time in our analysis. Without loss of generality we may rewrite θ as follows:

$$\theta = \alpha K + p \qquad \alpha \geq 0 \qquad 1 \leq p < K$$

The following result (whose proof is straight-forward) has been used in evaluating the complexity of $CowP$.

Lemma 13. *If $N/b^\alpha = O(1)$, then N is bounded by a constant.* $\Box$

A.1 Derivation of upperbound on D_θ

We shall show that for a cow using $CowP$, an upperbound on the value of D_θ (up to lower order terms) is as shown in expression 2.

Proof: Up to depth θ all nodes belong to S_N. Therefore at depth i, $i \leq \theta$, we have $CARD(S_N^i) = b^i$. Thus $DD_i = [2b/(b-1)]b^{i+1} + 2ib^{i-\theta_i}$. Therefore:

$$D_\theta = \sum_{i=1}^{\theta} \left(\frac{2b}{b-1} b^{i+1} + 2ib^{i-\theta_i} \right) \leq \frac{2b^2}{(b-1)^2} b^{\theta+1} + 2\theta \sum_{i=1}^{\theta} b^{i-\theta_i}$$

Let H_θ denote the second term on the right hand side. Rewriting θ as $\alpha K + p$, we have:

$$H_\theta = 2\theta \sum_{i=1}^{\alpha K + p} b^{i-\theta_i} \leq 2\theta \sum_{i=1}^{(\alpha+1)K} b^{i-\theta_i}$$

At a depth $i = Kd + j$, $0 \leq d \leq \alpha$ and $1 \leq j < K$, we have $\theta_i = d$. Thus:

$$H_\theta \leq 2\theta \sum_{d=0}^{\alpha} \sum_{j=1}^{K} b^{(K-1)d+j} \leq 2\theta b \left(\frac{b^K - 1}{b - 1} \right) \left(\frac{b^{(\alpha+1)(K-1)}}{b^{K-1} - 1} \right)$$

For $b \geq 2$, it is easy to see that $(b^k - 1)/(b^{k-1} - 1) \leq (b+1)$. Therefore:

$$H_\theta \leq \frac{2b}{b-1} \left(\theta(b+1)b^{(\alpha+1)(K-1)} \right) \leq \frac{2b}{b-1} b^{\theta+1} \left(\frac{\theta(b+1)}{b^{\alpha-K}} \right)$$

Returning to the derivation of D_θ, we now have:

$$D_\theta \leq \frac{2b}{b-1} b^{\theta+1} \left(\frac{b}{b-1} + \frac{\theta}{b^\alpha}(b+1)b^K \right) \leq \frac{2b}{b-1} b^{\theta+1} \left(\frac{b}{b-1} + \frac{N}{b^\alpha}(b+1)b^K \right)$$

From lemma 13 and the definition of worst case complexity it follows that the second term inside the brackets is a lower order term. The result follows. □

A.2   Derivation of upperbound on D_δ

We shall show that for a cow using $CowP$, an upperbound on the value of D_δ (up to lower order terms) is as shown in expression 3.
Proof: At a depth $i = \theta + j$, where $1 \leq j \leq \delta$, S_i^N contains b^θ nodes. Thus

$$DD_i = \left(\frac{2b}{b-1} b^{\theta_i+1} + 2i \right) \lceil b^{\theta-\theta_i} \rceil \leq \left(\frac{2b}{b-1} b^{\theta_i+1} + 2i \right) \left(b^{\theta-\theta_i} + 1 \right)$$

Without loss of generality, we may rewrite θ as $\alpha K + p$, where $\alpha \geq 0$ and $1 \leq p < K$. For any i greater than θ it is easy to see that $\theta_i \geq \alpha$. Also $i \leq N$. Therefore:

$$DD_i \leq \frac{2b}{b-1} \left(b^{\theta+1} + b^{\theta_i+1} \right) + 2N \left(b^{\theta-\alpha} + 1 \right)$$

We may now evaluate D_δ as follows:

$$D_\delta = \sum_{i=\theta+1}^{\theta+\delta} DD_i \leq \delta \left[\frac{2b}{b-1} b^{\theta+1} + 2N \left(b^{\theta-\alpha} + 1 \right) \right] + \sum_{i=\theta+1}^{\theta+\delta} \frac{2b}{b-1} b^{\theta_i+1}$$

Ignoring the lower order term $2N$, and simplifying, we have:

$$D_\delta \leq 2\delta b^\theta \left[\frac{b^2}{b-1} + \frac{N}{b^\alpha} \right] + \sum_{i=\theta+1}^{\theta+\delta} \frac{2b}{b-1} b^{\theta_i+1}$$

From lemma 13 and the definition of worst case complexity it follows that the term N/b^α inside the third brackets is a lower order term. The result follows. □

A.3 Derivation of upperbound on D_β

We shall show that for a cow using $CowP$, an upperbound on the value of D_β (up to lower order terms) is as shown in expression 4.

Proof: At a depth $i = \theta + \delta + d(K-1) + j$, where $0 \le d < \beta$ and $1 \le j \le (K-1)$, S_N^i contains $b^{\theta-d-1}$ nodes. Thus

$$DD_i = \left(\frac{2b}{b-1}b^{\theta_i+1} + 2i\right)\left\lceil b^{\theta-d-1-\theta_i}\right\rceil \le \left(\frac{2b}{b-1}b^{\theta_i+1} + 2i\right)\left(b^{\theta-d-1-\theta_i} + 1\right)$$

Without loss of generality, we may rewrite θ as $\alpha K + p$, where $\alpha \ge 0$ and $1 \le p < K$. For any i greater than θ it is easy to see that $\theta_i \ge \alpha$. Also $i \le N$. Therefore:

$$DD_i \le \frac{2b}{b-1}\left(b^{\theta-d} + b^{\theta_i+1}\right) + 2N\left(b^{\theta-d-1-\alpha} + 1\right)$$

Let $\eta = \theta + \delta + (K-1)\beta$. We may now evaluate D_β as follows:

$$D_\beta = \sum_{i=\theta+\delta+1}^{\eta} DD_i \le \sum_{d=0}^{\beta-1}\sum_{j=1}^{K-1}\left[\frac{2b}{b-1}b^{\theta-d} + 2N\left(b^{\theta-d-1-\alpha} + 1\right)\right] + \sum_{i=\theta+\delta+1}^{\eta}\frac{2b}{b-1}b^{\theta_i+1}$$

$$\le \frac{2b}{b-1}(K-1)\left(\frac{b^\beta-1}{b-1}\right)b^{\theta-\beta+1} + 2N(K-1)\left(\frac{b^\beta-1}{b-1}\right)b^{\theta-\alpha-\beta} + 2N\beta(K-1)$$

$$+ \sum_{i=\theta+\delta+1}^{\eta}\frac{2b}{b-1}b^{\theta_i+1}$$

Ignoring the lower order term $2N\beta(K-1)$, and simplifying, we have:

$$D_\beta \le \frac{2b}{b-1}(K-1)\left(1-\frac{1}{b^\beta}\right)b^\theta\left[\frac{b}{b-1} + \frac{N}{b^{\alpha+1}}\right] + \sum_{i=\theta+\delta+1}^{\theta+\delta+(K-1)\beta}\frac{2b}{b-1}b^{\theta_i+1}$$

From lemma 13 and the definition of worst case complexity it follows that the term $N/b^{\alpha+1}$ inside the third brackets is a lower order term. The result follows. □

A.4 Derivation of upperbound on D_q

We shall show that for a cow using $CowP$, an upperbound on the value of D_q (up to lower order terms) is as shown in expression 5.

Proof: At a depth $i = \theta + \delta + \beta(K-1) + j$, where $1 \le j \le q$, S_N^i contains $b^{\theta-\beta-1}$ nodes. Thus

$$DD_i = \left(\frac{2b}{b-1}b^{\theta_i+1} + 2i\right)\left\lceil b^{\theta-\beta-1-\theta_i}\right\rceil \le \left(\frac{2b}{b-1}b^{\theta_i+1} + 2i\right)\left(b^{\theta-\beta-1-\theta_i} + 1\right)$$

For any i greater than θ it is easy to see that $\theta_i \ge \alpha$. Also $i \le N$. Therefore:

$$DD_i \le \frac{2b}{b-1}\left(b^{\theta-\beta} + b^{\theta_i+1}\right) + 2N\left(b^{\theta-\beta-1-\alpha} + 1\right)$$

Let $\eta = \theta + \delta + (K-1)\beta + 1$. We may now evaluate D_q as follows:

$$D_q = \sum_{i=\eta}^{N-1} DD_i \le (q-1)\left[\frac{2b}{b-1}b^{\theta-\beta} + 2N\left(b^{\theta-\beta-1-\alpha} + 1\right)\right] + \sum_{i=\eta}^{N-1}\frac{2b}{b-1}b^{\theta_i+1}$$

Ignoring the lower order term $2N(q-1)$, and simplifying, we have:

$$D_q \leq \frac{2b}{b-1}b^\theta \left(\frac{(q-1)(b-1)}{b^{\beta+1}}\right)\left[\frac{b}{b-1} + \frac{N}{b^{\alpha+1}}\right] + \sum_{i=\theta+\delta+(K-1)\beta+1}^{N-1} \frac{2b}{b-1}b^{\theta_i+1}$$

From lemma 13 and the definition of worst case complexity it follows that the term $N/b^{\alpha+1}$ inside the third brackets is a lower order term. The result follows. $\square$

References

1. Baeza-Yates, R. A., J.C.Culberson, and G.J.E.Rawlins. Searching in the plane. *Information and Computation 106* (1993), 234–252.
2. Blum, A., P.Raghavan, and B.Schieber. Navigating in unfamiliar geometric terrains. In *STOC* (1991), pp. 494–504.
3. Dasgupta, P., P.P.Chakrabarti, and S.C.DeSarkar. Agent searching in a tree and the optimality of iterative deepening. *Artificial Intelligence 71* (1994), 195–208.
4. Dechter, R., and J.Pearl. Generalized best-first search strategies and the optimality of A^*. *JACM 32*, 3 (1985), 505–536.
5. Fiat, A., D.P.Foster, H.Karloff, Y.Rabani, Y.Ravid, and S.Vishwanathan. Competitive algorithms for layered graph traversal. In *FOCS* (1991), pp. 288–297.
6. Fiat, A., Y.Rabani, and Y.Ravid. Competitive k-server algorithms. In *FOCS* (1990), pp. 454–463.
7. Ghosh, S. K., and S.Saluja. Optimal on-line algorithms for walking with minimum number of turns in unknown streets. Tech. Rep. TCS-94/2, TIFR, Bombay, 1994.
8. Kao, M. Y., J.H.Reif, and S.R.Tate. Searching in an unknown environment: An optimal randomized algorithm for the cow-path problem. In *SODA* (1992), pp. 441–447.
9. Karp, R. M., M.Saks, and A.Widgerson. On a search problem related to branch-and-bound procedures. In *Proc. of 27th Annual Symp. on Foundations of Computer Science* (1986), pp. 19–28.
10. Klein, R. Walking an unknown street with bounded detour. *Computational Geometry: Theory and Applications 1* (1992), 325–351.
11. Kleinberg, J. M. On-line search in a simple polygon. In *Proc. of SODA '94* (1994), pp. 8–15.
12. Korf, R. E. Depth-first iterative deepening: An optimal admissible tree search. *Artificial Intelligence 27* (1985), 97–109.
13. Papadimitriou, C. H. Shortest path motion. In *Proc. FST-TCS Conference, New Delhi* (1987).
14. Papadimitriou, C. H., and M.Yannakakis. Shortest paths without a map. *Theoretical Computer Science 84* (1991), 127–150.
15. Pearl, J. *Heuristics: Intelligent Search Strategies for Computer Problem Solving.* Addison Wesley, 1984.

Efficient Algorithms for Vertex Arboricity of Planar Graphs

Abhik Roychoudhury and Susmita Sur-Kolay

Department of Computer Science and Engineering,
Jadavpur University, Calcutta 700032, India.
e-mail: dgd@jadav.ernet.in

Abstract : Acyclic-coloring of a graph $G = (V, E)$ is a partitioning of V, such that the induced subgraph of each partition is acyclic. The minimum number of such partitions of V is defined as the vertex arboricity of G. A linear time algorithm for acyclic-coloring of planar graphs with 3 colors is presented. Next, an $O(n^2)$ algorithm is proposed which produces a valid acyclic-2-coloring of a planar graph, if one exists, since there are planar graphs with arboricity 3.

Keywords : Vertex arboricity, planar graph, graph coloring, testing of sequential circuits.

1   Introduction

In this paper, the problem of determining the vertex arboricity of finite planar graphs efficiently is studied. Vertex arboricity of a graph is the minimum number of vertex-disjoint forests into which the graph can be decomposed. All vertices of a forest can be assigned the same color, thus producing no monochromatic cycles in the graph. This problem is also equivalent to a generalized vertex coloring problem [1] and will be henceforth referred to as *acyclic-coloring* [5].

Formally, let $G = (V, E)$ be a finite undirected (directed) graph. An acyclic-k-coloring *color* $: V \rightarrow \{1, 2, \ldots, k\}$ is a partitioning of V into $\{V_1, V_2, \ldots, V_k\}$ such that for all i, the subgraph G_{V_i} induced by V_i is a forest. Any vertex $v \in V_i$ is assigned color i, i.e. $color(v) := i$. The minimum value of k for which such partitioning exists, is called the *vertex arboricity* ρ of G.

The immediate motivation of this problem comes from the domain of *design for testability* in VLSI circuits [2]. A clocked sequential circuit is represented by a graph G whose vertices correspond to the flip-flops of the circuit and an $< i, j >$ edge implies the existence of a combinational path from flip-flop i to flip-flop j. The partitions obtained by acyclic-coloring of such a graph correspond to the independent clocks of the circuit. Each independent clock corresponds to a *test mode* of the circuit. In any particular test mode, the flip-flops in its own partition change states while the other flip-flops in the circuit maintain their states. An effective acyclic-coloring algorithm can minimize the number of partitions, and hence the number of test modes.

The acyclic-k-coloring problem for general digraphs is NP-complete [3]. A depth-first-search based simple heuristic for acyclic-coloring any digraph was suggested in [2]. Nevertheless, the fact that $\rho \leq 3$ for any planar graph [4, 5], leads us to the following pertinent question: is the acyclic-coloring problem polynomial-time solvable for planar graphs? No answers to this question are known to exist in literature to the best of the knowledge of the authors.

A simple linear time algorithm for acyclic-coloring of planar graphs using 3 colors is presented in Section 2. Section 3 provides a polynomial-time technique to determine whether it is possible to obtain acyclic-2-coloring of a planar graph. Experimental results and concluding remarks for future extensions appear in Section 4.

For graph-theoretic terminologies used without definition in this paper, the reader is referred to [6]. The induced subgraph formed by the neighbors of a vertex v is called the *Neighbor Induced Subgraph (NIG)* of v. With respect to a given acyclic-coloring, the number of vertices in $NIG(v)$ which have been colored i, is denoted by $N_i(v)$. A path or a cycle is said to be i-*monochromatic* if all the vertices on it are assigned the color i.

2 Acyclic-3-Coloring Algorithm for Planar Graphs

Since all planar graphs have $\rho \leq 3$ [4, 5], an acyclic-3-coloring exists always. In fact, it was proven in [5] that there exists an acyclic-3-coloring of any planar graph where each partition induces a linear forest. But no efficient algorithm follows from these results.

Our algorithm for acyclic-3-coloring a planar graph is a recursive one, similar to that for vertex-coloring [7]. In each recursive call, a vertex v of degree at most 5 (always exists in a planar graph) is deleted, until a sufficiently small graph G_s with at most 5 vertices is obtained. While deleting the vertex v, the vertices of $NIG(v)$ which have not yet been deleted, are stored.

The vertices of G_s are colored by procedure *Naïve_Color_3* as follows: (a) If $\Delta(G_s)$ is 0 or 1, all its vertices are assigned color 1. (b) Otherwise, it is shown below that a vertex $u \in G_s$ exists such that $2 \leq \deg(u) \leq 3$. Assign color 1 to u, color 2 to two neighbors of u and color 3 to all other vertices in G_s.

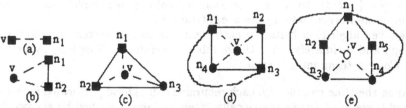

Fig. 1: Five cases of NIGs for acyclic-3-coloring [colors-- ○, ●, ■]

The recursive calls are completed by coloring the deleted vertex v in each call. The color i is given to v (for $i \in \{1, 2, 3\}$) such that there is at most one vertex in $NIG(v)$ with color i. Figs. 1a–1e show five NIGs and the corresponding acyclic-colorings; all possible other NIGs are subgraphs of one of these five. Since a vertex of degree at most 5 is always deleted, it is clear that $|V_{NIG}| \leq 5$. The complete algorithm is presented below in pseudocode.

procedure Acyclic_3_Color (G);

```
    begin
1.      if |V| ≤ 5 then
2.          Naïve_Color_3(G)
3.          else begin
4.              Find a vertex v in G with degree ≤ 5;
5.              Keep track of the neighbors of v;
6.              Acyclic_3_Color(G − v);
7.              Color v by considering the color of its neighbors;
            end;
    end;
```

Lemma 1 *Algorithm Naïve_Color_3 outputs an acyclic-3-coloring for any planar graph G_s with 5 or fewer vertices in constant time.*

Proof: If all vertices of G_s have degree 0 or 1, then G_s has no cycles. In the second case, if the existence of u is established, then the Lemma is proven since the color assignment does not produce any monochromatic cycles. Suppose contrary to the hypothesis, all vertices of G_s have degree 4 then G_s must have 5 vertices. But a planar graph of 5 vertices can have a maximum of 9 edges whereas the sum of the degrees of all vertices is 20, hence violating the Handshaking Lemma [6] for any graph. Therefore, vertex u exists. $\square$

Theorem 1 *Algorithm Acyclic_3_Color produces an acyclic-3-coloring for any planar graph with n vertices in $O(n)$ time.*

Proof: The correctness of the algorithm is established from the following observations:

i) Since 3 colors are available, by pigeon-hole-principle there exists no coloring of the vertices of $NIG(v)$ for which : $|N_i(v)| \geq 2, i = 1, 2, 3$. Thus, if $|N_j(v)| < 2$ for some $j \in \{1, 2, 3\}$, then v can be assigned color j to complete that recursive call and no monochromatic cycles are introduced.

ii) The given algorithm halts. In each recursive call, a vertex is removed from the graph and ultimately when the number of vertices is 5 or less, by Lemma 1 Naïve_Color_3 terminates correctly.

Regarding the time complexity, each recursive call can be performed in constant time and there are $O(n)$ recursive calls. Step 4 is accomplished by always choosing the minimum degree vertex. Before invoking *Acyclic_3_Color*, the vertices are kept sorted by bucket-sort in order of their degrees. Deletion of a vertex then causes minor adjustments. Step 5 takes constant time since $NIG(v)$ has at most 5 vertices. Step 7 involves computing $N_i(v)$ by counting the number of neighbors with color i $(i = 1, 2, 3)$ where $V_{NIG(v)}| \leq 5$. $\square$

3 Acyclic-2-Coloring Algorithm for Planar Graphs

An example of a planar graph with $\rho = 3$ is the geometric dual of the Tutte graph [5]. Determination of ρ of planar graphs is equivalent to testing whether an acyclic-2-coloring exists and producing it. The arboricity is 3 if the algorithm reports failure in yielding a valid acyclic-2-coloring.

3.1 Overview

The proposed algorithm colors the vertices of a planar graph G in a step-by-step fashion, coloring one vertex at each recursive call. At each stage, it is ensured that there exists no monochromatic cycle among the vertices already colored. A list of monochromatic paths for each color, obtained so far, is also stored. Initially, the two lists of monochromatic paths are empty. Since any cycle in G is entirely within a biconnected component (block) of G, the first step comprises in dividing G into its blocks. Next, the blocks of G are colored one by one in such a sequence that the cutpoints can be colored consistently. Each block is colored recursively by finding a vertex v of minimum degree which is certainly at most 5, keeping track of its not yet deleted neighbors, and deleting v. This continues until a sufficiently small graph G_s of at most 5 vertices is obtained, which is then colored naively. A recursive call is completed with coloring the deleted vertex by considering the colors of its neighbors and the set of monochromatic paths produced so far. This may induce limited backtracking if monochromatic cycles are obtained. The algorithm halts if elimination of these monochromatic cycles fails.

Naïve_Color_2 is very similar to *Naïve_Color_3* excepting in the case where there is a vertex u with degree 2 or 3. Here u is colored 1, two of its neighbors are assigned color 2 and all other vertices of G_s are given color 1.

It is to be noted that if the input graph G is divided into blocks $B_1, B_2, \ldots, B_n$, then while B_k is being colored, after the deletion of a finite number of vertices from B_k, the remaining graph could turn out to have cutpoints. The next recursive call will then again divide this graph into blocks $B_{k,1}, B_{k,2}, \ldots, B_{k,m}$. Then in our algorithm, coloring of $B_{k,i}, \{i = 1, 2, .., m\}$ are completed and thus the coloring of B_k is obtained. Only then can the coloring of B_{k+1} proceed.

The following sections discuss this algorithm in details. Section 3.2 gives a safe ordering of blocks for consistent coloring of cutpoints. The various cases of $NIG(v)$ during backtracking and the respective methods for elimination of monochromatic cycles appear in Section 3.3. The criterion for reporting failure, i.e., detecting the non-existence of an acyclic-2-coloring is given in Section 3.4. An idea about storing monochromatic paths and updating schemes is given in Section 3.5. Finally, Section 3.6 deals with the correctness proof and complexity analysis of the algorithm. The algorithm is presented below in pseudo code.

    BEGIN

        Read(G); % the input graph is read %

        Set list of monochromatic paths to ϕ;

        Acyclic_2_Color(G);

    END.

```
procedure Acyclic_2_Color(G);
    begin
1.        Divide G into blocks;
2.        Find a sequence σ in which the blocks are to be colored;
3.        For each block Bᵢ ∈ σ do
                begin
4.                Keep track of any already colored cutpoint c of Bᵢ;
5.                if |V_{Bᵢ}| ≤ 5 then
6.                    Näive_Color_2(Bᵢ)
7.                else begin
8.                    Find a vertex v in Bᵢ of minimum degree; % deg(v) ≤ 5%
9.                    Keep track of the neighbors of v;
10.                   Acyclic_2_Color(Bᵢ − v);
11.                   Color v by considering the color of its neighbors;
12.                   Eliminate monochromatic cycles by
                          first Toggle_Iterate and next Break_Cut_Cycle;
13.                   if failure occurs in Step 12 then report and exit;
14.                   Update the lists of monochromatic paths;
                  end;
15.               if color(c) is different now, then toggle all vertices on Bᵢ;
              end;
    end;
```

3.2 Safe Ordering of Blocks

Our strategy is to obtain acyclic-2-color for the blocks separately However, the cutpoints need to be consistently colored. An illustration of inconsistent coloring of cutpoints for the graph in Fig. 2a is given in Fig. 2b where a cutpoint may have been assigned different colors when the two blocks sharing it were colored separately. Changing the color of the cutpoint in any of these two blocks results in a monochromatic cycle. In order to ensure a consistent coloring the blocks should be colored in a particular sequence which is called a *safe ordering*. By this ordering, at the start of coloring any block, at most one cutpoint is previously colored. It is shown next that such an ordering always exists.

Lemma 2 *There exists a safe ordering of blocks of a graph such that when the coloring of the vertices of a particular block commences, there exists at most one vertex in that component which is already colored.*

Proof: Let us define an ordering σ as follows and then show that it is safe. For a graph G, its block-cutpoint tree [6] $BC(G)$ (Fig. 2c) has vertices corresponding to the blocks and the cutpoints of G and an undirected edge (c, B) exists if cutpoint c is on block B. Initially, the sequence σ is empty. A pruning step on $BC(G)$ involves (i) removing all its pendant vertices (these correspond to blocks of G only) from $BC(G)$, (ii) adding the list of these blocks to the beginning of σ, and (iii) removing the pendant vertices of the resultant graph (these correspond to cutpoints of G only). This pruning step is repeated until the resultant graph is empty or has one vertex which again corresponds to the central block of G. This block is added to the beginning of σ.

Since $BC(G)$ is a tree, it has no cycles and if the central block is treated as its root, any other block has a unique parent which is pruned after it and thus colored before it according to σ (Fig. 2d). Hence, when coloring of any block begins, only the cutpoint shared with its unique parent is previously colored. $\square$

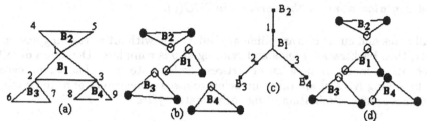

Fig. 2: (a) separable graph, (b) inconsistent coloring, (c) block-cutpoint tree, (d) consistent coloring.

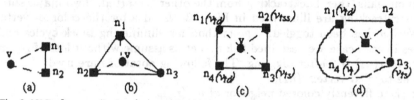

Fig. 3: NIGs for acyclic-2-coloring (a) with 2 vertices, (b) with 3 vertices, (c) C_4, (d) K_4-x.

3.3 Completion of Recursive Call by Coloring a Vertex

Before invoking the recursive call in step 10 of our acyclic-2-coloring algorithm, the NIG of the deleted vertex v is stored in Step 9 because on returning from the recursive call, the color of v in Step 11 depends on $NIG(v)$ (Fig.3) so that no monochromatic cycles are introduced.

If $N_k(v) < 2$ for some k, then $color(v) := k$ (Figs. 3a and 3b). If for both colors $k = 1, 2$, there are two vertices n_{i_k} and n_{j_k} in $NIG(v)$ which have been colored k but there is no k-monochromatic path in $G - v$ between n_{i_k} and n_{j_k} for at least some k, then $color(v) := k$.

In the remaining case, for both k there is *exactly one* k-monochromatic path of length 1 (i.e., direct edge) or more between n_{i_k} and n_{j_k} due to the assumption that the graph $G - v$ already has a valid acyclic-2-coloring when v is being colored. There is no option but to backtrack and $NIG(v)$ is referred to as the *backtracking NIG*. The major intricacy of our algorithm lies in effective backtracking. The operation starts with setting $color(v)$ to $color(n_i)$ where n_i is in $NIG(v)$, and changing $color(n_i)$ to $3 - color(n_i)$; this is called *toggling* the color of n_i. The choice of n_i depends on the backtracking $NIG(v)$. But toggling of n_i may give rise to monochromatic cycles now. Monochromatic cycles obtained while backtracking can be classified as follows:

Definition 1 *A block cycle of a toggled vertex n_i in $NIG(v)$ is a monochromatic cycle containing n_i and at least one other vertex in $NIG(v)$.*

Definition 2 *A* **cut cycle** *of a toggled vertex n_i in $NIG(v)$ is a monochromatic cycle containing n_i and no other vertex in $NIG(v)$.*

Definition 3 *A* **cut-chain cycle** *of a toggled vertex n_i in $NIG(v)$ is a monochromatic cut cycle which is formed by breaking a cut cycle of n_i and contains none of the vertices in $NIG(v)$.*

If all block and cut cycles are eliminated efficiently without re-toggling any vertex in G, then backtracking ends and coloring of v is complete. Detection of failure necessitates maintaining a list of vertices toggled upto now. The salient cases of backtracking $NIG(v)$ are taken up below to discuss the choice of n_i to be toggled and subsequent elimination of monochromatic cycles.

Case 1: $|V_{NIG}| = 4$.
Here $N(c_1) = N(c_2) = 2$ and color of any vertex of minimum degree in $NIG(v)$ can be toggled. If there are more than one (at most two) choices and backtracking from one fails, then backtracking from the other is started. Two major subcases for backtracking are illustrated in Figs. 3c & 3d after the color of vertex n_4 in $NIG(v)$ has been toggled. The method for eliminating block cycles and cut cycles in each case are described below. Let us assume without loss of generality that $color(n_4)$ is 2 after toggling. The following notations are used:
v_t: the toggled vertex, (n_4),
v_{td} : the differently colored neighbor of v_t, (n_1),
v_{ts} : the similarly colored neighbor of v_t, (n_3),
v_{tss} : the other similarly colored neighbor of v_t adjacent to v_{td}, (n_2),
$C_i(v_t)$: the i-th monochromatic cycle through v_t,
$v_i(v_t)$: the neighbors of v_t on a cycle $C_i(v_t)$, excepting v_{ts} if $C_i(v_t)$ is a block cycle of v_t in which case it contains the edge (v_t, v_{ts});
togg_set : set of vertices whose colors were toggled earlier.

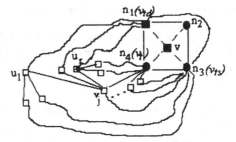

Fig. 4: Elimination of block cycles
[□ --- toggled vertices]

Case 1.1 : $NIG = K_4 - x$.
(a) Elimination of block cycles (Fig. 4): The key idea behind eliminating 2-colored block cycles due to toggling of v_t is to toggle the colors of $v_i(v_t)$ for all $i = \{1, 2, \ldots, p\}$ to color 1. It is to be observed that G being planar, for any plane embedding of it, only one of these, say $v_j(v_t)$ — either the first or the last one in clockwise order in the embedding of G — can create one or more 1-colored cut cycles through v_{td}. Once again, planarity guarantees that only u_l and u_r, the first and the last neighbor of $v_j(v_t)$ in clockwise order, may form 2-colored cut cycles through v_t or v_{ts}. In the next iteration, all the neighbors of

$v_j(v_t)$ excepting one of u_l and u_r (any one if neither has been toggled before) are toggled to color 2. However, if either of these two distinct vertices has a direct edge to v_t or v_{ts}, as the case may be, then all neighbors of $v_j(v_t)$ on 1-colored paths to v_{td} have to be toggled. Ultimately, if no more monochromatic cycles are created, the procedure halts successfully. It exits unsuccessfully if at any stage, an already toggled vertex must be toggled for elimination of a monochromatic cycle. These ideas could be presented in a semi-formal manner as follows.

Toggle_Iterate
(Eliminates block cycles due to toggling of $color(v_t)$)
Input : $G, v_t, v_{ts}, v_{td}, NIG(v)$, monochromatic paths formed by S, the set of already colored vertices, togg_set.
Output : A 2-coloring of $S \cup v$, possibly including cut-cycles, and the set of toggled vertices.
Technique:

T1. If v_t has one or more $color(v_t)$ monochromatic paths to v_{ts} then
      begin
T2.   *for all paths do toggle $color(v_i)$;*
T3.   *$u_l :=$ first just toggled neighbor of v_t in clockwise order;*
                $u_r :=$ last just toggled neighbor of v_t in clockwise order;
                (in the embedding of G)
      end
T4. for other_node := $\{v_t, v_{ts}\}$ do
      begin
            check_node :=v_{td} ; clr := $3 - color(v_t)$;
T5.   *while (either u_l or u_r has two or more clr-paths to check_node) do*
                %these are cut cycles%
                begin
T6.   *u := vertex having two or more clr-paths to check_node;*
T7.   *$U :=$ neighbors of u on clr-paths to check_node*
                    in clockwise order ;
T8.   *$u_l :=$ first element of U; $u_r :=$ last element of U;*
T9.   *if check_node $\in U$ then toggle all vertices in U - check_node*
                    % toggling check_node is avoided%
T10.   *else toggle all vertices in U except either u_l or u_r %not both%*
T11.   *Swap check_node and other_node; clr := $3 - clr$;*
            end
      end

Lemma 3 Toggle_Iterate *runs in $O(n)$ time when a vertex v in a graph of n vertices with $NIG(v) = K_4 - x$ is to be acyclic-2-colored and breaks all block cycles if it does not have to re-toggle.*

Proof: In the recursive algorithm for acyclic-2-coloring, there are no monochromatic cycles before v_t is toggled. Any block cycle resulting from toggling of v_t in $NIG = K_4 - x$, must have v_t and v_{ts} on it. All such block cycles are eliminated in Step T2. The remaining steps are for eliminating cut cycles arising due to toggling in Step T2. This procedure stops when either no more new cut cycles are created or some vertex has to be re-toggled. It will be established in Section 3.4 that this re-toggling condition occurs only when G has no acyclic-2-coloring.

In a planar graph, the number of distinct paths between two vertices is $O(n)$. So Step 2 may need $O(n)$ time. There are two iterations of the *for* loop in Step T5. Each iteration involves a *while* loop but since this inner loop toggles already colored vertices, it can do so at most $O(n)$ times over all its iterations. Hence, the total time complexity in the worst case is linear in n, the number of vertices in G. $\square$

(b) Elimination of cut cycles: Fig. 5a. gives a glimpse of the possible cut-cycles created due to toggling $color(v_t)$. It should be noted that many of the cut cycles get eliminated in the process of eliminating block-cycles, e.g. if in Fig. 5a, the path p_1 (_ _ _ _ _ _) is 2-monochromatic, then the cut cycle $(n_4, v_i, \ldots, v_j, n_4)$ gets broken as well. $(v_i, \ldots, v_j)$ implies the existence of a monochromatic path between v_i and v_j) But the ones in Fig. 5b need separate treatment because cut-chain cycles appear.

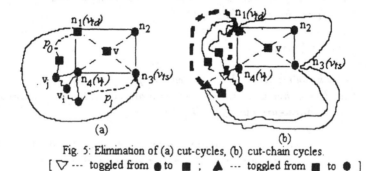

Fig. 5: Elimination of (a) cut-cycles, (b) cut-chain cycles.
[$\triangledown$ --- toggled from $\bullet$ to $\blacksquare$; $\blacktriangle$ --- toggled from $\blacksquare$ to $\bullet$]

Break_Cut_Cycle;
Inputs : $G, v_t, v_{ts}, v_{td}, NIG(v)$, monochromatic paths formed by S, the set of vertices colored so far, toggset;
Output : Valid acyclic-2-coloring of $S \cup \{v\}$
Technique :

$B1.\, just\_toggled := v_t; \; clr := color(v_t); \; check\_node := v_{td}; \; other\_node := v_{ts};$
$B2.\, for \; all \; v_i \in just\_toggled \; do$
  $begin$
$B3.$ $U_{v_i} := list \; of \; neighbors \; of \; v_i \; having \; color \; clr \; and$
              $clr\text{-}colored \; paths \; among \; them, \; in \; clockwise \; order;$
$B4.$ $if \; (\forall i, U_{v_i} = \phi, \; then \; return; \; \%no \; more \; cut \; or \; cut\text{-}chain \; cycles\%$

B5.   *else toggle all u_i except either the first or last in U_{v_i};*
         % Taking caution not to re-toggle any vertex %
   end;
B6. just_toggled := $\bigcup_k U_{v_k}$;
B7. if there is a cut cycle through any u_i and check_node then
    toggle neighbors of u_i as in Steps T7 to T11 in Toggle_Iterate;
B8. Swap check_node and other_node; clr := 3 - clr;
B9. Go to Step B2 for checking any further cut-chain-cycles.
 end;

Lemma 4 Break_Cut_Cycle *runs in $O(n)$ time when a vertex v in a graph of n vertices with $NIG(v) = K_4 - x$ is to be acyclic-2-colored and breaks all cut cycles which are not broken by Toggle_Iterate if it does have to re-toggle any vertex.*

Proof: Similar arguments as in Lemma 3. □

Case 1.2 : $NIG = C_4$ or its subgraphs.
(a) Elimination of block-cycles: The technique adopted is similar to *Toggle_Iterate* with the addition that 2-monochromatic paths between neighbor of a just toggled vertex to both v_{ts} and v_{tss} are to be detected [8]. Thus the for loop in Step T5 is for v_t, v_{ts}, v_{tss}. However, if a 2-monochromatic cycle containing v_{tss} is found at any stage, then in the subsequent iterations it is not necessary to search for cycles containing v_{ts} because of planarity.

(b) Elimination of cut-cycles: The technique is similar to the algorithm *Break_Cut_Cycle*.

Case 2: $|V_{NIG}| = 5$ (Fig. 6).
Here $N_i(v) \geq 2 : i = 1, 2$. For each color, a monochromatic path with endpoints e_1 and e_2 where $e_1, e_2 \in V_{NIG}$ is tested for. If such paths exist for both the colors, then obviously backtracking is needed. The criteria for selecting the vertex $n_i \in V_{NIG}$ to be toggled differs from case to case, and is included in the discussion for each individual case. The elimination of cut-cycles for this case is similar to the algorithm *Break_Cut_Cycle* discussed earlier. Therefore only the block cycles need to be discussed. The different NIGs in this case are as follows.

Case 2.1: $NIG = K_1 + (C_4 - x)$. (Fig. 6a)
Here, the vertex $n_i \in V_{NIG}$ to be toggled must satisfy :
a) $\deg(n_i) = 2$ in $NIG(v)$;
b) the neighbors of n_i are *differently* colored;
c) $N_{color(n_i)} = 2$.
If there is no vertex satisfying all three conditions enlisted above, then it has to be checked whether the degree 4 vertex (n_1 in Fig.6a) has at least two same colored neighbors. If so, then n_1 is toggled otherwise any vertex satisfying conditions

(a) and (b) but not (c) is toggled. The elimination of block-cycles here is exactly similar to the case $NIG = K_4 - x$.

Case 2.2 : $NIG = (K_1 + (C_4 - x)) - x$. (Fig. 6b)
The three conditions for selecting of $n_i \in V_{NIG}$ to be toggled for proper coloring of v in case 2.1 above are also applicable here. But if no vertex satisfying all three conditions exists, then n_i is a vertex meeting conditions (a), (b) and also the condition:
c') the two neighbors of n_i have no edge between them.

Possible subcases :
Case 2.2.1 : There is the edge (v_{ts}, v_{td}) (Fig. 7a).
Elimination of block cycles is done by algorithm *Toggle_Iterate* discussed earlier in Case 1.1.

Case 2.2.2 : No edge between v_{ts} and v_{td} (Fig. 7b).
Some modifications in *Toggle_Iterate* algorithm are required to tackle the situation. These are same as those mentioned in Case 1.2.

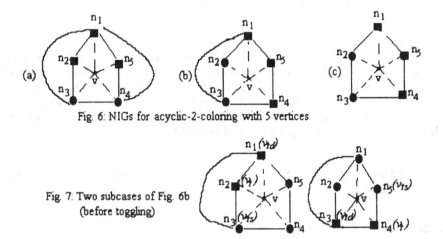

Fig. 6: NIGs for acyclic-2-coloring with 5 vertices

Fig. 7: Two subcases of Fig. 6b
(before toggling)

Case 2.3 : $NIG = C_5$ or its subgraphs. (Fig. 6c)
This is the most difficult situation with largest number of possibilities. To simplify the situation, the conditions for choosing a vertex $n_i \in V_{NIG}$ whose color is to be toggled are the following:
i) $\deg(n_i) = 2$;
ii) neighbors of n_i are *similarly* colored;
iii) $N_{color(n_i)} = 3$.
It can be easily seen that in the C_5 backtracking NIG, such a vertex n_i will always exist. This choice gives us an additional advantage as well in shortening the monochromatic paths obtained after coloring of v. The elimination of block cycles is done by making a few additions to the procedure *Toggle_Iterate* as follows [8]. This procedure is invoked twice, once with v_{td1} as check_node and then with v_{td2}. The basic strategy is similar to that for Case 1.1 but here monochromatic paths between the toggled vertex v_t and both of its neighbors v_{ts1} and v_{ts2} are relevant for creation of block cycle. If in this process, there is a vertex

u which has edges to both v_{ts1} and v_{ts2} and is toggled to $color(v_t)$, then vertex v_{ts2} (v_{ts1}) is toggled where check_node is v_{td1} (v_{td2}) and invoke Toggle_Iterate with $v_t := v_{ts2}$ (v_{ts1}); $v_{td} := u$; $v_{ts} := v_{td1}$ (v_{td2}).

3.4 Detecting non-existence of acyclic-2-coloring

When backtracking condition holds, some neighbor of v has to be toggled. Depending on the backtracking $NIG(v)$, there is a set of neighbors $S_{Toggle} \subseteq V_{NIG}$, any of which could be toggled. The algorithm fails if and only if $\forall n_i \in S_{Toggle}$, n_i has already been toggled once earlier in a previous recursive step (no vertex can be re-toggled). This is called the *re-toggling stopping condition*.

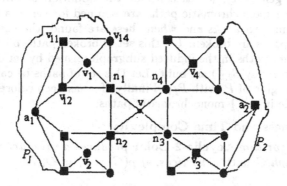

Fig. 8: Re-toggling stopping condition.

Lemma 5 *For a planar graph G having $\rho = 2$, algorithm* **Acyclic-2-Color** *does not encounter re-toggling stopping condition.*

Proof: Let us consider the situation shown in Fig. 8. While coloring v, it has neighbors n_1, n_2, n_3, n_4 each of which were toggled previously during the coloring of v_1, v_2, v_3, v_4 respectively. The neighbors of v_1, while it was colored were $v_{11}, v_{12}, v_{13}, v_{14}$; similarly for v_2, v_3, v_4. Since at all times, blocks are considered and there were no monochromatic cycles before the coloring of v, there are paths p_1 and p_2 as shown in Fig.8a. Also, as v is deleted before any of the above mentioned vertices, the minimum degree of the graph from which v was deleted is 4. So, there are at least two other vertices a_1, a_2 (refer Fig. 8a). Now, as $v_{11}, v_{12}, v_{13}, v_{14}$ are neighbors of v_1 while it is colored, therefore they were deleted after v_1. Similarly for v_2, v_3 and v_4. Now, after deleting v, either (i) a_1 or a_2 is deleted, or,
(ii) v_1 or v_2 or v_3 or v_4.
(1) If a_1 is deleted, then some v_{1i} is deleted before v_1, which is a contradiction.
(2) If v_1 is deleted after v, this is followed by the deletion of some v_{1i}, the deletion of a_1, then the deletion of some v_{2k}, after which v_2 is deleted (the minimum degree vertex of the residual graph is always deleted). A contraction is again obtained. Thus the situation in Fig. 8 cannot occur in a acyclic-2-colorable graph and some n_i can always be toggled to enable coloring of v and thereby get a valid 2-coloring. □

3.5 Monochromatic Path Management

The lists of monochromatic paths formed by vertices colored so far are consulted during the vital operation of backtracking. Here the storage and updation of monochromatic paths are discussed briefly. This covers Step 14 of our acyclic-2-coloring algorithm.

For both the colors, a set S of existing monochromatic paths for which the following invariant holds, is stored: For any path $p_i \in S$, there exists no $p_j \in S$, $(j \neq i)$, such that $p_i \subseteq p_j$.

Without loss of generality, let us consider the updation of 1-monochromatic paths. First, the 1-monochromatic paths are scanned for vertices that have just been toggled from 1 to 2. As and when these are found, the corresponding 1-monochromatic paths are broken. Let this set of broken paths be denoted as P_1. Then, the different paths in the induced subgraph formed by set of vertices just toggled from 2 to 1(say $S_{2,1}$) are built. Let this set of paths be called $P_{2,1}$. This is followed by merging of P_1 with $P_{2,1}$. Finally, if v has been colored 1, necessary changes are made in the 1-monochromatic paths.

3.6 Correctness and Time Complexity

Theorem 2 *Algorithm* **Acyclic-2-Color** *produces a valid acyclic-2-coloring of a planar graph G with n vertices, if $\rho(G) = 2$, in $O(n^2)$.*

Proof: The proof is based on the following facts.

(a) The 2-acyclic-coloring algorithm halts.
In each recursive call, a vertex is deleted from a biconnected component of the input graph. Ultimately, when the number of vertices ≤ 5, *Näive_Color_2* is invoked. Then, there are two options :
i) the algorithm reports failure while trying to color a vertex and thereby complete a recursive call;
ii) the recursive calls are completed for each block, after which the algorithm terminates successfully.
In either case, the 2-coloring algorithm halts.

(b) Algorithm Näive_Color_2 introduces no monochromatic cycles.
Let G_s be the small graph with which *Näive_Color_2* is invoked. Then, the only possibility of monochromatic cycles exists when for a color, say 1, N_1 is 3. If v_{start} is the first vertex in G_s to be colored, then $color(v_{start}) := 1$. Since, two of the neighbors of v_{start} are colored with 2, v_{start} can have at most one neighbor with color 1. Hence, no cycles are introduced.

(c) Given the coloring of the neighbors of a deleted vertex v, if no failure is reported, the coloring of v introduces no monochromatic cycles.
This follows from Lemmata 3 and 4.

(d) Given a valid coloring of the biconnected components, a valid coloring of the entire graph can be obtained.
This is based on the existence of safe ordering in Lemma 2.

(e) If the input graph is not 2-colorable, then a failure is reported.

From facts (a), (b), (c) and (d) above, it follows that if no failure is reported, the algorithm produces a valid 2-coloring. But if no failure is reported for an acyclic 3-colorable graph, then one of the two partitions produced by the algorithm must include a monochromatic cycle, i.e the coloring must be invalid. The proof follows by contradiction.

Finally, the correctness of our 2-coloring algorithm follows from Lemma 5 and observation (f) above.

The 2-acyclic-coloring algorithm presented in this section has a time complexity $O(n^2)$, where n is the number of vertices present in the input graph. There are $O(n)$ recursive calls. In each call, steps 1 which basically depth-first search on a planar graph, takes $O(n)$ time.Step is also linear time in the worst case and so are steps 12, 14 and 15. The remaining steps require constant time. □

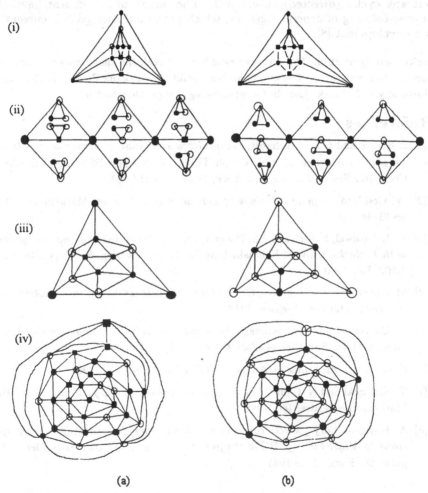

Fig. 9: Experimental results for acyclic-coloring with (a) 3 colors, (b) 2 colors on
(i) icosahedron, (ii) separable graph, (iii) maximal planar graph, (iv) 5-connected graph.

4 Experimental Results and Future Extensions

Both the algorithms discussed in this paper have been coded in PASCAL. Some of the results obtained on various types of planar graphs are shown in Fig. 9. The times on PC/AT-286 required for acyclic-3-coloring and acyclic-2-coloring these graphs are respectively less than 0.01 and 0.1 seconds. The algorithm Acyclic_2_Color reported failure on the geometric dual of the Tutte graph.

The acyclic-2-coloring algorithm can be improved by using a more efficient data structure for storing monochromatic paths and reducing the overhead of re-determining blocks after a vertex is deleted. There is scope for parallelising the algorithm since in the safe ordering of blocks, all blocks which are siblings (having same parent) in the block-cutpoint tree can be processed independently.

The algorithms described in this paper evidently work for planar digraphs as well. It may be worth investigating the class of planar graphs (digraphs) without any cycles (directed) of length 3. The design of an efficient heuristic for acyclic-coloring of general digraphs, which is the ultimate goal, is currently under development [8].

Acknowledgements: The authors would like to thank the anonymous referees for their constructive comments. The second author would like to thank Prof. J. S. Deogun of the University of Nebraska-Lincoln for introducing the problem to her.

References

[1] I. Broere and C.M. Mynhardt, *Generalized colorings of planar and outerplanar graphs*, Y. Alavi et al., eds., Graph Theory with Applications to Algorithms and Computer Science(Wiley, New York, 1985), pp 151-161.

[2] W. Goddard, *Acyclic colorings of planar graphs*, Discrete Mathematics 91, 1991, pp 91-94.

[3] V.D. Agrawal, S. Seth and J.S. Deogun, *Design for testability and test generation with 2 clocks*, Univ. of Nebraska-Lincoln, Dept. of Comp. Sc. Engg, Report series #102, Feb 1990.

[4] M. Garey and D.S. Johnson, *Computers and Intractability: A complete guide to NP-completeness*, Freeman, 1979.

[5] G. Chartrand and H.V. Kronk, *The point arboricity of planar graphs*, Journal of London Math. Soc., Vol 44, Part 4, Oct 1969.

[6] F. Harary, *Graph theory*, Addison Wesley Publishing Co., 1969.

[7] T. Nishizeki and N. Chiba, *Planar graphs: theory and algorithms*, North Holland Mathematics Studies 140, 1988.

[8] A. Roychoudhury , *Cycle-coloring of planar and general graphs with applications to VLSI testing* , BCSE Project Report, Jadavpur University, Dept. of Computer Sc. Engg., May 1995.

A Condition for the Three Colourability of Planar Locally Path Graphs

Ajit A. Diwan and N. Usharani

Department of Computer Science and Engineering,
Indian Institute of Technology, Bombay 400076.
email: {aad,usha}@cse.iitb.ernet.in

Abstract. A graph G is said to be locally path if for every vertex the subgraph induced by its neighbours is a path. Planar locally path graphs are a natural generalization of maximal outerplanar graphs. We show that they have a recursive construction which generalizes that of maximal outerplanar graphs. Using this characterization we obatin a 'local' condition for the three colourability of planar locally path graphs. As a corollary, we show that maximal planar graphs are three colourable iff every vertex has even degree. We also derive a structural property of maximal planar graphs.

1 Introduction

A graph G is said to satisfy a property locally if for every vertex v the neighbourhood of v, that is the subgraph induced by the set of vertices adjacent to v, satisfies the property. An important class of graphs are the locally cyclic graphs in which the neighbourhood of every vertex is a cycle. A locally cyclic graph is a triangulation of a surface in which every three cycle forms the boundary of a face. Thus planar locally cyclic graphs are precisely *4-connected* maximal planar graphs [2].

Locally cyclic triangulations of surfaces have been studied in [5], where it has been shown that all such triangulations of a given surface can be obtained from a finite number of basic triangulations using the operation of vertex splitting. Recent results in [7] have shown that for any $\epsilon > 0$, there are locally cyclic graphs with n vertices and $> n^{2-\epsilon}$ edges and this is the best possible. Locally cyclic triangulations have also been called 'Clean triangulations' in [4].

Locally path graphs, in which the neighbourhood of every vertex is a path, are triangulations of surfaces with 'holes' in which every three cycle is a face. A maximal outerplanar graph, which is locally path, is a triangulation of the sphere with one 'hole'. In general, planar locally path graphs are triangulations of the sphere with $k \geq 1$ 'holes'. Thus they generalize maximal outerplanar graphs in a natural way.

Maximal outerplanar graphs have a well known recursive construction which follows from the fact that they are chordal graphs [1, 3]. Any maximal outerplanar graph with > 3 vertices has at least two non-adjacent vertices of degree two. Deleting any vertex of degree two in a maximal outerplanar graph gives a smaller maximal outerplanar graph. This characterization also proves that maximal outerplanar graphs are three colourable and has been used in a simple proof of the *Art Gallery Theorem* [6].

In this paper we generalize the recursive construction for maximal outerplanar graphs to all planar locally path graphs. We show that in any planar locally path graph we can delete one of three possible subgraphs to obtain a smaller planar locally path graph. We use this recursive construction to find a 'local' characterization of three colourable planar locally path graphs. Intuitively, this characterization shows that a planar locally path graph is three colourable iff every 'hole' is locally similar to the single hole in a maximal outerplanar graph.

The results for planar locally path graphs can be generalized to planar graphs in which the neighbourhood of every vertex is either a path or a cycle and, in particular, for planar locally cyclic graphs. We show that a planar locally cyclic graph is three colourable iff every vertex has even degree. This result extends to all maximal planar graphs as well. An interesting structural property of maximal planar graphs is obtained as a consequence. We show that there is no maximal planar graph with exactly two vertices of odd degree such that they are adjacent.

The results in this paper are applicable only for planar graphs. They cannot be extended directly to all locally path or locally cyclic graphs. In fact, there are toroidal locally cyclic graphs in which every vertex has even degree but they are not three colourable. The recursive construction described for planar locally path graphs also does not hold for non-planar locally path graphs.

The paper is organized as follows. In section 2 we give the basic characterization of planar locally path graphs and sketch the proofs. In section 3 we consider the three colourability of planar locally path graphs. In section 4 we consider the generalization to planar locally cyclic and maximal planar graphs. Finally, in section 5 we conclude with some possible generalizations.

2 Characterization

In this section we show that planar locally path graphs have a recursive construction which generalizes that of maximal outerplanar graphs. We only sketch the proofs here, and the details can be found in [8]. We assume that all graphs considered are connected. The terminology used is standard except that we refer to cycles of length three as triangles. Before coming to the main result, we state some properties of planar locally path graphs which follow from the definition.

Let G be a planar locally path graph with > 3 vertices. Then

1. Every edge in G is contained in at least one and at most two triangles. This follows from the fact that the number of triangles containing an edge uv is just the degree of u in the neighbourhood of v, which is the same as the degree of v in the neighbourhood of u. This is at most 2 since the neighbourhoods are paths. The edges of G contained in exactly one triangle will be called *boundary* edges, while those in exactly two triangles will be called *internal* edges. A triangle will be called an *internal* triangle if all its edges are internal, else it will be called a *boundary* triangle.

2. Every vertex has exactly two boundary edges incident on it. Thus the set of boundary edges form a *2-factor* of G with k connected components, $k \geq 1$, where each component is a cycle of length ≥ 4. We refer to these cycles as *holes*.

3. There is a planar embedding of G such that every face is either a hole or a triangle and every hole and triangle in G is a face. This can be obtained from a natural cyclic ordering of the edges incident on each vertex, defined by the order in which the neighbours appear in the path induced in the neighbourhood. We call this embedding a *standard* embedding of G. As a consequence of this, if G has k holes, $k \geq 2$, then there are exactly $2k - 4$ internal triangles in G. Further, if every vertex in G has degree ≥ 3, then there are exactly n boundary triangles in G and $m = 2 * n + 3 * k - 6$. This follows from *Eulers formula*.

Henceforth, when we consider a planar locally path graph, we will actually consider a standard embedding of the graph and also assume that the external face is a triangle. The interior and exterior of a cycle will be defined with respect to this embedding.

Definition 2.1 *An* alternating cycle *in a standard embedding of a planar locally path graph G is an even cycle in G consisting of alternately boundary and internal edges, such that all boundary edges belong to different holes and all vertices not in the cycle are in one of the regions bounded by the cycle (either the interior or the exterior), with the other region containing no vertices.*

Note that in a standard embedding of G, the interior(exterior) of any alternating cycle will be triangulated. If the boundary edges and the interior (exterior) edges of an alternating cycle are deleted, the resulting graph is a planar locally path graph. Also, deleting a vertex of degree 2 gives a smaller locally path graph.

Theorem 2.1 *In a standard embedding of a locally path graph G with > 3 vertices, there is either a vertex of degree two or an alternating cycle of length four or six.*

Proof. We give a sketch of the proof here, the detailed version may be found in [8]. Assume that every vertex in G has degree ≥ 3. The main idea behind the proof is to do certain reductions on the graph which preserve the existence of alternating cycles. When an irreducible graph is obtained, a simple counting argument is used to show the existence of an alternating cycle. The reductions done are as follows

1. Let uvw be a boundary triangle with uv as the boundary edge. Suppose vertex w belongs to the same hole as the edge uv. Then the graph obtained by contracting the edge uv is also a planar locally path graph and an alternating cycle in the contracted graph is also an alternating cycle in G.

2. Let u be a vertex of degree 3. Let uv and uw be the two boundary edges incident on u and uz be the third edge. Delete the vertex u and all edges

incident on it and add the edge vw if it is not already present, so that vwz is a face in the resulting embedding. Mark this edge as a boundary edge. Note that the graph so obtained may not be locally path, and the edge added may be contained in more than one triangle but is still treated as a boundary edge. However, an alternating cycle in the reduced graph gives rise to an alternating cycle in the original graph.

3. Let u be a vertex of degree 4 and let $v\ w\ x\ y$ be the induced path in its neighbourhood. Suppose the edges vw, wx and xy are all internal edges. Note that all four vertices must be distinct, for otherwise we get an internal triangle which is not a face. However, all internal triangles in the original graph are faces, and the reductions performed do not create any internal triangle which is not a face. Delete vertex u and add the edges vy (if not present already) and either wy or vz so as not to create an internal triangle which is not a face. One of the two edges can always be added. Mark the edge vy as a boundary edge. Again, it can be verified that an alternating cycle in the reduced graph gives an alternating cycle in the original graph.

The reductions may create boundary triangles which are not faces. Since we are only interested in boundary triangles which are faces, we will refer to them as boundary triangle faces. The reductions ensure that if in the reduced graph two boundary triangle faces have an edge in common, they form an alternating cycle of length 4. Also if an internal triangle (which is necessarily a face) has no edge in common with another internal triangle, then it forms an alternating 6 cycle along with the boundary triangle faces which include its edges. The remaining step is to show that one of the two must exist.

Note that in the reduced graph a hole may be reduced to a single boundary edge. The number of boundary triangle faces is the same as the number of vertices in the reduced graph. The number of holes is still k and the number of internal triangles is $2k-4$, since the reductions preserve the number of holes and internal triangles. Since every hole has size at least 2, the number of boundary triangle faces is at least $2k$.

Every boundary triangle face has two other triangular faces which have an edge in common with it, while every internal triangle has three such faces. If any two boundary triangle faces have an edge in common they form an alternating cycle of length 4. If there is no such pair of boundary triangle faces, we consider an auxiliary graph whose vertices are the internal triangles. Two vertices in the auxiliary graph are adjacent iff there is a boundary triangle face which has edges in common with the internal triangles corresponding to the two vertices. The auxiliary graph thus has $2k-4$ vertices and at least $2k$ edges since there is an edge corresponding to every boundary triangle face. Therefore there must be at least 8 vertices of degree 3 in the graph. A vertex of degree 3 in the auxiliary graph corresponds to an internal triangle which has no edge in common with any other internal triangle. This gives an alternating cycle of length 6.

In fact, we can strengthen the statement of the above theorem. Given a standard embedding of a planar locally path graph with the external face being a triangle, there is either an interior vertex of degree 2 or there is an alternating cycle of length 4 or 6 whose interior does not contain any vertex. Thus we can always delete a vertex or an appropriate set of edges to get a standard embedding of a smaller planar locally path graph. This extension is required for the generalization of these results to planar locally connected graphs [8].

As a consequence of Theorem 2.1, we obtain the following recursive construction for planar locally path graphs.

Theorem 2.2 *G is a planar locally path graph iff there is a sequence of planar locally path graphs $G_1, G_2, \ldots, G_l$ such that G_1 is a triangle, $G_l = G$, and G_{i+1} is obtained from G_i by one of the following three operations, $1 \leq i < l$.*

1. *Add a new vertex in the interior of some hole in G_i and join it to the endpoints of some boundary edge of that hole.*
2. *Let uv and xy be boundary edges belonging to some hole in G_i such that when the boundary is traversed in a clockwise direction the vertices are encountered in the order u, v, x, y. Add the edges uy, vx and either ux or vy in the interior of the hole provided the distance in G_i between the endpoints of the added edges is ≥ 3.*
3. *Let uv, xy and wz be boundary edges belonging to some hole in G_i such that when the boundary is traversed in a clockwise direction the vertices are encountered in the order u, v, x, y, w, z. Add the edges vx, yw, zu and either the set of edges { ux, xw, wu } or the set of edges { vy, yz, zv } in the interior of the hole, provided the distance in G_i between the endpoints of all the added edges is ≥ 3.*

Note that in step 1 of Theorem 2.2 the length of an existing hole is increased by one. This is the only step necessary for constructing maximal outerplanar graphs. In step 2 an existing hole is split into two smaller holes while in step 3 it is split into 3 smaller holes. Also when step 2 is applied an alternating cycle of length 4 is created while when step 3 is applied an alternating cycle of length 6 is created. None of the three steps is redundant, since there are planar locally path graphs which have exactly one of the three possible configurations.

3 Three Colourability

In this section we derive a characterization of three colourable planar locally path graphs. Note that there is a simple sequential algorithm to test whether a given planar locally path graph is three colourable. We start with any edge and colour its endpoints by two different colours arbitrarily. Once a vertex and one of its neighbours has been assigned a colour, the colour of all its neighbours is determined. This follows from the fact that the neighbourhood is a path which is uniquely two colourable. We continue colouring the vertices in this way till either all vertices have been properly coloured or some vertex cannot be coloured. This is a sequential algorithm which can also be parallelised to obtain a fast parallel algorithm. Construct an auxiliary graph with the same set of vertices. For every internal edge uv in the original graph we add an edge wz in the auxiliary graph where w and z are the other vertices in the two triangles which contain uv. Thus in any proper three colouring of the original graph w and z must get the same colour. It can now be verified that the original graph is three colourable iff the auxiliary graph has exactly three connected components. These algorithms do not depend on the planarity of the graph and work correctly for all locally path (or cyclic) graphs.

Our characterization of three colourable planar locally path graphs is local in nature, and gives more insight into their structure. In a sense, we identify forbidden configurations which cause the three colouring algorithm to fail. This is similar to two colouring a connected graph, where an odd cycle is the only obstruction to two colouring. In the case of planar graphs, the graph is bipartite iff every face is of even length. We show that holes of a particular type are the only obstructions to three colouring planar locally path graphs. Our characterization also leads to an alternative parallel algorithm for deciding three colourability without actually finding a three colouring if one exists. However, in view of the fact that there is a simple parallel algorithm for this problem, we omit the discussion of this.

We first describe the intuitive idea behind the characterization. Consider a vertex u whose neighbourhood is a path with endpoints v and w. The edges uv and uw are the two boundary edges incident on u. If u has odd degree then in any proper three colouring of the graph v and w must have the same colour, while if u has even degree they must have different colours. This motivates the following definition.

Definition 3.1 *A hole is said to be* consistent *if the vertices of the hole can be properly three coloured so that the two vertices adjacent to a vertex u have the same colour iff u has odd degree in the graph. Any three colouring of the hole satisfying this condition will be called a* consistent *colouring of the hole.*

Clearly, if the graph is three colourable every hole must be consistent. The main result of this section is to show that the converse is also true.

Theorem 3.1 *A planar locally path graph G is three colourable iff every hole in G is consistent. Moreover, if G is not three colourable there are at least two inconsistent holes in G.*

Before proving the main theorem, we introduce some formalism which will help in the proof.

Definition 3.2 *Let $v_1, v_2, \ldots, v_l$ be the vertices on the boundary of some hole H in clockwise order. We associate a word w with the hole defined by $w = a_1 a_2 \ldots a_l$ where $a_i = o$ if v_i has odd degree and $a_i = e$ if it has even degree, $1 \leq i \leq l$, and denote this word by $s(H)$. Note that a word associated with a hole is uniquely defined upto a cyclic permutation. For any two vertices v_i, v_j on the boundary of H, we denote by $s(v_i v_j)$ the substring of $s(H)$ $a_i \ldots a_j$.*

Definition 3.3 *We define a finite automaton* $F(Q, \Sigma, \delta)$, *where* $Q = \{$ *ab, ac, ba, bc, ca, cb* $\}$ *is the set of states,* $\Sigma = \{o, e\}$ *is the alphabet, and the transition function* $\delta : Q \times \Sigma \to Q$ *is defined by* $\delta(xy, o) = yx$ *and* $\delta(xy, e) = yz$, *where xyz is any permutation of abc. As usual, we extend the transition function to* $Q \times \Sigma^*$ *and denote it by* δ^+. *We define a congruence relation* $\equiv$ *on* Σ^* *as* $w_1 \equiv w_2$ *iff* $\delta^+(q, w_1) = \delta^+(q, w_2)$, *for all* $q \in Q$.

We now state a few simple properties of the automaton F which follow from the definition.

Lemma 3.1 $w_1 \equiv w_2$ *iff* $\delta^+(q, w_1) = \delta^+(q, w_2)$ *for some* $q \in Q$. *w is accepted by the automaton iff* $w \equiv \lambda$. *Every word* $w \in \Sigma^*$ *is congruent to one of the words* λ, *o, e, oe, eo, ee.*

Lemma 3.2 *Let uv and xy be two edges on the boundary of some hole H such that the vertices are encountered in the order* u, v, x, y *in a clockwise traversal of the boundary. Suppose in some consistent colouring of the hole u and v are coloured a and b respectively. Then x and y must be coloured p and q respectively,* $p, q \in \{a, b, c\}$, *where* $\delta^+(ab, s(vx)) = pq$.

As a corollary of Lemma 3.2, we have

Lemma 3.3 *A hole H is consistent iff* $s(H) \equiv \lambda$.

We now come to the proof of Theorem 3.1. The proof uses the recursive construction for planar locally path graphs and the characterization of consistent holes obtained above.

Proof. (of Theorem 3.1).

It is sufficient to prove that if a planar locally path graph G is not three colourable there are at least two inconsistent holes in G, since if G is three colourable any three colouring of G is a consistent colouring of every hole in G.

Let G be a planar locally path graph. By Theorem 2.2, we can find a sequence of planar locally path graphs $G_1, G_2, \ldots G_l$ such that G_1 is a triangle, $G_l = G$ and G_{i+1} is obtained from G_i using one of the steps given in Theorem 2.2. The proof will be by induction on l. The base case is when G is a triangle, it has exactly one hole whose associated word is *eee* which is accepted by the automaton and the graph is clearly three colourable.

Now assume that the result is true for G_i. We consider three cases depending on the operation used to obtain G_{i+1}. In each of the three cases, there is only one hole H in G_i which is not a hole in G_{i+1}. The hole H in G_i is replaced by either one, two or three holes in G_{i+1}, depending on which operation is used to obtain G_{i+1} from G_i. We denote the new holes created in G_{i+1} by H_1, H_2 and H_3.

1. Suppose G_{i+1} is obtained from G_i by applying step 1 to hole H. Then s(H_1) is obtained from s(H) by changing two consecutive letters in s(H) and inserting an *e* between them. It can be easily verified that s(H_1) $\equiv$ s(H), hence hole

H_1 is consistent iff H is consistent. Thus if G_{i+1} is not three colourable then G_i is also not three colourable, and by induction, there are at least two inconsistent holes in G_i. These holes, possibly with H replaced by H_1, give two inconsistent holes in G_{i+1}.

2. Suppose G_{i+1} is obtained from G_i by applying step 2. Let u, v, x, y be the relevant vertices of the hole H and assume that the edges added are vx, yu and vy (the other case can be argued symmetrically). Then $s(H_1)$ is obtained from $s(vx)$ by changing the last letter and $s(H_2)$ is obtained from $s(yu)$ by changing the last letter. Thus $s(H_1) \equiv \lambda$ iff $s(vx) \equiv oe$ and $s(H_2) \equiv \lambda$ iff $s(yu) \equiv oe$. Suppose G_{i+1} is not three colourable. If G_i is also not three colourable then there are at least two inconsistent holes in G_i. If H is one of them then either H_1 or H_2 must be inconsistent, for otherwise, $s(H_1) \equiv \lambda$ and $s(H_2) \equiv \lambda \Rightarrow s(vx) \equiv oe$ and $s(yu) \equiv oe \Rightarrow s(H) \equiv s(vu) \equiv oeoe \equiv \lambda$. Thus there are at least two inconsistent holes in G_{i+1}. On the other hand, suppose G_i is three colourable. If $s(H_1) \equiv \lambda$ then $s(vx) \equiv oe$. This implies (by Lemma 3.2) that in a three colouring of G_i if vertex u and v are coloured a and b respectively, vertices x and y have colours a and c respectively. This means that the three colouring of G_i is also a three colouring of G_{i+1}, a contradiction. Thus $s(H_1) \not\equiv \lambda$ and also $s(H_2) \not\equiv \lambda$ by a symmetrical argument. Thus both H_1 and H_2 are inconsistent holes in G_{i+1}.

3. Suppose G_{i+1} is obtained from G_i by applying step 3 and let u, v, x, y, w, z be the relevant vertices of hole H. Assume the edges added are $\{vx, yw, zu, vy, yz, zv\}$. Then $s(H_1)$ is obtained from $s(vx)$ by changing the first and last letter, and $s(H_2)$ and $s(H_3)$ are obtained in a similar way from $s(yw)$ and $s(zu)$. Thus $s(H_1) \equiv \lambda$ iff $s(vx) \equiv ee$. Now the argument is very similar to case 2. If H_1, H_2, H_3 are all consistent, then $s(H) \equiv eeeeee \equiv \lambda$. Hence if H is inconsistent then at least one of H_1, H_2, H_3 must be inconsistent. Thus if G_i is not three colourable, two inconsistent holes in G_i give at least two inconsistent holes in G_{i+1}. Suppose G_i is three colourable, and hence $s(H) \equiv \lambda$. Suppose both $s(H_1)$ and $s(H_2)$ are $\equiv \lambda$. Then $s(vx) \equiv ee$ and $s(yw) \equiv ee$. Thus in the three colouring of G_i if u and v are coloured a and b respectively, x and y must be coloured c and a respectively while w and z must be coloured b and c respectively and the three colouring of G_i is also a three colouring of G_{i+1}, a contradiction. Thus both H_1 and H_2 cannot be consistent, and by a symmetric argument at least two of the three holes must be inconsistent.

This completes the proof of the main theorem.

4 Generalizations

In this section we briefly mention some generalizations of the results. In particular, for planar graphs in which the neighbourhood of a vertex is either a path or a cycle, we have a similar recursive construction which uses one more operation [8]. If uv and vw are two consecutive edges on the boundary of some hole and u

and w have no common neighbour other than v, then the edge uw can be added in the interior of the hole. This is the only operation which creates a cycle in the neighbourhood of a vertex (v). Arguing exactly as before we get the following results.

Theorem 4.1 *Let G be a planar graph in which the neighbourhood of every vertex is either a cycle or a path. Then G is three colourable iff every hole in G is consistent and every vertex whose neighbourhood is a cycle has even degree. In particular, a locally cyclic planar graph is three colourable iff every vertex has even degree. Moreover, if such a graph is not three colourable, there are at least two inconsistent components, where an inconsistent component is either an inconsitent hole or a vertex of odd degree whose neighbourhood is a cycle.*

A locally cyclic planar graph is a *4-connected* maximal planar graph. The above result can also be generalized to all maximal planar graphs.

Theorem 4.2 *A maximal planar graph is three colourable iff every vertex has even degree.*

Proof. It is sufficient to prove that if a maximal planar graph is not three colourable it must have a vertex with odd degree. We prove this by induction on the number of triangles which are not faces in some embedding of the maximal planar graph. Such triangles are separating triangles, since deleting the vertices in these triangles disconnects the graph. Our induction hypothesis is stronger than the statement of the theorem. We show that there is no maximal planar graph with exactly two vertices of odd degree such that the two odd degree vertices are adjacent. The base case is when there are no separating triangles, in which case the statement of the theorem follows from Theorem 4.1. The second assumption is also true for if there is a locally cyclic graph with exactly two odd degree vertices which are adjacent, then by deleting the edge joining them we get a graph with one hole whose associated word is $eeee$. Hence this graph is not three colourable and must have another inconsistent component (Theorem 4.1), which must be a vertex of odd degree.

Now consider a maximal planar graph G with $k \geq 1$ triangles which are not faces. Let T be any one of these triangles. The subgraphs induced by the vertices in the interior(exterior) of T together with the vertices in T are also maximal planar graphs with fewer number of separating triangles. Denote these by G_i and G_e respectively. G is three colourable iff both G_i and G_e are three colourable. Thus if G is not three colourable, at least one of G_i or G_e is not three colourable. Without loss of generality, assume that G_i is not three colourable. Thus, by induction hypothesis, there are at least two vertices of odd degree in G_i. If there are exactly two odd degree vertices, they are not adjacent and hence cannot both be in T. Thus at least one of them is an odd degree vertex in G. If there are more than two odd degree vertices in G_i, then there are at least four and at least one of them is not in T and is thus an odd degree vertex in G. To prove the second assumption, note that if both G_i and G_e are not three colourable, by the same argument as above, there is an odd degree vertex in both the subgraphs which is not in T, and hence there are two non-adjacent odd degree vertices in G. On the other hand, if G_e is three colourable, every vertex in it must have even degree. Hence a vertex has odd degree in G iff it has odd degree in G_i. Thus if G has exactly two odd degree vertices which are adjacent, then G_i also has exactly two odd degree vertices which are adjacent, contradicting the induction hypothesis. This completes the induction step and the proof follows.

The second assumption in the proof of Theorem 4.2 is of independent interest. We state it here as a theorem, whose proof follows from the proof of Theorem 4.2.

Theorem 4.3 *There is no maximal planar graph with exactly two vertices of odd degree such that the two odd degree vertices are adjacent.*

5 Conclusions

A natural question to ask would be whether these results can be generalized to all locally path graphs. Unfortunately, the results do not hold even for toroidal graphs. In particular, there are toroidal locally path graphs which are not three colourable even though all holes are consistent. Also there are toroidal locally cyclic graphs which are not three colourable even though all vertices have even degree. There also does not appear to be any recursive construction using only deletions of a finite family of subgraphs for general locally path graphs. An interesting question would be to find a family of forbidden structures (like the odd wheel for the planar case) for the three colourability of locally cyclic graphs.

References

1. M. C. Golumbic. *Algorithmic Graph Theory and Perfect Graphs*, Academic Press, New York, 1980.
2. J. Gross and A. Tucker. *Topological Graph Theory*, Wiley, New York, 1987.
3. F. Harary. *Graph Theory*, Addison-Wesley, Reading, MA, 1969.
4. N. Hartsfield and G. Ringel. Clean Triangulations, *Combinatorica*, Vol. 11, 1991, pp 145-155.
5. A. Malnic and B. Mohar. Generating locally cyclic triangulations of surfaces, *Journal of Combinatorial Theory Ser B*, Vol. 56, Nov. 1992, pp 147-164.
6. J. O'Rourke. *Art Gallery Theorems and Algorithms*, Oxford Press, New York, 1987.
7. A. Seress and T. Szabo. Dense Graphs with Cycle Neighbourhoods, *Journal of Combinatorial Theory Ser B*, 63(2), Mar. 1995, pp 281-293.
8. N. Usharani. A Study of Planar Locally Connected Graphs, Technical Report, Department of Computer Science and Engineering, I.I.T. Bombay, July 1995.

A Framework for the Specification of Reactive and Concurrent Systems in Z

Peter Baumann and Karl Lermer

Institut für Informatik der Universität Zürich
Winterthurerstr. 190, 8057 Zürich, Switzerland

Abstract. The formal specification language Z is used to specify transformational programs. We show in analogy to [2] that specifying concurrent systems in Z means conjoining the individual specifications and defining the input-output relations for the processes. The TLA approach to fairness is adapted to define liveness conditions for concurrent systems. Machine closure of the specifications will be proved. As a case study we apply our framework to the Bakery Algorithm where a complete Z specification and a correctness proof are presented.
Keywords: Z, formal methods, TLA, reactive systems

1 Introduction

Over the years Z has been successfully applied in numerous application areas and it has proved to be expressive enough for the specification of transformational programs, i.e. programs that can be specified by input/output relations [9]. Since more complex programs can be adequately specified only by describing their behaviors as infinite sequences of states it is reasonable to investigate whether and how Z can be used in this case, without extending existing Z notation and conventions.

In this sense we present techniques for the specification of complex systems in Z notation. This is done using concepts of nondeterministic finite state machines and temporal logic, especially TLA. Given a standard Z specification we first describe the underlying abstract state machine as a set of admissible sequences of operations and according states respecting the Z intuition of operation composition. We call it the according Z Machine. Then, we introduce fairness properties as operators on sequences of operations inspired by the strong and weak fairness approach of TLA [7]. A Z Machine together with strong and weak fairness assumptions on operations is called a Z System. It proves to be adequate for the description of complex systems which are not ruled by an ongoing interaction with their environment.

These investigations automatically lead to the question if a similar theory for the specification of reactive and concurrent systems could be developed, adhering to the existing Z notation and conventions. Familiar with the schema calculus [10] one could guess that this can be achieved by conjoining Z Systems which in addition would be in analogy to the TLA approach for conjoining specifications in [2].

Having this in mind we improve our specification theory in order to handle reactive systems [9] which communicate with their environment by input and output variables. The main problem is to introduce dependence on the environment in an adequate way. A generalized Z Machine takes into account not only operation and state sequences but also sequences of input values which simulate the information flow coming from the environment. Of course, liveness conditions for the process can be influenced by the underlying environment, too. Thus, we also involve dependency on the input flow in the definitions of weak and strong fairness. This allows us to define generalized Z Systems which now are adequate for the specification of the process components of reactive systems. Another beautiful aspect is the machine closure of generalized Z Systems.

In a next step we investigate how the underlying Z Machines can be joined in order to describe the whole reactive system. We derive in analogy to [1] that specifying a concurrent system of several processes means conjoining their generalized Z Machines (systems) and defining their input-output relations. An important characteristic of those so-called concurrent Z Machines is that they specify safety properties in a classical sense and that concurrent Z Systems inherit the property of being machine closed from their underlying generalized Z Systems.

L. Lamport's Bakery Algorithm [6] is a well known, concise example for a concurrent system and there exist a lot of investigations for several variants including specification and verification. Inspired by an evolving algebra approach [3] we apply our techniques and derive a complete Z specification for the Bakery Algorithm using a concurrent Z System. This enables us to provide yet another correctness proof.

1.1 Organization of the Paper

The next section introduces Z Machines and Z Systems for the specification of complex systems. How these techniques can be improved to specify process components of reactive systems will be shown in section three. It includes the generalized notions of Z Machines and Z Systems. Conjoining generalized Z Machines leads to complete specifications of reactive systems and will be presented in section four. That Z machines bear natural topologies and that they specify safety properties in a classical sense will be explained in section five. We prove the somewhat surprising fact that Z Systems are machine closed with respect to our notion of convergence. In the final part of our paper we present L. Lamport's Bakery Algorithm as a concurrent Z System and we use this Z specification to prove correctness and deadlock freedom.

We assume that the reader is familiar with the basic concepts of Z [10].

2   Z Machines and Z Systems

In the case of composition of nondeterministic operations, Z requires a demonic interpretation. A very clear motivation of demonic interpretations is given in [13]. In this section, we show how to obtain demonic interpretations of sequences of operations, thereby making explicit the abstract machine underlying a Z specification.

Given a Z specification with state schema $State$ and operation schemas $init, op_1, ..., op_n$ we define

$$OP == \{init, op_1, ..., op_n\} \cup \{(\Xi State)\}$$

as the set of admissible operations of the machine. $\Xi State$ will be used for the simulation of stuttering steps.

The composition of operations which are defined in the underlying Z specification can be interpreted as a sequence of operations in OP starting with $init$. In this sense, the following so-called simple machine SM defines sequences of operations and sequences of states requiring that the sequences of states could have been produced by the corresponding operation sequence[1].

SM

$\delta : \mathbf{N} \to OP; \ \sigma : \mathbf{N} \to State$

$\delta 0 = init \wedge (\forall n : \mathbf{N} \bullet csb(\sigma(n), \sigma(n+1)) \in \delta\, n)$

Introducing stuttering steps has the additional advantage that terminating computations can be modeled by sequences where there exists an index k such that $\delta(i) = \Xi State$ for all $i \geq k$ ([1], [9]).

A demonic interpretation requires control over all possible state sequences for a given operation sequence. This can be achieved by using a standard trick from the theory of nondeterministic finite state machines:

---

[1] The concatenation of state bindings (csb) is not a standard Z operation but can easily be defined by $csb : State \times State \to \Delta State \mid csb = \{\Delta State \bullet ((\theta State, \theta State'), \theta(\Delta State))\}$

$$\Lambda : (\mathbf{N} \rightarrowtail OP) \rightarrow (\mathbf{N} \rightarrow \mathbf{P}\ State)$$

$$\forall \delta : (\mathbf{N} \rightarrowtail OP) \bullet (\Lambda\delta)0 = State \land$$
$$(\forall n : \mathbf{N} \bullet (\Lambda\delta)(n+1) = \{t : State \mid \exists s : (\Lambda\delta)n \bullet csb(s,t) \in \delta n\})$$

The set *Cones* encompasses those sequences of operations which can be performed demonically irrespective to the nondeterministically chosen states [2].

$$Cones == \{\delta : \mathbf{N} \rightarrow OP \mid \delta 0 = init \land (\forall n : \mathbf{N}_1 \bullet (\Lambda\delta)n \subseteq Pre(\delta n))\}$$

The above introduced notions enable us now to state the central concept of a Z Machine *ZM* of a Z specification. *ZM* describes operation and state sequences capturing the intuitive meaning of executing the operations of the Z specification.

$$ZM \cong [SM \mid \delta \in Cones]$$

Liveness conditions can be specified in a systematic way as in the TLA approach [7]. The central idea is to incorporate conjunctions of weak and strong fairness conditions for operations into our specifications. An operation sequence fulfils the strong fairness condition with respect to an operation if either this operation occurs infinitely many times in the sequence or it is impossible to apply this operation from a certain point on.

$$SF : OP \rightarrow \mathbf{P}(\mathbf{N} \rightarrow OP)$$

$$\forall op : OP \bullet SF\ op =$$
$$\{\delta : \mathbf{N} \rightarrow OP \mid (\forall n : \mathbf{N} \bullet \exists k : \mathbf{N} \mid k \geq n \bullet \delta k = op) \lor$$
$$(\exists n : \mathbf{N} \bullet \forall k : \mathbf{N} \mid k \geq n \bullet \neg ((\Lambda\delta)k \subseteq Pre\ op))\}$$

An operation sequence fulfils the weak fairness condition with respect to an operation if either this operation occurs infinitely many times in the sequence or it is infinitely often impossible to apply this operation.

$$WF : OP \rightarrow \mathbf{P}(\mathbf{N} \rightarrow OP)$$

$$\forall op : OP \bullet WF\ op =$$
$$\{\delta : \mathbf{N} \rightarrow OP \mid (\forall n : \mathbf{N} \bullet \exists k : \mathbf{N} \mid k \geq n \bullet \delta k = op) \lor$$
$$(\forall n : \mathbf{N} \bullet \exists k : \mathbf{N} \mid k \geq n \bullet \neg ((\Lambda\delta)k \subseteq Pre\ op))\}$$

Directly from the definitions we derive that strong fairness implies weak fairness, i.e. $SF(op) \subseteq WF(op)$ for all $op \in OP$.

---

[2] We need the following extended version of the "pre operator": $Pre : \mathbf{P}\ \Delta State \rightarrow \mathbf{P}\ State \mid \forall A : \mathbf{P}\ \Delta State \bullet PreA = \{s : State \mid \exists t : State \bullet csb(s,t) \in A\}$

Z Systems are defined by Z Machines together with weak and strong fairness conditions for operations in the set OP.

$$ZS \mathrel{\widehat{=}} [ZM \mid \delta \in WF \; op_j \wedge \delta \in SF \; op_i \wedge \ldots]$$

An important feature of Z Machines and Z Systems is the so-called stuttering invariance, i.e. if we choose any binding zm in ZM, respectively ZS and add or cancel stuttering steps (except for index 0) in the operation sequence $zm.\delta$ and simultaneously in the state sequence $zm.\sigma$ we get a binding in ZM, respectively ZS again.
Z Systems proved to be appropriate for the description of complex systems which are independent from their environment. However, this concept often fails when processes have to be specified which live in an ongoing interaction with their environment.

3 Generalized Z Machines and Z Systems

A classical reactive program *runs concurrently with its environment* and maintains an *ongoing interaction* ([9]). Hence, studying those programs involves the analysis of interaction, between a program and its environment. Our goal will be to generalize the above stated model for processes communicating with their environment by input and output variables. Apart from sequences of operations and states we have to involve components which simulate the information flow between the process and its environment. Thus, a generalized Z Machine for a reactive system combines sequences of operations, states, input and output values and determines their relation.
We assume a Z specification for a process component of a given reactive system where the state space schema *State* is devided in input (Input), output (Output) and the internal variables (Intern).

$$State \mathrel{\widehat{=}} Intern \wedge Input \wedge Output$$

As in the previous section we first define all operations which determine the finite state machine: $init, op_1, \ldots, op_n$. For the simulation of stuttering steps it is inconvenient to take $\Xi State$. We have to ensure that the environment is allowed to change input values while the process itself is stuttering. This causes that we define

$$stut \mathrel{\widehat{=}} \Xi(Output \wedge Intern) \wedge \Delta State$$

and we get $OP == \{init, op_1, \ldots, op_n\} \cup \{(stut)\}$ for the operation set. The environment provides the underlying process with input values. Thus, when specifying any operation op in OP we are not allowed to

67

restrict the primed variables of the input part. This can be expressed equivalently by assuming $(\exists\, Input' \bullet op) \wedge Input' = op$.

Under this circumstances, a demonic interpretation requires flexibility with respect to the input. We assume that a process may get new input values in each step and we simulate this with a sequence $\iota \in (\mathbb{N} \to Input)$. Now, the demonic performance of operations depends on the input values ι and the original definition of $\varLambda$ has to be modified in this sense. [3]

$$\varLambda : (\mathbb{N} \to OP) \times (\mathbb{N} \to Input) \to (\mathbb{N} \to \mathbf{P}\, State)$$

$$\forall \delta : (\mathbb{N} \to OP);\ \iota : (\mathbb{N} \to Input) \bullet (\varLambda(\delta,\iota))0 = State \wedge$$
$$(\forall n : \mathbb{N} \bullet (\varLambda(\delta,\iota))(n+1) =$$
$$\{t : State \mid (\exists s : (\varLambda(\delta,\iota))n;\ int : Intern;\ out : Output \bullet$$
$$cib(int, \iota(n+1), out) = t \wedge csb(s,t) \in \delta n)\})$$

The original formation of cones was oriented on the process itself. The introduction of information flows from outside requires the formation of the set *Cones* with respect to the input sequence ι.

$$Cones : (\mathbb{N} \to Input) \to \mathbf{P}(\mathbb{N} \to OP)$$

$$\forall \iota : \mathbb{N} \to Input \bullet Cones(\iota) = \{\delta : \mathbb{N} \to OP \mid \delta 0 = init \wedge$$
$$(\forall n : \mathbb{N}_1 \bullet (\varLambda(\delta,\iota))n \subseteq Pre(\delta n))\}$$

Any sequence σ of states created by a sequence of operations δ can be devided in a sequence σ_{Input} with values in *Input*, a sequence σ_{Output} with values in *Output* and a sequence σ_{Intern} with values in *Intern*. The following function *Sel* selects all state sequences having a given input projection ι and a given output projection ω.

$$Sel : (\mathbb{N} \to Input) \times (\mathbb{N} \to Output) \to \mathbf{P}(\mathbb{N} \to State)$$

$$\forall \iota : \mathbb{N} \to Input;\ \omega : \mathbb{N} \to Output \bullet Sel(\iota,\omega) =$$
$$\{\sigma : \mathbb{N} \to State \mid \forall n : \mathbb{N} \bullet \exists s : Intern \bullet cib(s, \iota(n), \omega(n)) = \sigma(n)\}$$

The concept of a Z Machine of a Z specification can now be generalized. A specification of a process in a reactive system has to define the admissible operation, state, input and output sequences[4].

---

[3] For the concatenation of internal, input and output variables we use the function cib : $Intern \times Input \times Output \to State \mid cib = \{State \bullet ((\theta Intern, \theta Input, \theta Output), \theta State)\}$

[4] We state the components ι, ω explicitly, although they are determined by σ.

$$
\begin{array}{|l}
\hline
\_ZM _____ \\
\delta : \mathsf{N} \to OP;\ \sigma : \mathsf{N} \to State;\ \iota : \mathsf{N} \to Input;\ \omega : \mathsf{N} \to Output \\
\hline
\delta \in Cones(\iota) \wedge \sigma \in Sel(\iota, \omega) \\
\forall\, n : \mathsf{N} \bullet csb(\sigma(n), \sigma(n+1)) \in \delta\, n \\
\hline
\end{array}
$$

As in the preceding section we introduce liveness conditions on operations in OP. For the generalization of strong and weak fairness it is again necessary to introduce dependence on the input sequence ι. We state the generalized definition of weak fairness, the corresponding generalization of strong fairness (SF) should be obvious.

$$
\begin{array}{|l}
\hline
WF : OP \times (\mathsf{N} \to Input) \to \mathsf{P}(\mathsf{N} \to OP) \\
\hline
\forall\, op : OP;\ \iota : (\mathsf{N} \to Input) \bullet WF(op, \iota) = \\
\quad \{\delta : \mathsf{N} \to OP \mid (\forall\, n : \mathsf{N} \bullet \exists\, k : \mathsf{N} \mid k \geq n \bullet \delta k = op) \vee \\
\quad\quad (\forall\, n : \mathsf{N} \bullet \exists\, k : \mathsf{N} \mid k \geq n \bullet \neg\, ((\Lambda(\delta, \iota))k \subseteq Pre\ op))\} \\
\hline
\end{array}
$$

A Z Machine ZM conjoined with weak and strong fairness conditions on operations in OP is called a (generalized) Z System.

$$
ZS \mathrel{\widehat{=}} [ZM \mid \delta \in WF(op_i, \iota) \wedge \delta \in SF(op_j, \iota) \wedge ...]
$$

Z Machines and Z Systems as defined above generalize the corresponding notions presented in the preceding section. If the schema $Input$ contains no variable we would get the original definitions by hiding the output sequence ω.

The hiding concept of Z makes it easy to express the interior sight of the specified system. The schemas $ZM \setminus (\iota)$, $ZS \setminus (\iota)$ determine all admissible sequences of operations and states and hide the according input sequences. The external view of the machine is given by $ZM \setminus (\sigma)$, etc.

4   Z Machines and Z Systems for Reactive Systems

In the preceding section we presented a specification model for processes which live in interaction with their environment using input and output variables. Additionally having a specification, i.e. a Z Machine (system) of the environment we could ask for a specification of the whole concurrent machine (system). In this section we develop a technique for the specification of concurrent systems by conjoining the individual (generalized) Z Systems. In the following we often use subscripts for processes, schemas and definitions. We think that convenience and clarity of the presented techniques justify this violation of the syntactic rules of Z. Z syntax would request to specify any process explicitly and in a concrete specification, this could be done without any problem.

Let us assume N processes $P_1, ..., P_N$ forming a concurrent system. The environment of a concrete process P_i can be seen as a certain subset $\{P_{j_1}, ..., P_{j_{k_i}}\} \subset \{P_1, ..., P_N\}$. Thus, a specification of the whole system has to join the according Z Machines (systems) ZM_i, $i \in 1..N$[5] and to define their input-output relations. The input values for any process P_i are computed from the output of the surrounding processes. In our model this will be done by an abstract function F_i which determines ι_i with respect to $\omega_{j_1}, ..., \omega_{j_{k_i}}$. We have to guarantee that F_i delivers an input in every step and that the input value is only computed from the output variables of the processes.

$$F_i : Output_{j_1} \times ... \times Output_{j_{k_i}} \to \mathbf{P}\ Input_i \mid \emptyset \notin ran\ F_i$$

We get the Z Machine for the concurrent system by joining the Z Machines of the processes P_i, $i \in 1..N$ and by fixing the input-output relations with the help of the functions F_i, $i \in 1..N$. This is in analogy to the approach of M. Abadi and L. Lamport [2] for the parallel composition of TLA formulas: Concurrent composition of processes means conjoining their specifications.

$$
\begin{array}{l}
\hline
CZM \underline{\hspace{8cm}} \\
\quad ZM_1 \wedge .. \wedge ZM_N \\
\hline
\quad \forall i : 1..N;\ n : \mathbf{N} \bullet \iota_i(n) \in F_i(\omega_{j_1}(n), ..., \omega_{j_{k_i}}(n)) \\
\hline
\end{array}
$$

The notion of fairness which was introduced in the previous sections can be generalized without any problem. We define the so-called concurrent Z System by

$$CZS \mathrel{\widehat{=}} CZM \wedge ZS_1 \wedge .. \wedge ZS_N$$

Finally, the external sight of the concurrent system can be expressed by hiding the state sequences σ_i using the hiding concept of Z:

$$CZM \setminus (\sigma_1, ..., \sigma_N)\ or\ CZS \setminus (\sigma_1, ..., \sigma_N)$$

---

[5] We assume that all operations and sets which are necessary for the definition of ZM_i have as subscript i, i.e. $State_i$, OP_i, $Input_i$, σ_i,...

Note, that the conjunction of concurrent machines (systems) leads again to a concurrent machine (system).

5 A Topology for Z Machines

The spaces $sequ(X) == (\mathbf{N} \rightarrow X)$, ZM and CZM bear natural topologies which are generated by sequences of discrete metrics. At first we define the notion of sequential convergence for any of these spaces and then we will explain how this can be interpreted in a concrete specification.

A sequence $(f_n)_n$ in $sequ(X)$ converges to $f \in sequ(X)$ if for every coordinate i there exists n such that $f_n(i)$ equals $f(i)$ from n on.

A sequence of bindings $(zm_n)_n$ of a Z Machine ZM converges to a binding zm if the sequences $(zm_n.\delta)_n$, $(zm_n.\sigma)_n$ converge in the above defined sense to $zm.\delta$, respectively $zm.\sigma$ [6].

A sequence of bindings $(czm_n)_n$ of a concurrent Z Machine CZM converges to a binding czm if the according sequences of bindings of the ZM machines are convergent.

It is easy to see that the set of bindings given by a Z Machine ZM or CZM is a closed set with respect to the formation of limit points, i.e. if the sequences $(zm_i.\delta)_i$ and $(zm_i.\sigma)_i$ which belong to a sequence of bindings $(zm_i)_i$ of a Z Machine ZM have limit points $\delta \in (\mathbf{N} \rightarrow OP)$ and $\sigma \in (\mathbf{N} \rightarrow State)$ then we may conlude that δ and σ belong to a binding in ZM, too. In other words that δ and σ belong not to a binding of a given Z Machine can be computed in a finite number of steps, simply by looking at the prefixes of the sequences. This implies that the property specified by such a machine is a safety property (see [1]) with respect to cones.

A liveness property is called machine closed if it does not rule out any finite behavior (see [1]) of the corresponding state machine. This means that any finite behavior of the state machine is a prefix of a behavior satisfying the liveness condition.

An important feature of specifications using Z Systems is that they are machine closed with respect to our notion of convergence. This means that any finite sequence of operations respecting the cone condition determined by a given Z Machine ZM is a prefix of a cone which satisfies the fairness conditions defined by ZS.

---

[6] Note, that the according input, output sequences $(zm_n.\iota)_n$, $(zm_n.\omega)_n$ converge as well to $zm.\iota$, respectively $zm.\omega$.

Proposition 1. *ZS and CZS are machine closed.*

Before showing the proofs we need first some auxiliary notations. The embedding of finite sequences into $(\mathsf{N} \to OP)$ and $(\mathsf{N} \to Input)$ can be done by applying stuttering steps from a certain point on:

$$
\begin{array}{|l}
\_^E : (\mathsf{N} \twoheadrightarrow OP) \to (\mathsf{N} \to OP) \\
\hline
\forall f : (\mathsf{N} \twoheadrightarrow OP); \; n \in \mathsf{N} \bullet n \in \mathrm{dom}(f) \Rightarrow f^E(n) = f(n) \\
\qquad\qquad n \notin \mathrm{dom}(f) \Rightarrow f^E(n) = stut
\end{array}
$$

The operation $\_^I : (\mathsf{N} \twoheadrightarrow Input) \to (\mathsf{N} \to Input)$ is defined similarly. Important for the next proof is the following restriction of Λ to finite sequences.

$$
\begin{array}{|l}
\Gamma : (\mathsf{N} \twoheadrightarrow OP) \times (\mathsf{N} \twoheadrightarrow Input) \to (\mathsf{N} \to \mathbf{P}\, State) \\
\hline
\forall \delta : (\mathsf{N} \twoheadrightarrow OP); \; \iota : (\mathsf{N} \twoheadrightarrow Input) \bullet \Gamma(\delta, \iota) = \Lambda(\delta^E, \iota^I)
\end{array}
$$

Proof of: ZS is machine closed: We give the proof for the special case where *ZS* is defined by two strong fairness conditions. The general case for more strong fairness conditions can be done in similar way. In case of weak fairness conditions one may use the result for strong fairness conditions together with the closedness of *ZM*.

Let $zm \in ZM$ and $ZS \mathrel{\widehat{=}} [ZM \mid \delta \in SF(op_1, \iota) \wedge \delta \in SF(op_2, \iota)]$ with $op_1, op_2 \in OP$. Then we define a sequence $(zs^i)_i$ in *ZS* with limit point zm by the following algorithm:

We fix $i \in \mathsf{N}$ and construct sequences $\delta \in (\mathsf{N} \to OP)$ and $\sigma \in (\mathsf{N} \to State)$, $\iota \in (\mathsf{N} \to Input)$ as follows. We set $\delta(k) := zm.\delta(k)$ for $k \leq i$ and $\sigma(k) := zm.\sigma(k)$, $\iota(k) := zm.\iota(k)$ for $k \leq i + 1$ and $flag := 1$ [7]. Then, we proceed by recursive definition over $n \in \mathsf{N}$ using the following program.

$$
\begin{aligned}
&\textit{if } (\Gamma(\delta, \iota))(i + n + 1) \subseteq \textit{pre } op_1 \;\wedge \\
&\qquad\qquad (\Gamma(\delta, \iota))(i + n + 1) \subseteq \textit{pre } op_2 \wedge \textit{flag} = 1 \\
&\qquad \textit{then } \delta(i + n + 1) := op_2 \wedge \textit{flag} := 2 \\
&\textit{else if } (\Gamma(\delta, \iota))(i + n + 1) \subseteq \textit{pre } op_1 \;\wedge \\
&\qquad\qquad (\Gamma(\delta, \iota))(i + n + 1) \subseteq \textit{pre } op_2 \wedge \textit{flag} = 2 \\
&\qquad \textit{then } \delta(i + n + 1) := op_1 \wedge \textit{flag} := 1 \\
&\textit{else if } (\Gamma(\delta, \iota))(i + n + 1) \subseteq \textit{pre } op_1 \textit{ then } \delta(i + n + 1) := op_1 \\
&\textit{else if } (\Gamma(\delta, \iota))(i + n + 1) \subseteq \textit{pre } op_2 \textit{ then } \delta(i + n + 1) := op_2 \\
&\textit{else } \delta(i + n + 1) := stut;
\end{aligned}
$$

---

[7] We use the variable $flag \in \{1, 2\}$.

It remains to choose $\sigma(i + n + 2)$ in the range of $\delta(i + n + 1)$ applied to $(\Gamma(\delta, \iota))(i + n + 1)$ and to define $\iota(i + n + 2)$ as the input projection of $\sigma(i + n + 2)$. Thus, we get δ, ι, σ which determine $\omega \in (\mathbf{N} \to Output)$ and belong according to the construction to a binding $zs \in ZS$. We set $zs^i = zs$ and since any finite prefix of the operation and state sequence of zm appears as prefix of some zs^i we have $\lim zs^i = zm$.

Proof of: CZS is machine closed: We state the proof for $N = 2$ under the assumption that each of the systems ZS_1, respectively ZS_2 is defined by one strong fairness condition $SF(op_1, \iota_1)$, respectively $SF(op_2, \iota_2)$.
For $czm \in CZM$ we define a sequence $(czs^i)_i$ in CZS with limit point czm using the following algorithm.
We fix $i \in \mathbf{N}$ and define δ_1, δ_2, ι_1, ι_2 and σ_1, σ_2 by $\delta_j(k) := czm.\delta_j(k)$ for $k \leq i$ and $\sigma_j(k) := czm.\sigma_j(k)$, $\iota_j(k) := czm.\iota_j(k)$ for $k \leq i + 1$, $j = 1, 2$ and

$$\text{if } (\Gamma_1(\delta_1, \iota_1))(i + n + 1) \subseteq \text{pre } op_1 \text{ then } \delta_1(i + n + 1) := op_1$$
$$\text{else } \delta_1(i + n + 1) := stut_1;$$
$$\text{if } (\Gamma_2(\delta_2, \iota_2))(i + n + 1) \subseteq \text{pre } op_2 \text{ then } \delta_2(i + n + 1) := op_2$$
$$\text{else } \delta_2(i + n + 1) := stut_2;$$

Then, we choose s_1 in the range of $\delta_1(i + n + 1)$ applied to $\sigma_1(i + n + 1)$, respectively s_2 in the range of $\delta_2(i + n + 1)$ applied to $\sigma_2(i + n + 1)$ and by s_1^o, s_2^o we denote the according output projections. For $\iota_j(i + n + 2)$ we take any value in $F_j(s_1^o, s_2^o)^8$, $j = 1, 2$ and we may choose $\sigma_j(i + n + 2)$ in the range of $\delta_j(i + n + 1)$ such that the projection to $Input_j$ equals $\iota_j(i + n + 2)$ and the projection to $Output_j$ equals s_j^o, $j = 1, 2$. Directly from the construction follows that δ_1, δ_2, ι_1, ι_2 and σ_1, σ_2 belong to a binding $czs^i \in CZS$.
For more than two processes P_i, $i \in 1..N$, we have to increase the number of if-clauses according to N. The general case for more than one strong fairness condition per process can be done by combining the algorithm of the first part of the proof of proposition 1 with the one stated above.

6 The Bakery Algorithm

We introduce L. Lamport's improved version of the Bakery Algorithm [6] which we cite from [3] in the below stated form. Let $P_1, ..., P_N$ be processes which want from time to time to access a critical section where simultaneous access is not allowed. This can be done by supplying each

---

[8] W.l.o.g. we may assume that F_j is a binary function.

process P_i with a shared register R_i and internal variables $A_1, ..., A_N$. R_i is used by P_i for reading and writing and any of the remaining processes may only read R_i. An individual process can be in six consecutive phases: *start, doorway, ticket, wait, critical section, finale*. More details can be seen from the following algorithm which describes the consecutive phases for process P_i.

(*Start*) $A_i := 1$; $write(R_i, A_i)$

(*Doorway*) *for all* $j \neq i$, $read(R_j, A_j)$

(*Ticket*) $A_i := 1 + max_j A_j$; $write(R_i, A_i)$

(*Wait*) *for all* $j \neq i$, *repeat* $read(R_j, A_j)$

$\quad$ *until* $A_j = 0$ *or* $A_j > A_i$ *or* ($A_j = A_i$ *and* $j > i$)

(*Critical Section*)

(*Finale*) $R_i := 0$

Let us point out how this can be transformed in a Z specification using the above framework for the specification of concurrent processes.

$Mode ::= started \mid doorway \mid ticket \mid wait \mid critical\_section \mid finale$

$Input_i \mathrel{\widehat=} [R_1, ... R_{i-1}, R_{i+1}, ..., R_N : \mathbf{N}]$; $Output_i \mathrel{\widehat=} [R_i : \mathbf{N}]$

$Intern \mathrel{\widehat=} [mode : Mode;\ A_1, ..., A_N : \mathbf{N}]$

$State_i \mathrel{\widehat=} Input_i \wedge Output_i \wedge Intern$

Let us take $ProcState_i \mathrel{\widehat=} Output_i \wedge Intern$ for abbreviation. A process P_i can only change his internal (*Intern*) and his output (*Output_i*) variables. Stuttering steps for P_i are steps where only the input values (*Input_i*) can be changed by the environment, namely the processes $P_1, ..., P_{i-1}, P_{i+1}, ..., P_N$. We define $stut_i \mathrel{\widehat=} \Xi ProcState_i \wedge \Delta State_i$.

For each process P_i we assume an initializing operation which sets the register R_i and the internal variables $A_1, ..., A_N$ to zero.

$$init_i \mathrel{\widehat=} [\Delta State_i \mid mode' = finale \wedge R_i' = A_1' = ... = A_N' = 0]$$

The following schemas correspond to the consecutive phases of the Bakery Algorithm for process P_i. The actual status of P_i is stored in the variable mode.

SetStart_i

$\Delta State_i$; $\Xi(ProcState_i \setminus (mode, A_i, R_i))$

$mode = finale \wedge mode' = started \wedge A_i' = R_i' = 1$

SetDoorway_i

$\Delta State_i$; $\Xi(ProcState_i \setminus (mode, A_1, ... A_{i-1}, A_{i+1}, ..., A_N))$

$mode = started \wedge mode' = doorway \wedge \forall j : 1..N \mid j \neq i \bullet A_j' = R_j$

$$\boxed{\begin{array}{l} \underline{SetTicket_i} \\ \Delta State_i; \ \Xi(ProcState_i \setminus (mode, A_i, R_i)) \\ \hline mode = doorway \wedge mode' = ticket \\ A'_i = R'_i = 1 + \max\{A_1, ..., A_N\} \end{array}}$$

We divide the wait section into two parts: Reading ($SetWait_i$) the registers $R_1, ..., R_{i-1}, R_{i+1}, ..., R_N$ and computing ($SetWait1_i$) the predicate which regulates the repeat loop.

$$\boxed{\begin{array}{l} \underline{SetWait_i} \\ \Delta State_i; \ \Xi(ProcState_i \setminus (mode, A_1, ...A_{i-1}, A_{i+1}, ..., A_N)) \\ \hline mode = ticket \wedge mode' = wait \wedge \forall j : 1..N \mid i \neq j \bullet A'_j = R_j \end{array}}$$

$$\boxed{\begin{array}{l} \underline{SetWait1_i} \\ \Delta State_i; \ \Xi(ProcState_i \setminus (mode)) \\ \hline mode = wait \wedge mode' = ticket \\ \neg \ (\forall j : 1..N \mid j \neq i \bullet (A_j = 0 \vee A_j > A_i \vee (A_j = A_i \wedge j > i))) \end{array}}$$

$$\boxed{\begin{array}{l} \underline{SetCS_i} \\ \Delta State_i; \ \Xi(ProcState_i \setminus (mode)) \\ \hline mode = wait \wedge mode' = critical\_section \\ \forall j : 1..N \mid j \neq i \bullet (A_j = 0 \vee A_j > A_i \vee (A_j = A_i \wedge j > i)) \end{array}}$$

$$\boxed{\begin{array}{l} \underline{SetFinale_i} \\ \Delta State_i; \ \Xi(ProcState_i \setminus (mode)) \\ \hline mode = critical\_section \wedge mode' = finale \wedge R'_i = 0 \end{array}}$$

$$OP_i == \{init_i, SetStart_i, SetDoorway_i, SetTicket_i, SetWait_i, \\ SetWait1_i, SetCS_i, SetFinale_i\} \cup \{(stut_i)\}$$

For each process P_i we assume the following operations which were defined and explained in the previous sections: csb_i, cib_i, Λ_i, $Cones_i$, Sel_i, WF_i. The generalized Z Machine ZM_i specifies process P_i. Weak fairness conditions on the operations in OP_i guarantee that process P_i can not stay forever in one of the sections. Hence, the generalized Z System for process P_i is

$$\begin{array}{|l}
\hline
ZS_i \\
ZM_i \\
\hline
\delta_i \in WF_i(SetStart_i, \iota_i) \cap WF_i(SetDoorway_i, \iota_i) \\
\delta_i \in WF_i(SetTicket_i, \iota_i) \cap WF_i(SetWait_i, \iota_i) \\
\delta_i \in WF_i(SetWait1_i, \iota_i) \cap WF_i(SetCS_i, \iota_i) \cap WF_i(SetFinale_i, \iota_i) \\
\hline
\end{array}$$

Computing the input for process P_i is the task of the function F_i.

$$\begin{array}{|l}
\hline
F_i : Output_1 \times \ldots \times Output_N \rightarrow \mathbf{P}\ Input_i \\
\hline
\forall \omega_1 : Output_1; \ldots; \omega_N : Output_N \bullet F_i(\omega_1, \ldots, \omega_N) = \\
\quad \{b : Input_i \mid (b.R_1 = \omega_1.R_1 \wedge \ldots \wedge b.R_{i-1} = \omega_{i-1}.R_{i-1} \wedge \\
\quad b.R_{i+1} = \omega_{i+1}.R_{i+1} \wedge \ldots \wedge b.R_N = \omega_N.R_N)\} \\
\end{array}$$

Conjoining all Z Machines ZM_i and defining the input-output relations for the process communication leads to the specification of the whole system CZM. Now, the whole concurrent system expressing liveness conditions and the shared variable conditions can be stated as CZS.

For $i \in 1..N$ and $b \in State_i$ we abbreviate the predicates which guard the execution of $SetWait1_i$ and $SetCS_i$ by

$$p_i(b) = \forall j : 1..N \mid j \neq i \bullet b.A_j = 0 \vee b.A_j > b.A_i \vee (b.A_j = b.A_i \wedge j > i)$$

Let $b_1 \in State_1; \ldots; b_N \in State_N$ such that $b_i.A_k = b_j.A_k$, for all $i, j, k \in 1..N$. Then, a little computation gives us the existence of at least one $i \in 1..N$ such that $p_i(b_i)$ becomes true.

Having a complete Z specification in hand the next step is to prove correctness properties of the Bakery Algorithm. We will do this under the usual assumptions: We initialize the machine and investigate the resulting run which should only depend on the machine itself. In other words only bindings $czs \in CZS$ with $czs.\delta_i(n) \neq init_i$, for all $i \in 1..N$ and $n \in \mathbf{N}_1$ are interesting for the following observations. We claim the following properties:

(1.) *The deadlock freedom of the concurrent system CZS.*

(2.) *The correctness of the system CZS, which means that critical sections of different processes can not overlap.*

Sketch of the proofs: (1.) can be seen in the following way: We argue by contradiction and assume that there exists $czs \in CZS$ and $i_0 \in 1..N$, $n \in \mathbf{N}$ with $\forall k \geq n \bullet czs.\delta_{i_0}(k) \neq SetCS_{i_0}$.

The fairness assumptions on the operations imply that there exists $m \in \mathbf{N}$ such that

$$\forall k \geq m \bullet czs.\delta_{i_0}(k) \in \{SetWait_{i_0}, SetWait1_{i_0}, stut_{i_0}\}.$$

Now, we derive from the following lemma

Lemma 2. *Let $czs \in CZS$, $i,j \in 1..N$ and $n \in \mathbb{N}$ such that $czs.\delta_i(n) = SetWait_i$ and $czs.\delta_i(k) \in \{SetWait_i, SetWait1_i, stut_i\}$ for all $k \geq n$. Then, there must be $M \in \mathbb{N}$ such that $czs.\delta_j(k) \in \{SetWait_j, SetWait1_j, stut_j\}$ and $\neg\, p_j(czs.\sigma_j(k+1))$ for all $k \geq M$.*

that there must be $M \in \mathbb{N}$ with $\forall\, i \in 1..N\, \forall k \geq M \bullet$
$czs.\delta_i(k) \in \{SetWait_i, SetWait1_i, stut_i\} \wedge \neg\, p_i(czs.\sigma_i(k))$.
Thus, there exists a moment greater than M such that every process P_i computes its predicate p_i on the same values A_k, $k \in 1..N$ from this moment on. This causes that there is at least one index $i \in 1..N$ for which the according predicate p_i becomes true. Now, the weak fairness assumption on $SetCS_i$ supplies us with a contradiction.

Proof of lemma 2: We argue by contradiction. Because of the fairness assumptions we may conclude that $SetCS_j$ appears infinitely often in $czs.\delta_j$. Therefore, we may find $k_0 > n$ with $czs.\delta_j(k_0) = SetDoorway_j$. Then, the next ocurrence of $SetTicket_j$ in $czs.\delta_j$ after k_0 causes that process P_j gets a ticket $A_j \geq czs.\sigma_i(n+1).A_i + 1 \geq 2$. Hence, there exists $k_1 > k_0$ with $czs.\delta_j(k_1) = SetWait_j$. Thus, the predicate $p_j(czs.\sigma_j(k))$ gets false for $k \geq k_1 + 1$ and we get a contradiction.

(2.) follows from

Lemma 3. *Let $czs \in CZS$ and $n \in \mathbb{N}$, $i \in 1..N$ such that $czs.\delta_i(n) = SetWait_i$, $p_i(czs.\sigma_i(n+1))$. Let m be the minimal number greater than n with $czs.\delta_i(m) = SetFinale_i$. Then for all $n \leq k < m$ and $j \neq i$: $(czs.\delta_j(k) \neq SetWait_j \vee \neg\, p_j(czs.\sigma(k+1)))$.*

Proof. Let $czs.\delta_i(n) = SetWait_i$ and $p_i(czs.\sigma_i(n+1))$. We argue by contradiction and assume $n \leq k_0 < m$ minimal with $czs.\delta_j(k_0) = SetWait_j$ and $p_j(czs.\sigma_j(k_0 + 1))$. The case $j > i$ will be proved. The remaining case $j < i$ can be settled in similar way. $p_i(czs.\sigma_i(n+1))$ together with lemma 4 imply

$$czs.\sigma_i(k+1).A_j = 0 \vee czs.\sigma_i(k+1).A_j \geq czs.\sigma_i(k+1).A_i$$
$$\geq 2 ,\ n \leq k < m \qquad (1)$$

The definition of $SetWait_i$, $SetWait_j$ and the shared variable assumption leads to

$$czs.\sigma_i(n+1).A_j = czs.\sigma_j(n).R_j \ , \ czs.\sigma_j(k_0+1).A_i = czs.\sigma_i(k_0).R_i \quad (2)$$

Because of $p_j(czs.\sigma_j(k_0 + 1))$ and (1), (2) we may state

$$czs.\sigma_j(k_0 + 1).A_i > czs.\sigma_j(k_0 + 1).A_j \ . \quad (3)$$

We show that $czs.\sigma_j(n).R_j$ must be greater or equal than 2. If $czs.\sigma_j(n).R_j = 0$ then we may find $SetDoorway_j$, $SetTicket_j$ in $\{czs.\delta_j(n + 1), ..., czs.\delta_j(k_0)\}$, lets say $czs.\delta_j(k_1) = SetTicket_j$ and $czs.\delta_j(k_2) = SetDoorway_j$ with $n + 1 \leq k_2 < k_1 < k_0$. Thus, using lemma 4 we derive

$$
\begin{aligned}
czs.\sigma_j(k_0 + 1).A_j &= czs.\sigma_j(k_1 + 1).A_j > czs.\sigma_j(k_1).A_i \\
&= czs.\sigma_j(k_2 + 1).A_i = czs.\sigma_j(k_2).R_i \\
&= czs.\sigma_i(k_2).R_i = czs.\sigma_i(k_0).R_i \\
&= czs.\sigma_j(k_0).R_i
\end{aligned}
$$

which is impossible because of (2) and (3). The case $1 = czs.\sigma_j(n).R_j$ is not possible because of (1) and (2).

Hence, we may assume $czs.\sigma_j(n).R_j$ greater or equal than 2. Then, $SetTicket_j$ has been performed in $czs.\delta_j$ before the nth moment. Let l_0 be the last ocurrance of $SetTicket_j$ in $czs.\delta_j$ before k_0. Then, $l_0 < n$ and we derive using lemma 4 that the values of $czs.\sigma_j(k + 1).A_j$ and $czs.\sigma_j(k + 1).R_j$ are constant and equal for $l_0 \leq k \leq k_0$. Applying (2), (3) and lemma 4 again we get

$$
\begin{aligned}
2 \leq czs.\sigma_j \ (n + 1).A_j &= czs.\sigma_j(n).R_j = czs.\sigma_j(k_0 + 1).A_j \\
&< czs.\sigma_j(k_0 + 1).A_i = czs.\sigma_i(k_0).R_i = czs.\sigma_i(n + 1).A_i
\end{aligned}
$$

which is not possible because of (1).

Lemma 4. *Let $czs \in CZS$ and $n \in \mathbf{N}, i \in 1..N$ with $p_i(czs.\sigma_i(n + 1))$ and $czs.\delta_i(n) = SetWait_i$. For minimal $m > n$ and maximal $p < n$ such that $czs.\delta_i(m) = SetFinale_i$ and $czs.\delta_i(p) = SetTicket_i$ we get that $czs.\sigma_i(k).A_i$ and $czs.\sigma_i(k).R_i$ are constant and greater than 2 with $czs.\sigma_i(k).A_i = czs.\sigma_i(k).R_i$ for all $p + 1 \leq k \leq m$. If $j \neq i$ then $czs.\sigma_i(k + 1).A_j$ are constant for $n \leq k < m$. This implies that $p_i(czs.\sigma_i(k + 1))$ is true for $n \leq k < m$.*

Proof. Because of the fairness assumptions we may take $m \geq n$ minimal and $p \leq m$ maximal such that $czs.\delta_i(m) = SetFinale_i$ and $czs.\delta_i(p) = SetTicket_i$. Thus, $czs.\delta_i(k) \in \{stut_i, SetWait_i, SetWait1_i, SetCS_i, SetFinale_i\}$ for $p < k \leq m$ and a glance at those operations is enough to prove that $czs.\sigma_i(k).A_i$ and $czs.\sigma_i(k).R_i$ are constant and equal for $p < k \leq m$. $SetStart_i$ and $SetTicket_i$ have to appear before the performance of $SetWait_i$ (R_i, A_i may only be changed by $Init_i$, $SetStart_i$, $SetTicket_i$ and $SetFinale_i$) and so we get $czs.\sigma_i(k).R_i \geq 2$ for $n \leq k \leq m$.

7 Conclusions and Related Work

In this paper, we have shown how Z can be systematically used to specify complex systems which are independent from their environment. This is then generalized to a specification technique for processes that maintain an ongoing interaction between their environment by input and output variables, so called reactive systems.

To this end, we have introduced the notion of Z Machines which captures the ideas of demonic computations of standard Z specifications. Adding fairness conditions to Z Machines yields Z Systems which form the main concepts for the specification of complex reactive systems.

Our systematic use of standard Z for describing complex reactive systems by including fairness conditions is new. All other approaches which try to use Z for specifying reactive systems effectively extend Z in one way or the other. Leslie Lamport proposed a language TLZ [8] which is Z extended by temporal formulae à la TLA. In [4], Z is extended to include temporal operators. This extension is used to describe the alternating bit protocol. Particular solutions to specifying complex systems are reported in [12] and [5].

The paper that comes closest to our approach is [11], where Z is used for process modeling in a process algebraic style.

We wanted to develop a simple and yet general approach to systematically use Z for complex systems. We share the doubts of Leslie Lamport [8] that temporal formulae are easy to understand and work with in nontoy examples. Therefore, we do not want to introduce temporal operators as used in temporal logics [9]. Our operators SF and WF are easy to understand and they do not have to be further combined in specifications of complex reactive systems other than by conjunctions. The development of a Z System specification is systematic. Z Systems have similar expressive power as TLA and there is an underlying theory that guarantees important properties of specifications such as being machine closed.

We are currently investigating refinement techniques for Z Systems with the help of the schema calculus.

References

1. Martin Abadi and Leslie Lamport. *The Existence of Refinement Mappings.* In Proceedings of the 3rd Annual Symposium on Logic in Computer Science, p. 165-175. IEEE, 1988.
2. M. Abadi and L. Lamport. *Conjoining Specifications.* Digital Technical Report. December 7, 1993.
3. E. Börger and Y. Gurevich and D. Rosenzweig. *The Bakery Algorithm: Yet another specification and verification.* To appear in: E. Börger (Ed.). Specification and Validation Methods. Oxford University Press. 1995.
4. R. Duke and G. Smith. *Temporal logic and Z specifications.* The Australian Computer Journal, 21(2):62-66,1989.
5. I. Houston and M. Josephs. *Specifying distributed CICS in Z; accessing local and remote resources.* Formal Aspects of Computing, 6(6),1994.
6. L. Lamport. *A new Approach to Proving the Correctness of Multiprocess Programs.* ACM Transactions on Programming Languages and Systems, vol.1.1, July 1979, 84-97.
7. L. Lamport. *The temporal logic of actions.* ACM Transactions on Programming Languages and Systems, 16(3):872-923, 1994.
8. L. Lamport. *TLZ.* In J. Bowen and A. Hall, editors, Proceedings of the 1994 Z User Workshop, 267-268, Springer Verlag, 1994
9. Z. Manna and A. Pnueli. *The temporal logic of reactive and concurrent systems.* Springer Verlag, 1992.
10. J.M. Spivey. *The Z Notation: A Reference Manual.* Prentive-Hall, second edition, 1992.
11. B. Sufrin and He Jifeng. *Specification, analysis and refinement of interactive processes.* In M. Harrison and H. Thimbleby, editors, Formal Methods in Human-Computer Interaction, volume 2, chapter 6, 153-200. Cambridge University Press, 1990.
12. K.T. Narayana and S. Dharap. *Invariant properties in a dialog system.* ACM SIGSOFT Software Engineering Notes, 15(4), 67-79, 1990
13. J.C.P. Woodcock. *The rudiments of algorithm refinement.* The Computer Journal, 35(5):441-450, 1992.

Synthesizing Different Development Paradigms:
Combining Top-Down with Bottom-Up Reasoning About Distributed Systems[1]

J. Zwiers[2], U. Hannemann, Y. Lakhneche & W.-P. de Roever [3]

Abstract. Our goal is the presentation of a uniform framework for compositional reasoning about the development of distributed processes and data structures. This framework should be a synthesis because, depending on the structure of the processes involved and the verification steps required, *different* formalisms are most suitable for carrying out one's reasoning. We illustrate this uniform framework by presenting a methodology for reasoning about refinement of distributed data structures, i.e., data structures implemented by means of distributed networks. Our synthesis is compositional, state-based, history-based, and contains **sat** style, Hoare style , trace-invariant reasoning and assumption/commitment style specifications as dialects. The resulting formalism can be unfolded as if it were a portable telescope, yielding the style required according to its degree of unfolding.

1 Introduction

Many useful formal techniques for program development are well known by now, and are more and more accepted by software developers. For instance, one sees specifications in Z [S92] or VDM [J86] style, Hoare style pre/postconditions for verification of sequential programs, various logics and proof methods for concurrent programs, data refinement, etc. In practice, various techniques must be *combined*, as it is usually not the case that one single technique covers all aspects, and all stages, of development. There seems to be in particular a substantial gap between *sequential programming* on the one hand, and *concurrency* on the other. Our first contribution is that we show how to incorporate concurrency in the development of systems that start out as purely sequential programs. We are faced here with a so called *specification adaptation problem:* how to switch between a VDM or Z style specification and a trace based specification.

Specification adaptation does not only arise when stepping from sequential to concurrent programming. It also arises within a top-down development of some (sequential or concurrent) system S when one would like to reuse some available component C [H71]. Then C's specification may not immediately fit the required

---

[1] The collaboration of the authors has been partially supported by the European Community ESPRIT Basic Research Action Project 6021 (REACT)
[2] Twente University, P.O. Box 217, 7500 AE Enschede, The Netherlands, zwiers@cs.utwente.nl
[3] Institut für Informatik und praktische Mathematik, Christian-Albrechts-Universität zu Kiel, Preußerstraße 1-9, 24105 Kiel, Germany, {uha,yl,wpr}@informatik.uni-kiel.d400.de

"top-down"specification, and needs to be adapted to this new specification -this reflects the particular interface of C within S. Specification adaptation now consists of proving that C's original specification, which we assume to be given once and for all (e.g., a manufacturer's description of a chip interface), implies this new top-down specification. In case of a relational framework this merely boils down to proving inclusion (or implication, in case of a logical framework). However in case of Hoare-style specification formalisms for sequential systems, such as Z or VDM, and their more complicated versions for concurrent and distributed systems, in which trace-invariants and assumption/commitment pairs are used to characterize the communication interface, the problem becomes how to formulate and *prove* that one specification "implies" another one (in a relatively complete manner). A second contribution is that we show how this can be done; this is a generalization to concurrency of Olderog's masterful "*Note on the rule of adaptation*" [O83], which describes specification adaptation of sequential programs.

The context of these two contributions is our third contribution: a *uniform* framework for *compositonal* reasoning about the development of distributed processes and data structures. Why *uniform*? As already remarked, the choice of the best formalism for carrying out one's formal reasoning depends, among others, on the structure of the eventual implementation and the nature of the particular development step made. For instance, when sequential operations are emphasized, as in the field of data structure implementations, Hoare logics are preferable. Yet when given ready-made modules with given specifications have to be adapted to a given application, i.e., top-down reasoning meets bottom-up reasoning, a **sat** style *relational* formalism is better suited. Consequently a synthesis is called for in which both Hoare style and **sat** style reasoning can be represented [ZdeR 89, HHS87, vKH95].

And why should this framework be *compositional*? Because, following Dijkstra, distributed networks are derived from specifications in a hierarchically structured step-wise process, in which pure specification-oriented terms are gradually replaced by more implementation-oriented structures, as formalized in the so called *mixed term* approach [Wirth 71, Olderog 86, Z89]. In such a mixed term approach the distinction between specifications and programs becomes blurred because program operators may be applied to specifications, and specification operations to programs (e.g., the conjunction of programs $P_0 \wedge P_1$, or the sequential composition of specifications $S_0; S_1$). Now if this set-up is going to work, refining a (sub)specification S by one mixed term satisfying S or by another one, which equally satisfies S, should not cause any difference w.r.t. satisfaction of the original top-level specification one started out with. This implies that S should characterize the *complete interface* of that mixed term (incl. all of its refinements) with its environment in the context of that particular refinement step. This calls for a *compositional* framework.

We illustrate this uniform framework by presenting a methodology for reasoning about refinement of distributed data structures, i.e., data structures implemented by means of distributed networks. This imposes the following *requirements* to our framework: it should incorporate the mixed term approach, easily deal with real life state-based implementations, concurrency, and, as said before, adaptation. Consequently, our synthesis should be, respectively, compositional, state-based, trace-based (to deal compositionally with parallelism), and contain both sat style and Hoare style reasoning as dialects.

But this list doesn't cover all desiderata. As mentioned before, a promising way to specify and verify distributed algorithms compositionally in given programming languages for parallel computing is through the use of *assumption/commitment* based [MC81, PJ91] or trace-invariant based [CH81, Z89] formalisms and these methods should also find their place within our approach. The solution to all these at first sight conflicting requirements is a uniform formalism which is a synthesis of all of the above, specialized styles. This synthesis, rather than leading to a system whose complexity is the sum of its subsystems, forces us to adapt the *simplest* formalism available, that of (second order) predicate logic to our purposes by simulating the different styles of reasoning required for each development step within this single framework.

After this introduction we sketch our second order predicate logic formalism and explain how to embed sequential specification styles and CSP structures in it. In section 3 we discuss the relation between trace-based reasoning and state-based reasoning and how to integrate these approaches within a sat style formalism or Hoare logics. An approach to a formal solution of the adaptation problem within extensions of Hoare logics for reactive systems is presented in section 4. In order to illustrate these contributions we explain our techniques on a shortest path algorithm and its underlying data structure which is implemented as a network of processes.

2 A uniform framework

We introduce a general logical framework underlying the specification formalisms in this paper. We show how to embed Hoare logic, trace invariant logic, assumption/commitment and sat style logics. Thereafter we consider the adaptation problem for these logics. It is shown that adaptation can be done via translation, back and forth, between various logics.

Predicate formulae

We assume as given sets of typed first order variables $x \in Var$, and of typed second order relation variables $X \in VAR$. The exact typing scheme is not important for our goal. Expressions e are built up as usual. Relation constants R and relation variables X are interpreted as $n - ary$ typed relations. We assume that among these relational constants we have at least the equality relation "$=$" for each first order type τ. Within the syntax of formulae below, it is tacitly assumed that typing constraints are satisfied. The class of predicate formulae ϕ is defined by:

$$\phi ::= e_1 = e_2 \mid R(e_1, \ldots, e_n) \mid X(e_1, \ldots, e_n) \mid \phi_1 \wedge \phi_2 \mid \neg\phi \mid \forall x.(\phi) \mid \forall X.(\phi)$$

Formulae like $\phi_0 \rightarrow \phi_1$ are seen as standard abbreviations.

The satisfaction relation

In the sequel we show how, e.g., VDM, Z or Hoare style formulae for CSP processes can be embedded within predicate logic. This requires some notation. All these styles have in common that they are based on a general *satisfaction relation* "sat". A formula of the form ϕ **sat** ψ is defined here as a predicate formula $\forall \bar{x}.(\phi \rightarrow \psi)$. The variables $\bar{x}$ are called *base variables* of the specifications ϕ and ψ, expressed by $base(\phi, \psi)$. They are the subset of the free variables of ϕ, ψ consisting of those variables that actually refer to the *observable behaviour* of ϕ and ψ, and correspond in general to program variable, trace variables, ready-set variables, termination flags, and the like [P88]. The remaining first order variables 0f ϕ, i.e. the non-base ones, are denoted by $lvar(\phi)$, and are called the *logical variables* of ϕ, in the literature also referred to as "freeze" or "rigid" variables.

Sequential systems

A *relational specification* is a predicate formula with occurrences of unprimed and primed versions of variable names. In the examples we use a VDM style notation of the form:

require $pre(x_0, \ldots, x_n)$
ensure $post(x_0, \ldots, x_n, x_0\prime, \ldots, x_n\prime)$

which, for our purpose, abbreviates the predicate:

$pre(x_0, \ldots, x_n) \Rightarrow post(x_0, \ldots, x_n, x_0\prime, \ldots, x_n\prime)$

A *Hoare specification* is of the form $\{pre(\bar{x}, \bar{v})\}\ S\ \{post(\bar{x}, \bar{v})\}$, where we have pre/postconditions with free program variables $\bar{x}$ and logical variables $\bar{v}$. This formula can now be seen as an abbreviation that S satisfies the relational predicate $\forall \bar{v}.(pre(\bar{x}, \bar{v}) \Rightarrow post(\bar{x}\prime, \bar{v}))$. A convenient adaptation rule for turning relational specifications into Hoare style pre-postconditions is based on weakest preconditions or strongest postconditions defined for relational predicates. Let $\psi(\bar{x}, \bar{x}\prime)$ be a relational predicate, and let $pre(\bar{x}, \bar{v}), post(\bar{x}, \bar{v})$ be a state predicates, with possible occurrences of logical (i.e. non-state) variables $\bar{v}$. We define:

$$wp(\psi, post) \stackrel{\text{def}}{=} \forall \bar{x}\prime.(\psi \Rightarrow post[\bar{x}\prime/\bar{x}])$$
$$sp(pre, \psi) \stackrel{\text{def}}{=} \exists \bar{x}\prime\prime.(pre[\bar{x}\prime\prime/\bar{x}] \wedge \psi[\bar{x}\prime\prime/\bar{x}, \bar{x}/\bar{x}\prime])$$

The adaptation of Hoare formulae to relational ones is given above. Conversely, if a program S has been specified by a relational predicate ψ, then for any desired postcondition *post*, we know that S satisfies the formulae $\{wp(\psi, post)\}\ S\ \{post\}$.

Trace logic and CSP

As an application of the formalism above we recall how to deal with trace based specifications for CSP style processes [Hoare 85]. Trace based specifications are assertions of the form $S(h)$ with free occurrences of a trace typed variable h, denoting the communication history of a process. A *trace* is a finite sequence of communications taken from a given set of communication actions *Comm*. We employ standard notations and operations for trace expressions t and finite alphabets $\alpha \subseteq Comm$. We assume that α contains the special symbol "$\sqrt{}$", which by convention is used to signal termination. Furthermore, we need the following trace operations:

ε	the empty trace, also written as $<>$
$< a >$	a trace with a single communication $a \in \alpha$
$t_0 \frown t_1$	concatenation of t_0 and t_1
$t \restriction \alpha$	projection onto α
$t[b/a]$	renaming occurrences of a into b within t
$last(t)$	the last communication of t
$rest(t)$	t except for $last(t)$
$fin(t)$	abbreviates $last(t) = \sqrt{}$
$t_0; t_1 = t_0$	if $\neg fin(t_0)$,
$t_0; t_1 = rest(t_0) \frown t_1$	if $fin(t_0) \wedge t_1 \neq <>$
$t_0; <> = t_0$	
$t_0 \preceq t_1$	t_0 is a *prefix* of t_1

Note that when $t_1 \neq <>$, no $\sqrt{}$-symbol occurs *within* $t_0; t_1$, even if t_0 originally ended with a $\sqrt{}$. Our basic programming language consists of atomic communications, parallel and sequential composition of processes, choice of processes, hiding and renaming of channels:

$$P ::= stop \mid skip \mid a \mid P_0 ; P_1 \mid P_0 \| P_1 \mid P_0 \sqcap P_1 \mid P \backslash c \mid P[d/c]$$

It is known from the literature [CH81, Z89] how to define for a given CSP process P its *alphabet* $\alpha(P)$ and an equivalent trace specification $S(h)$. We give a brief sketch of this.

$stop(h)$	$h = \varepsilon$
$skip(h)$	$h = \varepsilon \vee h =< \sqrt{} >$
a	$h \preceq < a > \frown < \sqrt{} >$
$P_0 ; P_1$	$\exists t_0 \exists t_1 . P_0(t_0) \wedge P_1(t_1) \wedge h = t_0; t_1$
$P_0 \| P_1$	$P_0(h \restriction \alpha(P_0)) \wedge P_1(h \restriction \alpha(P_1)) \wedge h \restriction \alpha = h,$
	where $\alpha = \alpha(P_0) \cup \alpha(P_1)$
$P_0 \sqcap P_1$	$(P_0(h) \wedge h \restriction \alpha - \alpha(P_0) = \varepsilon) \vee (P_1(h) \wedge h \restriction \alpha - \alpha(P_1) = \varepsilon)$
	where $\alpha = \alpha(P_0) \cup \alpha(P_1)$
$P \backslash c$	$\exists t . P(t) \wedge h = t \restriction (\alpha(P) - \{c\})$
$P[d/c]$	$\exists t . P(t) \wedge (h = (t \restriction \alpha(P))[d/c]$

Thus we may regard CSP processes as abbreviations for certain trace specifications within our framework. The special variable h is the only base variable of such process specifications, i.e. all other free variables are logical variables. This implies that a relation of the form "P sat Q" must be read as "$\forall h.(P(h) \rightarrow Q(h))$". Formally speaking, both P and Q are trace formulae. However, it useful to distinguish the cases where P or Q are actually written down as CSP processes. Formulae that do not contain CSP operators at all will be called (logical) specifications.

3 Integrating trace- and state- based reasoning

How to incorporate concurrency into existing methods for development of sequential systems? It is well known that within process languages like CSP, *data* can be represented by *processes* with appropriately named channels for the operations that are allowed on the data. One can use a trace logic based on the **sat** relation to specify and verify such data processes. The problem then is: how to incorporate these results into a *sequential* program context? One might switch here completely to a formalism for concurrency, since data can now be seen as processes. However, it is advantagous to work at a more abstract level, relying on so called *virtual states*. The solution for this problem is related to techniques for data refinement. The step from traces to states is made formal by means of *retrieve functions* similar to those used in [J86]. Occasionally, one also needs *retrieve relations*, like those used in [Reynolds 81]. The basic idea is to introduce a process T for each abstract datatype T. The operations $op_1, \ldots, op_n$ defined for T data are taken as the names of channels of this T process. A declaration of the form: **var** x:T; S is then refined into $(x.T \mathbin{/\!/} S)\backslash x.op_0, \ldots, x.op_n$. As usual, x.T is the process T with the channels $op_1, \ldots, op_n$ renamed into $x.op_1, \ldots, x.op_n$. Also, x.T $\mathbin{/\!/}$ S denotes parallel composition where x.T is *subordinate* to S, in the sense of [Hoare 85]. That is, termination is determined by process S, and process x.T stops as soon as process S terminates.

What should be the specification for the T process? We derive it from the specification of the abstract datatype. The main stepping stone here is the definition of a retrieve relation retr_T that maps traces of a T process to states of an abstract T object. In general we assume that the T operations can be classified as either *constructor operations* $cons_i : T \times T_1 \times \cdots \times T_n \rightarrow T$, for $i \in I$, or as *selector operations* $sel_j : T \times T_1 \times \cdots \times T_m \rightarrow T'$, for $j \in J$. (The types T_i and T' are assumed to be different from T.)

The operations are specified by relational predicates of the form: $\psi_i(a, x_1, \ldots, x_n, a')$ (for constructors) and $\phi_j(a, x_1, \ldots, x_m, y)$ (for selectors). We define, in a canonical fashion, a retrieve relation $retr\_T(h, a)$. Informally, $retr\_T(h, a)$ holds iff a is one of the possible abstract values for T which result from applying the list of operations recorded in trace h. Assume that objects of type T have an initial value val_{init}. $retr_T$ is determined by the following:

$retr\_T(\langle\rangle, val_{init}) \stackrel{\text{def}}{=} \textbf{true}$,

$retr\_T(h \frown \langle (cons_i, v_1, \ldots, v_n) \rangle, a) \stackrel{\text{def}}{=} \exists \tilde{a}.(retr\_T(h, \tilde{a}) \wedge \psi_i(\tilde{a}, v_1, \ldots, v_n, a))$,

$retr\_T(h \frown \langle (sel_j, v_1, \ldots, v_m) \rangle, a) \stackrel{\text{def}}{=} retr\_T(h, a)$.

The next step is to define a trace specification $\chi_T(h)$ for the T process, based upon this retrieve function. Informally, it states that whenever selector operation sel_j is applied, the value returned is defined by ϕ_j, where the current *state* a is determined by $retr\_T(h, a)$:

$\chi_T(h) \stackrel{\text{def}}{=} \forall v_1, \ldots, v_m, y.(\forall t \leq h. \bigwedge_j (last(t) = (sel_j, v_1, \ldots, v_m, y) \Rightarrow \exists \tilde{a}.(retr\_T(rest(t), \tilde{a}) \wedge \phi_j(\tilde{a}, v_1, \ldots, v_m, y))))$

While this is a purely trace oriented style we preferably use formalisms where we can refer to data objects explicitly in our assertions. Thus, we incorporate these techniques into a state-trace formalism which in fact denotes an abbreviation of pure trace expressions. Let $p(h, x)$ be a predicate formula with free occurrences of the trace h and of variable x of type T. We define the meaning of such predicates by a translation back to (pure) trace formulae: $p(h, x)$ abbreviates $\tilde{p}(h)$, defined by: $\tilde{p}(h) \stackrel{\text{def}}{=} \exists a : T. (retr\_x.T(h, a) \wedge p(h, a))$, where

$retr\_x.T(\langle\rangle, val_{init}) \stackrel{\text{def}}{=} \textbf{true}$,

$retr\_x.T(h \frown \langle (x.cons_i, v_1, \ldots, v_n)\rangle, a) \stackrel{\text{def}}{=} \exists \tilde{a}.(retr\_x.T(h, \tilde{a}) \wedge \psi_i(\tilde{a}, v_1, \ldots, v_n, a))$,

$retr\_x.T(h \frown \langle (x.sel_j, v_1, \ldots, v_m)\rangle, a) \stackrel{\text{def}}{=} retr\_x.T(h, a)$, the retrieve function for variable x of type T. In the sequel we will regard all predicates as predicates on traces and such state variables unless stated otherwise.

Specification adaptation and adaptation completeness

Modular design of systems is often a combination of top-down global design and bottom-up reuse of existing modules. A pure top-down approach starts with a first specification S_0, which is then transformed gradually, via a series of intermediate designs $S_1, \ldots S_{n-1}$, into a final program text S_n. An intermediate design, say S_i, is built up from a number of (logical) specifications $\phi_0, \ldots, \phi_m$ by means of programming language constructs, such as sequential composition, iteration, etc. Since they combine programming constructs with logic specifications, the S_i are called *mixed terms*. Mixed term S_i is obtained from S_{i-1} by replacing some sub-term T in S_{i-1} by an implementing mixed term R, i.e. R must be such that R **sat** T holds. When T actually has the form of a logic specification and R the form of a program, then this describes a classical, top-down development step, an *implementation relation*. When both T and R are programs, we have a classical program transformation step, a *refinement relation*. Finally, when both T and R are logic specifications, we are dealing with *specification adaptation*. Such an adaptation step in the development of an implementation is in general necessary when one would like to implement T by some already available module M, where M is known to satisfy some given specification R. We do not want to verify *directly* that M **sat** T, since this would force us to inspect the internal structure of M. In fact, our programming language might include encapsulation constructs that do not even allow us to inspect this internal structure! And,e.g., chip manufacturers often produce manuals to describe the structure of their product; here such manuals fulfill the rôle of R. Therefore, we have to check indirectly that M **sat** T, by showing that R **sat** T, since M **sat** R is given once-and-for-all. Because T and R are specifications created by different designers, there might be a substantial "gap" between the two, and consequently verifying that R **sat** T then becomes a non-trivial design step (as illustrated in section 5).

In order to close this gap between two specifications one may apply rules that do not refer to the internal structure of their process objects (programs or formulae), but express the closure properties of the domain these objects belong to. In the terminology of Zwiers [Z89], a **sat** proof system is *adaptation complete* if, whenever for specifications P, Q,

$$\models \forall X.((X \textbf{ sat } P) \rightarrow (X \textbf{ sat } Q))$$

holds, then this implication is provable within this proof system. This is a simple task for a purely predicate based **sat** style, for it is clear that $\models \forall X.((X \textbf{ sat } P) \rightarrow (X \textbf{ sat } Q))$ holds iff $\models P \textbf{ sat } Q$, which for a **sat** style system is a simple verification condition. However in Hoare, trace-invariant based and assumption/commitment style formalisms this becomes a problem which is solved in the sections below.

Transformations between specification styles

Simple proof systems have been built for the **sat** relation for CSP [H69, Z89, Olderog 86]. Specification adaptation is rather straightforward here, as argued above. This is certainly one of the advantages of such **sat** systems. Yet, **sat** based logics do not treat sequential composition very well. For instance, consider the following proof rule for **sat** logic:

Rule 1 (Sequential Composition Sat style)

$$\frac{S_0 \text{sat } Q_0, \ S_1 \text{sat } Q_1}{S_0; S_1 \text{ sat } Q_0; Q_1}$$

Note that when Q_0 and Q_1 are logical specifications, then $Q_0; Q_1$ still contains a sequential operator that can be eliminated only at the expense of introducing an existential quantifier, i.e. in this style a proper rule for sequential composition doesn't eliminate this operator but reduces it to a new one at the level of specifications, as e.g., observed by Lamport [L91]. A nicer rule can be obtained by embedding Hoare logic within the **sat** system. For any trace predicate $S(h)$, define a *relation* on traces "; S" as follows:

$$(; S)(h, h') \stackrel{\text{def}}{=} \exists h''.(S(h'') \wedge h' = h; h'')$$

Now Hoare logic is embedded in the **sat** system by $\{P\} ; S \{Q\} \stackrel{\text{def}}{=} P; S \text{ sat } Q$

Note that the ";" on the right hand side of this definition is the operator defined in section 2. The main advantage of this approach is that now the following simple composition rule holds for these **sat** style formulae in Hoare style disguise:

Rule 2 (Sequential Composition Hoare style)

$$\frac{\{P\} ; S_0 \{R\}, \{R\} ; S_1 \{Q\}}{\{P\} ; (S_1; S_2) \{Q\}}$$

Similar Hoare style rules for these disguised **sat** style formulae can be given for other sequential constructs, such as the rule for iteration based on loop invariants. Dealing with relations on traces requires similar conventions on the use of variables as in section 2. We use h as trace variable in preconditions and h' for the free trace variable in postconditions. A novel aspect is that within $\{P\} ; S \{Q\}$ the pre and postconditions P and Q are *trace* specifications and may be CSP processes themselves. Transformation rules between **sat** specifications and Hoare style formulae are:

Rule 3 (Hoare- sat)

$$\frac{\forall \bar{u}.(\{P\} ; S \{Q\})}{S \text{ sat } \forall \bar{u} \exists t.(P(t) \to Q)}$$

Rule 4 (sat - Hoare)

$$\frac{S \text{ sat } Q}{\{wp(Q, R)\} ; S \{R\}} \qquad\qquad \frac{S \text{ sat } Q}{\{R\} ; S \{sp(R, Q)\}}$$

Here Q' abbreviates $Q[h'/h]$, $\bar{u}$ expresses the set of logical variables of P and Q, and $wp((;S),Q)$ and $sp(P,(;S))$ are trace logic equivalents to the definitions from section 2 (for variables $\bar{x}$).

As for classical Hoare formulae, one can define adaptation rules here too, as in , e.g., [O83, D92] for sequential constructs and [Z89] for distributed ones. Yet it appears to be simpler to translate back and forth between Hoare formulae and "sat" style formulae, due to the fact that adaptation is so simple to express in sat style systems. After all Hoare style formulae can be seen as mere abbreviations of "sat" formulae, and conversely, if S sat R, then S satisfies the Hoare formula $\{P\}\,;S\,\{P;R\}$.

4 Specification adaptation for extensions of the Hoare style system for CSP processes

For practical verification of reactive distributed systems, the Hoare style system for CSP is still not very convenient. Consider a formula $\{P\}\,;S\,\{Q\}$. The postcondition Q is required to hold for all traces of the system $P;S$, that is, both for traces that end in a "$\sqrt{}$" and for those that do not. Intuitively, the traces ending with a "$\sqrt{}$" correspond to a control point immediately after process S, whereas traces without such a "tick" correspond to intermediate stages of the execution of process S. In [ZRE85] a class of formulae was introduced where pre/postconditions P and Q describe only traces ending with a "$\sqrt{}$", corresponding to snapshots before and after execution of S, and where the specification of traces without a "$\sqrt{}$" is delegated to a *trace invariant* I which refers to the communication history only, but not to state variables:

$$I\,:\,\{P\}\,S\,\{Q\}\,\overset{\text{def}}{=}\,\{I\wedge(fin\to P)\}\,;S\,\{I\wedge(fin\to Q)\}$$

The formula above is valid, whenever: if (i) the computation starts in a terminated state such that P holds, then (ii) at every moment of execution of S the invariant I holds, and (iii) Q holds in case, and when, this execution terminates. This specification style separates local conditions such as pre- and postcondition upon the initial and final states from the communication interface w.r.t. the entire system, without losing the simplicity of the rules of Hoare logic which are slightly modified to fit this specification style.

A further extension of the invariant style splits up the invariant in two parts by distinguishing between the communication events of the environment and the communication events of the specified process. This assumption / commitment style was introduced by Misra and Chandy [MC81] in order to find a practically convenient formalism for distributed processes. The assumption refers to the expected communication behaviour of the environment and the commitment refers to the communications of the specified module, provided that the assumption has not been violated before by the environment. The attraction of this style (and of the related rely/guarantee formalism of [J81]) is that it enables the formulation of a proof rule for parallelism which is closest to Hoare's invariant rule for iteration in that both incorporate an induction argument on the length of the computation of which the soundness is established *once and for all* in their soundness proofs, and which, therefore, needn't be justified upon every application of these rules, as it is the case with the above trace invariant formalism. Proof rules for a state-trace based model were given in [ZdeBdeR 83] as derivations of rules of the invariant system. A relatively complete compositional proof system in this assumption/commitment style has been given in [P88], called *P-A Logic*.

Let A, C be trace assertions that do not refer to state variables. We define $\bullet A$ and $\mathcal{K}ern(A)$ by

$\bullet A(h) \stackrel{\text{def}}{=} (h \neq \langle\rangle \rightarrow A(rest(h)))$,

$\mathcal{K}ern(A)(h) \stackrel{\text{def}}{=} \forall h'(h' \leq h.(A(h')))$, intuitively denoting the largest prefix closed subset of A. An assumption/commitment specification is then defined by

$(A, C) : \{P\} S \{Q\} \stackrel{\text{def}}{=} \mathcal{K}ern(\mathcal{K}ern(\bullet A) \rightarrow C) : \{P\} S \{\mathcal{K}ern(A) \rightarrow Q\}$

and can therefore be regarded as a special case of the trace invariant formalism. Note that $\mathcal{K}ern(\mathcal{K}ern(\bullet A) \rightarrow C)$ expresses a trace invariant because of the outermost occurrence of the prefix closure operator $\mathcal{K}ern$. The general idea behind this style is that this correctness formula holds, whenever if **(i)** P holds for the initial terminated state and trace, then **(ii)** commitment C holds initially and after every communication provided A holds after all preceeding communications and **(iii)** if S terminates and A holds after all communications, then Q holds for the final state. Observe that $\mathcal{K}ern(\bullet A) \rightarrow C)$ is a literal translation of **(ii)** and $\mathcal{K}ern(A) \rightarrow Q$ a literal translation of **(iii)**, where $A(<>)$ holds by definition. While again the proof rules of Hoare logics have to be adapted slightly, an interesting rule for parallel composition is applicable: if we have specifications of the form $(A_i, C_i) : \{P_i\} S_i \{Q_i\}$, for $i = 0, 1$ such that for some assertion A, $(A \wedge C_i) \rightarrow A_{i-1}$ for $i = 0, 1$ hold, then the rule yields a specification for $S_0 \| S_1$ of the form:

$(A, C_0 \wedge C_1) : \{P_0 \wedge P_1\} S_0 \| S_1 \{Q_0 \wedge Q_1\}$.

As remarked before, the advantage of this rule is that it incorporates an induction argument on the length of the communication trace which is proved sound once and for all in the soundness proof of this rule, and needn't be repeated upon every application, as demonstrated in [ZdeBdeR 83].

Solving the adaptation problem for the trace invariant formalism

As a result of the different approaches towards the design of a process we may have a gap between the specification of the already developed module and the requirements as given by the proof obligations of the application process. This adaptation problems turns out to be more complicated: assuming that $\models I_1 : \{p_1\}X\{q_1\} \rightarrow I_2 : \{p_2\}X\{q_2\}$ holds, we have to prove formally that $I_2 : \{p_2\}X\{q_2\}$ is derivable from $I_1 : \{p_1\}X\{q_1\}$. Clearly this proof is more difficult than the one for the **sat** style and requires some more subtle rules than just an application of the consequence rule. Within this distributed setup, the semantics of programs (or specifications) involve more than state-based logic, in particular specific closure properties which are a consequence of the trace-based approach in the present representation, e.g., a program P is prefix-closed, hence $P = \mathcal{K}ern(P)$. Concerning the invariant formalism we sketch two different ways to include this additional information into the formal system. By expanding the definitions and applying some logical simplifications, one can show how to transform these formulae into equivalent **sat** style formulae, i.e., that $I : \{ P \} X \{ Q \}$ is equivalent to X **sat** $I : P \rightsquigarrow Q$, where

$$(I : P \rightsquigarrow Q)(h) \stackrel{\text{def}}{=} \forall t. fin(t) \wedge P(t) \rightarrow (I(t; h) \wedge (fin(t; h) \rightarrow Q(t; h)))$$

The adaptation problem for trace invariant formulae can thus be reduced to that between **sat** style formulae and is therefore in principle solved. This approach introduces a certain amount of complexity due to the transformations involved. A more promising way is therefore to adapt the specification directly within the invariant proof system itself. For a given precondition R and characterization $(I : P \rightsquigarrow Q)$ of a given black box module X both the *strongest invariant* and the strongest postcondition are formulated:

$$sinv(R, (I : P \rightsquigarrow Q)) \stackrel{\text{def}}{=} R; (\forall t. (P(t) \rightarrow I(t; h)))$$
$$sp(R, (I : P \rightsquigarrow Q))) \stackrel{\text{def}}{=} R; (\forall t. (P(t) \rightarrow I(t; h) \wedge Q(t; h))).$$

Using these constructs, an intuitivly appealing adaptation rule can be added to the proof system:

Rule 5 (Adaptation rule)

$$\frac{I : \{ P \} S \{ Q \}}{sinv(P_1, (I : P \rightsquigarrow Q)) : \{ P_1 \} S \{ sp(P_1, (I : P \rightsquigarrow Q)) \}}$$

Since $\models (I : P \rightsquigarrow Q) \rightarrow (I_1 : P_1 \rightsquigarrow Q_1)$ holds, we can conclude that $\models sinv(P_1, (I : P \rightsquigarrow Q)) \rightarrow I_1$ and $\models sp(P_1, (I : P \rightsquigarrow Q)) \rightarrow Q_1$ hold. Adaptation completeness of the enlarged invariant style proof system now follows immediately. This rule is related to the adaptation rules for sequential Hoare style systems that have been discussed by Olderog [O83].

A similar rule can be formulated for the assumption/commitment style by transforming a formula into its invariant style equivalent, applying the rule above and lifting the resulting formula up to its original formalism: Abbreviate the invariant style equivalent of $(A, C) : \{P\} S \{Q\}$ by $((A, C) : P \rightsquigarrow Q)$. Within the invariant style we can apply the adaptation rule 5 to obtain

$$sinv(R, ((A, C) : P \rightsquigarrow Q)) : \{ R \} S \{ sp(R, ((A, C) : P \rightsquigarrow Q)) \} \qquad (\dagger)$$

This formula can be interpreted both as an invariant formula and as an assumption/commitment formula, since every invariant formula $I : \{P\}S\{Q\}$ trivially denotes the assumption/commitment formula $(\textbf{true}, I) : \{P\}S\{Q\}$. In this derivation of an adaptation rule we introduce a new assumption A' in a straightforward way: If the formula $(\dagger)$ holds, we conclude that also for every A' the assumption/commitment formula

$$(A', \mathcal{K}ern(\bullet(A')) \wedge (sinv(R, ((A, C) : P \rightsquigarrow Q)))) :$$
$$\{ R \} S \{ \mathcal{K}ern(A') \wedge sp(R, ((A, C) : P \rightsquigarrow Q)) \}$$

holds (Note that A' on the assumption position corresponds to $\mathcal{K}ern(\bullet A)$ at the commitment position). Thus, one can formulate a rather complex adaptation rule within the assumption/commitment formalism itself:

Rule 6 (Adaptation Rule)

$$\frac{(A, C) : \{ P \} S \{ Q \}}{(A', \mathcal{K}ern(\bullet(A')) \land (sinv(R, ((A, C) : P \rightsquigarrow Q)))) : \{ R \} S \{ \mathcal{K}ern(A') \land sp(R, ((A, C) : P \rightsquigarrow Q)) \}}$$

The adaptation completeness proof of an assumption/commitment proof system including this rule is completely analogous to that for the trace invariant system, using a rule of consequence adapted to the assumption/commitment framework [Ha94]. Thus we are able to prove all valid implications between specifications of black box modules within this special framework as well as within the other formal systems discussed in this paper, be it the trace invariant framework, Hoare logic or the **sat** style proof system.

5 Integration of techniques

Our various adaptation problems are illustrated by developing a shortest path algorithm. This algorithm makes essential use of a priority queue. During the sequential part of its development, an a priori given state-based specification of this priority queue is used, illustrating the kind of specification adaptation required in case of modular reasoning, i.e., when a "clash" occurs between top-down and bottom-up development. When implementing this priority queue by a distributed network whose compositional trace-based specification needs to be adapted to show that it satisfies the state-based specification, we take again advantage of our uniform framework and present a canonical solution for these kind of problems.

Example - sequential part
Our example starts out with an abstract algorithm for calculating shortest paths from a given start node x, originally given in [D59]. Since we are aiming at *modular* development ,i.e. based on a priori given progam modules, we use a programming notation that is in the style of object oriented languages like Eiffel [M88]. We represent a directed graph by a set Node, where each Node y has a some attributes: "y.name" is a unique node identifier, and attribute "y.edges" is a set of pairs (z,d) where z is a Node and d is a real number. Informally, when $(z,d) \in$ y.edges, then there is an edge from y to z, with distance d. By "reachable(x,y)" we denote the predicate:

$$\exists x_0, d_1, x_1, \ldots, d_n, x_n.(x = x_0 \land y = x_n \land \bigwedge_{i=1,n}(x_i, d_i) \in x_{i-1}.edges)$$

In a similar style, one defines the partial function "distance(x,y)", denoting the distance of a shortest path from x to y, provided that reachable(x,y) is true. The algorithm assumes two more Node attributes: "y.reached" is a boolean indicating whether a path to y has been found. If y.reached is true, then "y.distance" contains the distance of some path from x to y. The procedure "path" itself is specified by a Hoare style pre-postcondition pair.

procedure path(x:Node)
pre $\forall$y:Node.($\neg$y.reached) $\land$ $\forall$(z,d) $\in$ y.edge. d $\geq$ 0
post $\forall$y:Node. (y.reached $\Longleftrightarrow$ reachable(x,y)) $\land$
$\qquad\qquad\qquad$ (y.reached $\Rightarrow$ y.distance = distance(x,y))

The algorithm itself, although not long, is quite intricate. A Hoare style correctness proof can be found in [Reynolds 81]. Rather than redoing that proof, we would like to focus on some relevant aspects for *modularity*, i.e. reasoning on the basis of program modules. The algorithm, as shown below, uses a so called *priority queue variable*, named "unexplored". Nodes y in "unexplored" have been found to be reachable, but the value of y.distance does not need to be the *shortest* distance from x to y as yet. *New* paths are found by extending those paths that end in some "unexplored" node. The correctness of the algorithm below crucially depends on the *order* of processing nodes from "unexplored" ones: nodes y are processed in the order of their y.distance attribute. The procedure "getmin" yields an element that is *minimal* among the nodes currently in the priority queue and deletes that element.

```
    procedure path(x:Node)
var   unexplored: Prioqueue; y,z: Node; d: Real
begin x.reached := true; x.distance:=0; unexplored.insert(x,0) ;
    while not unexplored.isempty do
        y:=unexplored.getmin;
        for (z,d) in y.edges do
            if not z.reached then z.reached := true;
                z.distance := y.distance + d; unexplored.insert(z);
            elsif y.distance + d < z.distance then
                z.distance := y.distance + d; unexplored.insert(z);
            end ;
        end ;
    end ;
end path
```

The correctness proof of procedure path in [Reynolds 81] is based on a number of state invariants, for instance:

$$\text{geninv} \stackrel{\text{def}}{=} (\forall v,w:\text{Node}.(v.\text{reached} \wedge v \notin \text{unexplored} \wedge w.\text{reached} \wedge w \in \text{unexplored})$$
$$\Rightarrow v.\text{distance} \leq w.\text{distance})$$

In order to check invariance, one should consider each program step in turn, but for now, let us focus on the "unexplored.getmin" step. Reynolds' proof includes the following Hoare formula:

$$\{\text{geninv} \wedge \text{unexplored} \neq \emptyset\} \ y:=\text{unexplored.getmin} \ \{\text{geninv}\} \qquad (*)$$

(Similar formulae are present for the steps of the form "unexplored.insert(z)".)

Now, let us assume that we do have an already existing "Priority queue" module. After a simple syntactic renaming, such that it actually operates on "Nodes", we have a specification as listed below.

```
class Prioqueue
export insert, getmin, isempty;  var queue: Set[Node]

procedure insert(x: T);  ensure queue/ = queue ⊎ {x}

function getmin: T
require queue ≠ ∅
ensure queue = queue/⊎{getmin} ∧ ∀x ∈ queue. (getmin.distance≤ x.distance)
```

function isempty: Bool; **ensure** isempty $\Leftrightarrow$ (queue = $\emptyset$)

end Prioqueue

In essence, this is a specification in VDM or Z style, so there is a gap between the Hoare formula (∗) for "getmin" above, and the specification within the Prioqueue class (At the end of this section we show how (∗) can be derived from this specification). One might remark that we should have specified the Prioqueue class with Hoare formulae, too. For instance, using this formula:

$\forall q.(\{$queue = q$\}$ y:=unexplored.getmin $\{q = $ queue $\uplus$ y $\wedge$

$\forall x \in$ queue. (y.distance $\leq$ x.distance)$\})$

Even here we have an adaptation problem: the desired specification does not follow from the one above by means of the rule of consequence, because it quantifies over all v s.t. $v \notin unexplored \wedge v.reached$ and not just y.

Now, specification adaptation for *sequential* programs is not difficult, because, as argued above, it can always be reduced to checking logical implication between relational predicate formulae. Z style specifications are by nature relational formulae, **require** **-ensure** specifications have been reduced to such formulae above, and VDM specifications can be treated analogously.

For example, from the Prioclass specification it follows that the assignment y:=unexplored.getmin satisfies the relational predicate $\psi(queue, queue\prime, y\prime)$, defined by:

queue $\neq \emptyset \Rightarrow$ (queue = queue$\prime$ $\uplus$ $\{y\prime\}$ $\wedge$ $\forall x \in$ queue. (y$\prime$.distance$\leq$ x.distance))

Hoare formula (∗) can now be verified by calculating the weakest precondition with respect to postcondition "geninv", and by checking that this is implied by "geninv $\wedge$ unexplored $\neq \emptyset$"

Example - concurrent part

For our shortest path example, the Prioqueue can play the role of type T. The corresponding Prioqueue *process* could be the following:

process Prioqueue **chan**   insert,getmin: Node; delete: Node; isempty: Bool
begin **do** insert?x $\rightarrow$ (Buffer(x) \\
                    Prioqueue[down/insert,up/getmin,ise/isempty,
                        del/delete])\ise,del

   or isempty!true $\rightarrow$ **skip**
   or delete?x $\rightarrow$ **skip**
   od
end Prioqueue

This process represents the empty queue, which is extended with a buffer process for each inserted element. Its communication interface satisfies the canonically constructed specification $\chi_{Prioqueue} \stackrel{def}{=}$

$\forall t \leq h.(last(t) = (getmin, x) \Rightarrow \exists v.(retr_{Prioqueue}(t, v) \wedge getmin(v, min(v))$

$\wedge last(t) = (isempty, \textbf{true}) \Rightarrow retr_{Prioqueue}(t, \emptyset)$

towards its environment, while it internally uses a Prioqueue process itself, and a Buffer process as data processes.

process Buffer(x:Node);

chan insert, getmin, up, down:Node; delete,del: Node; isempty,ise: Bool
begin   empty := false;

```
    while not empty do
        if insert?y        → if x.name=y.name then x.distance:=y.distance
                              elsif x.distance<y.distance then down!y
                              else down!x; x:=y fi
        or delete?y        → del!y;
                              if x.name=y.name then ise?empty
                                if not empty then up?x fi fi
        or getmin!x        → ise?empty; if not empty then up?x fi
        or isempty!false   → skip
        fi
    od
end Buffer
```

For the actual verification of this process an invariant specification is more convenient, e.g., $\chi_{Prioqueue}[down/insert, up/getmin] \rightarrow \chi_{Prioqueue}$:

$$\{retr_{Bag} = [x]\} \; Buffer \; \{retr_{Bag} = \emptyset\}$$

This specification states that whenever the *Buffer*'s internal data structure behaves as a priority queue, its communication to the environment will satisfy the priority queue specification itself.

Now assume that a *Buffer* process did already exist within some library, proven correctly w.r.t. a *different* specification, i.e., we are not given the program text above , but an invariant specification of some black box object *Buffer*. Since it is not possible to verify that library text again against the desired specification, our only possibility is to use the formal adaptation rules to derive the desired specification from the given one. As a result of section 4, we get that whenever one specification implies another one this can be proven within the proof system that was shown to be adaptation complete [Ha94](containing also a worked-out example of specification adaptation in case of a buffer, similar to the one above).

6 Conclusion

Integrating various techniques for software development is highly desirable. The need for integration arises in particular in a *modular* aproach, where existing software modules might have been specified in a different formalism than the one used by the developers of the system. This is even unavoidable when during the development process a switch is made from sequential programming to a parallel implementation. We have shown how to achieve this integration, by the technique of *specification adaptation*. Various forms of adaptation, as well as the issue of adaptation completeness, have been discussed. Some of these techniques have been illustrated by developing a shortest path algorithm. The technique used in this example for switching from sequential to concurrent systems is related to techniques for data refinement, in the style of [J86].

Finally, we remark that many of the techniques that we displayed in this paper could be formulated only because of the presence of an underlying uniform framework that is based on predicate logic.

Acknowledgement: We gratefully acknowledge the support of Paritosh K. Pandya when writing this paper.

References

[CH81]   Chen, Z.C. and Hoare, C.A.R., *Partial correctness of CSP*, Conf. on Distr. Comp. Sys., 1981

[D92]   Dahl, Ole-Johan, *Verifiable Programming*, Prentice Hall, 1992.

[D59]   Dijkstra,E.W., *A note on two problems in connexion with graphs*, Numerische Mathematik 1, 1959

[Ha94]   Hannemann, U., *Modular complete proof systems for distributed processes.*, Kiel, 1994.

[H69]   Hoare, C.A.R., *The axiomatic basis of programming*, CACM, 1969.

[H71]   Hoare, C.A.R., *Procedures and parameters: An axiomatic approach*, LNM,1971.

[Hoare 85]   C.A.R. Hoare: *Communicating Sequential Processes.* Prentice-Hall, 1985.

[HHS87]   Hoare, C.A.R., He Jifeng and Sanders,J.W.,*Prespecification in Data Refinement*, IPL 25, 1987.

[J81]   Jones, Cliff B., *Development methods for computer programs including a notion of interference*, Oxford, 1981.

[J86]   Jones, Cliff B., *Systematic software development using VDM*, Prentice-Hall, 1986.

[vKH95]   B. v.Karger and C.A.R. Hoare:*Sequential Calculus*, IPL 53, 1995

[L91]   Lamport, L.,*The Temporal Logic of Actions*, DEC, Systems Research Center, 1991.

[M88]   Meyer, B.,*Object-Oriented Software Construction*, Prentice-Hall, 1988.

[MC81]   Misra, J. and Chandy, K.M., *Proofs of networks of processes.* IEEE TSE, 7, 1981.

[O83]   Olderog, E.R., *On the Notion of Expressiveness and the Rule of Adaptation*, TCS 24, 1983.

[Olderog 86] E.-R. Olderog: *Process theory: Semantics, specification and verification.* LNCS 224, 1986.

[P88]   Pandya, P. *Compositional Verification of Distributed Programs*,Bombay, 1988.

[PJ91]   P. Pandya and M. Joseph: *P-A logic – a compositional proof system for distributed programs.* Distributed Computing 5, 1991.

[Reynolds 81] J.C. Reynolds:*The craft of programming.* Prentice-Hall, 1981.

[S92]   Spivey, Mike, *The Z notation: A reference manual*, Prentice-Hall, 1992.

[Wirth 71]   N. Wirth: *Program development by stepwise refinement.* Communications of the ACM, Vol. 14, No. 4, pp. 221 – 227, 1971.

[ZdeR 89]   J. Zwiers, W.-P. de Roever: *Predicates are Predicate Transformers: a Unified Compositional Theory for Concurrency.* Proceedings of the 8th Symposium on Principles of Distributed Computing, pp. 265 – 279, 1989.

[ZdeBdeR 83] Zwiers,J., de Bruin, A. and de Roever, W.-P., *A proof system for partial correctness of Dynamic Networks of Processes.* , LNCS 164, 1984.

[Z89]   Zwiers, J. *Compositionality, Concurrency and Partial Correctness*, LNCS 321, 1989.

[ZRE85]   Zwiers, J., de Roever, W.-P. and van Emde Boas,P. *Compositionality and concurrent networks: soundness and completeness of a proof system.* , LNCS 194, 1985.

Verifying Part of the ACCESS.bus Protocol Using PVS

Jozef Hooman

Dept. of Mathematics and Computing Science
Eindhoven University of Technology
P.O. Box 513, 5600 MB Eindhoven, The Netherlands
e-mail: wsinjh@win.tue.nl

Abstract. Based on a compositional framework for the formal specification of distributed real-time systems, we present a method for protocol verification. To be able to deal with realistic examples, the method is supported by the interactive proof checker PVS. In this paper we illustrate our approach by a protocol of the ACCESS.bus which is used for the communication between a computer host and its peripheral devices (e.g., keyboards, mice, joysticks, etc.). The bus supports dynamic reconfiguration while the system is operating. We specify and verify a safety property and a real-time progress property of this industrial example.

1 Introduction

In previous work we have addressed the formal specification and verification of distributed real-time systems (see, e.g., [Hoo91]). A framework based on Hoare triples has been applied to several examples such as a water level monitoring system [Hoo93] and a chemical batch processing system [Hoo94c]. In the current paper we only consider the basic ideas behind this formalism and concentrate on the application of such a method to protocol verification.

Essentially, we use an assertional method in which a system is characterized by expressing its properties in a particular logic. Such a method basically deals with formulas of the form P **sat spec**, denoting that process P satisfies specification **spec**. For a formal derivation of a program satisfying a certain specification, a proof system is given, i.e., a set of axioms and rules axiomatizing the programming constructs. To be able to reason with the specifications of components, without knowing their implementation, we use a compositional verification method. This means that in a rule for a compound programming construct only the specifications of its components are used. Compositionality makes it possible to verify design steps during the process of top-down system development. A large number of compositional proof systems have been developed, for instance, for sequential systems [Hoa69], untimed systems [Zwi89, PJ91], and real-time systems [Hoo91].

The only programming construct used in this paper is parallel composition. Under certain conditions we have the following compositional rule.

Rule 1 (Parallel Composition).
$$\frac{P_1 \text{ sat spec}_1, \quad P_2 \text{ sat spec}_2}{P_1 \| P_2 \text{ sat spec}_1 \wedge \text{spec}_2}$$

The main restriction on the application of this rule is that the specification $\mathbf{spec}_i$ of process P_i should only refer to the observable actions of process P_i itself, for $i = 1, 2$ (see, e.g., [Hoo91, Hoo94c] for more details).

We also need a *consequence rule* to weaken a specification.

Rule 2 (Consequence). $\dfrac{P \text{ sat } \mathbf{spec}_0, \mathbf{spec}_0 \rightarrow \mathbf{spec}_1}{P \text{ sat } \mathbf{spec}_1}$

To verify a protocol for a network of processes $P_1 \| \cdots \| P_n$, we first model the application domain and then proceed as follows:

1. Formulate a top-level specification $\mathbf{spec}_{tl}$ for the network $P_1 \| \cdots \| P_n$.
2. Axiomatize the communication mechanism between the processes $P_1, \ldots, P_n$ by a set of axioms $COMMAX$.
3. Find a suitable specification for each process P_i of the form P_i **sat** $\mathbf{spec}_i$, where $\mathbf{spec}_i$ refers to the external communication interface of P_i only, for $i = 1, \ldots, n$.
4. Prove $\mathbf{spec}_1 \wedge \cdots \wedge \mathbf{spec}_n \rightarrow \mathbf{spec}_{tl}$, using axioms of $COMMAX$ and perhaps other assumptions needed for correctness.
5. Derive a correct implementation of process P_i according to its specification P_i **sat** $\mathbf{spec}_i$, for $i = 1, \ldots, n$.

By the parallel composition rule and the consequence rule, this leads to a proof of $P_1 \| \cdots \| P_n$ **sat** $\mathbf{spec}_{tl}$. Hence we have derived a distributed program satisfying the top-level specification. Observe that in the fourth step the protocol is verified on an abstract level, independent of the implementation of the processes. The correctness of this step is based on compositionality of the parallel composition rule.

The method described above has been used to specify and verify a distributed real-time arbitration protocol [Hoo94a] in our Hoare-style framework. This protocol is inspired by an algorithm of the IEEE 896 Futurebus specification. In [ZH95] the first four steps of the method have been applied to an atomic broadcast protocol which involves both real-time and fault-tolerance. Whereas these examples have been verified manually, it is clear that the application of our formal method to large realistic systems requires some form of mechanical support. One should have a tool to construct proofs, to discharge simple verification conditions automatically, and to check proofs mechanically (e.g., after small changes in the specifications).

Therefore we investigate the use of the verification system PVS[1] (Prototype Verification System) [ORS92] for our method. The PVS specification language is a strongly-typed higher-order logic. Specifications can be structured into a hierarchy of theories. There are a number of built-in theories (e.g., reals, lists, sets, ordering relations, etc.) and a mechanism for automatically generating theories for abstract datatypes. The PVS system contains an interactive proof checker

---

[1] PVS is free available, see WWW page http://www.csl.sri.com/sri-csl-pvs.html

with, for instance, induction rules, automatic rewriting, and powerful decision procedures.

The first four steps of the method described above, supported by PVS, are applied to a part of the ACCESS.bus [ACC94]. ACCESS.bus is used for the communication between a computer host and its peripheral devices, such as keyboards, mice, joysticks, modems, monitors and printers. The ACCESS.bus protocol includes a physical layer and several software layers, including a base protocol, device driver interface, and several specific device protocols. The base protocol defines standard messages for device initialization, device identifications, address assignment, and a message envelope for device reports and control information. Usually the host computer is simply called the *computer*, all other partners on the bus are called *devices*, and together they are called *components*. A bus transaction is called a *message*.

The bus supports dynamic reconfiguration while the system is operating. The base protocol contains a configuration process which is used to assign a unique address to each device. Configuration shall occur at system start-up, reinitialization, or at any time when the computer detects the addition or removal of a device. At system start-up or reinitialization the computer shall send a Reset message to all addresses. When a device is reset or powered-up (henceforth we often refer to this as the *start* of the device), it shall always revert to the power-up default address. Then the device shall send an Attention message to the computer which shall reply with an Identification Request. This shall lead to an Identification Reply message of the device containing its identification string. Finally, the computer shall send an Assign Address message to the device containing its identification string and a new address.

Since there is a possibility that two devices are assigned to the same address, a device first transmits a Reset message to its assigned address before it may send data. This self-addressed Reset message forces other devices at the same address back to the power-up default address. We say that a process is in *data mode* if it may send data (i.e. it has obtained a new address, transmitted a self-addressed Reset message, and has not received any Reset to its new address during some period of time).

We investigate two properties of this protocol:

- A safety property which expresses that if two devices are in data mode then their addresses will be different.
- A progress property, expressing that a powered-up device will be in data mode within a bounded amount of time, provided the host does not send any Reset message.

Since program design is not relevant in the example of the ACCESS.bus protocol, we do not consider the fifth step of the method described above, but concentrate on the specifications. The main point here is step four of this method and we focus on the proof of the required implication in PVS. Therefore all primitives and specifications are defined in the PVS specification language. Although PVS has a facility to generate LATEX output of theories and proofs, we present our material in the original PVS format to give an impression of the user interface of this tool. First we formulate a few simple real-time primitives in Sect. 2. The configuration of the ACCESS.bus is defined in Sect. 3. Next the safety property and the progress property are specified and verified in, respectively, Sect. 4 and Sect. 5. Concluding remarks can be found in Sect. 6.

2   Real-Time Primitives

To reason about timing properties, we use a time domain which equals the real numbers. In the PVS theory rt below this is specified by defining the type Time to be equal to the built-in type real. Also nonnegative time points are needed, represented by the type NonNegTime. Further we define time intervals, using co to represent left-closed right-open intervals, etc. (Comments appear after '%'.) The types setof[Time] and pred[Time] are equivalent to the type [Time -> bool] denoting functions from Time to the built-in type bool.

The standard PVS operator NOT on bool is overloaded in rt such that it is defined on predicates over Time, too. Finally it is defined when a time predicate holds inside or during an interval.

```
rt   : THEORY
BEGIN
Time            : TYPE = real
NonNegTime      : TYPE = {t : Time | 0 <= t}
Interval        : TYPE = setof[Time]

t, t0, t1   : VAR Time
cc( t0, t1 ) : Interval = {t | t0 <= t AND t <= t1} %closed-closed
co( t0, t1 ) : Interval = {t | t0 <= t AND t < t1}  %closed-open
oc( t0, t1 ) : Interval = {t | t0 < t AND t <= t1}  %open-closed
oo( t0, t1 ) : Interval = {t | t0 < t AND t < t1}   %open-open

P           : VAR pred[Time]
NOT ( P ) : pred[Time] = (LAMBDA t : NOT P(t) ) ;

I                : VAR Interval
inside( P, I ) : bool = (EXISTS t : I(t) AND P(t))
dur( P, I )   : bool = (FORALL t : I(t) IMPLIES P(t))
END rt
```

3   Configuration of the ACCESS.bus

To apply the method of Sect. 1 to the ACCESS.bus protocol, first the application domain is modelled. The configuration of the ACCESS.bus is defined in theory config which imports the theory rt (i.e., the definitions given in rt are also available in config). Two uninterpreted types Components and Addresses are defined (for our purpose it is irrelevant how addresses are represented in the ACCESS.bus). The + behind TYPE indicates that the types are assumed to be nonempty. The host computer hc is defined as a constant of type Components and all other components are represented by the type Devices. Further, assume given two functions addr and datamode to denote, respectively, the address of a component at a certain point in time and whether a device is in data mode.

```
config : THEORY
BEGIN
IMPORTING rt
Components, Addresses : TYPE+

hc        : Components            % host computer
Devices : TYPE = {c : Components | c /= hc}

addr       : [ Components -> [Time -> Addresses] ]
datamode : [ Devices -> pred[Time] ]
```

In the ACCESS.bus, components communicate by means of messages which consist in our model of five fields: message identification, source address, destination address, identification string, and new address. Identification strings of devices are defined as an uninterpreted nonempty type. To define message identifications we use an enumeration type with five elements. Such an enumeration type is in fact an abstract datatype in PVS and by typechecking it the PVS system internally stores the information that all elements are different (so, e.g., message identification Reset is different from Attention). This information is used by the decision procedures of PVS. Messages are represented by a record with five fields. Primitives send and rec express that a component sends, respectively receives, a message at a particular point in time.

```
IdStrings  : TYPE+
MessageIds : TYPE = {Reset, Attention, IdReq, IdReply, AssignAddr}
Messages   : TYPE = [# mid      : MessageIds,
                       source : Addresses,
                       dest   : Addresses,
                       idstr  : IdStrings,
                       newaddr : Addresses
                     #]

send, rec : [ Components, Messages -> pred[Time] ]
```

Henceforth, in this and subsequent theories, the following variables are used:

```
t, t0, t01, t02, t1, t2, t3    : VAR Time
c, c0                          : VAR Components
d, d0, d1, d2                  : VAR Devices
a, a1, a2, newa, newa1, newa2 : VAR Addresses
```

```
mi                        : VAR MessageIds
istr, istr1, istr2        : VAR IdStrings
m                         : VAR Messages
```

By overloading the equality symbol, we can write (addr(d) = a)(t) instead of addr(d)(t) = a.

```
A              : VAR [Time -> Addresses]
= ( A, a ) : pred[Time] = (LAMBDA t : A(t) = a);
```

Since usually not all five fields of a message are relevant, we introduce a number of abbreviations. Similar abbreviations are defined (but not shown here) for **rec**.

```
send(c,mi)(t) : bool = (EXISTS m: send(c, m WITH [mid := mi])(t) )

send(c,mi,a)(t) : bool = (EXISTS m:
    send(c, m WITH [mid := mi, dest := a])(t) )

send(c,mi,a1,a2)(t) : bool = (EXISTS m:
    send(c, m WITH [mid := mi, source := a1, dest := a2])(t) )

send(c,mi,a1,a2,istr)(t) : bool = (EXISTS m:
    send(c, m WITH [mid := mi, source := a1,
                dest := a2, idstr := istr])(t) )

send(c,mi,a,istr,newa)(t) : bool = (EXISTS m:
    send(c, m WITH [mid := mi, dest := a, idstr := istr,
                newaddr := newa])(t) )

send(c,mi,a,istr)(t) : bool =(EXISTS m:
    send(c, m WITH [mid := mi, dest := a, idstr := istr] )(t) )
```

4   A Safety Property

In this section we verify a safety property, expressing that never two devices will be in data mode with the same address. This is done in theory **access1**, following the first four steps of the method presented in Sect. 1. Theory **access1** imports the theory **config**, which means that all the defintions of **config** and **rt** (which is imported by **config**) are available.

```
access1 : THEORY
BEGIN
IMPORTING config
```

4.1   Top-level Specification (spec$_{tl}$)

The first step in our approach is the definition of the top-level specification **top_level1**, expressing that when two different devices are in data mode they have different addresses.

```
top_level1 : bool = (FORALL d1, d2, t:
                datamode(d1)(t) AND datamode(d2)(t) AND d1 /= d2
                IMPLIES
                addr(d1)(t) /= addr(d2)(t))
```

4.2 Axiomatization of the Communication Mechanism (*COMMAX*)

The second step requires the axiomatization of the communication mechanism. The physical layer of the ACCESS.bus is based on the I^2C serial bus, for which the following holds:

> The same message sent at the same time by one or more entities is always correctly received once by every entity with a matching address that does not belong to the set of the senders (of the same message).

To be able to prove real-time properties, we use a nonnegative constant TD, representing an upper bound on the transmission time. It is not completely clear at which point in time a receiver should have the matching address; here we take a strong assumption (thus obtaining a weak axiom) by requiring that the receiver has this address during the whole receiving interval of TD time units.

```
TD : NonNegTime            % Transmission Delay

comm_ax1 : AXIOM (FORALL c, m, t:
        send(c,m)(t)
        IMPLIES
        (FORALL (c0 | NOT send(c0,m)(t) AND
                    dur( addr(c0) = dest(m) , co(t,t+TD) ) ) ):
            inside( rec(c0,m) , co(t,t+TD) ) ) )
```

4.3 Specification of the Processes (spec$_i$)

In step three we specify the components, as far as needed to prove top_level1. For this safety property we only have to specify devices. For a device d, assertion DS1(d) expresses that d is in data mode only if it has sent a self-addressed Reset at least DD time units earlier and since then no Reset has been received.

```
DD : NonNegTime            % Data mode Delay

DS1(d) : bool = (FORALL t : datamode(d)(t)
                    IMPLIES
                    (EXISTS (t0 | t0 < t - DD), a :
                        dur( addr(d) = a, cc(t0,t) ) AND
                        send(d,Reset,a)(t0) AND
                        dur( NOT rec(d,Reset,a), cc(t0,t) ) ) )
```

4.4   Correctness Proof

The goal of step four is to prove that the conjunction of the properties of all devices (i.e. (FORALL d : DS1(d))) implies top-level specification top_level1. First observe that DD >= TD is required to make sure that the self-addressed Reset of a starting device is received by all possible receivers before the device goes to data mode.

This, however, is not sufficient to prove top_level1. If two devices have the same identification string, they obtain the same address, and if they send their Reset messages simultaneously, then none of them will receive it and both can start sending data from the same address. The documentation [ACC94] does not give a solution for this case, but it seems reasonable to assume that the probability that this situation occurs is very small. To avoid reasoning with probabilities, we assume here that the situation can be neglected, i.e. that simultaneous resets do not occur, as expressed by not_sim_reset.

Then we can prove theorem cortl1, based on lemma cortl1lem which represents the situation that d1 has received an address before d2 sends a Reset to its own address.

```
not_sim_reset : bool = (FORALL t, d1, d2, a:
                        send(d1,Reset,a)(t) AND send(d2,Reset,a)(t)
                        IMPLIES
                        d1 = d2)


cortl1lem : LEMMA DD >= TD AND t01 <= t02 AND d1 /= d2 AND
                  t01 < t - DD AND
                  dur( NOT rec(d1,Reset,a1), cc(t01,t) ) AND
                  dur( addr(d1) = a1, cc(t01,t) ) AND
                  t02 < t - DD AND
                  send(d2,Reset,a2)(t02) AND addr(d2)(t) = a2 AND
                  not_sim_reset
                  IMPLIES
                  addr(d1)(t) /= addr(d2)(t)


cortl1 : THEOREM DD >= TD AND (FORALL d : DS1(d)) AND
                 not_sim_reset
                 IMPLIES
                 top_level1
END access1
```

The keywords LEMMA and THEOREM are equivalent in PVS. All lemmas and theorems presented here have been proved by means of the PVS interactive prover.

For instance, for cortl1lem the application of axiom comm_ax1 leads to three subgoals. One subgoal corresponds to the case that the Reset message of d2 is received by d1. Then we can derive a contradiction with the assumption that d1 does not receive Resets from t01 till t. The other two subgoals are related to the conditions of comm_ax1. Using not_sim_reset we can easily show that d1 does not send a Reset at t02. Finally we prove that d1 has address a2 during a certain interval by simply expanding a few definitions and providing the proper instantiations. In general, the user of the PVS prover gives the main outline of the proof, whereas details can be proved automatically (often by the powerful command "grind" which combines, e.g., automatic rewriting, skolemization, instantiation, and propositional simplification).

5 A Progress Property

In theory access2, which imports access1, we specify and verify a real-time progress property. To express when a device is powered-up we introduce a time dependent predicate up. Predicate start is used to express when a device starts (i.e. becomes powered-up or is reset).

```
access2 : THEORY
BEGIN
IMPORTING access1
up, start : [ Devices -> pred[Time] ]
```

5.1 Top-level Specification (spec$_{tl}$)

Specification top_level2 expresses that a device which starts at t1 and remains up till t2 will be in data mode during [t1+JD,t2), unless it receives a host Reset between t1 and t2, that is, as long as the host does not send a Reset between t1-TD and t2 (note that a transmission might take less than TD).

```
JD : NonNegTime      % Join Delay

top_level2 : bool = (FORALL d, t1, t2:
                        start(d)(t1) AND
                        dur( up(d), co(t1,t2)) AND
                        dur( NOT send(hc,Reset), co(t1-TD,t2))
                        IMPLIES
                        dur( datamode(d), co(t1+JD,t2) ))
```

5.2 Axiomatization of the Communication Mechanism (COMMAX)

Since access2 imports access1, axiom comm_ax1 is still available. We add an axiom to express that any message received by a component has been sent before by another component.

```
comm_ax2 : AXIOM (FORALL c, m, t:
      rec(c,m)(t)
      IMPLIES
      (EXISTS (c0 | c0 /= c) : inside( send(c0,m), oc(t-TD,t) )))
```

5.3 Specification of the Processes (spec$_i$)

In the third step we specify the components of the system. First the behaviour of devices is specified by a number of properties. We define a few constant addresses and a function **devid** to denote the identification string of a device.

```
pda     : Addresses      % power-up default address
hca     : Addresses      % host computer address

devid   : [ Devices -> IdStrings ]
```

A starting device has the power-up default address and sends an Attention message to the host computer address within **AD**, for some timing constant **AD**, provided this device stays up during at least **AD** time units.

```
AD : NonNegTime          % Attention Delay

DS2(d) : bool = (FORALL (t | dur( up(d), co(t,t+AD) )):
            start(d)(t)
            IMPLIES
            inside( send(d,Attention,pda,hca), co(t,t+AD) ) ) )
```

If a device receives an Identification Request message at the power-up default address, it responds within **IRD** by sending its device identification string to the computer address.

```
IRD : NonNegTime          % Id Reply Delay

DS3(d) : bool = (FORALL (t | dur( up(d), co(t,t+IRD) )):
                rec(d,IdReq,pda)(t)
                IMPLIES
                inside( send(d,IdReply,pda,hca,devid(d)),
                        co(t,t+IRD) ) )
```

If device **d** receives an Assign Address message containing its own device identification string then it will be in data mode after **DD** time units as long as it did not receive any Reset messages to the newly received address. Recall that **DD** has been defined in theory **access1**.

```
DS4(d) : bool = (FORALL t1, t2, newa:
            dur( up(d), co(t1,t2) ) AND
            rec(d,AssignAddr,pda,devid(d),newa)(t1) AND
            dur( NOT rec(d,Reset,newa), co(t1,t2) )
            IMPLIES
            dur( datamode(d), co(t1+DD,t2) ) ) )
```

Further a device only sends a Reset message to an address if it has received this address in an Assign Address message recently.

```
DS5(d) : bool = (FORALL t, a:
                        send(d,Reset,a)(t)
                        IMPLIES
                        inside( rec(d,AssignAddr,pda,devid(d),a),
                                cc(t-DD,t) ) )
```

Devices never send Identification Request or Assign Address messages.

```
DS6(d) : bool = (FORALL t, m:
                        send(d,m)(t)
                        IMPLIES
                        mid(m) /= IdReq AND mid(m) /= AssignAddr)
```

When a device starts, it has the power-up default address pda as long as it did not receive an Assign Address message at least TD time units before.

```
DS7(d): bool = (FORALL t0, t1:
    start(d)(t0) AND dur(up(d), cc(t0,t1)) AND
    dur( NOT rec(d, AssignAddr, pda, devid(d)), cc(t0,t1-TD))
    IMPLIES
    addr(d)(t1) = pda
```

The delay TD is related to comm_ax1 in access1, which requires that the receiver has the matching address during the whole receiving interval of TD time units. Properties DS6 and DS7 are often used in combination with comm_ax1 to show that a message sent to a device is also received (because the device cannot be a sender of this message and it will have the right address).

```
DS(d) : bool = DS2(d) AND DS3(d) AND DS4(d) AND
               DS5(d) AND DS6(d) AND DS7(d)
```

Next we specify the behaviour of the host computer. Property CS1 expresses that it responds to an Attention message by sending an Identification Request to the address of the sender within HCD1.

```
HCD1, HCD2 : NonNegTime    % Host Computer Delays
```

```
CS1 : bool = (FORALL t, a:
                        rec(hc,Attention,a,hca)(t)
                        IMPLIES
                        inside( send(hc,IdReq,a), co(t,t+HCD1) ) )
```

After receiving an Identification Reply, the computer returns an Assign Address message with a new address in less than HCD2 time units.

```
CS2 : bool = (FORALL t, a, istr:
    rec(hc,IdReply,a,hca,istr)(t)
    IMPLIES
    (EXISTS newa :
      inside( send(hc,AssignAddr,a,istr,newa), co(t,t+HCD2) ) ) )
```

The properties specified describe the exchange of messages between the computer and a starting device, resulting in the receipt of an Assign Address message by this device. As expressed by DS4, this only leads to data mode if the device does not receive a Reset to its new address. Therefore we assume that two Assign Address messages which contain different device identification strings, also contain different addresses. Then in the proof of top_level2 it is required that different devices have different identification strings.

```
CS3 : bool = (FORALL t1, t2, istr1, istr2, newa1, newa2:
                  send(hc,AssignAddr,pda,istr1,newa1)(t1) AND
                  send(hc,AssignAddr,pda,istr2,newa2)(t2) AND
                  istr1 /= istr2
                  IMPLIES
                  newa1 /= newa2 )
```

In property CS3 we made a simplification by not considering the re-use of addresses (in contrast with the real protocol which detects the absence of devices by Presence Check messages).

Similar to the devices, we have two properties about messages that are not sent by the computer and about its address. They are used in combination with comm_ax1 to derive that certain messages are received by the computer.

```
CS4: bool = (FORALL t, m:
                 send(hc,m)(t)
                 IMPLIES
                 mid(m) /= Attention AND mid(m) /= IdReply)

CS5: bool = (FORALL t: addr(hc)(t) = hca)

CS : bool = CS1 AND CS2 AND CS3 AND CS4 AND CS5
```

5.4 Correctness Proof

The correctness of the protocol described above with respect to specification top_level2 is proved, provided different devices have different identification strings (as expressed by id_diff). Further there is a constraint on the timing constants, using abbreviation RAB to represent the maximal delay between the start of a device and the moment it receives a new address.

```
id_diff : bool = (FORALL d1, d2: d1 /= d2 IMPLIES
                                    devid(d1) /= devid(d2))

RAB : Time = AD + 4*TD + HCD1 + IRD + HCD2 % Receive Address Bound

cortl2 : THEOREM id_diff AND JD >= RAB + DD AND CS AND
                 (FORALL d: DS(d))
                 IMPLIES
                 top_level2
```

In the proof of this theorem we use a number of lemmas (in the PVS file they occur before the theorem). It is rather straightforward to show that a starting device d receives an Assign Address message in less than RAB time units.

```
recAssignAddrlem : LEMMA
    DS2(d) AND DS3(d)AND DS6(d) AND DS7(d) AND
    CS1 AND CS2 AND CS4 AND CS5
    IMPLIES
    (FORALL (t | dur( up(d), co(t,t+RAB) )):
        start(d)(t)
        IMPLIES
        inside( rec(d,AssignAddr,pda,devid(d)), co(t,t+RAB) ) ) )
```

Next we show that this leads to datamode if d does not receive any Reset message to its new address and provided JD >= RAB + DD.

```
datalem : LEMMA DS4(d) AND dur( up(d), co(t1,t2) ) AND
                (EXISTS newa:
                    inside(rec(d,AssignAddr,pda,devid(d),newa),
                        co(t1,t1+RAB)) AND
                    dur(NOT rec(d, Reset, newa), co(t1,t2))) AND
                JD >= RAB + DD
                IMPLIES
                dur( datamode(d), co(t1+JD,t2) )
```

Hence it remains to prove that d does not receive any Reset message to its new address. This follows from the fact that the host computer does not send Reset messages and assuming no other device sends Reset messages to this new address.

```
notrecResetlem : LEMMA dur( NOT send(hc,Reset), co(t1-TD,t2)) AND
                        (FORALL (d0 | d0 /= d):
                            dur( NOT send(d0,Reset,newa), co(t1-TD,t2) ))
                        IMPLIES
                        dur( NOT rec(d,Reset,newa), co(t1,t2) )
```

Finally, we show that if the host computer sends a new address to device d then any other device will not send Reset messages to this new address. Note that id_diff is used here in combination with CS3.

```
notsendResetlem : LEMMA CS3 AND id_diff AND
                        (FORALL d: DS5(d)) AND
                        (FORALL d: DS6(d)) AND
                        (EXISTS t:
                            send(hc,AssignAddr,pda,devid(d),newa)(t))
                        IMPLIES
                        (FORALL (d0 | d0 /= d):
                          dur( NOT send(d0,Reset,newa), co(t1-TD,t2) ))
```

6 Concluding Remarks

Part of the ACCESS.bus protocol has been specified and verified using an assertional method supported by PVS. The protocol has real-time aspects and allows dynamic reconfiguration. Recent overviews on industrial applications of formal methods can be found in [Rus93] and [GCR94]. The main aim here was to investigate the support of the interactive theorem prover PVS for our protocol verification method.

An important part of the work described here concerns the tranformation from an informal description [ACC94] into a formal specification. In this paper we have extracted the essential points needed to prove a few properties, ignoring irrelevant implementation details, thus clarifying the protocol and making assumptions for its correctness explicit. For instance, to show the safety property (devices in data mode have different addresses) we had to introduce an assumption about the impossibility of simultanous Reset messages. Further we have slightly generalized the formulation in [ACC94] by introducing timing constants instead of concrete numbers and by deriving constraints on these constants that ensure correctness of the protocol.

Since the step from informal document to formal specification is crucial, we have tried to define our specifications close to the informal formulation. It turns out that the PVS specification language is very suitable for this purpose and by defining appropriate abbreviations the gap with the informal text is reasonably small. Clearly the transformation from informal to formal is an iterative process, since ambiguities have to be resolved, mistakes have to be corrected, etc. To our experience the use of a tool such as PVS increases the speed of this process to a large extent, since it is easy to incorporate changes, to rerun the proofs and to modify proofs interactively.

In the ACCESS.bus example we have not worked out step five of our approach in which a program for the components is derived. An example of program derivation by means of PVS can be found in [Hoo94b] where a mixed formalism, in which programs and assertional specifications are combined, has been defined in the PVS specification language.

Related to the work described here is the use of PVS [ORS92] and its predecessor EHDM for a number of applications, such as an interactive convergence clock synchronization algorithm [RvH93], a Byzantine fault-tolerant clock synchronization [Sha93], and an algorithm for interactive consistency [LR93]. We also mention the specification language TLA (Temporal Logic of Actions) which has been applied to a large number of examples. See, e.g., the specification and the hierarchically structured proof of a Byzantine generals algorithm [LM94]. Another nice example of protocol verification can be found in [BPV94] where an industrial protocol is specified and verified based on timed I/O automata.

Acknowledgements

The ACCESS.bus protocol has been proposed by Ron Koymans (Philips Research, Eindhoven) to a number of academic researchers as an example of an industrial protocol. He is also thanked for valuable comments on preliminary versions of this paper. We are grateful to the anonymous referees for useful suggestions.

References

[ACC94] *ACCESS.bus*™, *Specifications – Version 2.2*. Sunnyvale, California, 1994.

[BPV94] D. Bosscher, I. Polak, and F. Vaandrager. Verification of an audio control protocol. In *Formal Techniques in Real-Time and Fault-Tolerant Systems*, pages 170–192. LNCS 863, 1994.

[GCR94] S. Gerhart, D. Craigen, and T. Ralston. Experience with formal methods in critical systems. *IEEE Software*, 11(1):21–39, 1994.

[Hoa69] C.A.R. Hoare. An axiomatic basis for computer programming. *Communications of the ACM*, 12(10):576–580,583, 1969.

[Hoo91] J. Hooman. *Specification and Compositional Verification of Real-Time Systems*. LNCS 558, Springer-Verlag, 1991.

[Hoo93] J. Hooman. A compositional approach to the design of hybrid systems. In *Workshop on Theory of Hybrid Systems*, pages 121–148. LNCS 736, 1993.

[Hoo94a] J. Hooman. Compositional verification of a distributed real-time arbitration protocol. *Real-Time Systems*, 6(2):173–205, 1994.

[Hoo94b] J. Hooman. Correctness of real time systems by construction. In *Formal Techniques in Real-Time and Fault-Tolerant Systems*, pages 19–40. LNCS 863, 1994.

[Hoo94c] J. Hooman. Extending Hoare logic to real-time. *Formal Aspects of Computing*, 6(6A):801–825, 1994.

[LM94] L. Lamport and S. Merz. Specifying and verifying fault-tolerant systems. In *Formal Techniques in Real-Time and Fault-Tolerant Systems*, pages 41–76. LNCS 863, 1994.

[LR93] P. Lincoln and J. Rushby. The formal verification of an algorithm for interactive consistency under a hybrid fault model. In *Computer Aided Verification '93*, pages 292–304. LNCS 697, Springer-Verlag, 1993.

[ORS92] S. Owre, J. Rushby, and N. Shankar. PVS: A prototype verification system. In *11th Conference on Automated Deduction*, volume 607 of *Lecture Notes in Artificial Intelligence*, pages 748–752. Springer-Verlag, 1992.

[PJ91] P. Pandya and M. Joseph. P-A logic – a compositional proof system for distributed programs. *Distributed Computing*, 4(4), 1991.

[Rus93] J. Rushby. Formal methods and the certification of critical systems. Technical Report CSL-93-7, SRI International, November 1993.

[RvH93] J. Rushby and F. von Henke. Formal verification of algorithms for critical systems. *IEEE Transactions on Software Engineering*, 19(1):13–23, 1993.

[Sha93] N. Shankar. Verification of real-time systems using PVS. In *Computer Aided Verification '93*, pages 280–291. LNCS 697, Springer-Verlag, 1993.

[ZH95] P. Zhou and J. Hooman. Formal specification and compositional verification of an atomic broadcast protocol. *Real-Time Systems*, 9(2):119–145, 1995.

[Zwi89] J. Zwiers. *Compositionality, Concurrency and Partial Correctness*. LNCS 321, Springer-Verlag, 1989.

Reusing Batch Parsers as Incremental Parsers

Luigi Petrone

Università di Torino, Torino, Italy

ABSTRACT. *A simple solution to the incremental parsing problem is found by mapping sentential forms of a context-free grammar G over terminal strings of the placeholder language and then sub-tree forests, i.e. sequences of syntax sub-trees, over sequences of pairs (placeholder-token, subtree-pointer). Parsing such sequences-of-pairs/ placeholder-strings is equivalent to parsing and therefore to embedding sub-tree-forests into a global syntax tree. Using this placeholder-for-tree techmnique we may easily implement several incremental bottom-up parsing policies by re-using standard (batch) parsers, such as those generated by Yacc, or any LARL or LR(k) parser.*

KEY WORDS AND PHRASES: parsing, parse trees, LR parsers, incremental parsing, programming environments

CR CATEGORIES: 4.12, 4.42, 5.23

1.Introduction

One of the goals of Software Engineering during the last ten years has been the establishment of interactive programming environments. The need to reduce the delay between program editing time and detection of errors has led to the study of incremental parsing techniques [5, 6]. The need to assist the programmer in the task of handling programs as structured objects has led to the invention and later the improvement of structured editors [7, 9, 13]. The functionality of (hybrid) structured editors is fascinating and has attracted the interest of software engineers. Structured editors not only reduce editing time, a feature of less importance in these days of ever increasing machine performance, but also provide a high level structured view of program code which greatly facilitates the programmer's understanding and handling of the program text.
  The techniques presented in this paper rely upon the simple but powerful idea of mapping syntax subtrees over terminal strings of the placeholder language so that any standard batch parser can be used for parsing not only strings of characters, but subtree forests too. Such techniques can be used to solve several general problems of incremental parsing as well as the specific parsing problems described in [3]. Our methods are simpler and more flexible than other methods known in the literature and moreover have the important advantage that they can be used with any standard batch parser, whereas [5,6] require the use of a special purpose parser and need to store additional auxiliary information in the parse tree. Our placeholder-for-tree technique can also be used to solve the specific parsing problems of structured editors, see [3], but in a much simpler and more efficient way.

2. Intermixing analytic and synthesizer modes.

All hybrid editors, like Dual [9], KeyOne [10] or [13], see also [3], leave the user free to define programs by successive expansions in the synthesizer mode interleaved with replacements of placeholders with some text in the analytic mode or corrections (or updates) in text mode. Consequently the user is allowed to store and, later, load and parse programs still containing placeholders, which must therefore be represented in some concrete form by extending the base language. Usually placeholders are represented by tokens such as <declaration> or <expression> and the parser must be capable of parsing "terminal" strings which are essentially representations of sentential forms of the language, such as

```
while <expression> loop <statement-list> end loop;
```

3. The incremental problem

  Let $G=(N, \Sigma, P, A_1)$ be a proper unambiguous grammar defining a context-free language $L(G)=\{ x \mid A_1 \Rightarrow^* x \in \Sigma^* \}$, where as always $N =\{A_1, A_2, ... , A_k\}$ is the set of nonterminals, Σ is the terminal alphabet, P is the set of productions and A_I is the start symbol. Our starting point is an input string x and its corresponding parse tree T. Each internal node of the parse tree is labeled with a nonterminal symbol A_i - or with a code such as the production number allowing instant access to A_i - and if $X_1, X_2, \dots, X_r$ are labels of the descendant nodes of A_i then

$A_i \Rightarrow X_1 X_2 \dots X_r$. So our parse tree is not an abstract tree, see Figures 1 and 2. The nodes X_1, $X_2, \dots, X_r$ are also called constituents of A_i and r is called *arity* of the node.

As is usual in most implementations of structured editors, the output of the parser is stored as a two-way parse tree T: in other words, from any node of T of arity r you can either move uptree going through its parent node or downtree going through any child out of its r children. The presence of upward links in the tree can be used for other purposes as well: for instance it is essential for implementing a syntactic browser giving instant access to the declaration of any identifier of a program from its point of use (see [9] or [10]).

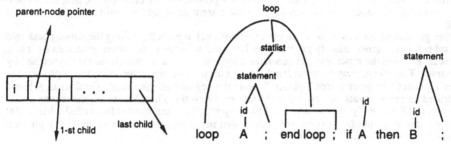

"i" is the production #

Fig. 1   A tree-node   Fig. 2.   A tree-row

In general, an editing action over a string $x=ysz$ amounts to replacing a substring s of x with a second string s' so that x is updated into $x'=ys'z$. If s' is the empty word then the action is called a *cut* or a *delete* operation; if s is the empty word then we say that we have *inserted* the string s'; if s is a placeholder then the replacement of s with s' is called a *placeholder expansion*.

In order to eliminate tedious details from our treatment let us make the simplifying assumption that any cut point in our text is always located between two successive tokens.

The string x (or the tree T) and the string x' (or the tree T') are called the old string (or old tree) and the updated string (or updated tree), respectively. The incremental problem consists in building the new tree T' making as much use as possible of the subtrees of the old tree T.

In order to solve the general case of insertion of unstructured objects or of unrestricted editing followed by incremental analysis we need to introduce the concept of *parse-subtree-forest*.

First, let us call *forest* a sequence of trees and *frontier of the forest* the catenation of the strings which are the frontiers of each tree of the sequence. Given a string $x=ysz$ and its parse tree T, we call *parse-subtree-forest* based on a substring s of ysz, and we write $PSF(s,ysz)$, that sequence of subtrees and/or tokens which are part of T, have highest profile, and whose frontier is exactly the substring s. Obviously $PSF(x,x)=\mathsf{T}$. If τ denotes the mapping replacing each subtree of the parse-subtree-forest with the nonterminal labelling its root, then the string

$y \ \tau(PSF(s,ysz)) \ z$ is a sentential form of $L(G)$ and $\tau(PSF(y,yz))$ is the stack contents of any bottom-up parsing of the string yz when the substring y of yz has been processed, i.e.

$A_I \Rightarrow^* \tau(PSF(y,yz))z$ by a rightmost derivation of the grammar G and $A_I \Rightarrow^* y \ \tau(PSF(z,yz))$ by a leftmost derivation of G.

4. Two-level parsers

A parser (P) of a language $L(G)$ is an algorithm that, for any string x, outputs a yes or no-answer to the question whether x belongs to $L(G)$. A structural parser (SP) is a parser, complemented with semantic routines, mapping any string x of $L(G)$ into a tree, its parsing tree, having as frontier the input string x.

Most current implementations of parsers P (or structural parsers SP) find it convenient to make use of an input function called *Fetch-char* (or respectively *Fetch-this-char*). *Fetch-char* is a function, used by P, returning, at each call, an element of the terminal alphabet which is the next character of the string-to-be-parsed. *Fetch-this-char* is a function used by SP returning, at each call, an item made up of two objects: a character of the alphabet and a pointer to the physical location (i.e. occurrence) of this character in the input text. Pointers are then handled by the semantic routines allocating tree nodes and mutually linking lower nodes, or frontier tokens, with the upper nodes of the parse tree, as follows. A semantic routine *MakeNode(i)* is associated to each *i-th* grammar production of arity r

$$A_j \rightarrow X_1 X_2 \ldots X_r \quad \{ MakeNode(i) \}$$

and activated when the parser reduces a handle. Using a table that for each i gives both the arity r and the left-hand nonterminal A_j of the *i-th* grammar production, a possible implementation of the routine *MakeNode(i)* will allocate a parent-node (having as node-attributes the production number, a pointer to the parent-node, and an array of r pointers for the r children), retrieve and pop from the top r positions of the semantic stack r pointers to the child-nodes, link them both-ways to the parent-node and eventually push the parent-node address on top of the semantic stack.

Parser generators such as Yacc make use of a two-level approach dividing the parsing task into two subtasks: a scanner, usually generated by Lex, and a "high-level" parser generated by Yacc. The scanner scans the input text and generates a sequence of tokens which are then processed by the parser. The tokens may be considered for all purposes as characters of a new alphabet: the token alphabet.[1] In other words a parser P treats the scanner as the routine Fetch-char, while a structured parser SP treats the scanner as the routine Fetch-this-char. Therefore, the tokens which are input to SP are really items made up of two objects: the token itself (often called token code) and a pointer to the physical occurrence of the token in the input string (or a pointer to a physical replica of the token itself: its value).

5. The placeholder languages

Given a grammar $G = (N, \Sigma, P, A_1)$, we must first eliminate an important source of confusion by making a clear distinction between nonterminal symbols of N and placeholders, which are additional symbols of an extended terminal alphabet used in the *placeholder* grammars G_i'. The placeholder grammar of a grammar $G_i = (N, \Sigma, P, A_i)$ is the grammar $G_i' = (N, \Sigma \cup T, P \cup K, A_i)$ where $T = \{t_1, t_2, \ldots, t_k\}$, $card(T) = card(N)$, and K is the set of additional productions:

$$\{ A_1 \rightarrow t_1, A_2 \rightarrow t_2, \ldots, A_k \rightarrow t_k \}$$

Let us shorten G_1' into G': the language $L(G')$ is called the *placeholder* language.

In order to avoid introducing too many additional terminal characters, the placeholders are often coded over new tokens such as <statement>, <declaration>, <identifier> etc. which, for this purpose, are added to the language provided no clash takes place with the tokens of the original language; this is an easy matter: it is sufficient to select, as placeholder-tokens, k strings not belonging to $L(G)$.

We must keep in mind that a placeholder string is not a sentential form of the grammar but just a terminal string of the placeholder language and therefore may naturally be parsed by any of its batch parsers. Let σ denote the *(1-1)*-mapping sending nonterminal symbols A_i into the corresponding nonterminals t_i, and let SF denote the set of sentential forms of the original grammar, then the placeholder language has the interesting *closure* property of being the union of $L(G)$ and of $\sigma(SF)$. A string of the placeholder language either belongs to the original language or is the image of one of its sentential forms. For example, if $x_1 A_2 x_3 A_4$ is a sentential form of G - and obviously also a sentential form of G' - then $x_1 t_2 x_3 t_4$ is a terminal string of $L(G')$; incidentally let us note that $x_1 t_2 x_3 A_4$ is a sentential form of G' but obviously not of G.

Let us now introduce the concept of *incremental syntax tree* (IST) of a string x belonging to the placeholder language $L(G')$. The IST of a string x is the tree obtained from the syntax tree for x by replacing all nonterminals labelling its internal nodes with the corresponding placeholders. An IST *looks like* a normal syntax tree except that any cross-section of a syntax

---

[1] Henceforward we shall use indifferently the terms token and character. The term *token* is more appropriate when the grammar is a real grammar with hundreds of productions, tens of characters and you really need a scanner routine, the term *character* is more appropriate when the grammar is a toy-grammar with 4-5 productions and 2 or 3 characters.

tree is a sentential form of the language, while the corresponding cross-section of the IST is a placeholder string, image of the corresponding sentential form. Let us denote with $A_i\text{-}IST(x)$ the incremental syntax tree of a string x deriving from the nonterminal A_i.

6. The batch parser of the incremental language

We want to be able to use a single parser for parsing k placeholder languages L_i' $\{i=1, ... , k\}$, defined by k distinct grammars $G_i' = (N, \Sigma \cup T, P \cup K, A_i)$, having as start symbol in turn each of the k nonterminal symbols of the original grammar G. In practice, k may be a fairly large number (for the Ada grammar, for instance, k is greater than 100). The solution adopted in [3] is rather complex and requires a special parser operating on k distinct parsing tables. Our solution is much simpler and requires a single standard parser operating on a slightly modified grammar.

For this purpose let us introduce a sort of universal grammar: the *incremental* grammar $G'' = (N \cup A_0, \Sigma \cup T \cup D, P \cup K \cup H, A_0)$ where D is a set of additional terminal symbols $\{d_1, d_2, ... , d_k\}$ and H contains the following additional productions

$$\{ A_0 \rightarrow d_1 A_1 , A_1 \rightarrow d_2 A_2 , ... , A_1 \rightarrow d_k A_k \}$$

It is easily seen that the grammar G'' generates the incremental language $L(G'')$ [2] which is the union of all the placeholder languages, each pre-concatenated with the driving symbol d_i:

$$L(G'') = d_1 L(G_1') \cup d_2 L(G_2') \cup ...... \cup d_k L(G_k')$$

The driving symbols can be defined as virtual tokens, i.e. tokens having a token code but not having a concrete representation in the input language and therefore not recognized by the scanner; in some cases the driving symbols may be identified with the placeholder tokens: the additional shift-reduce conflicts of the grammar can be handled by tricks such as the Yacc default rule.

The single parser of the incremental grammar may perfectly replace the k parsers for the various different placeholder languages. In fact, it may easily be verified that

$$A_i \Rightarrow^* x \qquad \text{if and only if} \qquad A_0 \Rightarrow^* d_i x$$

Therefore, in order to check whether x is derivable from A_i it is sufficient to check whether $d_i x$ is derivable from A_0.

The incremental language is the key element of our development since *a batch parser of the incremental language may be used as the key tool for building the incremental parser for the original language* (see also [12]).

Let us give an example.

Let the productions of the grammar $G_1 = (N, \Sigma, P, procedure)$ be the following:

procedure	→	**procedure** *id* **is begin** *statlist* **end** ;
statlist	→	*statlist statement* | *statement*
statement	→	**if** *id* **then** *statlist* **end if** ;| *iteration loop* | *loop* | *id* ;
iteration	→	**while** *id* | **for** *for-clause*
loop	→	**loop** *statlist* **end loop** ;
for-clause	→	*id* **in** *id* .. *id*
id	→	*id letter* | *letter*
letter	→	A | B | ... | Z

The tokens of this grammar are the keywords, the separator ";", and the identifiers *id* [3]; the nonterminal symbols are *procedure, statlist , statement, iteration, loop, for-clause* (the last two productions and *letter* are handled by the scanner) and since the tokens <procedure>, <statlist>,· <statement>, etc. do not belong to the language we may safely add them to G_1' to denote the corresponding placeholders.

The incremental grammar G_1'' associated with the grammar G_1 is obtained from the latter by adding a new starting symbol *procedure '*, 6 driving symbols, 6 placeholder tokens and the following twelve productions:

---

[2] One should not confuse the incremental grammar G'' with the *naive* method mentioned in [3] of adding to P all productions $A_1 \rightarrow A_i$ which is both *wrong* (any string derivable from any nonterminal can be substituted for the start symbol) and liable to originate several additional conflicts!

[3] To simplify drawings of parse trees we will consider both sequences "end if ; " and "end loop ;" as single tokens.

```
procedure'  →   <1> procedure | <2> statlist | <3> statement | <4> iteration
                                            | <5> loop | <6> for-clause

procedure   →   <procedure>
statlist    →   <statlist>
statement   →   <statement>
iteration   →   <iteration>
loop        →   <loop>
for-clause  →   <for-clause>
```

The following string

<3> **if** BETA **then** DELTA ; <statement> GAMMA ; **end if** ;

made up of the driving symbol <3> concatenated with a placeholder string derived from the nonterminal *statement* is a typical example of the kind of string generated by G_1" and, incidentally, let us mention that is not the image of a rightmost sentential form of G_1".

Let us suppose now that by using Yacc, or a similar tool, we have built a structural parser for the language $L(G_1")$ that we will call *ISP*. Let us point out that *ISP* has no more conflicts than the original parser *SP* of $L(G)$ and it is only slightly more complex than the latter. [4]

It is a simple matter to modify *ISP* into a parametrized routine A_i-*ISP* that will parse only the strings derivable from the nonterminal A_i *according to the grammar G'* . First, we parametrize the *Fetch-this-character* routine (in other words the scanner) into *Fetch-this-character(i)* in such a way that the first time it is called it will behave *as if* the driving symbol d_i had been read from the input file; then we make the semantic routines associated with the productions $A_1 \to d_i A_i$ $(i=1, .., k)$ return in A_1 the pointer to the A_i subtree as the symbol d_i *does not exist in the input string* ; finally, if we replace in *ISP* all calls to *Fetch-this-character* with calls to *Fetch-this-character(i)*, then we get A_i-*ISP*. Since we may want to A_i-parse several times higher forests of the same IST , it is convenient to implement both A_i-*IP* and A_i-*ISP*. [5] First we will check whether A_i generates x by A_i-parsing x with the procedure A_i-*IP* ; only when we are sure that A_i generates x will we call A_i-*ISP* to build the A_i-$IST(x)$. Note that this precaution is common in structured editing. Since we do not know in advance whether an update is correct, we cannot structure-parse it right away because doing so would risk partially corrupting the syntax tree.

7. A placeholder-for-tree technique for parsing forests

In the approach followed in [6], the incremental parsing algorithm recreates the stack contents of the parser by suitably threading the nodes of the parse tree and by storing some additional information in them (*LR* tables for instance).

In this paper we follow a different approach, which we may shortly describe as a *placeholder-for-tree* technique and which is based on the following

Remark 1. *If the string* α, *obtained by replacing each subtree of a forest with the corresponding placeholder, belongs to the placeholder language* $L(G')$ *and* $IST(\alpha)$ *is its incremental parse tree, then the given forest can be embedded into a parse tree obtained by putting* $IST(\alpha)$ *on top of the forest.*

Let us suppose that the output of our *ISP* is expressed in the form of an incremental syntax tree. Once a certain cross-section of the old IST has been made and the *subtrees that can be integrally reused in the new parsing* have been identified, we are confronted with a sequence of trees or perhaps tokens followed by a sequence of characters, the modified text, followed by another sequence of trees or tokens. Each tree is labeled with the placeholder corresponding to its syntactic category. Let us imagine we have run the scanner over the modified text so that we may

---

[4] A comparison related to the Yacc generated parser for the grammar used in KeyOne to describe the full Ada language gives the following numbers: the original (placeholder) Ada grammar with some 551 rules has been modified into an incremental grammar having 674 rules (for simplicity, the driving symbols have been identified with the placeholder tokens; the additional shift-reduce conflicts are handled by Yacc by the default rule). The number of table entries has increased from 2464 to 3151 but the total occupancy of the Yacc generated parser has increased from 123,100 bytes to 127,100 bytes. The increase in size, 3%, is almost negligible and compares very favourably with the burden of generating and managing more than one hundred different parsing tables as proposed in [3].

[5] In practice, A_i-*IP* is derived from A_i-*ISP* by replacing all the semantic routines with no operations.

suppose that the sequence of characters has been replaced by a sequence of tokens. We are now confronted with a forest.

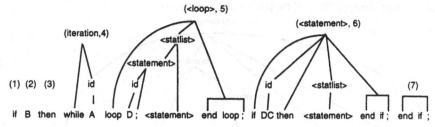

Fig. 3

For instance, with reference to G_1 grammar, our forest may consist, see Figure 3, of the three tokens "if", B and "then", followed by the IST of the following while-clause

while A

labelled by the placeholder <iteration>, followed by the IST of the following text

loop D ; <statement> end loop ;

labelled with the placehoder <loop>, followed by the IST of the following text

if DC then ; <statement> end if ;

labelled with the placehoder <statement>, followed by token "end if ;". The frontier of this seven-item forest is the if-statement:

(1)   **if B then while A loop D ; <statement> end loop ; if DC then ; <statement> end if ; end if ;**

A forest may be represented as a sequence of pairs, each pair being made up either of a placeholder (represented by a certain token code) and a pointer to the syntax subtree that should be re-used in the new parsing, or of a token-code and a pointer to the token value or the token occurrence in the input file. Being a list of pairs (token-code, token-pointer), a forest may be analysed by any standard two-level structural parser. The only difference is that the second item of some pairs may not point to a token value but to a syntax subtree of *identical syntactic category*. Let us call *tokenized forest* such a list of pairs of tokens and their associated pointers (token-code, token-pointer).

A tokenized forest is a two-face object: from the parser viewpoint it is a sequence of tokens, from the viewpoint of its semantic routines it appears as a sequence of subtrees or tokens. A tokenized forest may be given in input to a standard structural parser A_i-ISP by replacing the usual scanner with a routine returning the next item of the tokenized list at each call. In case of success the *ISP* will build an incremental syntax tree having the given forest as *one of its cross-sections and the entire input string as frontier*. So batch parsing a sequence of a few tokens is equivalent to incremental parsing the frontier of the entire forest. Note that if the tokenized forest is part of a preexisting parse tree, then the *ISP* will give rise to a new tree which will be new only in its upper section and will share the initial forest with the old parse tree. With reference to the above example the parser of *ISP* will parse the seven roots of the forest of Figure 3, i.e. the seven token sequence[6]

(2)   **if B then <iteration> <loop> <statement> end if ;**

while at the same time the semantic routines of *ISP* will build a tree for the entire statement (1), which is the frontier of this forest. We might say that our *placeholder-for-tree* technique cheats the structural parser: the syntactic part of the structural parser believes it is parsing the seven token sequence (2), but since this sequence is a forest, the pointers associated to some placeholder tokens of its frontier are not pointing to placeholder values but to syntax subtrees. *The net effect will be the implicit embedding of the syntax subtrees of the given forest (of Figure 3) into the newly built syntax tree of Figure 4 while the incremental parsing activity is performed by a standard batch parser, such as a Yacc generated parser, operating, as usual, on "lexical" tokens.*

---

[6] For the sake of simplicity, here and hereafter we will describe tokenized tree-rows by explicitly listing only their token-codes and intentionally omitting their associated pointers. It is up to the reader to figure out which pointers actually point to syntax subtrees. But so as not to become completely unreadable we will write the identifier names instead of the always identical token-code *id*.

In case of failure we may reject the input string *only if* we are sure that the subtrees occurring in the input *must be part of every parse tree* for its frontier. From now on, by *A*-parsing a given forest we mean the application of the process described above.

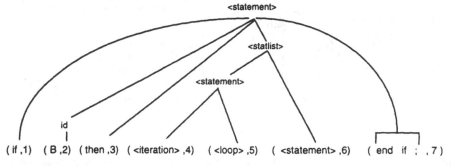

Fig. 4

Starting from an $S\text{-}IST(x)$, i.e. starting from an incremental syntax tree whose root is labeled S and whose frontier is the string x, we may easily produce several forests which we may call *cross-sections* of the $S\text{-}IST(x)$. The first cross-section one may think of is the forest obtained by breaking the root of the S-tree into the sequence of its constituent subtrees. We may recursively obtain other cross-sections by arbitrarily breaking any of the subtrees of a given cross-section into its constituent subtrees. The lowest cross-section of an $S\text{-}IST(x)$ is obviously the string x. Conversely, any cross-section may be generated from a bottom-up parsing process that sometimes omits the step of reducing a handle on top of the stack. Cross-sections correspond to sentential forms of the grammar.

Remark 2. *A bottom-up parsing of a given forest, which happens to be the cross-section of an $S\text{-}IST(x)$, will rebuild a tree isomorphic to the original $S\text{-}IST$ (x).*

The proof is based on the simple observation that if you modify any bottom-up parsing algorithm sometimes forgetting to reduce a handle lying on top of the stack, you will not reduce the input string to S but to a certain cross-section of the $S\text{-}IST(x)$. If you find a way to memorize the reduce actions that you have omitted and start a second parsing pass over that cross-section all over again, making exclusive use of the memorized reduce actions, you will reduce it to S and the net effect will be only a delayed execution of the missing reductions. In other words, S-parsing any forest, which is a cross-section of a given $S\text{-}IST(x)$, will rebuild a set of nodes isomorphic to that part of the $S\text{-}IST(x)$ which, having initially been *neglected* while creating the given cross-section, was considered missing from the forest.

This result allows us to start any bottom-up parsing process, such as shift-reduce or $LR(h)$, from forests which *are not necessarily rightmost sentential forms of the grammar.*

8. The incremental parsing algorithm

The previous sections allow us to solve the incremental parsing problem provided we know which subtrees of the old parse tree can be re-used in the new parsing. If the grammar G is $LR(k)$, then the placeholder grammar G' is also $LR(k)$, see [5]. Let $x=ysz$ be the old string, $x'=ys'z$ the updated string and let us suppose we have built the $A_1\text{-}IST(ysz)$. Let us suppose that G' is an $LR(k)$ grammar; this means that if $x=yy'sz$ and $x'=yy's'z$ and $|y'|=k$, then $PSF(y,yy'sz)=PSF(y,yy's'z)$. In other words we may re-use for the parsing of the updated string x' the parse-subtree-forest whose frontier is the substring y of the old string x ending k tokens (the look-ahead tokens) to the left of the modification point. In the particular case of $LR(0)$ grammars we have $PSF(y,ysz)=PSF(y,ys'z)$, and therefore the old parse-subtree-forest based on the string y lying on the left of the modification point may be integrally re-used for building the new parse tree.

Theorem *2.4* of [5] can be re-phrased by saying that if G' is an $LR(k)$ and also an $RL(h)$ grammar[8] and $x=yy'sz'z$ and $x'=yy's'z'z$ with $|y'|=k$ and $|z'|=h$, then we may re-use for the new parse tree both the left parse-subtree-forest based on y and the right parse-subtree-forest based on z provided that such left (right) parse-subtree-forest is followed (preceded) by k (h) unmodified look-ahead (or look-backwards) tokens.

While k also influences the type of parsing algorithm, and consequently values of k greater than *1* are unpractical, h only determines the number of tokens to be excluded from the frontier of the right parse-subtree-forest and therefore fairly high values of h are acceptable. Our trial algorithm will work, perhaps less efficiently, even for an unlimited value of h, i.e. also for an $LR(k)$ grammar which for this purpose can be considered as an $LR(k) \wedge RL(\infty)$ grammar. For the sake of simplicity, we will outline the algorithm for the $LR(0) \wedge RL(0)$ case, since the general case of $LR(k) \wedge RL(h)$ grammars differs only in the number of tokens, k and h respectively, to be systematically excluded from the frontier of the left (and right) parse-subtree-forest that will be reused in the new parse tree.

Remark 3. *The A_1-$IST(ys'z)$ of the updated string x' can be constructed by A_1-parsing the tokenized forest made up of $PSF(y,ysz)$ concatenated with the string s' concatenated with $PSF(z,ysz)$, i.e.*
$$(4) \qquad\qquad PSF(y,ysz) \; s' \; PSF(z,ysz)$$

Note that forest (4) may not be a rightmost sentential form of the grammar; however the $LR(0) \wedge RL(0)$ property of the grammar guarantees that both the left and right forests, $PSF(y,ysz)$ and $PSF(z,ysz)$, must be part of every parse-tree for the string $ys'z$ and therefore, if $ys'z \in L(G)$ then forest (4) must necessarily be a cross-section of a parse-tree. Remark 2 of section 7 allows us to build the parse tree for the string $ys'z$ by any bottom-up parsing algorithm starting from any forest that happens to be a cross-section of the parse tree.

8.1. An improvement: matching old nodes rather than building new ones

The systematic exclusion of h or k tokens from the left (right) parse-forest may cause the exclusion of some perfectly re-usable subtrees of the old parse tree. The situation is even worse for $LR(k) \wedge RL(\infty)$ grammars for which we cannot guarantee the reuse of any subtree of the right parse-forest. A simple modification of the tree-building routine *MakeNode(i)*, that we will from now on rename *MakeOrMatchNode(i)*, will solve the problem in the following manner: if, at the time of reducing an r-items handle, the r pointers on top of the semantic stack are old [9] child-nodes, pointing to an already existing parent-node, and such parent-node has the right node-code, there is no need to destroy the old node to rebuild a node of same type. We may suppose that the routine *MakeOrMatchNode(i)* does exactly this check before building a new tree-node; the amount of checking seems considerable but actually it is not because it amounts to a series of and-conditions and failure of one of them will prevent evaluation of all the others. Note that in terms of tree re-using this algorithm outperforms [6] as it re-uses *all re-usable subtrees* of the old parse tree both for $LR(0) \wedge RL(0)$ grammars as well as for $LR(0)$ grammar. Obviously, in the case of $LR(0) \wedge RL(\infty)$ grammar the gain is more important because only in this way are we able to make use of the right parse-forest.

8.2. Another improvement: parsing only a subtree

But we can do much better. Significantly better when the reconstruction of the tree may be done within a possibly small subtree of A_1-$IST(ysz)$, while the rest of the old tree may be kept unaltered. This case is rather frequent since an update operation will seldom affect ancestor nodes older than two or three generations. We must find the smallest A_i-IST such that its frontier contains the substring s and the following three *subtree* conditions are fullfilled:

---

[8] G is RL(k) iff its reversal G^R, i.e. the grammar whose productions have right-hand sides which are the reversals of those of G, is LR(k).

[9] A node is a new-node if its pointer-to-parent attribute is undefined.

$$
\begin{array}{lll}
i) & A_1 \Rightarrow^* \alpha_1 A_i \beta_1 & \Rightarrow^* y_1 A_i z_1 \\
ii) & A_i \Rightarrow^* \alpha_2 s \beta_2 & \Rightarrow^* y_2 s z_2 \\
iii) & A_i \Rightarrow^* \alpha_2 s' \beta_2 & \Rightarrow^* y_2 s' z_2
\end{array}
$$

where $\quad x=ysz$, $\qquad x'=ys'z$, $\qquad y=y_1 y_2$, $\qquad$ and $\qquad z=z_2 z_1$.

Finding the smallest matching subtree will be done in parallel with the left-to-right parsing of forest (4) and will be the task of a new version of the semantic routine *MakeOrMatchNode(i)* which we will rename *MakeOrMatchNodeOrStop(i)*. Each time we perform a handle reduction, the semantic routine will also check whether the subtree conditions are fulfilled; if they are the partially new subtree whose frontier includes s' will be linked to the old tree and we will immediately stop the left-to-right parsing process of forest (4).

Before describing this second extension let us partially interpret the subtree-conditions in terms of nodes belonging to the old tree and to the newly built subtree. Condition *i)* is related to the upper part of the old tree and states the possible existence of an old node, labelled A_i : condition *ii)* states that the frontier of that old node, labelled A_i , that is going to be replaced by the newly built subtree, must include the updated substring s and that such node of the old tree must be found amongst the ancestor nodes of the string $y_2 s z_2$; conditions *ii) and iii)* state that the syntax category of the newly built subtree, i.e. A_i , must match the category of the root of the replaced subtree; condition *iii)* says that the frontier of the newly built subtree must include the new string s'.

The routine *MakeOrMatchNodeOrStop(i)* will operate in two states. In the first state it will find the smallest subtree embeddding the substring s' in its frontier. Remember that the semantic routine handles trees (or zero-trees, i.e. tokens) via their pointers stored in the semantic stack of the parser. In order to know when a parse subtree, pointed to by a pointer stored in the semantic stack, has completely embedded the modified substring s' in its frontier, we may devise a *boundary* computation by associating a *negative* sign both to the first and last token-pointer of the updated substring s'.

Whenever a handle reduction takes place, the synthesized parent pointer becomes a *boundary* subtree, being given a negative sign if just one of its children is associated with a negative pointer, i.e. is a *boundary* subtree. As soon as two of the pointers, which happen to be associated with the same handle, both have a negative sign, then we know that the corresponding parent-node E is the minimal node embedding the modified substring s' within its frontier; therefore the third subtree-condition *iii)* is satisfied and the routine may now switch to its second state. We anticipate that in the second state a negative pointer, stored in the semantic stack, will be interpreted as denoting an embedding node.

In the second state we will stop the parsing process as soon as a reduction involving an embedding node E would create a new node N which can fit directly in the old parse tree because it satisfies the first two subtree conditions *i)* and *ii)*.

More precisely, let us suppose that we are reducing a handle

$$
A_i \rightarrow X_1 X_2 \ldots X_r
$$

of arity r , such that one node amongst the nodes pointed to by the r pointers $P_1,...,P_r$ lying on top of the semantic stack, is an embedding node E. Let us suppose we have already created, as usual, a new parent-node N of syntax category A_i and not yet linked it both ways to its children $X_1 X_2 \ldots X_r$.[10]

We now circularly scan all the r nodes of the handle looking for a node [11] of syntax-category A_i : it is to the parent-node P of such node that we will glue node N. In fact, subtree condition *ii)* requires that the old node to which we could possibly glue node N must be found amongst the ancestor nodes of $y_2 s z_2$ and therefore amongst the parent-nodes (or perhaps their ancestors) of the old nodes of the handle $X_1 X_2 \ldots X_r$, provided the corresponding child-node has syntax category A_i.

If all the above conditions are satisfied then the newly-built node N may be down-linked both-ways to its r children $X_1 X_2 \ldots X_r$ and up-linked both-ways to the parent-node P. All the conditions defining a syntax tree are verified and we may stop the parsing process. If no brother of syntax category A_i is found, node N is down-linked both-ways to its children, N is defined to be the new embedding node and the semantic routine exits.

---

[10] After this linking in fact all the old nodes of the reduced handle will be pointing to N and reference to their old parent-nodes will be lost.

[11] If $r=1$ and the embedding node E is an old node then we glue E to its parent-node and we may stop the parse algorithm; if $r=1$ and E is not an old node, the search fails and we exit the semantic routine; the effective creation of the unary node N is, as usual, an irrelevant implementation decision.

In order to stop parsing from within the semantic routine even if the end-of-file is a long way ahead we may force an exit by causing a sort of error exception: indeed there is no other way of stopping the standard batch parser before the end of the string (or better of the forest) is reached.

Let us now consider the same example discussed in [6]. Let G_2 be the grammar with productions:

$$
\begin{aligned}
S &\rightarrow ABC \\
A &\rightarrow aAa \mid b \\
B &\rightarrow AB \mid cBc \mid d \\
C &\rightarrow fAB
\end{aligned}
$$

The productions of the incremental grammar $G_2{}''$ are

$$
\begin{aligned}
S_0 &\rightarrow d_S\, S \mid d_A\, A \mid d_B\, B \mid d_C\, C \\
S &\rightarrow ABC \mid t_S \\
A &\rightarrow aAa \mid b \mid t_A \\
B &\rightarrow AB \mid cBc \mid d \mid t_B \\
C &\rightarrow fAB \mid t_C
\end{aligned}
$$

where S_0 is the new starting symbol ; t_S , t_A ; t_B and t_C are the newly added placeholder tokens and d_S, d_A , d_B , d_C are the driving symbols. G_2 is an $LR(O) \wedge RL(1)$ grammar, but for a simpler comparison with [6] let us ignore its right hand property and consider it only as an $LR(O)$ grammar.

Let x be the string $x_1 x_2 x_3 x_4 x_5 x_6$ in $L(G_2)$, $x_i \in \Sigma^*$, where x_1 , x_2 , x_3, x_4, derive from A, x_5 derives from B and x_6 derives from C. If we ignore its dashed arcs and consider the cut-arc between node $(t_B, 7)$ and node $(t_B, 6)$ valid, then figure 7.a may be seen as representing the parse tree $S-IST(x)$. Let $x'=x_1 x_2 x_3\, aba\, x_4 x_5 x_6$ be a modification of x obtained by inserting the string aba between $y= x_1 x_2 x_3$ and $z= x_4 x_5 x_6$. Since we suppose G_2 to be only an $LR(O)$ grammar, its right PSF is flattened down to its frontier $x_4 x_5 x_6$ and the tokenized forest we need to S-parse is $t_A t_A t_A abax_4 x_5 x_6$ which we represent as a sequence of pairs of tokens and subtree references:

$$(t_A ,1)\,(t_A,2)\,(t_A,3)\,(a,-)\,(b,+)\,(a,-)\,(x_4,+)\,(x_5,+)\,(x_6,+)$$

The minus sign denotes the boundaries of the replaced string and is used for computing the embedding node. In order to understand how our algorithm works we must keep in mind that, up to the time when the semantic routine explicitly updates their parent-attribute, all the roots of this forest, as well as all the tokens of $x_4 x_5 x_6$, are still pointing to their respective parent-nodes in the old tree. When the input still to be processed is the string $x_4 x_5 x_6$, the stack contents will be:

parser stack:	A	A	A	a	A	a
semantic stack:	+1	+2	+3	-	1'	-

and the next reduction, which is $A \rightarrow aAa$, will involve a handle including two boundary pointers; the newly-built node 2, labeled t_A , will be the new embedding node and given a negative sign. The routine enters its second state. The next input $x_4 x_5$ will be syntactically parsed re-using nodes 4, 5 and 6 of the old parse tree as the conditions of section 8.1 hold here.

At this point the input still to be processed is x_6 and the stack contents will be:

parser stack:	A	A	A	A	B
semantic stack:	+1	+2	+3	-2'	6

The next reduction, $B \rightarrow AB$, will cause, as usual, the creation of a new parent node $(B, 3')$, but since it will involve $(A,-2')$, which is the embedding node E , we have to test whether the remaining subtree conditions $i)$ and $ii)$ are satisfied. On scannning the two nodes of the handle we find that $(B, 6)$, has the same syntax category as the left hand part of the current reduction $B \rightarrow AB$: all subtree conditions are now verified and the old parent, i.e. node$(B, 7)$, of node $(B, 6)$ may be made to become the parent of the newly built node $(B, 3')$ whose partially new subtree now becomes entirely glued to the old tree.[12] A pseudo-error condition is issued to stop parsing.

---

[12] The crossed out link in figure 5.a is meant to suggest that the old parent of the (old) node 6, i.e. node 7, previously had two children, node 3 and node 6; after the update node 7 still has two children: namely, old child 3 and the newly built node 3' (labelled B) which includes the old subtree rooted in node 6.

The effective output of the incremental algorithm are the dashed arcs of Figure *5.a*; the nodes enclosed in the dotted area are scanned and syntactically parsed but, being matched by

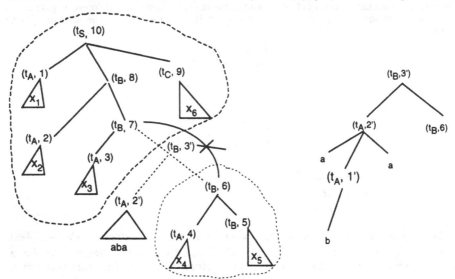

Fig. 5.a . The dotted (dashed) line encloses portion of the parse tree whose reconstruction is skipped by the MakeOrMatch routine (syntactic parser,).

Fig. 5.b

the semantic routine, they are not created anew; the nodes in the dashed area are skipped completely. The reconstructed portion of the parse tree is represented in Figure *5.b* and is equal to the tree portion reconstructed in [6]; in fact, the subtree conditions used by our algorithm and by the method presented in [6] are similar and it will easily be seen that both algorithms have a theorethical comparable time complexity, but in practice we believe our algorithm to be faster, simpler and requiring much less space. A minor drawback of our algorithm is the need to parse the left *PSF*, i.e. in the case of this example the *3*-tokens forest $t_A\ t_A\ t_A$, but since parsing this additional sequence consists only of multiple shifts with no reductions, we know it will take minimum time; on the other hand [6] can avoid such shifts at the cost of storing *LR* tables in the parse tree.

Note that our algorithm is tuned for structured editors which have the special need of parsing placeholder strings belonging to $L(G')$. But it is trivial (albeit amusing) to make it accept only strings of the base language $L(G)$. It is sufficient to break the agreement between scanner and parser by restricting the scanner to recognize only the tokens of the original grammar G. The placeholder tokens become virtual objects in the sense that they are still referenced in the productions of the grammar to allow parsing of the tokenized forests, they materialize whenever a forest is created as a cross-section of an *IST*, but they are not recognized as such by the scanner which would, for example, return three tokens for the sequence of characters *<statement>* , namely the operator "<" , the identifier "statement", and the operator ">".

9. Implementations

The method, described in section *8*, of using a single universal parser for parsing any incremental part of a source program containing placeholders has been used since *1985* by all versions of the Keyline/KeyOne series of PDL analysers and structured editors, built and commercially distributed in Europe and the US [13] by the LPS company (see also the reviews of KeyOne in [8]). KeyOne has been implemented for more than a dozen standard languages and dialects and, in particular, for full Ada, VMS Pascal and C.

---

[13] In the US the editor was better known under the name of Vads Edit as it was distributed under an OEM agreement by Verdix Corporation, the Ada compiler vendor.

The KeyOne version, mostly distributed in Europe, can be used as a design editor supporting step-wise refinements. The design description uses abstractions which can be written anywhere a syntax construct, such as a placeholder, can be written. KeyOne design descriptions are hypertexts expanding into other program refinements, perhaps containing other abstractions. A KeyOne user may edit his program at various levels of abstraction and the system will incrementally parse a given high level description replacing the abstraction name or the ellipsis [1] not with its refinement but with the corresponding placeholder. The idea of parsing sequences of text and parse subtrees, replacing the subtrees with the corresponding placeholders, is therefore already germinally present (even if not completely exploited) in the technology behind KeyOne.

10. Concluding remarks

At first sight it might seem rather peculiar that our incremental parsing algorithm is implemented tinkering with the semantic routines which are usually ignored as being too close to implementation details - and incidentally this may explain why we have spent half a page trying to redefine well known concepts such as the cooperation of scanners, *structural* parsers and semantic routines in building a parse tree -. In reality we had no other choice because we did not want to modify the parsing algorithm in the first place; indeed, we might consider it strange to deal with parse trees from within the parsing process; all syntax trees are built by semantic routines, such as our *MakeNode* routine, and keeping to deal with them from within the same process should give better results even in an incremental situation.

In this paper we have presented a general solution to the problem of incremental parsing, making an interesting *re-use* of standard batch parsers which nowadays can be easily generated by tools such as the very popular Yacc or the newly available Bison. We hope that the simple solution to this problem presented here as well as several other solutions that we have been able to obtain by applying our techniques (see [11] and [12]) will make the incremental approach to parsing more popular in fields other than structured (text) editing or incremental compilation, such as the newly emerging syntax directed graphical applications.

The main drawback of the other incremental parsing methods described in the literature, the considerable amount of storage needed in the parse-tree to record the configurations entered by the parser at each step of the analysis together with other auxiliary information such as special links to descendant nodes, has been eliminated, even *totally* eliminated, for those tools such as structured editors, which need to represent parse-trees as double-linked trees. However, the advantage of our solution should not be evaluated only in terms of bytes saved but also in the neatness of syntax tree design that may encourage re-use and standardization.

Acknowledgements

I wish to thank reviewer 2 for his in-depth reading and perceptive comments that have helped me to improve the final version of this paper.

REFERENCES

1. BARIOGLIO, M., CAPELLA, G., LUPO, I., AND PETRONE, L. Software Productivity Tools for Program Design, Implementation, Documentation. *Proc.s of the 1986 IFIP Congress . Elsevier Publ.* (North-Holland), IFIP 1986, 909-914.
2. BERGADANO, E. Parsificazione incrementale negli Editor Strutturati. *Dip. Informatica, Universita' di Torino* (1993)
3. DEGANO, P., MANNUCCI, S., AND MOJANA, B. Efficient Incremental LR Parsing for Syntax-Directed Editors. *ACM Trans. Prog. Lang. Syst. 10*, (July 1988), 345-373.
4. DONZEAU-GOUGE, V., HUET, G., KAHN, G, AND LANG, B. Programming Environments based on Structured Editors: the MENTOR Experience. *Workshop on Programm. Environments*, Ridgefield, CT (1980).
5. GHEZZI, C., AND MANDRIOLI, D. Incremental Parsing. *ACM Trans. Prog. Lang. Syst. 1*, (July 1979), 56-71.
6. GHEZZI, C., AND MANDRIOLI, D. Augmenting Parsers to Support Incrementality *Journal of the ACM* (July 1980), 564-579.
7. MEDINA MORA, R., AND FEILER, P. An Incremental Programming Environment. *IEEE Trans. Softw. Eng.* SE-7, 5 (Sept. 1981), 472-482.

8. PAPPAS, T. L., Designing Ada Programs. *(a Product Review of KeyOne) IEEE Computer* , (March 1990), 92-94.

9. PETRONE, L., et alii. DUAL: An Interactive Tool for Developing Documented Programs by Step-wise Refinements. *Proc.s Int. Conf. Software Engin.* , Tokyo (1982), 350-357.

10. PETRONE, L., et alii. KeyOne Manual. Copyright LPS. Torino, Italy (1985, and rev. 1990)

11. PETRONE, L., AND BERGADANO, E. The new Architecture of Free Structured Editors *(in preparation)*.

12. PETRONE, L. Incremental Parsers, Batch parsers and Free Structured Editors *(submitted for publication)* .

13. TEITELBAUM, T., AND REPS, T., The Cornell Program Synthesizer: A Syntax-directed Programming Environment. *Commun. ACM 24,9* (Sept. 1981), 563-573.

The Expressive Power of Indeterminate Primitives in Asynchronous Computation

Prakash Panangaden*

School of Computer Science
McGill University
Montréal, Québec, Canada

Abstract. It has long been realized that the exigencies of systems programming require primitives that behave indeterminately. The best-known dataflow primitive is the so called **fair merge** which abstracts aspects of fair resource allocation. It has been known for about two decades that fair primitives lead to unbounded indeterminacy. Around seven years ago E. W. Stark, Vasant Shanbhogue and I discovered that various variants of fair merge primitives, all manifesting unbounded indeterminacy, were provably different. These differences are based on simple monotonicity properties.

In the present paper I review these results and discuss some related phenomena involving a fair stack. I then describe results about fair splitting. These results are based on topological properties rather than simple order-theoretic properties. This gives some basic insight into what can and cannot be described by oracles and the relative power of various oracles. Finally I describe a result, implicitly due to Jim Russell, which establishes the most powerful possible oracle.

1 Introduction

In the past when I have discussed or mentioned results on the expressiveness of concurrent languages the typical reaction has been to say "but isn't everything there known by Turing completeness?" Indeed the success of classical recursion theory has led to the belief that all problems in this area are settled. On closer scrutiny it becomes clear that the usual picture of Turing machines, the halting problem and the expressive power of programming formalisms has several parameters and that interesting questions remain when these parameters are altered.

Of course an intellectually sterile twiddling of parameters in order to publish more papers is not worth the effort. There is, however, a fundamental reason to explore new notions of expressiveness. The new reason is, in a word, "parallelism". Imagine connecting the output of one Turing machine to the input of

---

* Research supported in part by Natural Sciences and Engineering Research Council of Canada and by Fonds pour la Formation de Chercheurs et a l'Aide à la Recherche du Québec.

another. As soon as we have two such machines functioning autonomously we are out of the realm of classical recursion theory.

But has anything important really changed? We can easily imagine a Turing machine simulating the two Turing Machines by interleaving the actions of each. The fundamental change that has occurred is the way in which data can now be presented to Turing Machines. One can imagine that the input tape of a Turing Machine is being written while the machine in question has begun examining the input. In short the input is *incomplete*. As soon as there are two loci of computational activity one can have the possibility that one machine consumed output from the other. Now the question is what does it do when it runs out of input? Another major point is that we can now talk sensibly of *infinite* input/output.

In the world of sequential computation if one subprogram produces infinite output this output is "useless". The subprogram will run forever and nothing will happen with the rest of the program. If, however, two subprograms run *concurrently* then one could be producing infinite output which the other examines *while it is being produced*. As a simple example suppose that a Turing Machine produces the infinite sequence of primes in increasing order. Another Turing Machine can examine this list and check for the occurrence of a specific number. From the traditional viewpoint of sequential computation this is a divergent program! Clearly from a parallel viewpoint the program is entirely nontrivial and does not deserve to be lumped together with a program that just loops without ever producing output.

In the present paper we review the dataflow paradigm as the arena for discussing expressiveness questions.

2 The Dataflow Paradigm

The basic ideas of the dataflow paradigm go back to Gilles Kahn [16]. His model of distributed systems is as follows. A system is viewed as a collection of autonomous computing processes communicating by one-way, unbounded data channels. Each computing process executes its own program and, from time to time, may communicate with other processes by sending data tokens along the channels or reading tokens from incoming channels. If a process attempts to read a token from an incoming channel that has no data, it suspends until data becomes available. If a datum arrives at the input channel of some process that is not interested in reading it, the datum is buffered indefinitely. Thus all processes are assumed *receptive*. An operational semantics, based on coroutines to simulate parallelism, was worked out by Kahn and McQueen [17]. Henceforth we refer to such networks as *Kahn networks*.

There is an extremely pleasant mathematical description of dataflow networks. First note that a process could have an internal state. Thus one cannot think of a process as being a function from input tokens to output tokens. For example consider a process that reads integer tokens and adds them to an internal integer variable which is initially zero. Every time it reads an input it outputs the updated value of its internal counter. If the input sequence $111\ldots$

were input the output would be 123.... One can, however, think of a process as being a function from input sequences to output sequences. In general these sequences could be infinite.

The collection of finite or infinite data sequences are called *streams*. Let V be a collection of data values, say integers or booleans or some other type. Then we write V^ω for finite or infinite sequences, V^∞ for infinite sequences and V^* for finite sequences of elements of V. The set V^ω comes equipped with the structure of a complete partial order with least element [14]. The ordering is the prefix ordering and the least element is the empty sequence. The infinite elements are infinite sequence and each is the least upper bound of a chain of finite sequences. Thus V^ω is an algebraic complete partial order.

Under the assumptions of the Kahn model, a process computes a *continuous* function from input token streams to output token streams. One can thus model processes as functions. If one has a network of such processes then one can write down a system of equations to model the behaviour of the network. If the network has feedback loops the system of equations will be recursive. Such systems of equations can be solved using elementary fixed-point theory. The correspondence between the operational semantics and the denotational semantics is known as Kahn's principle and has been proved many times[12, 21].

A basic question that arises is "what happens if we no longer have determinate primitives"? One can ask, in particular, what happens if the processes can test for the absence of data and thereby become sensitive to arrival times. Since this temporal information is not present in the streams of data we clearly have indeterminacy.

Much of the research in indeterminate dataflow is aimed at developing a theory of comparable elegance and utility as Kahn's was for determinate dataflow. In this paper fairness properties of indeterminate Kahn networks play a major role. Fairness has been the paradigmatic liveness property and has been studied extensively in a variety of formalisms with several different definitions; see, for example, the book by Francez [13]. It is known that fairness introduces unbounded indeterminacy (or countable nondeterminism), the proof is by a simple Koenig's lemma argument. Plotkin's pioneering study of powerdomains for indeterminacy included the observation that the powerdomain that he introduced had been specifically designed for bounded indeterminacy and therefore excluded the study of fair systems [32]. Several people have worked on generalizations of powerdomain techniques that would apply to unbounded indeterminacy [1, 3, 7, 30, 29, 33].

3 What is Expressiveness?

Three parameters are basic to any notion of expressiveness. These are

1. a notion of *observation*,
2. a set of primitives and
3. a collection of composition rules.

Thus when we ask whether **while loops** are more "powerful" than **do loops**, in the context of some language, we usually have in mind observing inputs and outputs *at the basic datatypes*; the primitives and composition rules being those associated with the language in question.

In the dataflow context one can have two quite different viewpoints; one could make *finite* observations or one could make *infinite observations*. The latter are of course an idealization of infinitely many finite observations. If one's attitude is that "observation" should correspond to some, in principle, physically feasible experiment then clearly the latter are not observations. However, the notion of idealized experiment is nevertheless valuable and serves to provide a conceptual basis to many theories. Consider physical theories couched in terms of differential equations based on notions of an underlying smooth continuum. Clearly the classical notion of real number does not correspond to anything directly measurable but such theories are nevertheless useful.

In the treatment of the denotational semantics of Kahn networks the basic observable is the stream. Even though this is an infinite object it can be viewed as the limit (lub) of finite streams and thus one is not entirely removed from notions of physical observability. We shall take the same view in the present paper. What one is really doing, of course, is providing imagery for the notion of semantic equality. To summarize, we take as observable the input-output relation of a network where the members of the relation are tuples of (possibly infinite) streams. A detailed study of dataflow from the viewpoint of finite observability was undertaken by Rabinovich and Trakhtenbrot [35, 36, 37].

The background primitives that we take are the determinate processes. More precisely we assume that any continuous function is available as a primitive. This means that we are not interested in the traditional recursion-theoretic notion of expressiveness. In later sections we will look at other possibilities, for example, allowing only sequential primitives. Finally the composition that we consider is the ability to connect subnetworks together by using the output channel of one as an input channel to another. In particular we permit the formation of feedback loops. Now we can define formally what it means to express a process.

Definition 1 *Suppose that M and N are (indeterminate) processes. We say that M is* **more expressive than** *or* **more powerful than** *N if it is possible to construct a network with arbitrary determinate processes and as many copies of M as is needed with the same IO relation as N.*

There is a crucial point to be made here. The notion of IO Relation is not *compositional*. This is of course the well-known Brock-Ackerman anomaly [6]. The following example, due to Russell [38], is more perspicuous than the original Brock-Ackerman example and relies on "weaker" primitives as we shall see later. Consider a process, shown in the left hand side of figure 3, with a single input channel and a single output channel; ignore, for the moment, the dotted line. Its behaviour is as follows. It makes a top-level choice between two behaviours. The first behaviour is to output a 0, then wait for an input token and then output a 0. We summarize this as $0; read; 0$. In the same notation the other possible behaviour is $read; 0; 1$. The input-output relation is

Input	Output
Empty	0 or Empty
Nonempty	00 or 01

Fig. 1. IO Relation of the Processes

Consider a second process with the same IO channels and with the following behaviour. It can make a three-way choice between the first two possible behaviours of the first process but has, in addition, the possibility of choosing the behaviour represented as 0; *read*; 1. It is easy to see that this has exactly the same IO relation as the first process. Now suppose that we form the feedback loop shown in the dotted line. The modified processes have only a single output channel. Such input as is available comes from the feedback. The first process can now produce either nothing or 00 while the first can produce either nothing, or 00 or 01. Thus we have two processes with the same IO relation but when placed in identical contexts (in this case the context is the feedback loop) have quite different IO relations.

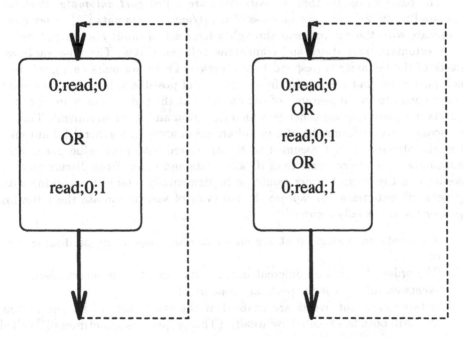

Fig. 2. The IO Relation is not Compositional

What we see is that the IO relation is not enough to give a compositional semantics; but the IO relation is our basic observable. Hence *any compositional semantics* must contain more information than the IO relation. Thus our notion of expressiveness is the most liberal possible. One can consider more refined notions of expressiveness, based on traces, for example, that would yield more distinctions between processes. The point is that if we prove negative results with the most liberal definitions then our results are *as strong as possible*. In other words, if we can show that, with our definitions, that M cannot express N it means that the result will be true in any compositional semantics of dataflow processes.

4 Automata and Processes

The operational semantics of Kahn networks are typically given in terms of concurrent automata [22, 21, 42]. In this brief section we cannot give a substantial account of these automata and their fascinating theory. We refer the reader to the articles of E. W. Stark for a thorough treatment [42, 43, 44] and to his papers with the author for the expressiveness results [28, 27]. The general theory of asynchronous systems is discussed in Arnold and Nivat [4], Mazurkiewicz [23], Pratt [34], Kok [18], Kwiatkowska [20], Jonsson [15] and Rabinovich and Trakhtenbrot [35].

The basic automata that we work with are called *port automata*. Port automata [21, 26, 27] are a special class of *input/output automata* [21]. They communicate with the environment through a fixed set of input and output ports. The automata have states and transitions between states. The observable aspects of the behavior correspond to the *events*. These are pairs (p, v), where p is a port name and v is a value from some set of possible values. Such an event corresponds to the appearance of the datum v at the port p and represents the observable communication between an automaton and its environment. The set of possible events forms the *alphabet* of an automaton. In a general IO automaton the alphabet is not assumed to be structured into port-value pairs. Port automata are more concrete than IO automata and none of the discussion that we have in the present paper would be fundamentally altered by working with general IO automata. To capture the behavior of such automata the following properties are usually assumed:

- All events on a single port are either causally dependent or disable each other.
- The order of events on different input ports cannot be observed, hence, all events on different input ports are concurrent.
- If two concurrent events are enabled at the same state of an automaton, they will both be executed eventually. (This property is sometimes [20] called *concurrency fairness*).
- The environment cannot be prevented from supplying a new input to an automaton. This is called *receptivity*.

The upshot of these assumptions is that processes are reactive and receptive. The input is assumed to be buffered, thus the fact that processes are receptive does not mean that a given process needs to react in a nontrivial way to all input. The basic causality and conflict relations between events are axiomatized below.

Definition 2 *A concurrent alphabet is a pair* $(E, ||)$, *where* E *is any set, and* $||$ *is a symmetric and irreflexive binary relation on* E, *called the concurrency relation.*

We will use the notation $\mathcal{E}$ for the concurrent alphabet $(E, ||)$.

Definition 3 *Let* V *be a set of data values, and* $P^{inp} \cap P^{out} = \emptyset$ *two finite sets of names. A port signature is a set* E *of the form* $E = E^{inp} \cup E^{out}$ *where* $E^{inp} = P^{inp} \times V$ *and* $E^{out} = P^{out} \times V$.

Elements of $E^{inp}, E^{out}, P^{inp}, P^{out}$ are called respectively, input events and output events, input ports and output ports. For any event $e = (p, v)$ we write $port(e) = p$ and $value(e) = v$ for its corresponding port and value component. We use the notation E^p to denote all events on the port p and we use a letter P for the union $P^{inp} \cup P^{out}$.

4.1 Port Automata

The following definition of port automata is similar to the one given in [21, 26, 27]. For the sake of simplicity, we omit the notion of internal events from the definition of port automata.

Definition 4 *A port automaton is a 4-tuple* $M = (\mathcal{E}, Q, q', \rightarrow)$ *where*

- E *is a port signature.*
- $\mathcal{E} = (E, ||)$ *is a concurrent alphabet, which satisfies the following properties:*

Single port dependency *For all* $e, e' \in E$: *if* $e || e'$ *and* $port(e) = port(e')$, *then* $e = e'$.
Buffered input *For all input events* e, e': *if* $port(e) \neq port(e')$, *then* $e || e'$.

- Q *is a set of states, and* $q' \in Q$ *is a distinguished initial state,*
- $\rightarrow$ *is a transition function; i.e. a partial function of type* $Q \times E \rightarrow Q$, *which satisfies the following properties:*

Receptivity *For any state* q *and any input event* a, *there exists a state* r *such that* $q \rightarrow ar$.
Commutativity *For any state* q *and any events* a, b *such that* $a || b$, *if* $q \rightarrow ar$ *and* $q \rightarrow br'$ *then, there exists a state* q' *such that* $r \rightarrow bq'$ *and* $r' \rightarrow aq'$.

We say that event e is *enabled* at the state q if there exists a state r, such that $q \rightarrow er$. In that case, we sometimes write qa for the state r. Note that if two events a and b are concurrent we **do not imply** that if one is enabled in a given state the other has to be as well. A key difference between asynchronous transition systems [20] and port automata can be indicated here. In ATSs one assumes that if a and b are concurrent and the sequence ab is possible in a given state then ba must also be possible. A simple example of a dataflow primitive that violates this is *left-strict OR*.

A very important point is the following. The concurrency relation is part of the definition of an automaton. It is not a derived concept obtained from the transition function. Thus, if, for example, there are two output events a and b such that when either one is executed both a and b are enabled in the resulting state it does not necessarily follow that a and b are concurrent. It could be the case that a conflict and that the choice between a and b is being made available in the new state. This makes phrases like "the same transition is enabled in the next state" quite subtle. In order to make such statements rigorous one must appeal to the concurrency relation. If a and b are concurrent then, in the situation just outlined, the "same" b transition is enabled after the a transition but not otherwise. A rigorous treatment is based on the notion of *residual* [28, 42] adapted from the lambda calculus [5]. For our purposes we will not need the formal definitions of residuals or the algebra of residuals but the concept is important.

We now introduce two important classes of automata. We say that a port automaton is *monotone* if it satisfies the following property:

Monotonicity If some output event b is enabled at the state q, then for any input event a, event b is also enabled at the state qa.

The intuition here is that "arrival of new input cannot disable an already enabled output transition". Thus monotone automata rule out timeouts and interrupts. These are important in real programs but are painful to reason about; for a very good reason, they introduce nonmonotonicity in a fundamental way.

We say that a port automaton is *determinate* if it is monotone and it satisfies the following property in addition:

Determinacy If two different output events b and b', are both enabled at the same state then, $b\|b'$.

This prevents the presence of internal choice in an automaton. It is straightforward to show that if an automaton satisfies this property the input-output relation is indeed a (continuous) function. From now on, we will use the word *automaton* to mean a *port automaton*.

Networks of automata are built by linking ports of different automata. Here (as in [26, 28]) we use three operations on automata to build more complex (networks of) automata.

Aggregation of automata M and N is a new automaton $M\|N$ obtained by putting M and N side by side (no shared ports).

Feedback of automata M is a new automaton $loop(M, p_1, p_2)$ obtained from M by linking its output port p_1 to its input port p_2.

Hiding Automaton $M\backslash X$ is obtained by hiding a set of output ports X of automaton M.

Ordinary sequential composition can be obtained by aggregation followed by feedback. For each of these three operations, certain compatibility conditions have to be defined, otherwise, the operation will not necessarily produce a port automaton. These compatibility conditions are spelled out, for example, in [26]. They merely serve to ensure that there are no conflicts in port names and that interconnection is port-to-port, i.e. there is no implicit merging of output ports.

From now on, we will use the term *network of automata* to denote an automaton resulting from composing finitely many automata, using the above three operations. We write $\prod M_i$ for a network consisting of finitely many automata M_i. It is important to keep in mind that a network of automata is just another automaton.

4.2   Completed Computations and Traces

In this subsection we give the basic definitions of computations and traces.

Definition 5 *A* computation sequence, γ, *of an automaton M is any finite or infinite sequence of transitions $q^\iota \to e_1 q_2 \to e_2 \dots$ where each transition is a transition of M.*

The set of all computations of an automaton M is written $Comps(M)$. A compositional description of the processes described by automata is given by ignoring the intermediate states and recording only the events.

Definition 6 *The* trace *of a computation sequence $\gamma = q^\iota \to e_1 q_2 \to e_2 \dots$ is a sequence $Tr(\gamma) = e_1 e_2 \dots$ consisting of the input and output events in γ.*

The set of all traces of an automaton M is written $Tr(M)$. If γ is a finite computation sequence of length n, we use the following notation: $q^\iota \to Tr(\gamma) q_n$ to indicate that γ starts at the state q^ι and finishes at the state q_n, having a trace $Tr(\gamma)$. [2] We say that some event e is enabled at the state r of some trace $t = Tr(\gamma)$ if e is enabled at r in the computation sequence γ. Similarly, we say that event e is executed in the trace t if it is executed in γ.

The following definition captures the notion of so-called "concurrency fairness". Notice that we do not expect fair behavior with respect to all continuously enabled events, but only for those which are continuously enabled **and concurrent with** all other enabled events.

Definition 7 *A computation sequence γ is* completed *iff it is either finite, and no output event is enabled at its final state, or it is infinite and for no event e, does there exist a finite computation sequence δ such that $\gamma = \delta\gamma'$ and e is enabled at every state of γ' but never executed or disabled in γ'.*

---

[2] We will make always clear from the context, whether the symbol $\to$ is used for a transition function, or for a finite derivation.

According to this definition it is possible that an output event a may be in conflict with an output event b on a different output port. It may happen that when a occurs the events a and b are both enabled in the new state and the choice between them has to be made afresh. If you just look at the transition table, it may seem that a and b are concurrent. In fact, the concurrency relation has to be given as primitive. Just because two transitions appear to commute does not mean they are concurrent. In this circumstance it is possible for a repeatedly to happen and for b never to happen and for the computation to be deemed completed. In this case we take the view that b is getting disabled and reenabled every time a is executed. It is only if a and b are deemed to be concurrent that b must happen eventually. An example of this phenomenon is a dataflow primitive called a *splitter*. Here one has a single input port and two output ports. Each input token is placed on one or other of the two output ports. It is entirely possible for the choice to be made consistently in favor of one of the two output ports.

Next, we define a *completed computation sequence of a network of automata*. We use the notation $\gamma|_X$ for the projection of the computation sequence γ onto the set of events X. We use the same notation for the projection of a trace. Also, we use the notation $t|_{\sim X}$ to stand for $t|_{E \setminus X}$. The set of all completed computations of an automaton M is written $FairComps(M)$.

Definition 8 *A computation sequence γ of a network of automata $\prod M_i$ is completed iff $\forall i \gamma|_{E_i} \in FairComps(M_i)$.*

Definition 9 *The* trace set *of an automaton M is the following set:*

$$Tr(M) = \{t \in E^\infty | t = Tr(\gamma), \gamma \in FairComps(M)\}.$$

4.3 The Permutation Preorder

In reasoning about computations or traces or concurrent alphabets in general the usual order structure on sequences, i.e. the prefix ordering, is not appropriate because of the ability to commute concurrent events. One can define a preordering (and induced equivalence relation) on sequences by using the permutation of concurrent events. In fact the actual sequences are not as important as the equivalence classes under permutation. The basic ideas here are due to Mazurkiewicz.

We use A^* and A^∞ to denote the set of all finite and respectively - finite and infinite words over the alphabet A.

We define *permutation equivalence* [23], on the set of finite strings over the concurrent alphabet $\mathcal{E} = (E, \|)$, to be the least congruence $\equiv$ with respect to concatenation, such that $ab \equiv ba$ iff $a\|b$, where $a, b \in E$. We say that two strings $w, w' \in E^*$ are permutation equivalent $w \equiv w'$ iff one can be obtained from the other by permuting adjacent, concurrent symbols. From now on, we use the notation $\sqsubseteq$ for the usual prefix ordering and we omit the subscript $\equiv$ when the concurrent alphabet is clear from context.

Definition 10 *We first define the* **permutation preorder** *on finite words as:* $w \preceq w'$ *iff* $\exists w''(ww'' \equiv w')$ *and then we extend it to infinite words as follows:* $w \preceq w'$ *iff* $\forall w_1 \sqsubseteq w(\exists w_1' \sqsubseteq w'(w_1 \preceq w_1'))$. *Now, we extend the relation* $\equiv$ *on infinite words:* $w \equiv w'$ *iff* $w \preceq w' \wedge w' \preceq w$

4.4 Traces are Fully Abstract

The fact that traces give a fully-abstract semantics for dataflow networks was first established by Jonsson in his doctoral dissertation but was largely overlooked until Kok developed a different fully-abstract semantics [18]. The result was proved independently by several other groups, for example Rabinovich and Trakhtenbrot proved it with a different notion of observation. Kok's contribution was a simple universal context, built using fair merge, that served to distinguish any two networks with different traces; this was used by Jonsson to simplify his original result. Jim Russell showed that one could construct a universal discriminating context using purely determinate primitives. In this section we quickly review the presentation from Shanbhogue's thesis which differs in minor details from Jonsson's presentation.

The key point is that in order to have full abstraction one should close the set of traces under permutations. Consider the following two primitives. In each case there is one input port and two output ports called b and c. The output ports each receive a copy of the input tokens. In the first case the automaton always performs the output event on port b first whereas in the second case the automaton performs the output event first on b or c alternately. These are both determinate and quite distinct as automata and they have different trace sets. However, a moment's thought shows that no context can possibly distinguish these two automata if we are observing the input-output relation. Thus the set of traces needs to be closed under certain permutations.

Definition 11 *Let* $\mathcal{E}$ *be a concurrent alphabet and let* $X \subseteq E^\infty$. *Then,* $Cl(X) = \{w \in E^\infty : w \equiv w'$ *for some* $w' \in X\}$.

It is not difficult to see that Cl is a closure operator on $(E^\infty, \subseteq)$. It has been proved in [41] that closure sets of traces are compositional.

The observable aspect of the behavior of automata is the *input-output relation*. A port history of M is a vector of values over the set V^∞, indexed by port names. Every trace t determines a port history H_t, where for each port p, the projection of H_t onto the port p is a sequence of values $H_t^p = value(t|_{E^p})$. We write H_t^{inp}, H_t^{out} for the port histories of t restricted to input and output ports respectively. Also, if $\mathbf{x} \in (V^\infty)^n$, and $b \in E$, we will often use the notation $\mathbf{x}b$ to denote the port history $\mathbf{x}'$ such that $\mathbf{x}_i' = \mathbf{x}_i$, for all $i \neq port(b)$, and $\mathbf{x}_j' = \mathbf{x}_j value(b)$ for $port(b) = j$.

If $L \subseteq E^\infty$, then we write $Rel(L)$ to denote the following set: $\{(H_w^{inp}, H_w^{out}) : w \in L\}$.

Definition 12 *The* **input-output relation** *of an automaton* M *is the following set:* $IO(M) = Rel(Tr(M))$

The following propositions are due to Lynch and Stark [21].

Proposition 13 *If M is a determinate automaton, then $IO(M)$ is continuous function.*

Proposition 14 *[27] If function $f : (V^\infty)^m \to (V^\infty)^n$ is continuous, then there exists a determinate automaton M such that $IO(M) = f$.*

It was shown in [18, 15, 35, 38] that two automata have the same input-output relation in every network context iff the closure of their trace sets are the same. This result is known as the *full-abstraction* theorem for trace semantics.

5 The Role of Monotonicity Properties

When one is working with determinate automata the IO relation computed is a Scott-continuous function on streams. In this section we see that the continuity and monotonicity properties can break down; indeed monotonicity properties will turn out to be a key tool for proving expressiveness theorems. Of course we have to define suitable analogues of these properties. We cannot just define monotonicity for the IO relation because, as we have seen, the IO relation is not compositional so we cannot reason easily about properties of networks if we only have information about the IO relation of the components of the network. The discussion in the first two subsections of this section is taken from [26].

We assume that the trace languages that we are interested in are over a concurrent alphabet that has the form $E^{inp} \otimes E^{out}$. The intention is that the trace languages are those associated with port automata and closed under the permutations that lead to a fully abstract model.

Definition 15 *Let L be a trace language. We say that L is* **Hoare-monotone** *if*

$$\forall w \in L. \forall u \in (E^{inp})^\infty. w|_{E^{inp}} \preceq \Rightarrow \exists w' \in L. w'|_{E^{inp}} \equiv u \wedge w \preceq w'.$$

If R is an IO relation we say that R is **Hoare-monotone** *if*

$$\forall (\alpha, \beta) \in R \forall \alpha'. \alpha \sqsubseteq \alpha' \Rightarrow \exists \beta'. (\alpha', \beta') \in R \wedge \beta \sqsubseteq \beta'.$$

What this says is that if we have any trace in L we can "add new input" and get a trace that is "bigger" than the original trace. The definition originally given by Panangaden and Shanbhogue [26] is essentially the same though very clumsily stated because they did not use the permutation preorder on traces. Clearly if the trace set computed by some network is Hoare monotone the associated IO relation will be as well. Note also that if the IO relation happens to be a function then Hoare-monotonicity is the same as ordinary monotonicity of functions.

The following definition and proposition say what happens to Hoare monotonicity if we "hide some of the channels."

Definition 16 *Suppose we are given a binary relation R on vectors of streams; $R \subseteq (\Sigma^\infty)^m \times (\Sigma^\infty)^n$. Suppose we also have the relation $S \subseteq (\Sigma^\infty)^j \times (\Sigma^\infty)^k$ with $j \leq m$ and $k \leq n$ related to R in the following way. We are given injections $\mu : \{1, \ldots, j\} \rightarrow \{1, \ldots, m\}$ and $\nu : \{1, \ldots, k\} \rightarrow \{1, \ldots, n\}$ such that*

$$S = \{((x_{mu(1)}, \ldots, x_{mu(j)}), (y_{nu(1)}, \ldots, y_{nu(k)}))$$
$$|where \;\; ((x_1, \ldots, x_m), (y_1, \ldots, y_n)) \in R\}.$$

*Then we say that S is obtained from R by **projection**.*

The following obvious proposition can now be stated.

Proposition 17 *Suppose that R is a Hoare-monotone IO relation and that R' is obtained from R by projection. Then R' is also Hoare monotone.*

It would be nice if we could have proved that Hoare monotonicity is a compositional property but unfortunately it says nothing about limiting behaviour of sequences of traces and some such property is needed for reasoning about feedback loops. Accordingly we define a property analogous to continuity and use it instead.

Definition 18 *We say that a trace language L is **limit closed** if it is Hoare monotone and the following property holds for all sequences of traces $\{t_i | i \in \omega\} \subseteq L$.*

$$\forall i \in \omega.t_i \preceq t_{i+1} \Rightarrow \exists t \in L.\forall i \in \omega.t_i \preceq t \land H_t = \sqcup H_{t_i}.$$

Note that if we take a limit-closed relation R and obtain R' by projection we might well obtain something that is not limit-closed. Examples of this possibility are in fact abundant. However, in view of the proposition above we know that the result will still be Hoare monotone. In [26] the following theorem is proved by structural induction and an appropriate fixed-point argument to deal with feedback.

Theorem 19 *Suppose that N_i is a family of networks with limit-closed trace sets then ΠN_i has a Hoare-monotone trace set.*

There is a dual property obtained by reversing the inequalities in the definition of Hoare monotonicity.

Definition 20 *A trace language L is said to be **Smyth monotone** if*

$$\forall w \in L.\forall u \in (E^{inp})^\infty.u \preceq w|_{E^{inp}} \Rightarrow \exists w' \in L.w'|_{E^{inp}} \equiv u \land w' \preceq w.$$

Analogously we define Smyth-monotone IO relations.

Once again ordinary monotonicity of functions implies Smyth monotonicity. But now one needs no limit-closed type property to reason about feedback. Here one is "descending" so well-foundedness of traces does the trick. In [26] the following theorem is proved by induction on the structure of networks.

Theorem 21 *Suppose that N_i is a family of networks with Smyth-monotone trace sets then ΠN_i has a Smyth-monotone trace set.*

5.1 The Expressive Power of Merge Primitives

In a prescient paper David Park [30] discussed a number of variants of fair merge. The fact that he named them differently clearly shows that he was aware that they were different but this was never mentioned.

Fair merge has two input channels and a single output channel. Every token arriving at an input is eventually put on the output channel in the same order, relative to other tokens on the same channel, as it arrived but interleaved with respect to the tokens on the other channel. Fair merge captures fair resolution of contention between two streams of requests.

So called *angelic merge* has two input channels, like fair merge, but only guarantees that it never gets stuck. It is thus fair on finite streams but may be unfair on infinite streams. The third variant is called *infinity-fair* merge. It is fair only on infinite streams. If one of the streams is finite it will get stuck.

We now give formal definition of the associated IO relations. One can define processes (i.e. trace sets) that give these IO relations as done in [26]. Since we are only trying to convey the idea rather than the proof we will content ourselves with the IO relations.

Definition 22 *The IO relation* **fairmerge** *is a set of triples of streams* (s_1, s_2, s_3) *such that* s_3 *can be split into two subsequences* s_3' *and* s_3'' *with* $s_1 = s_3'$ *and* $s_2 = s_3''$. *The IO relation* **amerge** *is the set of triples of the form* (s_1, s_2, s_3) *such that* s_3 *can be broken up into two subsequences* s_3' *and* s_3'' *such that*

1. s_3' *is a prefix of* s_1 *and* s_3'' *is a prefix of* s_2,
2. *if* s_1 *is finite then* $s_2 = s_3''$ *and if* s_2 *is finite then* $s_1 = s_3'$.

The IO relation **imerge** *is the set of triples of the form* (s_1, s_2, s_3) *such that* s_3 *can be broken up into two subsequences* s_3' *and* s_3'' *such that*

1. s_3' *is a prefix of* s_1 *and* s_3'' *is a prefix of* s_2,
2. *if* s_1 *is infinite then* $s_2 = s_3''$ *and if* s_2 *is infinite then* $s_1 = s_3'$.

Now we note that **fairmerge** is neither Hoare nor Smyth monotone. Indeed **fairmerge** has no total Hoare monotone subsets. The relation **amerge** is Hoare monotone but not Smyth monotone while **imerge** is both Hoare and Smyth monotone. A clever construction due to E. W. Stark shows how to express **imerge** with **amerge** and an easy argument shows that one can implement **amerge** with **fairmerge** [41]. The monotonicity properties noted and the theorems of the last subsection show that in the reverse direction we have impossibility results. Of course these are with our choice of processes corresponding to the IO relations. The "real" results here are about expressiveness of classes of processes and not about the expressiveness of relations. It is possible that the condition on limit closedness could be lifted by a careful analysis of the possible trace sets but none of my collaborators nor I have ever tried to do this with any seriousness.

Where do these primitives come from? and what do these results mean? The nonmonotonicity of **fairmerge** is a signal that the semantic theory associated with it will be subtle and this is indeed the case. The automata-theoretic results of Panangaden and Stark [28] show that fair merge requires automata that can be interrupted. One easy way to implement fair merge can be described informally as follows. Suppose that one can check whether a channel has data on it with committing to read it; this is usually called *polling*. Then one can just cycle periodically between the two input channels and poll them before attempting to read. It is trivial to emulate polling with a fair merge. Thus the nonmonotonicity phenomenon is present in polling, interrupts or timeouts. The expressiveness results are saying that there is no semantically more tractable primitive that has the same expressive power.

One can try to implement fair merge by using an "oracle". Let us suppose that there is a device that emits an arbitrary sequence of positive integers. One can use this as a driver of a merge as follows [3]. At every phase, read an integer, say n, and take n tokens from the current channel then switch channels and get another integer. Repeat this forever. This only works with infinite input streams; one has implemented **imerge**. This example shows how one may lose limit-closedness by hiding an internal channel. We will discuss the expressive power of oracles in a later section.

Angelic merge can be obtained in the following way. Suppose that one is working in a functional language with recursion, say pure Lisp. Now consider McCarthy's construct **amb**. What **amb** does is evaluate two integer expressions and pick one of them as its result. It will only diverge if *both* its subcomputations diverge. It is possible to implement **amerge** by using **amb** recursively but one cannot get fair merge this way [25]; even though **amb** does embody some sort of polling.

5.2 Full Abstraction Revisited

A proof of full abstraction has two parts. One has to show that the semantics is compositional and one must show that if two processes have different denotational semantics there must be a context that distinguishes them. The distinguishing contexts used by Kok [19] and Jonsson [15] use fair merge. This was of course before the expressiveness results were proved. The question immediately arises whether something other than traces turns out to be fully abstract if one restricts to networks that use less powerful primitives. In fact Russell [38] and Rabinovich and Traktenbrot [35] showed that one can construct contexts that distinguish traces using only determinate primitives. Thus traces *are* fully abstract even with the mildest form of indeterminacy.

5.3 Distinguishing Trace Sets: Fair Stack

So far we have taken the view that we should try to prove negative results with the mildest possible requirements on emulation. On the other hand we know that if one is observing the IO relation then (a suitable) trace equivalence corresponds to equality in the fully abstract model. It seems therefore that the next (and last!) type of expressiveness result one can reasonably try to prove is about whether *trace sets* can be emulated. The results of this section are hitherto unpublished results of Čubrić and Panangaden [45].

The first question we consider is automata theoretic. What characterizes the trace languages defined by monotone automata? We have seen that monotone automata have a Hoare monotone IO relation [28], in fact one can show without too much trouble that they have Hoare monotone trace sets. But, contrary to what I expected, the converse is not the case. My former student, Marija Čubrić, came up with an example of a process that has a Hoare monotone trace set but cannot be implemented by any monotone automaton.

We use the notation $\pi_\sqsubseteq(L)$ to stand for the prefixes of L.

Definition 23 *A trace language is said to be* **monotone** *iff*

$$\forall a \in V^{inp}.\forall b \in V^{out}.\forall w \in V^*.wb, wa \in \pi_\sqsubseteq(L) \rightarrow wab \in \pi_\sqsubseteq(L).$$

It is easy to show that this *is* the correct characterization of the trace language of a monotone automaton.

Theorem 24 Čubrić. *M is a monotone automaton iff $Tr(M)$ is a monotone trace language.*

It is also not too hard to show that

Proposition 25 *If $Tr(M)$ is a monotone trace language it is a Hoare monotone trace language.*

The results cited in the last paragraph allow us to discuss Marija Čubrić's example. The process is called a *fair stack*. Intuitively, it is a process that has an input channel and an output channel. The process consumes input tokens and pushes them on an internal stack. At arbitrary times it pops the stack, if it not empty. If the input stream is infinite the stack is cleared out infinitely often; this guarantees that no token sits on the bottom of the stack forever. We use the notation $\bar{a}$ to stand for an output event that produces the token a. Now we define a function *depth* acting on partial traces, as follows

$$depth(\bot) = 0$$
$$depth(ua\bar{a}v) = depth(uv)$$
$$depth(w) = |w|$$

where $u, v \in V^*$ and $w \in (V^{inp})^*$. Now we define the trace set of fair stack (FS) as follows

$$Tr(FS) = \{t \in V^*|depth(t) = 0\} \cup \{t \in V^\omega|\forall u \sqsubseteq t.\exists v.u \sqsubseteq v \sqsubseteq t.depth(v) = 0\}.$$

The fact that FS does not have a monotone trace language is immediate. The fact that it does have a Hoare monotone, and limit-closed, trace language is not hard. This shows that FS cannot be implemented (upto equality of trace sets) by any monotone automaton. On the other hand it can certainly be implemented upto IO equality since it *is* Hoare monotone.

The remarks at the end of the last paragraph show that FS cannot be implemented (upto trace equality) by angelic merge. Since FS does not have a Smyth monotone IO relation there is no hope of implementing it with **imerge** even upto just IO equality. On the other side a mild programming trick allows the trace language of FS to be implemented by fair merge.

What about implementing **amerge** with a fair stack? In fact this cannot be done even upto just IO. Thus from the viewpoint of IO equality FS lies strictly between **amerge** and **imerge**. If we consider emulation of trace languages it is incomparable to **amerge**. It remains to prove these assertions. The idea is to exhibit another property that is compositional, satisfied by FS but violated by **amerge**. This property is given the silly name[3] *f-continuity*.

Definition 26 *An IO relation R is said to be* **f-continuous** *(f-monotone) if there exists a continuous (monotone) function f such that* $graph(f) \subseteq R$.

The proof that this property is compositional should not come as a surprise; an explicit proof is given in [45]. Clearly the identity function is contained in the IO relation of FS because the stack can always push and pop immediately, thus appearing to be an identity buffer. On the other hand the IO relation of **amerge** is not even f-monotone. Suppose not and let f be the function that is contained in the IO relation of **amerge**. Note that $(a, \perp, a)$ and $(\perp, b, b)$ must both be in f, since f has to be total and be contained in the IO relation of **amerge**. But neither (a, b, ab) nor (a, b, ba) would be consistent with monotonicity of f yet these are the only choices. The structure of the expressiveness lattice is more complicated than we suspected before.

6 The Expressive Power of Oracles

A popular way of modeling indeterminacy is to use *oracles*. The idea is that every process really is determinate but some of the inputs are "hidden". Thus, an observer would record that, the response to the visible inputs is indeterminate. As we have already noted, processes like fair merge or angelic merge cannot be modelled this way. It is however worth understanding what different sorts of oracles are possible.

In this section we give new proofs of some of the results of a paper by McAllester, Panangaden and Shanbhogue [24]. The results of that paper expose some of the structure of the hierarchy of expressiveness but the proofs are very operational in nature. The present proofs are due to an anonymous referee of [24] and have not appeared in print before. An automata theoretic characterization of processes that are oracleizeble is given by E. W. Stark [43], he calls the resulting automata *semi-determinate*.

The intuitive notion of being oracleizable is formalized as follows.

Definition 27 *If D is a Scott domain, we say that an n-input, m-output relation R on streams,* $R \subseteq (V^\infty)^m \times (V^\infty)^n$, *is* **D-oracleizable** *if there is*

---

[3] I am responsible.

a *Scott-continuous* function $F : D \rightarrow [(V^\infty)^m \rightarrow (V^\infty)^n$ such that $R = \cup_{d \in D}\{(x, y)|F(d)(x) = y\}$.

In this point of view a process has a visible behaviour given by R. The function F is the "real behaviour" but since we do not see the D input we see only the relation R as the IO relation of the process. The domain D is an important parameter in the hierarchy of expressiveness properties. Our classification is based on the power of different choices for D. If $Unit$ is the one-point domain then $Unit$-oracleizable relations are just continuous functions.

In Plotkin's remarkable 1976 paper on powerdomains [32] he introduces the concept of finite generability to capture *finitely branching* indeterminacy. If we let $\mathbf{B}^\infty$ be the domain of boolean streams and $\mathbf{B}^\omega$ the infinite boolean streams then a subset G of a domain D is *finitely generable* if there is a continuous function f from $\mathbf{B}^\infty$ to D such that $f(\mathbf{B}^\omega) = G$. In short finite generability is the same concept as $\mathbf{B}^\omega$-oracleizable.

The basic domains that we consider are N^∞, the domain of streams of natural numbers, $N \times \mathbf{B}^\infty$, and $\mathbf{B}^\omega$. We will prove that these are strictly different from each other. To give concrete focus to these domains we will work with three basic primitives called **USS**, meaning *unfair splitter with signal*, **WFS** meaning *weakly fair splitter with signal* and **SFS** meaning *strongly fair splitter with signal*. Each has a single input channel and three output channels. The first output channel is called the *signal channel*. Tokens arrive at the input channel. The process makes an internal choice about whether to pass the token to the second or to the third output channel. If the token is placed on the second channel a 0 is sent along the signal channel, whereas if a token is sent along the third channel a 1 is sent along the signal channel. Thus the signal is used to "announce" the choices made. The three primitives differ in the fairness assumptions. The **USS** will make a binary choice each time with no attempt to be fair; it could put all the tokens on a single channel. It is clear that **USS** is $\mathbf{B}^\omega$-oracleizable and that it can generate any oracle value (just use the signal). The **WFS**, when given an infinite input will send at least one token on each output channel. It is clear that **WFS** is $N \times \mathbf{B}^\infty$-oracleizable and can generate any $N \times \mathbf{B}^\infty$ oracle. If infinitely many tokens arrive at an **SFS** each output channel gets infinitely many tokens. It is easy to see that **SFS** is N^∞-oracleizable and can generate any oracle in N^∞.

6.1 Topological Preliminaries

We assume that the reader is familiar with basic notions of topology and domain theory. We collect some basic facts about the metric properties of Scott domains and how these relate to the order structure. The results are primarily due to Comyn and Dauchet [9].

Recall that a Scott domain is a complete partial order which is required to be (i) *algebraic*, (ii) have a *countable* basis of finite (compact) elements and finally (iii) bounded-complete (consistent complete). The Scott topology is generated by taking as a subbase the sets of the form $N_e = \{u|e \sqsubseteq u\}$ with e a compact element. This topology is T_0 but not T_1. The specialization order reproduces the order structure of the domain. Functions that are continuous with respect to this topology are precisely the functions that preserve directed suprema. Given the basis of finite elements one can recover the whole domain by the familiar process of ideal completion. See the excellent recent book by Gunter [14] or the Pisa Notes [31] for an expository account of these concepts.

Given a Scott domain D one can define a metric as follows. Let D_0 be the countable set of compact elements that generate D via directed suprema. Fix an enumeration $\phi : N \to D_0$, where N is the natural numbers. Let x and y be elements of D. We define $d(x, y)$ to be 2^{-k} where k is the smallest natural number such that $\phi(k)$ is below exactly one of x and y. One can check that this is a metric. Comyn and Dauchet show that Cauchy completion of the compact elements gives the same result as ideal completion. The resulting metric space depends on the choice of the enumeration function ϕ, but the *topology* does not. The topology obtained is the well-known Cantor topology [32] (also known as the Lawson topology) They also prove the following very important fact.

Theorem 28 Comyn and Dauchet. *Suppose that D is a Scott domain equipped with a metric as above. Suppose that $(d_i | i \in N)$ is a Cauchy sequence converging to d. Then there is an em increasing sequence of compact elements $(e_j | j \in N)$ such that $\bigsqcup_j e_j = d$ and $\forall k. \exists l. \forall n \geq l. e_k \sqsubseteq d_n$.*

Thus every Cauchy sequence can be "emulated" by a directed sequence converging to the same limit and staying below the members of the Cauchy sequence.

In working with streams we use the following closely related, but not quite the same, metric. Given two streams s and t we look at the first position n where the streams differ. We then define $d(s, t) = 2^{-n}$. The topology defined is the same as with the enumerating functions if the alphabet is finite but not with a countable alphabet. The theorem above applies to this metric as well. If the alphabet is finite the space of all streams with this metric is *compact* but not when the alphabet is countable; the metric of Comyn and Dauchet is always compact. Whether the alphabet is countable or not the metric space of streams with the simple comparison metric is always a complete metric space as is also the subspace of infinite streams.

One key theorem that we use is the Baire category theorem, see, for example, the book by Dugundji [11]. We recall some definitions and the theorem. Suppose that X is a complete metric space. We say that a subset is *dense* if the closure of the subset is the whole space. We say that a subset Y of X is *nowhere dense* if every point in $X \setminus Y$ has an open ball around it that does not intersect Y.

Theorem 29 Baire Category Theorem. *A complete metric space cannot be the union of countably many nowhere dense sets.*

There are many equivalent ways of saying this; this is the form that we will use.

6.2   Expressiveness Theorems

We state the expressiveness results as theorems about the domain needed to serve as oracles for **USS**, **WFS** and **SFS**. We begin with an obvious proposition.

Proposition 30 *Any element of* $\mathbf{B}^\infty$ *can be generated by* **USS**. **USS** *is not determinate but is* $\mathbf{B}^\infty$-*oracleizable.*

USS is a typical finitely branching primitive. By contrast **WFS** requires countable branching.

Proposition 31 **WFS** *is* $N \times \mathbf{B}^\infty$-*oracleizable but is not* $\mathbf{B}^\infty$-*oracleizable. Furthermore any element of* $N \times \mathbf{B}^\infty$ *can be generated by* **WFS**.

Proof. To generate an element of $N \times \mathbf{B}^\infty$ from **WFS** we send the sequence of natural numbers in as input. By reading, say, the second output channel we get an arbitrary element of N. Suppose that this number is k, we ignore the first k inputs from the signal channel and read the rest of them to get an arbitrary member of $\mathbf{B}^\infty$. Note that if we just took the signal output we would never get an element like 1^∞. It is easy to write an oracle-program for **WFS**. We read the integer, say k and ensure that the first k elements go to one of the channels and that the $k + 1$st goes to the other channel. Then we use the boolean stream to drive an unfair splitter on the rest of the input.

Suppose that **WFS** were $\mathbf{B}^\infty$-oracleizable. Suppose that $F : \mathbf{B}^\infty \to [(V^\infty) \to (V^\infty)^2$ is a function such that

$$\mathbf{WFS} = \cup_{x \in V^\infty} \cup_{d \in \mathbf{B}^\infty} \{F(d)(x)\}.$$

Let us choose as x the stream $01234\ldots n\ldots$. For every integer $k \geq 0$ there is an *infinite* oracle d_k such that $F(d_k)(x) = (y_k, z_k)$ with the first element of z_k being *at least* k. Now since $\mathbf{B}^\infty$ is compact with the usual metric topology, the sequence of elements from $\mathbf{B}^\infty$ being used as oracles, viz. the d_k must have a convergent subsequence, d_{k_i} which converges to, say, d. Note that d must also be infinite. Now using the theorem of Comyn and Dauchet we can find an increasing sequence of *finite* elements of $\mathbf{B}^\infty$, e_i such that $\sqcup e_i = d$ and for all i there exists a j such that $\forall n \geq j.e_i \sqsubseteq d_{k_n}$. Since F is monotone $F(e_i)(x) \sqsubseteq F(d_{k_n})(x)$. Let us define (y_i, z_i) by the equation $F(e_i)(x) = (y_i, z_i)$. Now the z_i has to have a first value which is at least k_n *for all* n, but this is a strictly increasing sequence. This is only possible if $z_i = \bot$ for all i. But, since F is continuous, $F(d)(x) = \sqcup_i F(e_i)(x)$ we must have that $F(d)(x) = (x, \bot)$ which is impossible for **WFS**.

The above proof hinged on the compactness of the space $\mathbf{B}^\infty$. The space $N \times \mathbf{B}^\infty$ is not compact and an oracle for **WFS** is constructible using $N \times \mathbf{B}^\infty$, as we have seen. One might wonder whether it would make a difference if we guaranteed that each output got at least k elements rather than just one. A programming trick by Shanbhogue shows that one gets no additional expressive power this way. In any case, this should be intuitively clear.

Proposition 32 **SFS** *is* N^∞ *oracleizable but is not* $N \times \mathbf{B}^\infty$ *oracleizable. Using* **SFS** *one can generate any sequence of positive integers.*

Proof. We establish the second claim first. We send in any infinite input sequence and look at the signal generated. This signal must contain *infinitely many* 1s *and infinitely many* 0s. This is because of the fairness guarantee of **SFS**. Now we can easily write a determinate process that reports the sizes of successive blocks, giving us all possible positive integer sequences. In fact we will work with the process **NN**, which generates all possible positive integer sequences rather than with **SFS**, for the rest of this proof since the two are clearly interdefinable.

Suppose that **NN** has $N \times \mathbf{B}^\infty$ oracle. Then it must be the countable union of *finitely generable sets*, one for each value from N. Now **NN** computes N^ω, the set of all *infinite* integer sequences. With the usual metric, N^ω is a closed subspace of N^∞. Now we claim that any finitely generable subset, say A, of N^∞ that lies in N^ω must be *nowhere dense*. The result then follows from the Baire category theorem.

In order to prove the claim we proceed as follows. Let the set A be as in the previous paragraph. Let $F : \mathbf{B}^\infty \to N^\infty$ be a Scott-continuous function such that $A = F(\mathbf{B}^\omega)$. Using the same sort of argument as in the last proposition we can readily verify that A is closed. We shall prove that A cannot contain an open ball thus showing that A has empty interior and hence is nowhere dense. Let z be a *finite* element of N^∞. Consider the set of finite sequences z_k where k is a positive integer and $z_k = z \cdot k$, i.e. the sequence z followed by the integer k. Now we consider the collection of sets $z_k \uparrow = \{u | z_k \sqsubseteq u\}$. We claim that only finitely many of these sets can intersect A. Assume for the moment that this is true. Now an open ball in N^ω looks like $u \uparrow \cap N^\omega$ where u is a finite element of N^∞. If the claim we have just made is true, then clearly A cannot contain an open ball and hence A has empty interior and hence is nowhere dense. Thus if we establish the claim we will be done.

Suppose infinitely many of the sets $z_k \uparrow$ intersect A; we will write $z_{k_i} \uparrow$ for the sets that intersect A. Then we can find infinitely many infinite streams $b_i \in \mathbf{B}^\omega$ such that $F(b_i) \in z_{k_i} \uparrow$. Now, since $\mathbf{B}^\omega$ compact, we can find a convergent subsequence $b_{i'}$ which converges to b in $\mathbf{B}^\omega$. Using the theorem of Comyn and Dauchet we can construct a sequence of finite elements e_i of $\mathbf{B}^\infty$ such that $\bigsqcup_i e_i = b$ and with the property

$$\forall m. \forall m' \geq m. e_m \sqsubseteq b'_m.$$

Since F is monotone $\forall m. \forall m' \geq m. F(e_m) \sqsubseteq F(b_{m'}) \in z_{m'} \uparrow$. But the z_m are all inconsistent, so the only way to satisfy this condition is if $\forall m. F(e_m) \sqsubseteq z$. Since F is continuous $F(\bigsqcup e_i) = F(b) \sqsubseteq z$. But $F(b) \in A$ which contains only infinite sequences, so we have a contradiction.

The last paragraph is a rehash of the proof of the previous proposition. The use of the Baire category theorem conceals a diagonalization argument which is how we proved it in our paper [24]. What is interesting about these proofs is the interplay between the order structure and the metric structure. Both structures are essential for these proofs to go through.

6.3 Montonicity and Oracleizability

The process **SFS** is easily seen to be equivalent to **imerge** or to **NN**. One might be tempted to think that being both Hoare and Smyth monotone would precisely correspond to being oracleizable. A very clever example due to Jim Russell [39] shows that this is false. First note that if a process is oracleizable then the IO relation is the union of a collection of functional relations. This is one approach to generalizing Kahn's principle; think of a process as realizing a set of functions [2, 39].

Consider a process with 2 input and 4 output channels. We construct the following table. We assume that the only possible input is a 1.

$$
\begin{array}{ll}
\textit{Input} & \textit{Outputs} \\
(\bot, \bot) & (1, \bot, \bot, \bot), (\bot, \bot, \bot, 1) \\
(1, \bot) & (1, 1, \bot, \bot), (\bot, \bot, 1, 1) \\
(\bot, 1) & (1, \bot, 1, \bot), (\bot, 1, \bot, 1) \\
(1, 1) & (1, 1, \bot, 1), (1, \bot, 1, 1)
\end{array}
$$

It is easy to see that this is both Hoare and Smyth monotone. On the other hand it contains no total functional subrelations; thus it cannot be oracleizable.

6.4 The Most Powerful Oracle

Given the results of the last subsection and early experiences of recursion theory we might be tempted to think that the possible oracles form a rich collection like the lattice of degrees in recursion theory. The point is, however, that determinate primitives are very powerful and are capable of surprising encodings. In this section we show that N^∞ is in fact the most powerful oracle possible. In other words **imerge** or **SFS** or **NN** can all be viewed as being "complete" for the class of oracleizable primitives. We close with conjectures about the structure of the hierarchy of expressiveness among oracleizable primitives.

The result of this section is due to Jim Russell and the present author but in his thesis Russell did not take quite the same view and was led to the conclusion that **imerge** was not "complete" for this class. In the view of the present paper, oracles are used effectively. Thus an infinite stream used as oracle input must be examined incrementally. In Russell's thesis he allowed oracles to be read as atomic values even if they were infinite objects. With this view, of course, **imerge** cannot be universal.

Proposition 33 (Russell) *Given any oracleizable network its behaviour can be described by a set of continuous functions closed under the following operation. If F is the set of functions then $Cl(F) = \{f | \forall i_0 \sqsubseteq i_1 \sqsubseteq \ldots, \sqcup i_l = i,$ with the i_l finite, and for all chains of finite outputs $o_0 \sqsubseteq o_1 \sqsubseteq o_2 \ldots, f(i) = sqcup o_l, o_l \sqsubseteq f(i_l)$ for each l $\exists f' \in F.f'(i) = f(i), o_l \sqsubseteq f'(i_l)$ for each $l\}$. Furthermore two networks are indistinguishable in all contexts if and only if they have the same representation as such sets of functions.*

Remark This is the semantics of his thesis [39]. The nontrivial part of his semantics is coming up with the right closure conditions on the set of functions to get full abstraction. The proposition is trivial with out the closure condition.

Now we can prove the main result.

Theorem 34 Russell. *Given any oracleizable network one can construct a network that uses N^∞ as its oracle and computes the same relation.*

Proof. By the proposition it suffices to show that for a set of continuous functions F we can construct a network with N^∞ as an oracle, which computes the relation obtained by taking the union of the graph of all the functions in F. We use the notation $[s]_n$ to mean the length n prefix of the stream s. Given two functions f, f' we can define a family of equivalence relations $\equiv_n$ indexed by the natural numbers as follows. We say $f \equiv_n f'$ iff $\forall s.[f([s]_n)]_n = [f'([s]_n)]_n$. Note that there are only countably many equivalence classes of F for $\equiv_n$ so we can index them by the naturals. Let C_{k_1} be the k_1th equivalence class of $\equiv_1$. We can then consider the effect of the relation $\equiv_2$ on C_{k_1}. We get countably many subclasses which we denote $C_{k_1 k_2}$. In this way we can define C_r for any finite sequence r of natural numbers. For an infinite stream t of naturals we define C_t as the intersection of the C_r for all r, a finite prefix of t. Now C_t is either empty or contains a single function.

Because of the closure condition on the sets of functions we can tell at a finite stage whether the equivalence class is empty or not. Thus we can program a determinate device that examines the integer stream as it comes in and modifies it to ensure that it yields a stream that defines a nonempty equivalence class. Now we define the function P as follows

$$P(s, i) = \begin{cases} [f([i]_l)]_l, f \in C_s & \text{if } s \text{ is finite of length } n \\ f(i), f \in C_s & \text{if } s \text{ is infinite and } C_s \text{ is not empty} \\ \sqcup\{P(r, i) | r \text{ is a finite prefix of } s\} & \text{if } s \text{ is infinite and } C_s \text{ is not empty} \end{cases}$$

Clearly P is continuous and computes exactly the functions in F.

Note that the last clause makes P total but will never arise because the preprocessor makes sure that this case never arises. The closure condition is essential for this.

The basic conjecture is that we have a "gap" in the hierarchy.

Conjecture 35 *There is no oracleizable network strictly between $N \times B^\infty$ and N^∞.*

7 Conclusions

The main results described here show that the hierarchy of indeterminate primitives is rich and far from explored as yet. At the top level we have the **fair merge** primitive which manifests nonmonotonic behaviour. We have **angelic merge** and **infinity-fair merge** forming provably weaker primitives with **fair**

stack strictly between them. We have shown that nothing above **infinity-fair merge** can be described by oracles. We have showed that oracles can be classified by the structure of the domain from which the oracle values are taken. Finally we have discussed some provably distinct primitives at the level of oracles.

All the above results hold with arbitrary continuous functions taken as available as primitive determinate networks. If we look at what happens if we only allow sequential or stable functions then new differences appear [24]. For example with the primitives **WFS** or **USS** one cannot reconstruct the output on the signal channel without having it available at first. This is very closely linked to stability phenomena. In the purely determinate realm Sazanov [40] and more recently Bucciarelli [8] have shown that the hierarchy associated with sequential and stable functions (called "degrees of parallelism") forms a very rich structure.

The monotonicity arguments that we used can be used in other contexts as well. For example Critchlow and Panangaden [10] have used it to show that delay operators in SCCS are provably different. Similarly one can argue that one cannot implement fair merge in logic programming without some analogue of polling. Typically the primitive **var?**, which tests whether a variable has any bindings or not is used to implement fair merge. This is clearly the same as polling.

Is there any monotonicity property that **fair merge** *does* satisfy? Fair merge does satisfy Hoare monotonicity if restricted to finite inputs only. We conjecture that any primitive must satisfy this property. Indeed we conjecture that *there is no indeterminate dataflow primitive* strictly more powerful than **fair merge**. This is more a "thesis" about what is physically realizable than a mathematical conjecture that can be proved. Of course it could be disproved by constructing an example that is strictly more powerful; assuming, of course, there is no controversy about whether the proposed primitive is physically realizable or asynchronous.

Acknowledgments

I have benefitted from many discussions including with the following: Samson Abramsky, David Benson, Steve Brookes, Carol Critchlow, Marija Čubrič, Radha Jagadeesan, Bengt Jonsson, Joost Kok, David McAllester, Albert Meyer, Catuscia Palmidessi, Alex Rabinovich, Jim Russell, Jan Rutten, Vasant Shanbhogue, Boris Traktenbrot and especially Gene Stark. I have used results of Samson Abramsky, Marija Čubrič and Jim Russell that have not appeared in print before. I would like to acknowledge NSERC for financial support. Finally I am delighted to thank the program committee and its chairman, P. S. Thiagarajan, for inviting me to talk at FSTTCS.

References

1. S. Abramsky. On semantic foundations for applicative multiprogramming. In J. Diaz, editor, *Proceedings of the Tenth International Conference On Automata, Languages And Programming*, pages 1–14, New York, 1983. Springer-Verlag.

2. S. Abramsky. A generalized kahn's principle. In M. Mislove, editor, *Proceedings of the Fifth Workshop on Mathematical Foundations of Programming Semantics.* Springer-Verlag, 1990. Lecture Notes In Computer Science 442.

3. K. R. Apt and G. D. Plotkin. Countable nondeterminism and random assignment. *Journal of the ACM*, 33(4):724–767, 1986.

4. A. Arnold and M. Nivat. Formal computations of nondeterministic recursive program schemes. *Math. Systems Theory*, 13:219–236, 1980.

5. H. P. Barendregt. *The Lambda Calculus, Its Syntax and Semantics.* Studies in Logic. North-Holland, Amsterdam, revised edition edition, 1984.

6. J. Dean Brock and W. B. Ackerman. Scenarios: A model of non-determinate computation. In J. Diaz and I. Ramos, editors, *International Colloquium on Formalization of Programming Concepts*, pages 252–259. Springer-Verlag, 1981. Lect. Notes in Comp. Sci. 107.

7. M. Broy. Fixed point theory for communication and concurrency. In *Formal Description of Programming Concepts II*, pages 125–148. North-Holland, 1983.

8. A. Bucciarelli. Degrees of parallelism in the continuous type hierarchy. Workshop on Full Abstraction of PCF and Related Languages, University of Aarhus,, April 1995.

9. G. Comyn and M. Dauchet. Metric approximations in ordered domains. In M. Nivat and J. Reynolds, editors, *Algebraic Methods in Semantics*, pages 251–276. Cambridge University Press, 1985.

10. C. Critchlow and P. Panangaden. The expressive power of delay operators in sccs. *Acta Informatica*, 28:447–452, 1991.

11. J. Dugundji. *Topology.* J. Wiley, 1966.

12. A. A. Faustini. An operational semantics for pure dataflow. In *Proceedings of the Ninth International Colloquium On Automata Languages And Programming*, pages 212–224. Springer-Verlag, 1982. Lecture Notes in Computer Science 140.

13. N. Francez. *Fairness.* Springer-Verlag, 1986.

14. Carl A. Gunter. *Semantics of Programming Languages: Structures and Techniques.* MIT Press, 1992.

15. B. Jonsson. Fully abstract trace semantics for dataflow networks. In *Proceedings of the Sixteenth Annual ACM Symposium on Principles of Programming Languages*, 1989.

16. G. Kahn. The semantics of a simple language for parallel programming. In *Information Processing 74*, pages 993–998. North-Holland, 1977.

17. G. Kahn and D. MacQueen. Coroutines and networks of parallel processes. In Gilchrist, editor, *Proceedings of Information Processing*, pages 993–998. North-Holland, 1977.

18. J. Kok. Denotational semantics of nets with nondeterminism. In *Proceedings of the 1986 European Symposium on Programming*, pages 237–249, 1986.

19. J. Kok. A fully abstract semantics for dataflow nets. In *Proceedings of Parallel Architectures And Languages Europe 1987*, pages 351–368, Berlin, 1987. Springer-Verlag.

20. M. Kwiatkowska. *Categories of Asynchronous Systems.* PhD thesis, University of Leicester, May 1989.

21. N. A. Lynch and E. W. Stark. A proof of the Kahn principle for input/output automata. *Information and Computation*, 82(1):81–92, 1989.

22. N. A. Lynch and M. Tuttle. Hierarchical correctness proofs for distributed algorithms. Technical Report MIT/LCS/TR-387, M. I. T. Laboratory for Computer Science, April 1987.

23. A. Mazurkiewicz. *Advanced Course in Petri Nets*, volume 255 of *Lecture Notes In Computer Science*, chapter Trace Theory, pages 279–324. Springer-Verlag, 1986.

24. D. Mcallester, P. Panangaden, and V. Shanbhogue. Nonexpressibility of signaling and fairness. *J. Comput. Syst. Sci.*, 47(2):287–321, 1993.

25. P. Panangaden and V. Shanbhogue. Mccarthy's amb cannot implement fair merge. In K. V. Nori, editor, *Proceedings of the 8th Conference on Foundations of Software Technology and Theoretical Computer Science*, pages 348–363. Springer-Verlag, 1988. Lecture Notes in Computer Science 338.

26. P. Panangaden and V. Shanbhogue. The expressive power of indeterminate dataflow primitives. *Information and Computation*, 98(1):99–131, 1992.

27. P. Panangaden, V. Shanbhogue, and E. W. Stark. Stability and sequentiality in dataflow networks. In M. S. Paterson, editor, *Proceedings of the Seventeenth International Colloquium On Automata Languages And Programming*, pages 308–321. Springer-Verlag, 1990. Lecture Notes in Computer Science 443.

28. P. Panangaden and E. W. Stark. Computations, residuals and the power of indeterminacy. In T. Lepisto and A. Salomaa, editors, *Proceedings of the Fifteenth International Colloquium on Automata Languages and Programming*, pages 348–363. Springer-Verlag, 1988. Lecture Notes in Computer Science 317.

29. D. Park. On the semantics of fair parallelism. In *Proceedings of the Winter School on Formal Software Specification*, pages 504–526, New York, 1980. Springer-Verlag. Lecture Notes In Computer Science 86.

30. D. Park. The "fairness problem" and non-deterministic computing networks. In J. de Bakker and L. van Leeuwen, editors, *Proceedings of the Fourth Advanced Course on Foundations of Computer Science - Distributed Systems*, pages 133–161. Mathematisch Centrum, 1982.

31. G. D. Plotkin. Lecture notes on domain theory. The Pisa Notes.

32. G. D. Plotkin. A powerdomain construction. *SIAM Journal of Computing*, 5(3):452–487, 1976.

33. G. D. Plotkin. A powerdomain for countable nondeterminism. In *Proceedings of the Ninth ICALP*. Springer-Verlag, 1982. LNCS 140.

34. V. Pratt. Modeling concurrency with partial orders. *International Journal Of Parallel Programming*, 15(1):33–71, 1986.

35. A. Rabinovich and B. A. Trakhtenbrot. Nets of processes and dataflow. To appear in Proceedings of ReX School on Linear Time, Branching Time and Partial Order in Logics and Models for Concurrency, LNCS, 1988.

36. A. Rabinovich and B. A. Trakhtenbrot. Nets and data flow interpreters. In *Proceedings of the Fourth Annual IEEE Symposium on Logic in Computer Science*, pages 164–174, 1989.

37. A. Rabinovich and B. A. Trakhtenbrot. Communication among relations. In M. S. Paterson, editor, *Seventeenth International Colloquium on Automata Languages and Programming*, number 443 in Lecture Notes In Computer Science, pages 294–307. Springer-Verlag, 1990.

38. J. R. Russell. Full abstraction for nondeterministic dataflow networks. In *Proceedings of the 30th Annual Symposium of Foundations of Computer Science*, pages 170–177, 1989.

39. J. R. Russell. On oracleizable networks and kahn's principle. In *Proceedings of the seventeenth Annual ACM Symposium on Principles of Programming Languages*, 1990.

40. *Degrees of Parallelism in Computations*, volume 45 of *Lecture Notes In Computer Science*. Springer-Verlag, 1976. Proceedings of the Conference on Mathematical

Foundations of Computer Science.

41. V. Shanbhogue. *The Expressiveness of Indeterminate Dataflow Primitives*. PhD thesis, Cornell University, 1990.
42. E. W. Stark. Concurrent transition systems. *Theoretical Computer Science*, 64:221–269, 1989.
43. E. W. Stark. Connections between a concrete and an abstract model of concurrent systems. In *Mathematical Foundations of Programming Language Semantics*, Lecture Notes in Computer Science 442. Springer-Verlag, 1990.
44. E. W. Stark. A simple generalization of kahn's principle to indeterminate dataflow networks. In M. Z. Kwiatkowska, M. W. Shields, and R. M. Thomas, editors, *Semantics of Concurrency, Proceedings of the International BCS-FACS Workshop*. Springer-Verlag, 1990. Available as SUNY Stonybrook TR 89-29.
45. Marija Čubrić and P. Panangaden. Monotone and nonmonotone dataflow networks. ACAPS Memo 81, McGill University, 1993.

The Transformation Calculus

Jacques Garrigue

Research Institute for Mathematical Sciences
Kyoto University, Kitashirakawa-Oiwakecho
Sakyo-ku, Kyoto 606-01 JAPAN
Tel ++81-75-753-7211
Fax ++81-75-753-7272
E-mail garrigue@kurims.kyoto-u.ac.jp

Abstract. The lambda-calculus, by its ability to express any computable function, is theoretically able to represent any algorithm. However, notwithstanding their equivalence in expressiveness, it is not so easy to find a natural translation for algorithms described in an imperative way.

The transformation calculus, which only extends the notion of currying in lambda-calculus, appears to be able to correct this flaw, letting one implicitly manipulate a state through computations.

This calculus remains very close to lambda-calculus, and keeps most of its properties. We proved confluence of the untyped calculus, and strong-normalization in presence of a typing system.

1 Introduction

Currying is as old as lambda calculus. For the simple reason that, in raw lambda calculus —without pairing or similar built-in constructs—, this is the only way to represent multi-argument functions. This just means that we will write

$$\lambda x. \lambda y. M[x, y]$$

in place of

$$(x, y) \mapsto M[x, y].$$

At this stage appears a first asymmetry: while in the pair (x, y) the two variables play symmetrical roles, in $\lambda x. \lambda y. M$ they don't. An implicit order was introduced. Materially this means that we can partially apply our function directly on x but not on y.

We now look at types. There, currying can be seen as isomorphism of types [3]:

$$(A \times B) \to C \simeq A \to B \to C.$$

Here comes another asymmetry: why don't we get any similar isomorphism for $A \to (B \times C)$.

The calculus we will present here generalizes currying to these two kinds of symmetries: between arguments, and between input and output. For the first one, we are just taking over the mechanism of *label-selective currying* developed previously [1, 11].

For the second one we develop a new notion of *composition*, which, contrary to the usual one, is compatible with currying.

The resulting system, *transformation calculus*, is a conservative extension of lambda calculus. Why such a name? Because this essentially syntactic extension —semantics remain very similar— provides us with a new way of representing state transformations, *i.e.* state being represented by labeled input parameters, that may get returned by our term. Handling state as a supplementary parameter that gets returned with the result is not new. But by extending currying we get more flexibility, in two ways. First, since a part of the state is no more than a labeled parameter, we can dynamically extend it by simply adding a new parameter at some point in our term. Second, selective currying lets a transformation ignore parts of the state it doesn't need. They will just be left unmodified.

To demonstrate our point, we introduce *scope-free variables*, which are trivially encoded in the transformation calculus, and can be used in place of usual scoped mutable variables, in the Algol tradition. Since they have no syntactic scope, scope-free variables respect dynamic binding rather than static binding; but they are more flexible than Algol variables, while simulating blocks and stack discipline.

The rest of this paper is composed as follows. In Section 2 we introduce progressively the different features which form the transformation calculus. Section 3 is devoted to the formal definition of the transformation calculus. Sections 4 and 5 respectively define and give the fundamental properties of scope-free variables and a simply typed transformation calculus. Related works are presented in Section 6. Finally, Section 7 concludes. For lack of space, no proofs ar given in this paper[1].

2   Composition and streams

We first introduce informally and progressively the features of our calculus. We start from the classical pure lambda-calculus, that is[2]

$$M ::= x \mid \lambda x.M \mid (M).M$$

with β-reduction

$$(N).\lambda x.M \rightarrow_\beta [N/x]M$$

and where terms are considered modulo α-conversion (renaming of bound variables).

2.1   Implicit currying

Currying is the fundamental transformation by which multi-argument functions are encoded in the lambda-calculus. It can appear in abstractions as well as applications. For instance $f(a,b)$ will be encoded as $(b).(a).f$, and $\lambda(x,y).M$ becomes $\lambda x.\lambda y.M$.

This operation does not modify the nature of calculations, since clearly $(a,b).\lambda(x,y).M$ and $(b).(a).\lambda x.\lambda y.M$ reduce to the same $[a/x,b/y]M$ (provided x and y are distinct variables).

---

[1] They can be found, together with denotational semantics for the simply typed calculus, in the report version [9].

[2] Application, denoted by a dot, is written postfix, and is left associative.

By implicit currying, we mean that we will write curried and uncurried versions of terms (for an arbitrary number of arguments) indifferently, always supposing that we reduce curried ones. Of course we work in the pure lambda-calculus without pairing, so that no confusion is possible. The new syntax becomes

$$M ::= x \mid \lambda(x, \ldots).M \mid (M, \ldots).M$$

where abstracted variables under the same λ should be distinct. Implicit currying is expressed by the two structural equivalences[3]:

$$\lambda(x_1, \ldots, x_n).M \equiv_\lambda \lambda(x_1, \ldots, x_k).\lambda(x_{k+1}, \ldots, x_n).M$$
$$(N_1, \ldots, N_n).M \equiv (N_{k+1}, \ldots, N_n).(N_1, \ldots, N_k).M$$

where all x_i's are distinct.

$\equiv$ is defined as the reflexive, symmetric and transitive closure of $\equiv_\varsigma$'s.

2.2 Composition

The next step is to introduce a binary *composition* operator[4] (";") and a *transformation constructor* ("$\downarrow$").

$$M ::= \ldots \mid \downarrow \mid M; M$$

A *transformation* is a term such that, provided enough input, it gets a transformation constructor at its head position.

Together we add a new reduction rule, and a new structural equivalence, to eliminate compositions. Some other equivalences are introduced in the actual calculus, to enable earlier flattening of terms, but we leave them for later.

$$\downarrow; M \rightarrow_\downarrow M$$
$$(N_1, \ldots, N_k).(M_1; M_2) \equiv (N_1, \ldots, N_k).M_1; M_2$$

We can see the sequencing role of composed pairs as follows: when we apply $(M_1; M_2)$ to a sufficient input tuple of arguments, we first apply M_1 to this tuple, get (hopefully) a *tuple-term* (term of form $(N_1, \ldots, N_k).\downarrow$) as result of its reduction, and apply M_2 to this result tuple.

It just looks like if we added a stack machine into the lambda-calculus. For instance, we can write the transformation that switches two terms on top of a stack as $sw = \lambda(x, y).(y, x).\downarrow$, and can apply it to an input tuple of any size:

$$
\begin{aligned}
& (a, b, c) \ .\lambda(x, y).(y, x).\downarrow \\
\rightarrow_\beta \quad & (b, c) \ .\lambda(y).(y, a).\downarrow \\
\rightarrow_\beta \quad & (c) \ .(b, a).\downarrow \\
\equiv \quad & (b, a, c) \ .\downarrow
\end{aligned}
$$

---

[3] In fact, if we remember the habit many have of writing $(\lambda xy.M)\,a\,b$ for the above term, we have done absolutely nothing new. However this syntax lets us emphasize some natural groupings of values. For instance the encoding of pairs in lambda-calculus can be written as $\lambda f.(a, b).f$.

[4] Both the dots of abstraction and application bind tighter than composition.

Composed with another term, it plays the same role as the $C = \lambda fxy.fyx$ combinator; but in a postfix way.

$$
\begin{aligned}
& (a, b, c) \,.(sw; K) \\
\equiv_; \; & (a, b, c) \,.sw; K \\
\xrightarrow{*} \; & (b, a, c) \,.\!\downarrow; K \\
\equiv_; \; & (b, a, c) \,.(\!\downarrow; K) \\
\rightarrow_\downarrow \; & (b, a, c) \,.K \\
\xrightarrow{*} \; & \quad\;\; (c) \,.b
\end{aligned}
$$

Since we are in the lambda-calculus, we can define the fix-point operator Y. We just define then loops in terms of this operator. The functional for a while-do loop can be defined as

$$
\begin{aligned}
\text{while} = Y(&\lambda whl. \\
&\lambda(end, do).(end; \\
&\quad \lambda b.\text{if } b \text{ then } do; (end, do).whl \text{ else } \downarrow))
\end{aligned}
$$

The *end*-condition is a transformation that adds to its input a boolean b, false to end, true to go on, leaving the rest in position. *do* may change the values from the input, but not their number. Such a functional works on a state of any size.

An imperative version of Euclid's algorithm for the greatest common divisor can then be written

$$
\begin{aligned}
&(\lambda(x).(x \neq 0, x).\!\downarrow, \\
&\lambda(x, y).(y \bmod x, x).\!\downarrow).\text{while}; \\
&\lambda(x, y).y
\end{aligned}
$$

We notice here an important difference between this "while" functional and something equivalent written using pairing. Here our *end*-condition only uses x, whereas a functional using pairing would have required it to receive the whole state even though y is not needed. This remark will become even more important when we will add to our calculus the power of *selective currying*.

2.3 Selective currying

Combining lambda-calculus and a stack machine should be enough to express algorithms both in their functional and imperative form. Variables are denoted by positions on the stack. However, these positions being relative to the stop of the stack, they change if we add anything on top of if: *e.g.* c is the third element in $(a, b, c).\!\downarrow$, but it becomes the 4th in $(a, b, c).\!\downarrow; (d).\!\downarrow \rightarrow (d, a, b, c).\!\downarrow$, where d is either a parameter for the next function or a new variable. That makes it quite uneasy to actually write algorithm using variable in such a calculus.

The possibility of mixing parameters and variables is good, since it means that we can see everything with a functional insight. What we would like is to have a more uniform way to handle a position. The idea is to use several named stacks simultaneously. As stacks were represented by tuples, we now use sets of named tuples, or *streams*. For instance, $\{\epsilon \Rightarrow (a, b), p \Rightarrow c, q \Rightarrow (d, e)\}$ denotes a set of 3 independent stacks: (a, b) named ϵ[5], (c) named p, and (d, e) named q. Defining a label as being a pair (*stack name*,

---

[5] ϵ is a default name, such that a tuple (a, b, c) without name will be interpreted as the stream $\{\epsilon \Rightarrow (a, b, c)\}$.

stack index), we can also write a stream as a set of labeled values: $\{\epsilon 1 \Rightarrow a, \epsilon 2 \Rightarrow b, p \Rightarrow c, q1 \Rightarrow d, q2 \Rightarrow e\}$ [6].

Using this we can write a transformation incrementing the value labeled i as

$$\lambda\{\epsilon \Rightarrow x, i \Rightarrow y\}.\{i \Rightarrow x + y\}.\downarrow$$

Since i is a named position its index is not modified by operations on other names.

We see here a new version of Euclid's algorithm, using labels for both the function and the while functional.

while $= \lambda\{end \Rightarrow end, do \Rightarrow do\}.$
    $end; \lambda\{ok \Rightarrow ok\}.$
    if ok then $do; \{end \Rightarrow end, do \Rightarrow do\}.$while else $\downarrow$

gcd $= \lambda(x, y).\{m \Rightarrow x, n \Rightarrow y\}.\downarrow;$
    $\{end \Rightarrow \lambda\{m \Rightarrow m\}.\{ok \Rightarrow m \neq 0, m \Rightarrow m\}.\downarrow,$
    $do \Rightarrow \lambda\{m \Rightarrow m, n \Rightarrow n\}.\{m \Rightarrow n \bmod m, n \Rightarrow m\}.\downarrow\}.$while;
    $\lambda\{m \Rightarrow m, n \Rightarrow n\}.n$

On such an example the addition of labels may look as pure verbosity, but what we obtain here is very close to what we would write as an imperative algorithm. We only have to add trivial abstractions of the form $\lambda\{m \Rightarrow m\}$ in order to transform an assignment-like syntax (where $\{m \Rightarrow M\}$ is read as $m := M$) into functions.

2.4 Stream behavior

The examples we presented above worked all right, but what happens with "incorrect" terms, that are not well-behaved?

We had already such terms in classical lambda-calculus. For instance, if we encode an if-then-else by a pair $\lambda s.(s\ t\ e)$, where s is expected to be an encoded boolean, and t and e the two cases, we expect in most cases t and e to be well-behaved, that is if t encodes a pair, then e should also encode a pair. Otherwise, we will have unexpected behavior trying to apply a projection on it.

This problem of behavior is even more pernicious with the transformation calculus. Again in an if-then-else we expect the two branches to have similar behavior. But even if the second one gives back a stream with more labels than the first, it may well not appear, as long as we only use transformations that only access labels present in the first stream. This is still an incoherence.

So, by *well-behaved*, we will mean here that for any acceptable input with same stream structure, a transformation should give back a stream with same labels. That is, its *stream-behavior*, the stream structure of the result with respect to the stream structure of the input, should not be dependent on encoded values in the input.

For instance,

$$\lambda b.\text{if } b \text{ then } \{l \Rightarrow M\}.\downarrow \text{ else } \downarrow$$

---

[6] We allow omitting the index when there is only one value in the stack, as for p.

is not well behaved since it returns either a stream with label l or an empty stream, depending on the value of b.

This is difficult to give a precise definition of *well-behaved* terms in an untyped framework, since it depends on what encodings we use. In a typed framework that amounts to subject reduction, and we give in Section 5 a simply typed transformation calculus that satisfies it (*i.e.* all typable terms are well-behaved).

2.5 Scope-free variables

We have insisted on how this calculus was a potential basis for an integration of imperative and functional styles in the design of algorithms. Here we introduce a general method to directly map the imperative notion of variable into the transformation calculus.

In fact, what we mean by *scope-free variable* is slightly stronger than a mutable variable. We call it scope-free, since it is not syntactically scoped like in structured programming, neither is it global. We can say that it is local to a sequence of transformations, composed together.

A scope free variable is essentially a name v whose use in labels is exclusively reserved in the concerned sequence of transformations. This sequence is delimited by the creation of the variable with value a, encoded $\{v \Rightarrow a\}.\downarrow$, and its destruction by an abstraction, $\lambda\{v \Rightarrow x\}.\downarrow$. Between these, all transformations using or modifying this variable should once take it (through abstraction) and then put it back (by application), identical or modified. Typically a modification can be written $\lambda\{v \Rightarrow x\}.\{v \Rightarrow M\}.\downarrow$. That is, the sequence has form:

$$\{v \Rightarrow a\}.\downarrow; \ldots; \lambda\{v \Rightarrow x\}.\{v \Rightarrow M\}.\downarrow; \ldots; \lambda\{v \Rightarrow x\}.M$$

Since some transformations may be functionals, the recognition of such a structure is not immediate, but for instance m and n in the last version of Euclid's algorithm are scope-free variables.

The most interesting property of scope-free variables is that, like scoped variables, they have no effect outside of the sequence they are used in. That is, we can use the same label v outside of the sequence our scope-free variable is local to, without interference. A scope-free variable may even be used in a subsequence of another scope-free variable using the same label:

$$\{v \Rightarrow a\}.\downarrow; \ldots; \underline{\{v \Rightarrow b\}.\downarrow; \ldots; \lambda\{v \Rightarrow x\}.\downarrow}; \ldots; \lambda\{v \Rightarrow x\}.\downarrow$$

In the underlined subsequence the external scope-free variable is identifiable by the label $v + 1$ but comes back to v after.

Still, we must be careful that scope-free variables are not variables in the meaning of lambda-calculus: they appear on a completely different level, that of labels. Nor are they pervasive like would be references. We do not add side-effects to functions, but just provide some implicit way to manipulate a "stream" of arguments. That means that a function that is not called directly on this stream (through composition) will not access the scope-free variables it contains, and as such cannot have any imperative behavior with respect to this stream. This is this limitation which permits us to assimilate scope-free variables with arguments, and still be a conservative extension of lambda-calculus.

We give two examples of the use of scope-free variables. The first one is a simple encoding of an imperative programming language *a la* Algol. The second one shows how scope-free variable are stronger than scoped ones.

Here is the program and its translation.

```
begin
   var x=5, y=10;
   x := x+y;
   begin
      var x=3;
      y := x+y
   end
   x := x-y;
   return(x)
end
```

$$\{x \Rightarrow 5, y \Rightarrow 10\}.\downarrow;$$
$$\lambda\{x \Rightarrow x, y \Rightarrow y\}.\{x \Rightarrow x + y, y \Rightarrow y\}.\downarrow;$$
$$\{x \Rightarrow 3\}.\downarrow;$$
$$\lambda\{x \Rightarrow x, y \Rightarrow y\}.\{x \Rightarrow x, y \Rightarrow x + y\}.\downarrow;$$
$$\lambda\{x \Rightarrow x\}.\downarrow;$$
$$\lambda\{x \Rightarrow x, y \Rightarrow y\}.\{x \Rightarrow x - y, y \Rightarrow y\}.\downarrow;$$
$$\lambda\{x \Rightarrow x, y \Rightarrow y\}.x$$

We expect this program to evaluate to $5 + 10 - (3 + 10) = 2$.

$$\{x \Rightarrow 5, y \Rightarrow 10\}.\downarrow; \ldots$$
$$\{x \Rightarrow 5, y \Rightarrow 10\}.\lambda \cdots.\{x \Rightarrow x + y, y \Rightarrow y\}.\downarrow; \ldots$$
$$\{x \Rightarrow 15, y \Rightarrow 10\}.\{x \Rightarrow 3\}.\downarrow; \ldots$$
$$\{x1 \Rightarrow 3, x2 \Rightarrow 15, y \Rightarrow 10\}.\lambda \cdots.\{x \Rightarrow x, y \Rightarrow x + y\}.\downarrow; \ldots$$
$$\{x1 \Rightarrow 3, x2 \Rightarrow 15, y \Rightarrow 13\}.\lambda\{x \Rightarrow x\} \downarrow; \ldots$$
$$\{x \Rightarrow 15, y \Rightarrow 13\}.\lambda \cdots.\{x \Rightarrow x - y, y \Rightarrow y\}.\downarrow; \ldots$$
$$\{x \Rightarrow 2, y \Rightarrow 13\}.\lambda\{x \Rightarrow x, y \Rightarrow y\}.x$$
$$2$$

Note here that since we encode dynamic binding[7] for scope-free variables, we would get the same result even if the central part was a call to the same piece of code defined elsewere: with scope-free variable, even Basic's subprograms, which have no variable passing, would be a nice feature, since we can create a scope-free variable before the call to pass a parameter, and destroy it after.

The translation we propose here is a general one. By defining variables at the beginning of blocks and destroying them by abstractions at the end, we can translate any Algol-like program (with dynamic binding for mutable variables), even containing procedures and functions.

The above example still respects a scoping discipline: variables are created and destroyed in opposite order. To show the specificity of scope-free variables, we must disobey it.

Not respecting a scoping discipline seems quite dangerous for variables, and of little use in purely computing programs. However, if we think of IO's, then the situation is

---

[7] Dynamic binding is generally considered as bad, because destroying referential transparency. However, if we distinguish between static (defined only once, like λ-variables) and mutable variables, the notion of referential transparency for the last is not so clear. Since they are already not referentially transparent w.r.t. their values, the simpler modeling offered by dynamic binding can be seen as an advantage.

different. Consider a program with structure

$$\overline{A;\overline{B};C}$$

in which we want the console to be redirected in part $A;B$, and the screen to be changed in $B;C$. We suppose that we have mutable variables *con* and *scr* to indicate respectively which console and which screen should be used. Moreover we do not know which were the console and screen before entering A.

A dirty method is to use temporary variables c and s, to store the old values:

```
c:=con; con:=newc; A; s:=scr;
scr:=news; B; con:=c; C; scr:=s
```

The problem is that these temporary variables may be modified by error in A, B or C.

So a better solution is to use static variables, only set once:

```
let c = !con in
  con:=newc; A ;
  let s = !scr in
    scr:=newc; B ; con:=c; C ; scr:=s
end end
```

However, because of the scope discipline, c is still defined in C, whereas we do not need it anymore. We can see here an inconsistency between the scope of c, which is $A;B;C$, and its expected area of use, $A;B$.

We think that the scope-free variable way to do it is cleaner:

$$\{con \Rightarrow newc\}.\downarrow; A; \{scr \Rightarrow news\}.\downarrow; B;$$
$$\lambda\{con \Rightarrow c\}.\downarrow; C; \lambda\{scr \Rightarrow s\}.\downarrow$$

We didn't define any new variable, but did just temporarily hide the original value by the redirected one. And there are no "dangling" definitions (variables still defined out of their area of use).

3 Syntax of transformation calculus

In this section we define the untyped transformation calculus.

The definition is done in three steps. 1) We define streams[8], as a tool for defining the calculus. 2) We give a syntactic definition of terms in the transformation calculus, and add a structural equivalence on these terms[9]. 3) Then we define reduction rules for these equivalence classes.

---

[8] In [9], we give another definition of streams, permitting more commutations in the calculus, but here we use a simpler one.

[9] We could use all equivalences as directed reduction rules. This would result in a slightly more complicated system (*cf.* [1] for selective λ-calculus)

Definition 1 stream monoid. Let $\mathcal{L}$ be a set of names, $\mathcal{A}$ a set of values. $\mathcal{S}(\mathcal{A})$, the set of $\mathcal{L}$-streams on $\mathcal{A}$, is the set of the functions from $\mathcal{L}$ to the tuples of $\mathcal{A}$, such that only a finite number of tuples are not empty.

$$\mathcal{S}(\mathcal{A}) = \{s : \mathcal{L} \to \mathcal{A}^* \mid \sum_{l \in \mathcal{L}} |s(l)| \in \mathbb{N}\}$$

For a stream s, $\mathcal{D}_s = \{(l, n) \mid 1 \leq n \leq |s(l)|\}$ is the set of its defined positions. We write s as $\{l_1 n_1 \Rightarrow s(l_1)_{n_1}, \ldots, l_m n_m \Rightarrow s(l_m)_{n_m}\}$ by enumerating all its *defined positions*.

Concatenation on streams, the monoidal operation, is the name-wise concatenation of tuples:

$$(r \cdot s)(l) = r(l) \cdot s(l)$$

Notations In the following definitions we will use the abbreviations $A \mathbin{\varnothing} B$ for $A \cap B = \emptyset$, $FV(M)$ for the free variables of M, and $V(R)$ for the values contained in the stream R.

Definition 2. Terms of the transformation calculus, or Λ_T, are those generated by M in the following grammar, where variables should be distinct in abstractions. Composition has lower priority than dots.

$$
\begin{aligned}
M ::= \ &x &&\text{variable} \\
\mid \ &\downarrow &&\text{transformation constructor} \\
\mid \ &\lambda \mathcal{S}(x).M &&\text{abstraction} \\
\mid \ &\mathcal{S}(M).M &&\text{application} \\
\mid \ &M ; M &&\text{composition}
\end{aligned}
$$

They are considered modulo $\equiv$, the minimal congruence defined by the closure of the following equalities.

$$
\begin{aligned}
S.R.M &\equiv (R \cdot S).M \\
\lambda R.\lambda S.M &\equiv_\lambda \lambda(S \cdot R).M & V(R) \mathbin{\varnothing} V(S) \\
R.\lambda S.M &\equiv_\lambda \lambda S.R.M & FV(R) \mathbin{\varnothing} V(S), \mathcal{D}_R \mathbin{\varnothing} \mathcal{D}_S
\end{aligned}
$$

$$
\begin{aligned}
(R.M_1); M_2 &\equiv_; R.(M_1; M_2) \\
(\lambda R.M_1); M_2 &\equiv_{\lambda;} \lambda R.(M_1; M_2) & V(R) \mathbin{\varnothing} FV(M_2) \\
(M_1; M_2); M_3 &\equiv_; M_1; (M_2; M_3)
\end{aligned}
$$

Equalities $\equiv_.$ and $\equiv_\lambda$ are derived from the monoidal structure. $\equiv_{.;}$, $\equiv_{\lambda;}$ and $\equiv_;$ are intuitive. Equality $\equiv_\lambda$ is the "symmetrical" of β-reduction: when they have no common names, application and abstraction may switch, and permit earlier reductions.

Substitutions are done in the same way as for lambda-calculus, composition not interacting with variable binding. Terms will always be considered modulo α-conversion. That is $\lambda\{l \Rightarrow x\}.M \equiv \lambda\{l \Rightarrow y\}.[y/x]M$ when $y \notin FV(M)$.

Definition 3. "→" is defined on transformation calculus terms by β-reduction and $\downarrow$-elimination[10].

$$\{l \Rightarrow N\}.\lambda\{l \Rightarrow x\}.M \rightarrow_\beta [N/x].M$$
$$\downarrow; M \qquad \rightarrow_\downarrow \qquad M$$

$\xrightarrow{*}$ is the reflexive and transitive closure of →.

Selective λ-terms and β-reduction define the selective λ-calculus.

Theorem 4. *Transformation calculus is confluent.*

The proofs are given in [9]. Confluence of transformation calculus is obtained from selective λ-calculus through a translation into it.

4 Scope-free variable encoding

We cannot expect to give a precise definition of *scope-free variable* in the transformation calculus, where it is only encoded. It appears as an intuitive notion of a variable whose locality is not syntactical but operational. We will define it outside of the calculus.

For this we use a framework in which a program is a sequence of operations. Operations can themselves contain programs, but these are independent, and may not have side-effects on the external sequence.

Definition 5. A *scope-free variable* is some way to create, modify and destroy a value such that:

1. these operations may appear in different syntactic entities, which may be used independently.
2. a *closed use* of this variable is obtained when a creator, some modifiers, and a destructor result in a *modification sequence*.
3. its closed use in a modification sequence has no side-effect outside it.

A consequence of this definition is the hiding property we insisted on. The same variable may have several independent closed uses, with modification sequences included in one another, and there is no problem as long as we do not try to modify the value from one use inside another's modification sequence.

As we introduced in Section 2, elementary creators, modifiers and destructors in transformation calculus are respectively $\{v \Rightarrow M\}.\downarrow$, $\lambda\{v \Rightarrow x\}.\{v \Rightarrow M\}.\downarrow$ and $\lambda\{v \Rightarrow x\}.\downarrow$. But we can think of more complex ones, acting simultaneously on multiple variables, taking arguments, or returning results.

To ensure that we have a correct scope-free variable encoding here, we must verify the third point of the definition, which says that it has no effect outside the sequence it is used in.

Proposition 6. *The scope-free variable encoding into the transformation calculus ensures locality to the modification sequence.*

---

[10] We chose to make $\downarrow$-elimination a reduction rule rather than a structural equality because it reduces the size of terms, while the structural equalities of Definition 2 do not change it.

As we have seen, thanks to this property, scope-free variables are not only more flexible than classical scoped variables, but can replace them in most of their uses. Particularly, in functional language they can replace "disciplined" references (which do not go out of their scope), without the need of a specific evaluation strategy. Their only limitation is that —in the transformation calculus— one cannot export them like references, since they are linked to an explicit name. However, this is a limitation of the label system we use, and not of scope-free variable in themselves: one can add a syntactical scope to scope-free variables [10]. The real point about them is that the use of a (now scoped) scope-free variable is not restricted by that syntactical scope[11] (which is only a problem of naming), like with the stack discipline, but by its *life area*, or modification sequence (its real operational scope).

5 Simply typed transformation calculus

To obtain a simply typed form of transformation calculus, we annotate variables with some type in abstractions, just the same way it is done in lambda calculus. But first we must define what are these types.

The two most important novelties are that, first, stream types are introduced, and second, that function type are not from any type to any other, but only from stream types to stream or base types. This last particularity "flattens" types, but still contains as a subset all simple types of lambda-calculus.

Definition 7. Simple types in the transformation calculus are generated by t in the following grammar.

$$
\begin{aligned}
u &::= u_1 \mid \dots \text{ base types} \\
r &::= \mathcal{S}(t) \quad \text{ stream types} \\
w &::= u \mid r \quad \text{ return types} \\
t &::= r \to w \text{ types}
\end{aligned}
$$

The same label may not appear more than once in the same stream type; stream types are equal up to different orders, and $(\{\} \to \tau) = \tau$, for short.

Definition 8. A term in the simply typed transformation calculus is constructed according to the following syntax.

$$
M ::= x \mid \downarrow \mid \lambda \mathcal{S}(x{:}t).M \mid \mathcal{S}(M).M \mid M; M
$$

Finally the relation between terms and types is given in the following definition.

Definition 9. A *type judgement*, written $\Gamma \vdash M : \tau$, expresses that the term M has type τ in the context Γ. Induction rules for type judgements are given in figure 1.

Rules (I,II,III) are the traditional ones for typed lambda calculus, simply extended to streams. We can go back to it by limiting labels in streams to sequences of integers starting from 1 (that is, in the above rules, having only $l = \epsilon 1$).

---

[11] This is true with references too, but their operational scope is only defined by garbage collection.

$$\Gamma[x \mapsto \tau] \vdash x : \tau \tag{I}$$

$$\frac{\Gamma[x \mapsto \theta] \vdash M : \tau \to w}{\Gamma \vdash \lambda\{l \Rightarrow x{:}\theta\}.M : (\{l \Rightarrow \theta\} \cdot \tau) \to w} \tag{II}$$

$$\frac{\Gamma \vdash M : (\{l \Rightarrow \theta\} \cdot \tau) \to w \quad \Gamma \vdash N : \theta}{\Gamma \vdash \{l \Rightarrow N\}.M : \tau \to w} \tag{III}$$

$$\Gamma \vdash \downarrow : \{\} \tag{IV}$$

$$\frac{\Gamma \vdash M : \tau_1 \to \tau_2 \quad \Gamma \vdash N : \tau_2 \to w}{\Gamma \vdash M; N : \tau_1 \to w} \tag{V}$$

$$\frac{\Gamma \vdash M : \tau_1 \to \tau_2}{\Gamma \vdash M : (\tau_1 \cdot \tau) \to (\tau_2 \cdot \tau)} \tag{VI}$$

Fig. 1. Typing rules for simply typed transformation calculus

Rule (IV) types the constant $\downarrow$. However it will most often need the cooperation of rule (VI), transformation subtyping, which expresses that any transformation may be applied to labels it is not concerned with: they will simply be rejected to the result. For instance, it gives to $\downarrow$ any symmetrical type $(\tau \to \tau)$. Rule (V) types composition: M is applied to the result stream of N, and re-abstracted by its abstraction part. Here again, we need the collaboration of rule (VI) to extend the types of either M or N.

Proposition 10 subject reduction. *If $\Gamma \vdash M : \tau$ and $M \to N$ then $\Gamma \vdash N : \tau$.*

Proposition 11 strong normalization. *If $\Gamma \vdash M : \tau$ then there is no infinite reduction sequence starting from M.*

This last property is interesting, since it is general belief that introducing mutables suppresses strong normalization: we keep it here, because all values used by a term appear in its type.

6 Related works

Since transformation calculus only happens to be able to represent state, its origin is not to be found in the field of semantics of stateful languages. It is rather based on two independent threads of work. The first one is the Categorical Combinatory Logic [6], in which composition and currying play a central role. The direction seems opposed: one encodes lambda-calculus into CCL (or its abstract machine version, the CAM [5]), while transformation calculus extends lambda-calculus. But the intuition that algorithmicity can be found in the structures of the lambda-calculus itself is the same.

The second one is process calculi. Their use of names for communication is similar to the principle of the transformation calculus. In [2], Boudol proposes the γ-calculus. The base is lambda-calculus, but applications express emissions of messages and abstractions their reception, while multiple terms can be evaluated simultaneously. Milner's π-calculus [17] proceeds alike, and by labeling with names applications and abstractions, it allows the use of multiple channels. The fundamental difference with our calculus is that non-determinism of the receptor of a message make these calculi divergent, while our terms are syntactically sequenced in order to keep determinism.

After these somewhat different directions, our claims makes necessary to look at the larger literature concerning modeling of mutables in Algol, Lisp, and modern functional programming languages. Algol is the closest to our system, since scope-free variables cannot be used out of their *life area*, like with Algol's stack discipline, where a variable cannot be exported out of its scope.

Landin first proposed an encoding of Algol 60 into the lambda calculus [13]. But the problem was not solved: "The semantics of *applicative expressions* can be specified formally without the recourse to a machine. [. . .] With *imperative applicative expressions* on the other hand it appears impossible to avoid specifying semantics in terms of a machine".

After a number of attempts, including marked stores [16] and subtler models [15], a denotational semantics was finally obtained with Oles and Reynolds category-theoretic models [19, 21]. The essential idea is to define blocks as functions that can be applied to a range of states with various shapes, but do not change their shapes. However, inside the block, state is temporarily extended with local variables. Thus, they do not appear in its meaning. Our approach shares a lot with this view, since we syntactically "expand" and "shrink" our state when we create and delete a scope-free variable.

If we go out of the Algol tradition, we can forget about the stack discipline. As a result, most systems give a formalization of references. So does the λ_v-S-calculus [7] for Scheme, and a call-by-value reduction strategy. With effect inference [14], restrictions on the reduction strategy can be reduced, and, for instance, parallelism can be introduced.

Still, we feel more concerned by systems going the other way, starting without a specific reduction strategy. There are a number of them, which enforce single-threadedness of variables by various typing disciplines [12, 20, 22, 23, 24]. We can see an intuitive relation between the way scope-free variables are used and linear types, but still we are not relying to typing for single-threadedness.

We actually do it in a syntactical way. In that we are very close to λ_{var} [18, 4]. In fact, even the structures of the calculus have similarities: like us, they use the linear structure of spines to ensure single-threadedness. They have rules to propagate the values of mutable variables along the spine of a term, like does our structural equivalences for labeled arguments, and their **return**-elimination rule $((\textbf{return } N) \triangleright \lambda x.M \rightarrow (\lambda x.M)N$, *cf.* [18]) can be seen as a variant of $\downarrow$-elimination ($\downarrow; M \rightarrow M$) including value-passing. The essential difference is that, since we use the same mechanism for scope-free variables and value-passing, we obtain a more unified calculus. In particular, the fact they are encoding references means that they must do some kind of garbage collection (their **pure** construct) to convert a value obtained using mutable variables into a purely functional one. In the transformation calculus, since we explicitly delete variables, we do not need such an "impure" purifier.

7 Conclusion

We proposed the transformation calculus as an extension of currying in the lambda-calculus permitting both functional and imperative encoding of algorithms.

We think this gives interesting answers to the two sides of the relation function/algorithm: as a demonstration of the relation between lambda-calculus and algorithms, and as a basis for functional languages handling states and sequentiality problems.

Still there are many topics left to explore. Typing is one of them. If we are to write program in this calculus, it is even unavoidable, since we must be able to verify that scope-free variables are correctly used. We presented here a simply typed system. We propose in [8] a polymorphic version of it, extending that for selective λ-calculus [11]. This could be completed by the introduction of linear types [24]: in the transformation calculus, variables are single threaded (operations on them are sequenced), but there is no restriction to their duplication. This is particularly a problem with IO: we can semantically create fictitious worlds, but there is no way to implement them. Moreover, using linear types within the transformations calculus relieves the programmer of most of the grudge of the linear style, since sending back an argument is easy.

Compilation, which is easy with stores, is complex here. The final goal would be to eliminate label information, but the possibility to compose a term with a variable makes impossible to reach it in the general case. There is no problem if we use non-polymorphic typing, since everything can be compiled into tuples, but the polymorphic case is still open.

One strength of the transformation calculus is its system of labels. We may be interested in extending it. For instance, the possibility to generate new label names will give unique identifiers for scope-free variable, and avoid the hiding of a label in those subsequences which create new variables on this label. A more structured label space even enables the use of object-based techniques, and solve the restrictions of dynamic binding.

Last, the similarities between this calculus and process calculi suggest that it might be used to express some forms of parallelism. If one looks at the way data flows in our terms, lots of reminiscences of the dataflow model may be seen. A topic like compilation of the calculus into this model looks interesting.

References

1. Hassan Ait-Kaci and Jacques Garrigue. Label-selective λ-calculus: Syntax and confluence. In *Proc. of the Conference on Foundations of Software Technology and Theoretical Computer Science*, pages 24–40, Bombay, India, 1993. Springer-Verlag LNCS 761.
2. Gerard Boudol. Towards a lambda-calculus for concurrent and communicating systems. In *Proceedings of TAPSOFT '89*, pages 149–161, Berlin, Germany, 1989. Springer-Verlag, LNCS 351.
3. Kim Bruce, Roberto Di Cosmo, and Giuseppe Longo. Provable isomorphisms of types. Technical Report LIENS-90-14, LIENS, July 1990.
4. Kung Chen and Martin Odersky. A type system for a lambda calculus with assignments. In *Proc. of the International Conference on Theoretical Aspects of Computer Software*, pages 347–363, 1994.

5. Guy Cousineau, Pierre-Louis Curien, and Michel Mauny. The categorical abstract machine. *Science of Computer Programming*, 8, 1987.
6. Pierre-Louis Curien. *Categorical Combinators, Sequential Algorithms, and Functional Programming*. Progress in Theoretical Computer Science. Birkhauser, Boston, 1993. First edition by Pitman, 1986.
7. M. Felleisen and D.P. Friedman. A syntactic theory of sequential state. *Theoretical Computer Science*, 69:243–287, 1989.
8. Jacques Garrigue. *Label-Selective Lambda-Calculi and Transformation Calculi*. PhD thesis, University of Tokyo, Department of Information Science, March 1995.
9. Jacques Garrigue. The transformation calculus (revised version). Technical report, Kyoto University Research Institute for Mathematical Sciences, Kyoto 606-01, Japan, 1995. *cf.* http://wwwfun.kurims.kyoto-u.ac.jp/~garrigue/papers/.
10. Jacques Garrigue. Dynamic binding and lexical binding in a transformation calculus. In *Proc. of the Fuji International Workshop on Functional and Logic Programming*. World Scientific, Singapore, to appear.
11. Jacques Garrigue and Hassan Ait-Kaci. The typed polymorphic label-selective λ-calculus. In *Proc. ACM Symposium on Principles of Programming Languages*, pages 35–47, 1994.
12. J.C. Guzman and P. Hudak. Single threaded polymorphic calculus. In *Proc. IEEE Symposium on Logic in Computer Science*, pages 333–343, 1990.
13. P. J. Landin. A correspondence between ALGOL 60 and Church's lambda notation. *Communications of the ACM*, 8(2-3):89–101 and 158–165, February 1965.
14. John M. Lucassen and David K. Gifford. Polymorphic effect systems. In *Proc. ACM Symposium on Principles of Programming Languages*, pages 47–57, San Diego, California, 1988.
15. Albert R. Meyer and Kurt Sieber. Toward fully abstract semantics for local variables. In *Proc. ACM Symposium on Principles of Programming Languages*, pages 191–203, 1988.
16. R. Milne and C. Strachey. *A Theory of Programming Language Semantics*. Chapman and Hall, 1976.
17. Robin Milner. The polyadic π-calculus: A tutorial. In *Logic and Algebra of Specification*, pages 203–246. NATO ASI Series, Springer Verlag, 1992.
18. Martin Odersky, Dan Rabin, and Paul Hudak. Call by name, assignment, and the lambda calculus. In *Proc. ACM Symposium on Principles of Programming Languages*, pages 43–56, 1993.
19. F.J. Oles. Type algebras, functor categories, and block strcutures. In N. Nivat and J.C. Reynolds, editors, *Algebraic Methods in Semantics*, chapter 15, pages 543–573. Cambridge University Press, 1985.
20. Simon L. Peyton Jones and Philip Wadler. Imperative functional programming. In *Proc. ACM Symposium on Principles of Programming Languages*, pages 71–84, 1993.
21. John C. Reynolds. The essence of ALGOL. In de Bakker and van Vliet, editors, *Proc. of the International Symposium on Algorithmic Languages*, pages 345–372. North Holland, 1981.
22. Vipin Swarup, Uday S. Reddy, and Evan Ireland. Assignments for applicative languages. In John Hugues, editor, *Proc. ACM Symposium on Functional Programming and Computer Architectures*, pages 192–214. Springer Verlag, 1991. LNCS 523.
23. Philip Wadler. Comprehending monads. In *Proc. ACM Conference on LISP and Functional Programming*, pages 61–78, 1990.
24. Philip Wadler. Linear types can change the world! In M. Broy and C. B. Jones, editors, *Programming Concepts and Methods*. North Holland, 1990.

Equational Axiomatization of Bicoercibility for Polymorphic Types

Jerzy Tiuryn*

Institute of Informatics

Warsaw University

Banacha 2, 02-097 Warsaw, POLAND

(tiuryn@mimuw.edu.pl)

Abstract

Two polymorphic types σ and τ are said to be bicoercible if there is a coercion from σ to τ and conversely. We give a complete equational axiomatization of bicoercible types and prove that the relation of bicoercibility is decidable.

1   Introduction

The notion of a *subtype* of a type plays an important role in typed programming languages and it has been a subject of an intensive research recently [Ben93, BCGS91, CMSS94, CGL92, CP94, LMS95, Mit90, Tiu92, TW93]. There is a number of fundamental open problems which have to be solved in order to gain a better understanding of the notion of a subtype. One of such problems is the question of decidability of the relation of subtyping for second-order polymorphic types. This relation has been axiomatized by John Mitchell [Mit90]. The fact that a type σ is a subtype of a type τ can be established by finding a *coercion* from σ to τ. Coercions are denoted by terms, typable in system F of polymorphic second-order lambda calculus. The important feature of coercions is that after erasing all the type information they are $\beta\eta$ equal to the identity. Despite this simple computational behaviour it is not known how to decide for two given polymorphic types σ and τ whether there is a coercion from σ to τ. Part of the difficulty in proving this problem decidable is the axiom which says that the type $\forall X.\ \sigma$ is a subtype of each of its instances. In symbols:

*This work is partly supported by NSF Grant CCR-9113196, KBN Grant 2 P301 031 06 and by ESPRIT BRA7232 GENTZEN.

$$\forall X. \ \sigma \leq [\tau/X]\sigma$$

holds for every type τ. The above subtype relation is established by the coercion $(\lambda x : \forall X. \ \sigma \ x[\tau]) : \forall X. \ \sigma \rightarrow [\tau/X]\sigma$.

In the present paper we are concerned with the notion of a bicoercion. Call σ and τ *bicoercible* iff there are coercions from σ to τ and conversely from τ to σ. The main result of the paper is a complete equational axiomatization of the relation of bicoercibility. It turns out that this equational theory is decidable, implying therefore decidability of the problem of bicoercibility for polymorphic types.

The relation of bicoercibility for polymorphic types is a special kind of isomorphism. This observation follows from the *coherence property* (see [LMS95]), *i.e.* the property which says that if M and N are coercions from σ to τ, then they are provably equal in a certain extension of the system F. Bicoercibility of polymorphic types neither contains nor is contained in the notion of a provable isomorphism in system F. The reader is refered to [Dic95] for characterization of the latter notion. For example, $\forall Y. \ \forall X. \ X$ and $\forall X. \ X$ are bicoercible but they are not provably isomorphic in F. On the other hand, $X \rightarrow (Y \rightarrow Z)$ and $Y \rightarrow (X \rightarrow Z)$ are provably isomorphic in F but they are not bicoercible.

The paper is organized as follows. In Section 2 we recall the Mitchell's system of sybtyping for polymorphic types. We also define there a rewrite system which is very much related to the Mitchell's proof system. Rewriting σ into τ results in establishing that σ is a subtype of τ. In Section 3 we introduce the equational system for deriving bicoercions. This system is decidable. Sections 4 and 5 are devoted to the proof of its completeness. Section 4 collects some technical results which show some kind of a control which we have on moving quantifiers in types when rewriting them. This part of the proof deals with a more general situation of arbitrary coercions, rather than bicoercions. The proof of completeness is concluded in Section 5 where we characterize the so called *reversible rewrite steps* — steps which are used when deriving a bicoercion.

2 Subtyping for Polymorphic Types

First we present the Mitchell's system of subtyping for polymorphic types (see [Mit90]). The system derives formulas of the form $\sigma \leq \tau$, where σ and τ are polymorphic types.

Axioms:

(refl) $\sigma \leq \sigma$

(inst) $\forall X. \sigma \leq [\rho/X]\sigma$

(dummy) $\sigma \leq \forall X. \sigma$, X doesn't occur free in σ.

(distr) $\forall X. (\sigma \rightarrow \tau) \leq \forall X. \sigma \rightarrow \forall X. \tau$

Rules:

$$(\rightarrow) \quad \frac{\sigma' \leq \sigma \qquad \tau \leq \tau'}{\sigma \rightarrow \tau \leq \sigma' \rightarrow \tau'} \qquad\qquad (\forall) \quad \frac{\sigma \leq \tau}{\forall X. \sigma \leq \forall X. \tau}$$

$$(\text{trans}) \quad \frac{\sigma \leq \rho \qquad \rho \leq \tau}{\sigma \leq \tau}$$

We write $\vdash \sigma \leq \tau$ to indicate that $\sigma \leq \tau$ is derivable in the above system.

In the rest of the paper we will view polymorphic types as binary trees such that:

- inner nodes are labelled $\rightarrow$;
- leaves are labelled with type variables $X, Y, \ldots$;
- every node is labelled by a finite sequence of quantified variables $\forall X_1 \forall X_2 \ldots$.

A node in a type σ will be indentified with a path (a sequence of 0's and 1's) leading from the root to that node. A node w is said *positive* if the number of 0's in w is even. Otherwise w is called *negative*.

We define a rewrite system for polymorphic types, closely related to the above formal system. We say that σ *rewrites* into τ in one step, denoted $\sigma \sqsubset \tau$, if τ is obtained from σ by applying one of the following rewrite rules:

For a positive node perform one of the following steps:

(*p-inst*) $\forall X. \sigma \sqsubset [\rho/X]\sigma$

(*p-dummy*) $\sigma \sqsubset \forall X. \sigma$, X doesn't occur free in σ.

(*p-distr*) $\forall X. (\sigma \to \tau) \sqsubseteq \forall X. \sigma \to \forall X. \tau$

For a negative node perform one of the following steps:

(*n-inst*) $[\rho/X]\sigma \sqsubseteq \forall X. \sigma$

(*n-dummy*) $\forall X. \sigma \sqsubseteq \sigma,$ X doesn't occur free in σ.

(*n-distr*) $\forall X. \sigma \to \forall Y. \tau \sqsubseteq \forall Z. ([Z/X]\sigma \to [Z/Y]\tau),$ Z is a new variable.

When performing steps (*p-inst*) and (*n-inst*) a care has to be taken in order to avoid quantifier clashes — rename bound variables, when necessery, to prevent such clashes.

3   Bicoercion

Two types σ and τ are said to be *bicoercible* if $\vdash \sigma \leq \tau$ and $\vdash \tau \leq \sigma$ holds. We start with some examples of bicoercible types.

Lemma 1 *Types* $\forall X \forall Y. \sigma$ *and* $\forall Y \forall X. \sigma$ *are bicoercible.*

Proof: We prove the $\leq$ inequality.

$$\dfrac{\forall X \forall Y. \sigma \leq \forall Y \forall X \forall Y. \sigma \qquad \dfrac{\dfrac{\dfrac{\forall Y. \sigma \leq \sigma}{\forall X \forall Y. \sigma \leq \forall X. \sigma}\,(\forall)}{\forall Y \forall X \forall Y. \sigma \leq \forall Y \forall X. \sigma}\,(\forall)}{}}{\forall X \forall Y. \sigma \leq \forall Y \forall X. \sigma}\,(\text{trans})$$

The opposite inequality follows by symmetry of assumptions.   ∎

Lemma 2 *If every occurence of* X *in* σ *is positive, then* $\forall X. \sigma$ *and* $[(\forall X. X)/X]\sigma$ *are bicoercible.*

Proof: The $\leq$ inequality is just the (inst) axiom. For the proof of $\geq$ we first show the following fact which is very easy to establish by induction on σ and τ (we omit the proof). If X occurs only positively in σ and only nagatively in τ, then

$$[(\forall X.\ X)/X]\sigma \leq \sigma \quad \text{and} \quad \tau \leq [(\forall X.\ X)/X]\tau$$

Now, having the above, we proceed as follows.

$$\frac{[(\forall X.\ X)/X]\sigma \leq \forall X.\ [(\forall X.\ X)/X]\sigma \quad \dfrac{\dfrac{[(\forall X.\ X)/X]\sigma \leq \sigma}{\forall X.\ [(\forall X.\ X)/X]\sigma \leq \forall X.\ \sigma}\ (\forall)}{\ }(\text{trans})}{[(\forall X.\ X)/X]\sigma \leq \forall X.\ \sigma}$$

∎

Lemma 3 *If X doesn't occur free in σ, then $\forall X.\ (\sigma \to \tau)$ and $\sigma \to \forall X.\ \tau$ are bicoercible.*

Proof: For $\leq$ we have the following derivation.

$$\frac{\forall X.\ (\sigma \to \tau) \leq \forall X.\ \sigma \to \forall X.\ \tau \quad \dfrac{\dfrac{\sigma \leq \forall X.\ \sigma \quad \forall X.\ \tau \leq \forall X.\ \tau}{\forall X.\ \sigma \to \forall X.\ \tau \leq \sigma \to \forall X.\ \tau}\ (\to)}{\ }(\text{trans})}{\forall X.\ (\sigma \to \tau) \leq \sigma \to \forall X.\ \tau}$$

For the proof of $\geq$ we have the following derivation.

$$\frac{\sigma \to \forall X.\ \tau \leq \forall X.\ (\sigma \to \forall X.\ \tau) \quad \dfrac{\dfrac{\dfrac{\forall X.\ \tau \leq \tau \quad \sigma \leq \sigma}{\sigma \to \forall X.\ \tau \leq \sigma \to \tau}\ (\to)}{\forall X.\ (\sigma \to \forall X.\ \tau) \leq \forall X.\ (\sigma \to \tau)}\ (\forall)}{\ }(\text{trans})}{\sigma \to \forall X.\ \tau \leq \forall X.\ (\sigma \to \tau)}$$

∎

3.1   Proof System

The system given below is for deriving expressions of the form $\sigma \equiv \tau$, where σ and τ are polymorphic types.

Axioms:

(A1) $\sigma \equiv \sigma$

(A2) $\forall X \forall Y.\ \sigma \equiv \forall Y \forall X.\ \sigma$

(A3) $\forall X.\ \sigma \equiv [(\forall X.\ X)/X]\sigma$, $X \neq \sigma$ and all occurences of X in σ are positive.[1]

(A4) $\forall X.\ (\sigma \to \tau) \equiv \sigma \to \forall X.\ \tau$, X doesn't occur free in σ, and it has a negative occurence in τ.[2]

Rules:

$$\text{(arrow)}\quad \frac{\sigma \equiv \sigma' \qquad \tau \equiv \tau'}{\sigma \to \sigma' \equiv \tau \to \tau'} \qquad\qquad \text{(quant)}\quad \frac{\sigma \equiv \sigma'}{\forall X.\ \sigma \equiv \forall X.\ \sigma'}$$

$$\text{(trans)}\quad \frac{\sigma \equiv \rho \qquad \rho \equiv \tau}{\sigma \equiv \tau} \qquad\qquad \text{(symm)}\quad \frac{\sigma \equiv \tau}{\tau \equiv \sigma}$$

We write $\vdash \sigma \equiv \tau$ to indicate that there is a derivation of $\sigma \equiv \tau$ in the above proof system.

The main result of this paper is the following theorem. It shows that the above proof system captures bicoercibility.

Theorem 4 σ and τ are biceorcible iff $\vdash \sigma \equiv \tau$ holds.

Let us observe that the implication ($\Leftarrow$) is *soundness* of our proof system, while the implication ($\Rightarrow$) expresses its *completeness*. Proof of the above theorem will be given at the end of Section 5. We conclude this section with a corollary of the above result, and with a technical lemma which will be used later.

Corollary 5 *It is decidable for two given polymorphic types whether they are bicoercible.*

Proof: Let us identify types which differ only with respect to names of bound variables (α-conversion) and with respect to the order of quantifiers. Clearly, it is decidable for given two polymorphic types, whether they are identical in the above sense.

---

[1] If $X = \sigma$, then this case is handled by (A1).

[2] If X occurs in τ only positively, then this case is handled by (A3). We have added this constraint to (A4) in order to simplify the proof of Corollary 5 and Lemma 6.

Consider a rewrite system consiting of the rewrite rules being axioms (A3) and (A4), ordered from left to right. This system is clearly confluent and strongly terminating. Confluence of the system easily follows from the fact that the rules are non-overlaping. Strong termination follows as a result of introducing a measure on types, which to a given type σ assigns the sum, over all occurences of quantifiers in σ, a natural number equal 1 plus the size of the subtype below that occurence. Each of the rules decreases this measure.

Given two types σ and τ. Decidability of $\vdash \sigma \equiv \tau$ follows from the observation that $\vdash \sigma \equiv \tau$ holds iff σ and τ have the same (up to the above identification) normal form in the above rewrite system. The implication ($\Rightarrow$) is proved by induction on the length of derivation of $\vdash \sigma \equiv \tau$. The implication ($\Leftarrow$) is proved by induction on the sum of numbers of rewrite steps to normal form for σ and τ. Both proofs are completely routine and we leave them for the reader.

Thus, by Theorem 4, bicoercibility is decidable. ∎

Lemma 6 *If $\vdash \sigma \to \sigma' \equiv \tau \to \tau'$ holds, then both $\vdash \sigma \equiv \tau$ and $\vdash \sigma' \equiv \tau'$ hold.*

Proof: We prove the lemma by induction on the number of steps in derivation of $\sigma \to \sigma' \equiv \tau \to \tau'$. The only slightly non-trivial step is when the last rule used in the derivation is (trans). Then, for a certain type ρ we have

$$\vdash \sigma \to \sigma' \equiv \rho \quad \text{and} \quad \vdash \rho \equiv \tau \to \tau'$$

If ρ is of the form $\rho' \to \rho''$, then we are done (apply the induction assumption). Otherwise ρ starts with a quantifier.[3] Hence $\rho \equiv \tau \to \tau'$ and $\rho \equiv \sigma \to \sigma'$ are instances of an axiom (A3) or (A4). Since these axioms are applicable in disjoint situations, it follows that both $\rho \equiv \tau \to \tau'$ and $\rho \equiv \sigma \to \sigma'$ must be instances of the same axiom. Hence $\sigma \to \sigma' = \tau \to \tau'$, i.e. $\sigma = \tau$ and $\sigma' = \tau'$, and we are done. ∎

4 Auxiliary Results

We first introduce the concept of a *marking*. Let σ be a type and let X be a variable which occurs in σ. This occurence is marked 0 if it is free, it is marked -1 if it is bound by a quantifier which is at a negative node, and marked $+1$ if it is bound by a quantifier which is at a positive node.

The following result expresses the property that -1 marks are "easy to create" but impossible to "get rid off".

---

[3] Obviously it cannot be a type variable.

Lemma 7 *Let $\sigma \sqsubseteq^* \tau$ and let w be a leaf in σ. Assume that one of the following three conditions holds.*

(i) *w doesn't belong to τ.*

(ii) *w is marked in σ with -1.*

(iii) *w is marked in σ with 0, and it is a leaf in τ marked $+1$ or -1.*

Then there is $u \leq w$ such that u is a leaf in τ marked -1.

Proof: We prove the lemma by induction on the number of steps in deriving $\sigma \sqsubseteq^* \tau$. Let's consider first a one step $\sigma \sqsubseteq \tau$. If w doesn't belong to τ, then this must have resulted from an application of $(n\text{-}inst)$ rule at node $w' \leq w$. Then w' is a negative node and this step introduces a quantifier at this node, which binds a variable at a leaf $u \leq w$. Thus u is marked -1. Assume now that w belongs to τ. If w is marked in σ with 0 or -1, then w must be a leaf in τ (since no step substituting a type for w in σ can be performed in this case). It follows that if w is marked -1 in σ, then the same mark stays at w in τ. Now, if w is marked 0 in σ but w is bound in τ, then it means that this step must have introduced a quantifier at an existing node. This is only possible at $(n\text{-}inst)$ step and this means that w is marked -1 in τ. This completes the base step of induction.

Let's assume now that $\sigma \sqsubseteq \rho \sqsubseteq^* \tau$. If w doesn't belong to ρ, then by the previous analysis, it follows that there is $w' \leq w$ such that w' is a leaf in ρ marked -1. Thus, by induction assumption applied to $\rho \sqsubseteq^* \tau$ and the node w' we get the conclusion.

Assume now that w belongs to ρ. If w doesn't belong to τ, then take any w' which extends w and is a leaf in ρ. Clearly w' doesn't belong to τ either. By induction assumption applied to $\rho \sqsubseteq^* \tau$ and the node w' we get the conclusion (since w doesn't belong to τ, it follows that the leaf u in τ such that $u \leq w'$ must satisfy $u \leq w$).

To complete the proof let's assume that w is marked in σ with 0 or -1 and w belongs to ρ. It follows from the analysis of the base case that then w is a leaf in ρ marked 0 or -1. Thus, by induction assumption applied to $\rho \sqsubseteq^* \tau$ and the node w we get the conclusion. This completes the proof of lemma. ∎

The next result shows that moving a quantifier downwards from a positive node to a positive node can be done only in a very limited way.

Lemma 8 *Let $\sigma \sqsubseteq^* \tau$ and let X be a variable which occurs in both: σ and τ at the same node, say u. Let us assume that this occurence of X in σ is bound by* *a quantifier at a positive node w_σ and in τ at a positive node w_τ. If $w_\sigma \leq w_\tau$, then $w_\tau = w_\sigma 1^m$, for some $m \geq 0$.*

Proof: We prove the lemma by induction on the number of steps in getting from σ to τ. If $\sigma \sqsubset \tau$ and the quantifier moves down from a positive node to a positive node then it is only possible by performing (*p-distr*) step. Then clearly we obtain $w_\tau = w_\sigma 1$, as required.

Now, let us assume that $\sigma \sqsubset \rho \sqsubset^* \tau$, and let u be the node at which the variable X occurs. If u is not a leaf in ρ, then by Lemma 7 (i) and (ii) the node u in τ must have been marked -1, contradiction. Applying Lemma 7 again we conclude that u must be marked in ρ with $+1$. Let w_ρ be the node in ρ at which the binding quantifier for u is located. It follows from the analysis of the base case that either $w_\rho = w_\sigma$, or $w_\rho = w_\sigma 1$. If $w_\sigma \leq w_\tau$ and former possiblity holds, then, by induction assumption, we get that $w_\tau = w_\rho 1^m$, for some $m \geq 0$. Hence $w_\tau = w_\sigma 1^m$. If $w_\sigma \leq w_\tau$ and latter possibility holds, then either $w_\tau = w_\sigma$ and we are done, or else $w_\rho \leq w_\sigma$ (since $w_\rho \leq u$ and $w_\tau \leq u$), and again by induction assumption we conclude that $w_\tau = w_\rho 1^m$, for some $m \geq 0$, i.e. $w_\tau = w_\sigma 1^{m+1}$, for some $m \geq 0$. This completes the proof. ∎

Let u_1 and u_2 be leaves in a type σ. A positive occurence of a quantifier $\forall X$ in σ which binds a variable at nodes u_1 and u_2 is said to be *splitable* at nodes u_1 and u_2, if there exists a type τ and two positive nodes w_1 and w_2 in τ such that $\sigma \sqsubset^* \tau$ and u_1 and u_2 are leaves in τ bound by two different quantifiers: one at node w_1 and the other at node w_2. The case $w_1 = w_2$ is allowed. We call every pair (σ, τ) with the above properties *splitting pair.*

Lemma 9 *If $\forall X$ is splitable in σ at nodes u_1 and u_2, then u_1 and u_2 are positive.*

Proof: We prove the lemma by induction on the number of steps in deriving $\sigma \sqsubset^* \tau$ such that (σ, τ) is a splitting pair. By inspecting all the six cases it is easy to verify that in one step one cannot get a splitting pair.

Let us suppose that $\sigma \sqsubset \rho \sqsubset^* \tau$ and (σ, τ) is a splitting pair at nodes u_1 and u_2. If (ρ, τ) is also a splitting pair at nodes u_1 and u_2 (perhaps for a quantifier located in a position different than $\forall X$ in σ) then, by induction assumption, we conclude that u_1 and u_2 are positive.

Consider now the situation when (ρ, τ) is not a splitting pair for u_1 and u_2. If u_j is not contained in ρ (where $j = 1$ or 2), then by Lemma 7 (i), there is $u_j' \leq u_j$ such that u_j' is a leaf in ρ, marked with -1. Since $\rho \sqsubset^* \tau$, it follows Lemma 7 (ii) that there is $u_j'' \leq u_j'$ such that u_j'' is a leaf in τ marked -1. This is a contradiction since u_j is a leaf in τ marked $+1$. Thus both: u_1 and u_2 belong to ρ.

If u_j is not a leaf, then by Lemma 7 (ii) we would get a contradiction since the marking of u_j in τ is $+1$. Thus u_1 and u_2 are leaves in ρ.

If the variable at u_j is marked -1 or 0, then again, by Lemma 7 (ii) or (iii) we get a contradiction with the marking of u_j in τ. Thus both u_1 and u_2 are variables in ρ, marked $+1$, and they are bound by two different quantifiers.

Let's analyse now the first step $\sigma \sqsubset \rho$ to see when we can obtain ρ with the above properties. Clearly it cannot be $(p\text{-}dummy)$ or $(n\text{-}dummy)$. Let w be the node in σ at which the $\forall X$ occurs. Consider the following remaining four cases.

Case I: $(p\text{-}distr)$
The only possibility in this case is that $(p\text{-}distr)$ is performed at w for $\forall X$ and $u_1 \leq w0$ and $u_2 \leq w1$. Then u_1 would be marked -1 in ρ, a contradiction. Thus this case is impossible.

Case II: $(n\text{-}distr)$
This case is clearly impossible.

Case III: $(p\text{-}inst)$
Then the quantifier $\forall X$ at node w must have been instantiated. Obviously it must have been instantiated with the type of the form $\forall Y_1 \ldots \forall Y_n . Y_i$, where $1 \leq i \leq n$. Thus, in order to conform to the requirement that both u_1 and u_2 are marked $+1$ in ρ — it follows that both u_1 and u_2 must be positive.

Case IV: $(n\text{-}inst)$
If the quantifier introduced by this step binds u_i (for $i = 1$ or 2), then u_i would have been marked in ρ with -1. This is impossible.

This completes the proof of the lemma. ∎

5   Reversible Steps

Call one step $\sigma \sqsubset \tau$ *reversible* if $\tau \sqsubset^* \sigma$ holds. In the next sequence of lemmas we will chracterize all reversible steps.

Lemma 10 (p-dummy)
$(p\text{-dummy})$ *step is always reversible.*

Proof: Obvious. ∎

Lemma 11 (p-distr)

Let w be a positive path in σ and let $\sigma \sqsubset \tau$ by a (p-distr) step performed at a node w by moving the quantifier $\forall X$ to $w0$ and $w1$. This step is reversible iff X doesn't occur free at node $w0$.

Proof. If X occurs free below $w0$, then this occurence is marked -1 in τ, while it is marked $+1$ in σ. Thus, by Lemma 7 (ii), the relation $\tau \sqsubset^* \sigma$ is impossible. Obtained contradiction proves the result. ∎

Lemma 12 (p-inst)

Let w be a positive path in σ and let $\sigma \sqsubset \tau$ by a (p-inst) step performed at node w by instantiating a quantifier $\forall X$ with a type ρ. This step is reversible iff one of the following conditions holds:

(i) *$\forall X$ is dummy.*

(ii) *$\forall X$ is not dummy, all bindings of $\forall X$ are positive and ρ is of the form $\forall Y_1 \ldots \forall Y_n . Y_i$, for some $1 \leq i \leq n$.*

(iii) *$\forall X$ is not dummy, there is another quantifier $\forall Z$ at a node $u \leq w$, such that $w = u1^m$, for some $m \geq 0$; $\forall Z$ has only positive bindings; and $\forall X$ is in scope of $\forall Z$. Type ρ is of the form $\forall Y_1 \ldots \forall Y_n . Z$, where $Z \notin \{Y_1, \ldots, Y_n\}$. Moreover, if $\forall X$ has a negative binding, then $\forall Z$ is dummy.*

Proof. Let us first observe that each of the above conditions implies reversibility of the (*p-inst*) step. In case of (ii) as well as in case of (iii), when both $\forall X$ and $\forall Z$ have only positive bindings, then it follows from Lemma 2. In case of (iii) when $\forall Z$ is dummy one gets back from τ to σ by a sequence of (*p-distr*) steps.

We prove now that if the (*p-inst*) step is reversible then one of the (i)–(iii) must hold. Assume that $\forall X$ is not dummy. If depth of ρ is greater than 1, then, by Lemma 7 we cannot get $\tau \sqsubset^* \sigma$. Hence ρ is of one of the following two forms:

$$\rho = \forall Y_1 \ldots \forall Y_n . Y_i \tag{1}$$

or

$$\rho = \forall Y_1 \ldots \forall Y_n . Z \tag{2}$$

where $1 \leq i \leq n$ and $Z \notin \{Y_1, \ldots, Y_n\}$.

Suppose ρ is of the form (1) and $\forall X$ has a negative binding at node v. Then, after substituting ρ for X, we get that in τ the node v is marked -1, while it is marked $+1$ in σ. Thus, by Lemma 7, we cannot get $\tau \sqsubset^* \sigma$. The obtained contradiction proves (ii).

Next, let us suppose that ρ satisfies (2). If this substitution introduces a free variable Z, i.e. $\forall X$ is not in the scope of a quantifier $\forall Z$, then by Lemma 7 (iii), the relation $\tau \sqsubset^* \sigma$ would be impossible. Hence Z must be bound at some node $u \leq w$. Since $\tau \sqsubset^* \sigma$ holds, it follows from Lemma 7 that u must be positive. Hence, by Lemma 8, it follows that $w = u1^m$, for some $m \geq 0$. Let v_1 and v_2 be nodes in σ which are bound by $\forall Z$ and by $\forall X$, respectively. Since $\tau \sqsubset^* \sigma$ holds, it follows that $\forall Z$ is splitable in τ at nodes v_1 and v_2. Thus, by Lemma 9, we obtain that either $\forall Z$ is dummy in σ or $\forall Z$ and $\forall X$ have only positive bindings in σ. This completes the proof of (iii). ∎

Proposition 13 *For arbitrary types σ and τ, if $\sigma \sqsubset \tau$ by a reversible step, then $\vdash \sigma \equiv \tau$.*

Proof. We prove the conclusion assumimg first that the reversible step $\sigma \sqsubset \tau$ was performed at a positive node. If this was a (*p-dummy*) step then using (A3) we obtain the conclusion. If this was a (*p-distr*) step then, by Lemma 11, it is enough to use (A4) to obtain the conclusion. Finally, if this step was a (*p-inst*) step, then we apply Lemma 12. In case of (i) or (ii) described in Lemma 12 use (A3). In case of (iii), when $\forall Z$ is dummy (using the notation of this lemma), we use (A2) and (A4) to get the conclusion. In case of (iii), when both $\forall X$ and $\forall Z$ have only positive bindings we use (A3) to get the conclusion.

Now, if the reversible step $\sigma \sqsubset \tau$ was performed at a negative node, then it is easy to check that if X is a new type variable, then the step $(\tau \to X) \sqsubset (\sigma \to X)$ is also reversible and is performed at a positive node. Hence, by the first part of this proof we conclude that $\vdash (\tau \to X) \equiv (\sigma \to X)$ holds. By Lemma 6 it follows that $\vdash \tau \equiv \sigma$ holds. Thus $\vdash \sigma \equiv \tau$ holds as well. This completes the proof. ∎

Proof of Theorem 4: Now we can prove Theorem 4. It follows from Lemma 1, Lemma 2, and Lemma 3, by obvious induction, that if $\vdash \sigma \equiv \tau$ holds, then σ and τ are bicoercible. This proves soundness.

For the completeness part let us assume that σ and τ are bicoercible. Then $\sigma \sqsubset^* \tau$, and all these steps are reversible. Thus, by Proposition 13, we obtain $\vdash \sigma \equiv \tau$. This completes the proof of the theorem. ∎

Acknowledgments: The author would like to thank Pierre-Louis Curien, Roberto Di Cosmo and Giuseppe Longo for stimulating discussions on the topics presented in this paper, during his stay at Ecole Normale Superieure in 1994, supported by a grant of French Ministry of Research.

References

[Ben93] M. Benke," Efficient type reconstruction in the presence of inheritance", in: A. M. Borzyszkowski, S. Sokołowski (Eds.) *MFCS'93: Mathematical Foundations of Computer science, Proc. 18th Intern. Symp., Gdansk 1993*, Springer-Verlag LNCS **711**, (1993), 272–280.

[BCGS91] V. Breazu-Tannen, T. Coquand, C.A. Gunter, and A. Scedrov, "Inheritance as implicit coercion", *Information and Computation* **93**, (1991), 172-221.

[CMSS94] L. Cardelli, S. Martini, J.C. Mitchell, and A. Scedrov, "An extension of system **F** with subtyping", *Information and Computation* **94**, (1994), 4-56.

[CGL92] G. Castagna, G. Ghelli, and G. Longo, "A calculus of overloaded functions with subtyping", *Proceedings, ACM conference on LISP and Functional Programming*, San Francisco (1992). *Information and Computation* **117** (1995), 115-135.

[CP94]   G. Castagna, and B.C. Pierce, "Decidable bounded quantification", *21st Ann. ACM Symposium on Principles of Programming Languages*, (1994), 151-162.

[Dic95] R. Di Cosmo, *Isomorphisms of types: from λ-calculus to information retrieval and language design*, Birkhauser, 1995.

[LMS95] G. Longo, K. Milsteed, and S. Soloviev, "A logic of subtyping", *Proc. 10-th IEEE Symp. Logic in Computer Science*, San Diego (1995), 292-299.

[Mit90] J.C. Mitchell, "Polymorphic type inference and containment", in: G. Huet (Ed.), *Logical Foundations of Functional Programming*, Addison-Wesley (1990), 153-193.

[Pie92] B.C. Pierce, "Bounded quantification is undecidable", *19th Ann. ACM Symposium on Principles of Programming Languages*, (1992), 305-315. *Information and Computation* **112** (1994), 131-165.

[Tiu92] J. Tiuryn, "Subtype inequalities", *Proc. 7-th IEEE Symp. Logic in Computer Science*, Santa Cruz (1992), 308-315.

[TW93]   J. Tiuryn, and M. Wand, "Type reconstruction with recursive types and atomic subtyping", in: M.-C. Gaudel and J.-P. Jouannaud (Eds.)

TAPSOFT'93: Theory and Practice of Software Development, Proc. 4th Intern. Joint Conf. CAAP/FASE, Springer-Verlag LNCS **668**, (1993), 686-701.

From Causal Consistency
to Sequential Consistency
in Shared Memory Systems

Michel Raynal[1] and André Schiper[2]

[1] IRISA, Campus de Beaulieu, 35042 Rennes Cédex (France) raynal@irisa.fr
[2] Dpt d'informatique, EPFL, 1015 Lausanne (Switzerland) schiper@di.epfl.ch

Abstract. Sequential consistency and causal consistency constitute two of the main consistency criteria used to define the semantics of accesses in the shared memory model. An execution is sequentially consistent if all processes can agree on a same legal sequential history of all the accesses; if processes perceive distinct legal sequential histories of all the accesses, the execution is only causally consistent (legality means that a read does not get an overwritten value).

This paper studies synchronization constraints that, when obeyed by operations of a given causally consistent execution, make it sequentially consistent. More precisely, the paper introduces the MSC synchronization (mixed synchronization constraint) which generalizes (1) the known DRF (data race free) and CWF (concurrent write free) synchronizations and (2) a new one called CRF (concurrent read free). The MSC synchronization allows for concurrent conflicting operations on a same object, while ensuring sequential consistency; this is particularly interesting in the context of distributed systems (where objects are possibly replicated) to cope with partition failures: conflicting operations in two distinct partitions do not necessarily block processes that issue them (as it is the case of quorum based protocols). Technically, a tag (control type) is associated with each operation, and all operations endowed with the same tag obey the same synchronization constraint.

1 Introduction

For several years the shared memory model has become a pervasive concept in parallel and distributed systems. This is due to the universality of the model: processes distributed over a network and interacting through shared objects (objects distributed over the network, and possibly replicated), fit for example perfectly into this model. Moreover, the shared memory model[3] is the adequate framework for defining consistency criteria: a consistency criterion defines the value returned by every read operation invoked by a process on some object (or some variable). It is important to stress that the definition of a consistency criterion must be independent of the possible existence of multiples copies of

---

[3] A (logically) shared memory includes simultaneously all the objects with all their implicit mutual and causal dependencies.

objects, and must not rely on a particular protocol implementing the criterion; it must be based on a formal model and be as general as possible to make designers capable to study properties of consistency criteria, and to produce results not bound to particular implementations. As for abstract data types, such a classical approach distinguishes clearly the semantics offered to users from its particular implementations. Several authors have correctly claimed that a memory consistency criterion is a contract between the memory system and application programs [11].

Three main consistency criteria have been proposed in the literature: atomic consistency [9] (also called linearizability [7]), sequential consistency [8] and causal consistency [3]. In all three cases a read operation returns the *last* value assigned to the variable (or written into the object). The three consistency criteria differ however in the definition of the *last* write operation. Atomic consistency is the more restrictive of the three consistency criteria: it requires that all the processes agree on a total order including all the read/write operations that they have issued, and this total order has to respect real time (i.e. if op_1 precedes op_2 in real time, then all the processes have to agree that op_1 has occurred before op_2). With sequential consistency, the processes have also to agree on a total order of their read/write operations, but this total order does not have to respect their real time order. With causal consistency, processes can disagree on the way they totally order concurrent write operations.

Causal consistency includes sequential consistency (i.e. an execution that satisfies sequential consistency also satisfies causal consistency) and sequential consistency includes atomic consistency (i.e. an execution that satisfies atomic consistency also satisfies sequential consistency). If some relationships between atomic and sequential consistencies are well understood (e.g. [5] compares their respective powers), a unifying framework, that would allow for a better understanding of the links between sequential consistency and causal consistency criteria, is still missing. This is precisely the purpose of this paper, which shows that sequential consistency can be obtained from causal consistency by adding some appropriate synchronization constraints (a synchronization constraint orders some pair of operations). This is particularly interesting from methodological and implementation points of view as it means that a family of protocols implementing sequential consistency can be seen as consisting of two independent layers: a basic layer implementing causal consistency and, on top of it, another layer enforcing the chosen synchronization constraints (an interesting consequence of the approach is that only this second layer has to be changed when we want to replace a set of synchronization constraints by another one).

Our work can be seen as a continuation of the work started by Ahamad *et al.* [4]. These authors consider however only two types of synchronization constraints: *Data Race Free* (DRF) synchronization, and *Concurrent Write Free* (CWF) synchronization. We generalize their work in two directions. First we distinguish between two classes of synchronization constraints: (1) the *per object* synchronization constraints, which synchronizes operations on each object independently, and (2) the *inter-object* synchronization constraints, which synchronizes operations on distinct objects. The DRF synchronization fits into the per object synchronization class, and the CWF synchronization fits into the inter-object synchronization class. The per object synchronization class is particularly interesting, as it provides the *locality* property introduced by Herlihy and Wing [7]. The second, and most important contribution of the paper, is the introduction of two new synchronization constraints: the *Concurrent Read Free* (CRF) synchronization, and the *Mixed Synchronization Constraint* (MSC) which combines the DRF, CWF and CRF synchronizations. The MSC synchronization has the nice property to allow conflicting operations on the same object to proceed concurrently. This is particularly interesting in the context of distributed systems where objects are possibly replicated, to cope with partition failures. If sequential consistency is obtained by implementing the MSC synchronization on top of a causally consistent distributed shared memory, then conflicting operations issued from two distinct partitions do not necessarily block processes that issued them (when sequential consistency is ensured by quorum based protocols such blocking always occurs).

The paper is structured as follows. Section 2 formally defines causal and sequential consistencies. Section 3 introduces the two classes of synchronization constraints (per object synchronization and inter-object synchronization), and defines the DRF, CWF and CRF synchronization constraints. The mixed synchronization constraint MSC is introduced in Section 4. Due to page limitation, proof of a lemma, practical impact of the MSC synchronization constraint in the context of distributed systems, and a discussion about other consistency criteria from which sequential consistency can also be obtained, are not included in this paper; the interested reader can consult [12] where these points are addressed.

2 Shared Memory Model

2.1 Notations

We consider a finite set of sequential processes $P_1, \ldots, P_n$ that interact via a finite set X of shared objects. Each object $x \in X$ can be accessed by read and write operations. A write into an object defines a new value for the object; a read allows to obtain a value of the object. A write of value v into object x by process P_i is denoted $w_i(x)v$; similarly a read of x by process P_j is denoted $r_j(x)v$ where v is the value returned by the read operation; op will denote either r (read) or w (write). For simplicity, we assume all values written into an object x are distinct. Moreover, the parameters of an operation are omitted when they are not important. Each object has an initial value; it is assumed that this value has been assigned by an initial fictitious write operation.

2.2 Histories

The *local history* $\hat{h}_i$ of P_i is the sequence of operations issued by P_i. If $op1$ and $op2$ are issued by P_i and $op1$ is issued first, then we say $op1$ precedes $op2$ in P_i's process-order, which is noted $op1 \rightarrow_i op2$. Let h_i denote the set of operations executed by P_i; the local history $\widehat{h_i}$ is the total order $(h_i, \rightarrow_i)$.

An *execution history* (or simply a history) $\widehat{H}$ of a shared memory system is a partial order $\widehat{H} = (H, \rightarrow_H)$ such that :

- $H = \bigcup_i h_i$
- $op1 \rightarrow_H op2$ if :

  i) $\exists P_i : op1 \rightarrow_i op2$ (in that case, $\rightarrow_H$ is called *process-order* relation),

  or *ii)* $op1 = w_i(x)v$ and $op2 = r_j(x)v$ (in that case $\rightarrow_H$ is called *read-from* relation),

  or *iii)* $\exists op3 : op1 \rightarrow_H op3$ and $op3 \rightarrow_H op2$.

A read operation $r(x)v$ is *legal* if: $\exists w(x)v : w(x)v \rightarrow_H r(x)v$ and $\nexists op(x)u :$ $(u \neq v) \wedge (w(x)v \rightarrow_H op(x)u \rightarrow_H r(x)v)$. A history $\widehat{H}$ is legal if all its read operations are legal[4].

Two operations $op1$ and $op2$ are *concurrent* in $\widehat{H}$ if we have neither $op1 \rightarrow_H op2$ nor $op2 \rightarrow_H op1$.

2.3   Sequential Consistency

Sequential consistency has been proposed by Lamport in 1979 to define a correctness criterion for multiprocessor shared memory systems [8]. A system is sequentially consistent with respect to a multiprocess program, if "*the result of any execution is the same as if (1) the operations of all the processors were executed in some sequential order, and (2) the operations of each individual processor appear in this sequence in the order specified by its program*".

This informal definition states that the execution of a program is sequentially consistent if it is equivalent to a sequential execution[5]. More formally, we define sequential consistency in the following way.

Definition. Sequential consistency. A history $\widehat{H} = (H, \rightarrow_H)$ is *sequentially consistent* if it admits a linear extension[6] $\widehat{S}$ in which all reads are legal.

---

[4] Other definitions of legality [3, 4] eliminate only the possibility of an intervening write ($w(x)u$) between the writing of some value ($w(x)v$) and a reading of the same value ($r(x)v$).

[5] In his definition, Lamport assumes that the *process-order* relation defined by the program (see point *(2)* of the definition) is maintained in the equivalent sequential execution, but not necessarily in the execution itself. As we do not consider programs but only executions, we implicitly assume that the *process-order* relation displayed by the execution histories are the ones specified by the programs which gave rise to these execution histories.

[6] A linear extension of a partial order is a topological sort of this partial order, so it maintains the order of all ordered pairs of the partial order.

As an example let us consider the history $\widehat{H_1}$ (Figure 1)[7]. Each process P_i, $(i=1,2)$, has issued three operations on the shared objects x and y. The write operations $w_1(x)0$ and $w_2(x)1$ are concurrent. It is easy to see that $\widehat{H_1}$ is sequentially consistent by building a legal linear extension including first the operations issued by P_2 and then the ones issued by P_1. It is also easy to see that the history $\widehat{H_2}$ (Figure 2) is not sequentially consistent, as no equivalent legal sequential history can be built.

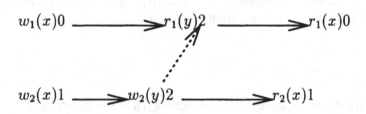

Fig. 1. A sequentially consistent history $\widehat{H_1}$

Various cache-based protocols implementing sequential consistency have been proposed in the context of parallel machines [1, 5, 10]. The protocols presented in [1, 10] allow several read operations and one write operation to concurrently access a same variable (reading of cached values and writing into the main memory) but do not allow concurrent write operations on a same variable. One of the protocols (called *fast write*) presented in [5] allows write operations on a same variable to proceed concurrently. However, these protocols do not assume an underlying causally consistent memory, and thus could not identify the two layers approach and the mixed synchronization constraint given in the paper.

In the context of distributed systems, where each object is supported by several permanent copies, non cache-based protocols implementing sequential consistency have been proposed. Usually these protocols use votes [14] or quorums [6] mechanisms and, consequently, implement actually atomic consistency which is stronger than sequential consistency.

2.4   Causal Consistency

Causal consistency has first been introduced by Ahamad *et al.* in 1991 [3], and then it has been studied by several authors [2, 4]; it defines a consistency criterion weaker than sequential consistency. Causal consistency allows for a wait-free

---

[7] In all figures, only the edges that are not due to transitivity are indicated (transitivity edges come from *process-order* and *read-from* relations). Moreover, (intra-process) *process-order* edges are denoted by continuous arrows and (inter-process) *read-from* edges by dotted arrows.

implementation of read and write operations in a distributed environment, i.e. causal consistency allows for cheap read/write operations (see [3, 4] for protocols implementing causal consistency).

With sequential consistency, all processes agree on a same legal linear extension ("legal sequential history") $\widehat{S}$. The agreement defined by causal consistency is weaker. Given a history $\widehat{H}$, it is not required that two processes P_i and P_j agree on the same ordering for the write operations which are not ordered in $\widehat{H}$. The reads are however required to be legal.

Definition. Causal consistency. Let $\widehat{H} = (H, \rightarrow_H)$ be a history. $\widehat{H}$ is *causally consistent* if all its read operations are legal.

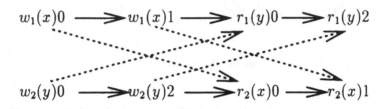

Fig. 2. A causally consistent history $\widehat{H_2}$

In a causally consistent history, all processes see the same partial order on operations but, as processes are sequential, each of them might see a different linear extension of this partial order.

So, in a causally consistent history, no read operation of a process P_i can get a value that, from P_i's point of view, has been overwritten by a more "recent" write. As an example consider history $\widehat{H_2}$ (Figure 2). This history is causally consistent as all its read operations are legal. The history $\widehat{H_3}$ (Figure 3) is not causally consistent as the read operation $r_3(x)1$ issued by P_3 is not legal: $w_1(x)1 \rightarrow_H r_3(x)2 \rightarrow_H r_3(x)1$. Said another way, when P_3 has issued its first read operation on x (namely $r(x)2$), it has got the value 2, and consequently for this process, the value 1 of x has logically been overwritten.

Remark. Our definition of causal consistency is akin to *causal reads* introduced in [2]. Another definition has been proposed in [4]. Let $\widehat{H_i}$ be the sub-history of $\widehat{H}$ from which all read operations not in the local history h_i of P_i have been removed (more formally, $\widehat{H_i}$ is the sub-relation of $\widehat{H}$ induced by the set of all the writes of H and all the reads issued by P_i). History $\widehat{H}$ is causally consistent if, for each process P_i, the associated history $\widehat{H_i}$ admits a legal linear extension $\widehat{S_i}$. From a formal point of view, this definition and ours are not strictly equivalent. Ours is slightly stronger but allows a simple statement of

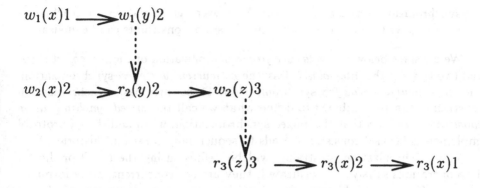

Fig. 3. A non causally consistent history $\widehat{H}_3$

causal consistency and, as shown in [12], it corresponds with the definition of causal order [13] in message passing systems. But, it is important to note that the main theorem stated in Section 4 is valid whatever definition of legality is chosen (ours or the one of [4]).

3 Basic Synchronization Constraints

3.1 Adding Synchronization Constraints

As mentioned in the introduction, this paper shows that a causally consistent history whose operations respect some synchronization constraints is a sequentially consistent history $\widehat{H}$: these synchronization constraints ascertain that $\widehat{H}$ admits a legal linear extension $\widehat{S}$. From an implementation point of view this means that a protocol implementing sequential consistency can be seen as consisting of two independent layers: (1) a first layer implementing causal consistency (i.e. basically ensuring the legality of reads), and (2) a second one enforcing the synchronization constraints.

A synchronization constraint orders some pairs of operations. Let $op(x)$ and $op'(y)$ be such a pair. Two classes of constraints can be defined, depending whether or not x and y are the same object.

1. *Per object synchronization.* In this case the synchronization constraint applies to each object x independently of the others. Two operations $op(x)$ and $op'(y)$, such that x and y are distinct, are never synchronized. This type of synchronization is particularly interesting as it provides the *locality* property introduced in [7].
2. *Inter-object synchronization.* In this case the synchronization constraint orders some pairs of operations $op(x)$ and $op'(y)$ on distinct objects x and y. This type of synchronization is more general (as it includes the *per object*

synchronization as a special case). Moreover, as *op* and *op'* can each be a read or a write operation, several subclasses of constraints can be envisaged.

We consider below the *data race free* synchronization of the *per object* class, and two types of the *inter-object* class: the *concurrent write free* synchronization, and the *concurrent read free* synchronization. These three basic synchronization constraints can be combined to define what we call the *mixed synchronization constraint*. We show that this mixed synchronization, when added to a protocol implementing causal consistency, leads to sequentially consistent histories.

The synchronization constraints will be defined using the ORD predicate. Two operations satisfy this predicate if they are not concurrent. More formally, let $\widehat{H} = (H, \to_H)$ be a history, and op, op' be two distinct operations of H:

$$ORD\,(op, op') \overset{def}{=} (op \to_H op') \; or \; (op' \to_H op)$$

3.2 Synchronization Tags

Each read or write operation of a history $\widehat{H}$ is associated with one and only one synchronization tag. The possible tags are *DRF (data race free)*, *CWF (concurrent write free)*, *CRF (concurrent read free)*. A read operation can be tagged *DRF* or *CRF*; a write operation can be tagged *DRF* or *CWF*. Operation op tagged T, and accessing object x, will be noted $op_T(x)$.

Informally, all operations endowed with the same tag T obey the same synchronization constraint (which is called T). Operations with distinct tags are not synchronized. The next two subsections define synchronization constraints based on tagged operations.

3.3 Per Object Synchronization

As indicated previously, the per object synchronization constraint is defined for each object independently of the others: it consists in ordering conflicting operations on each object x (two operations on an object x are conflicting iff one of them is a write). In other words, operations on each object obey the readers-writers synchronization. This constraint is usually called DRF *(data race free)*.

DRF synchronization. The DRF synchronization orders conflicting operations tagged *DRF*. Let $\widehat{H} = (H, \to_H)$ be a history and x an object. For any two distinct operations $op_{DRF}(x)$ and $op'_{DRF}(x)$, the DRF synchronization ensures that, if at least one of these operations is a write, then $ORD\,(op_{DRF}(x), op'_{DRF}(x))$ holds.

It has been proved in [4] that a causally consistent history $\widehat{H}$ in which (using our terminology) all operations are tagged *DRF* is sequentially consistent. This result is obtained here as a special case of a more general result (see Section 4.3).

3.4 Inter-Object Synchronization

The DRF synchronization does not order operations on distinct objects x and y. We introduce two synchronization constraints that order operations on distinct objects: one called CWF *(concurrent write free)* applies only to write operations, the other, called CRF *(concurrent read free)*, applies only to read operations.

In both definitions hereafter we consider a history $\widehat{H} = (H, \rightarrow_H)$ and a pair of (not necessarily distinct) objects x and y.

CWF synchronization. This synchronization orders write operations tagged *CWF*. For any two distinct write operations, $w_{CWF}(x)$ and $w_{CWF}(y)$, the CWF synchronization ensures $ORD\ (w_{CWF}(x), w_{CWF}(y))$.

CRF synchronization. This synchronization orders read operations tagged *CRF*. For any two distinct read operations, $r_{CRF}(x)$ and $r_{CRF}(y)$, the CRF synchronization ensures $ORD\ (r_{CRF}(x), r_{CRF}(y))$.

It has been shown in [4] that a causally consistent history $\widehat{H}$, in which all write operations are tagged *CWF*, is sequentially consistent. This result will also be proved in Section 4.3 as a special case of our more general result. Notice that the CRF synchronization (with all read operations tagged *CRF*) is not sufficient to ensure sequential consistency out of causal consistency. The CRF synchronization, together with the DRF synchronization for write operations, leads however to this result. This is included in our mixed synchronization constraint.

4 A Mixed Synchronization Constraint

4.1 The Mixed Constraint

The synchronization constraints DRF, CWF and CRF introduced so far can be combined to define a mixed synchronization constraint, noted MSC *(mixed synchronization constraint)*. MSC is a generalization of the previous synchronizations in the sense that it allows for combinations of synchronization constraints (either CWF with DRF, or CRF with DRF). In other words, MSC does not require that the same synchronization tag be attached to every operation of a history. We distinguish between two MSC constraints, MSC_W and MSC_R:

- A history $\widehat{H}$ satisfies the MSC_W constraint if and only if (1) all its operations are tagged either *CWF* or *DRF*, and (2) the *CWF* tagged operations obey the CWF synchronization, and the *DRF* tagged operations obey the DRF synchronization.
  Note that in this case all read operations are necessarily tagged *DRF*, while write operations are tagged *DRF* or *CWF*.
- A history $\widehat{H}$ satisfies the MSC_R constraint if and only if (1) all its operations are tagged either *CRF* or *DRF*, and (2) the *CRF* tagged operations obey the CRF synchronization, and the *DRF* tagged operations obey the DRF synchronization.
  Note that in this case all write operations are necessarily tagged *DRF*, while read operations are tagged *DRF* or *CRF*.

We say that the MSC synchronization is satisfied by a history $\widehat{H}$ if and only if $\widehat{H}$ satisfies either MSC_W or MSC_R. Notice that, in a history $\widehat{H}$ that satisfies the MSC constraint, the tags CWF and CRF are incompatible. If $w(x)$ is tagged CWF in $\widehat{H}$, then all read operations of $\widehat{H}$ are tagged DRF (constraint MSC_W). Similarly if a read is tagged CRF then all write operations are tagged DRF (constraint MSC_R).

In order to understand the MSC_W and MSC_R synchronizations, it is important to understand that they do not require conflicting operations on a given object x be ordered. Consider a history $\widehat{H}$ and the two cases of conflicting operations (namely, read/write conflict and write/write conflict). Consider first read/write conflicting operations. If $\widehat{H}$ satisfies the MSC_W constraint, and if a write operation w on some object x is tagged CWF, whereas a read operation r on x is tagged DRF, then these two operations are not ordered by the MSC_W synchronization constraint: hence $w_{CWF}(x)$ and $r_{DRF}(x)$ are not synchronized. Consider now the same conflict with $\widehat{H}$ satisfying the MSC_R constraint: if a write operation w on x is tagged DRF, whereas a read operation r on x is tagged CRF, then these two operations are not ordered by MSC_R: hence $w_{DRF}(x)$ and $r_{CRF}(x)$ are not synchronized.

The same result holds for write/write conflicts. Suppose the MSC_W constraint is satisfied by $\widehat{H}$; if a write operation w on some object x is tagged DRF whereas another write operation w' on x is tagged CWF, then these operations are not ordered by MSC_W: hence $w_{DRF}(x)$ and $w'_{CWF}(x)$ are not synchronized, i.e. concurrent conflicting writes are allowed!

Remark. To understand that the DRF and the CWF constraints mentioned in [4] are special cases of the MSC constraint, consider the following explanation. First, the DRF synchronization constraint in [4] is obviously a special case of either the MSC_W or of the MSC_R synchronization constraint, in which all the operations are implicitly tagged DRF. Second, the CWF synchronization constraint in [4] is a special case of the MSC_W synchronization constraint in which all the write operations are implicitly tagged CWF and obey CWF synchronization, and all the read operations are implicitly tagged DRF and obey DRF synchronization (in that case the DRF synchronization is actually a *nil* synchronization as it applies only to read operations of $\widehat{H}$ which are never conflicting!).

4.2 Deterministic Read Operations

Section 4.3 proves our main result, namely: a causally consistent history $\widehat{H}$, that satisfies either MSC_W or MSC_R, is sequentially consistent. Because MSC_W allows for concurrent writes on the same object, MSC_W requires an additional deterministic *read rule*, in order for our main result to hold. This rule defines which value has to be returned by a $r(x)$ operation in case the read operation is aware of two concurrent write operations.

Deterministic read rule. Consider $\hat{H}$ a causally consistent history obeying the MSC$_W$ synchronization, an object x, and two concurrent writes $w_{DRF}(x)u$ and $w_{CWF}(x)v$. Let $r(x)$ be a read operation such that both $r(x)u$ and $r(x)v$ are legal[8]. Then the read operation returns the value written by $w_{CWF}(x)$, i.e. $r(x)$ returns v.

4.3 The MSC Theorem

We prove in this Section our main result relating causal consistency to sequential consistency. We prove the results for MSC$_W$ and for MSC$_R$ together.

Theorem 4.1 *Let $\hat{H} = (H, \rightarrow_H)$ be a causally consistent history such that (1) either MSC$_W$ or MSC$_R$ is satisfied, and (2) in case of MSC$_W$ each read operation $r(x)$ follows the read rule (Sect. 4.2). Then $\hat{H}$ is sequentially consistent.*

Preliminary Definitions In order to prove the Theorem 4.1, we introduce the Lemma 4.2. This Lemma is based on two additional relations on the operations of a history $\hat{H}$: a *logical write-write precedence* relation (denoted $\rightarrow_w$) and a *logical read-write precedence* relation (denoted $\rightarrow_r$).

Logical write-write precedence. Let $\hat{H} = (H, \rightarrow_H)$ be a history. The write-write precedence relation $\rightarrow_w$ is defined on pairs of write operations, on a same object x, that are concurrent in $\hat{H}$. By definition, this can happen only if one of them is tagged *DRF* while the other is tagged *CWF* (two writes on a same object, both tagged either *DRF* or *CWF*, are ordered in $\hat{H}$). The logical write-write precedence relation states that the write tagged *DRF* is logically *before* the write tagged *CWF*.

Definition: Let $w_{CWF}(x)$ and $w_{DRF}(x)$ be two concurrent write operations in $\hat{H}$. Then we have: $w_{DRF}(x) \rightarrow_w w_{CWF}(x)$. (Moreover $\rightarrow_w$ holds only in these cases.)[9]

Logical read-write precedence. Let $\hat{H} = (H, \rightarrow_H)$ be a history. The logical read-write precedence relation $\rightarrow_r$ is defined on pairs of read and write operations for each object x.

Definition: Let $w(x)u$, $r(x)u$ and $w(x)v$ be three operations in $\hat{H}$ such that $w(x)u \rightarrow_H w(x)v$ or $w(x)u \rightarrow_w w(x)v$. Then we have: $r(x)u \rightarrow_r w(x)v$. (Moreover $\rightarrow_r$ holds only in these cases.)[10]

---

[8] This means the read is aware of both writes and there is no intervening operation $op(x)a$, with $a \neq u$ and $a \neq v$, in between $w_{CWF}(x)v$ and $r(x)$ and in between $w_{DRF}(x)u$ and $r(x)$.

[9] So, given $w(x)u$ and $w(x)v$, we necessarily have one of these four relations: $w(x)u \rightarrow_H w(x)v$, or $w(x)u \rightarrow_w w(x)v$, or $w(x)v \rightarrow_H w(x)u$, or $w(x)v \rightarrow_w w(x)u$.

[10] Note that in this case u and v are necessarily distinct.

Lemma 4.2 (Acyclicity) *Let $\hat{H} = (H, \to_H)$ be a causally consistent history that satisfies the MSC synchronization constraint. Let $\to_w$ and $\to_r$ be the two relations on H defined above. Then the relation $\to_H \cup \to_w \cup \to_r$ defines a partial order on H.*

PROOF. Let $\to$ be either $\to_H$ or $\to_w$ or $\to_r$, and let $\to_{w/r}$ be either $\to_w$ or $\to_r$. Consider the directed graph whose vertices are the operations of H, and whose edges are the $\to$ relation. We prove that, for any $n > 0$, there are no (directed) cycle of length n in this graph. The proof is by induction on m, where m is the number of $\to_{w/r}$ edges in the cycle.

i) **Base step** *($m=1$).* Let $op_1 \to op_2 \to \ldots \to op_n \to op_1$ be a simple cycle of length $n > 1$, and assume that one single edge of this cycle is of type $\to_{w/r}$. We show that this leads to a contradiction. Without loss of generality, let $op_1 \to op_2$ be the only $\to_{w/r}$ edge in the above cycle; so, as $\hat{H}$ is transitive, we have $op_2 \to_H op_1$. There are two cases to consider, numbered $i.1)$ and $i.2)$.

i.1) $op_1 \to_r op_2$.
From the definition of $\to_r$ it follows $op_1 \equiv r(x)u$ and $op_2 \equiv w(x)v$, and either (a) $w(x)u \to_H w(x)v$ or (b) $w(x)u \to_w w(x)v$.

(a) $w(x)u \to_H w(x)v$. As $op_2 \to_H op_1$, i.e. $w(x)v \to_H r(x)u$, we have $w(x)u \to_H w(x)v \to_H r(x)u$, which means that $r(x)u$ is not legal, in contradiction with the assumption that, because $\hat{H}$ is a causally consistent history, its read operations are legal.

(b) $w(x)u \to_w (x)v$. From the definition of $\to_w$ it follows that $w(x)u$ and $w(x)v$ are concurrent and respectively tagged DRF and CWF. So we get (1) $w_{CWF}(x)v \to_H r(x)u$ (because $op_2 \to_H op_1$), (2) $w_{DRF}(x)u \to_H r(x)u$ (read-from relation), and (3) $w_{CWF}(x)v$ and $w_{DRF}(x)u$ are concurrent. This is in contradiction with the read rule of Section 4.2 (namely $r(x)$ cannot read value u; it reads v or, if it exists, a more recent value v' such that $w(x)v \to_H w(x)v'$).

i.2) $op_1 \to_w op_2$.
By definition of $\to_w$, op_1 and op_2 are two concurrent write operations, but $op_2 \to_H op_1$ (because $op_1 \to op_2$ is the only $\to_{w/r}$ edge in the cycle): a contradiction.

ii) **Induction step** *($m > 1$).* Let $op_1 \to op_2 \to \ldots \to op_n \to op_1$ be a simple cycle of length n, and assume that there is no cycle with m or less than m edges $\to_{w/r}$ ($m < n$). We prove that there can be no cycle with $m + 1$ edges $\to_{w/r}$. The proof is again by contradiction.

Assume a cycle with $m + 1$ edges $\to_{w/r}$, and pick arbitrarily two of these $m + 1$ edges. Without loss of generality let op_1, op_2, op_t ($t \geq 2$), op_{t+1} be the four operations that are the endpoints of the two $\to_{w/r}$ edges: $op_1 \to_{w/r} op_2$ and $op_t \to_{w/r} op_{t+1}$. There are four cases to consider, numbered $ii.1)$ to $ii.4)$.

ii.1) $op_1 \to_r op_2$ and $op_t \to_r op_{t+1}$.
By definition of $\to_r$: op_1, op_t are read operations and op_2, op_{t+1} are write operations.

ii.11). If op_1 and op_2 (or op_t and op_{t+1}) are both tagged *DRF* we have $op_1 \to_H op_2$ or $op_2 \to_H op_1$ and consequently there exists a cycle of $\to_{w/r}$ with less than $m + 1$ edges.

ii.12). If both op_1 and op_2 are not tagged *DRF*, and the same holds for op_t and op_{t+1}, then due to the incompatibility of *CRF* and *CWF* tags, either both reads (op_1 and op_t) are tagged *CRF*, or both writes (op_2 and op_{t+1}) are tagged *CWF*. If the reads are tagged *CRF*, then as $\hat{H}$ satisfies MSC, either $op_1 \to_H op_t$ or $op_t \to_H op_1$. If the writes are tagged *CWF*, then as $\hat{H}$ satisfies MSC, either $op_2 \to_H op_{t+1}$ or $op_{t+1} \to_H op_2$. In all of these four cases, we are able to exhibit a cycle with no more than m edges $\to_{w/r}$, which is in contradiction with the induction hypothesis: if $op_1 \to_H op_t$ or $op_{t+1} \to_H op_2$ (respt. $op_t \to_H op_1$ or $op_2 \to_H op_{t+1}$) then there is a cycle not including the edge $op_1 \to_r op_2$ (respt. $op_t \to_r op_{t+1}$) and so including less than $m + 1$ edges $\to_{w/r}$.

ii.2) $op_1 \to_r op_2$ and $op_t \to_w op_{t+1}$.
By the definition of $\to_r$ and $\to_w$, op_1 is a read operation, op_2, op_t, op_{t+1} are write operations and op_{t+1} is tagged *CWF*. Because op_{t+1} is tagged *CWF*, there can be no read operations tagged *CRF*, i.e. op_1 is tagged *DRF* and op_2 is tagged *DRF* or *CWF*.

ii.21). If op_2 is tagged *DRF* then we have $op_1 \to_H op_2$ or $op_2 \to_H op_1$ and there is a cycle of $\to_{w/r}$ of less than $m + 1$ edges.

ii.22). If op_2 is tagged *CWF* then, as op_{t+1} is also tagged *CWF*, we have either $op_2 \to_H op_{t+1}$ or $op_{t+1} \to_H op_2$. In both cases, we can exhibit as previously a cycle with no more than $m \to_{w/r}$ edges, in contradiction with the induction hypothesis: if $op_2 \to_H op_{t+1}$ (respt. $op_{t+1} \to_H op_2$), then there is a cycle not including the edge $op_t \to_w op_{t+1}$ (respt. $op_1 \to_r op_2$).

ii.3) $op_1 \to_w op_2$ and $op_t \to_r op_{t+1}$.
By renaming op_1 to op_t, op_2 to op_{t+1}, op_t to op_1 and op_{t+1} to op_2, case *ii.3)* becomes identical to *ii.2)*.

ii.4) $op_1 \to_w op_2$ and $op_t \to_w op_{t+1}$.
By the definition of $\to_w$, op_2 and op_{t+1} are write operations tagged *CWF*, i.e. either $op_2 \to_H op_{t+1}$ or $op_{t+1} \to_H op_2$. In both cases, we can exhibit as previously a cycle with no more than $m \to_{w/r}$ edges, in contradiction with the induction hypothesis: if $op_2 \to_H op_{t+1}$ (respt. $op_{t+1} \to_H op_2$) there is a cycle not including the edge $op_t \to_w op_{t+1}$ (respt. the edge $op_1 \to_w op_2$).
(*end of the proof of Lemma 4.2*)

Proof of the MSC Theorem Lemma 4.2 has showed that $\to_H \cup \to_w \cup \to_r$ defines a (partial) order on H. The following Lemma 4.3 completes the proof of the MSC theorem by showing a legal linear extension $\hat{S}$ can be constructed by a topological sort of $(H, \rightarrow_H \cup \rightarrow_w \cup \rightarrow_r)$.

Lemma 4.3 (Legality) *A topological enumeration of* $(H, \rightarrow_H \cup \rightarrow_w \cup \rightarrow_r)$ *produces a legal linear extension* $\widehat{S}$ *of* $\widehat{H}$.

PROOF. See [12].

5 Conclusion

This paper has studied synchronization constraints that, when obeyed by operations of a given causally consistent execution, make it sequentially consistent. Such an approach is particularly interesting as, from methodological and implementation points of view, it means that a protocol implementing sequential consistency can consist of two independent layers: a basic one implementing causal consistency and, on top of it, another one implementing some synchronization constraints for the operations issued by processes.

The paper introduced the MSC synchronization (mixed synchronization constraint) which generalizes (1) the known DRF (data race free) and CWF (concurrent write free) synchronizations, and (2) a new one called CRF (concurrent read free). A main interest of this constraint lies in the fact it allows concurrent conflicting operations on a same object while ensuring sequential consistency; this is particularly interesting in the the context of distributed systems (where objects are possibly replicated) to cope with partition failures: conflicting operations in two distinct partitions do not necessarily block processes that issue them (as it is the case with quorum based protocols). Technically, a tag (control type) is associated with each operation, and all operations endowed with the same tag obey the same constraint.

The paper has also classified the synchronization constraints in two classes: the *per object* synchronization class, and the *inter-object* synchronization class (which includes the per object synchronization class). This classification allows to better understand linearizability (which has the nice locality property [7]) with respect to sequential consistency: linearizability is obtained by the *per object* synchronization. Finally, while the paper has identified MSC as a sufficient condition to get sequential consistency out of causal consistency, it would be interesting to identify a necessary condition to get sequential consistency out of causal consistency.

Acknowledgements

The presentation has benefited from valuable comments from the anonymous referees. We thank Ambuj Singh for pointing out the difference between the definition of causal consistency used in this paper and the one used in [4].

194

This work has been supported in part by the Commission of European Communities under ESPRIT Programme BRA 6360 (BROADCAST), by France Telecom under a CNET grant "Cohérence d'objets répartis", and by the "Fonds national suisse" and OFES under contract number 21-32210.91.

References

1. Y. Afek, G. Brown, and M. Merritt. Lazy caching. *ACM Trans. on Prog. Lang. and Systems*, 15(1):182–205, 1993.
2. D. Agrawal, M. Choy, H.V. Leong and A. Singh. Mixed consistency: a model for parallel programming. *In Proc. 13th ACM Symposium on Principles of Dist. Computing*, Los Angeles, pages 101–110, 1994.
3. M. Ahamad, J.E. Burns, P.W. Hutto, and G. Neiger. Causal Memory. In *Proc. 5th Intl. Workshop on Distributed Algorithms (WDAG-5)*, pages 9–30. Springer Verlag, LNCS 579, 1991.
4. M. Ahamad, P.W. Hutto, G. Neiger, J.E. Burns, and P. Kohli. Causal Memory: Definitions, Implementations and Programming. TR GIT-CC-93/55, Georgia Institute of Technology, July 94, 25p.
5. H. Attiya and J.L. Welch. Sequential Consistency versus Linearizability. *ACM Trans. on Comp. Systems*, 12(2):91–122, 1994.
6. H. Garcia-Molina and D. Barbara. How to assign votes in a distributed systems. *Journal of the ACM*, 32(4):841–850, 1985.
7. M. Herlihy and J. Wing. Linearizability: a correctness condition for concurrent objects. *ACM Trans. on Prog. Lang. and Systems*, 12(3):463–492, 1990.
8. L. Lamport. How to make a multiprocessor computer that correctly executes multiprocess programs. *IEEE Trans. on Computers*, C28(9):690–691, 1979.
9. J. Misra. Axioms for memory access in asynchronous hardware systems. *ACM Trans. on Prog. Lang. and Systems*, 8(1):142–153, 1986.
10. M. Mizuno, M. Raynal, and J.Z. Zhou. Sequential Consistency in Distributed Systems. *Proc. Int. Workshop "Theory and Practice in Dist. Systems"*, Dagstuhl, Germany, Springer-verlag LNCS 938 (K.Birman, F. Mattern and A. Schiper Eds), 1994, pp.224-241.
11. D. Mosberger. Memory consitency models. *ACM Operating Systems Review*, 27(1):18–26, 1993.
12. M. Raynal and A. Schiper. *From causal consistency to sequential consistency in shared memory systems*. INRIA Research Report 2557, May 1995, 27 pages.
13. M. Raynal, A. Schiper and S. Toueg. The causal ordering abstraction and a simple way to implement it. *Inf. Proc. Letters*, 39:343–350, 1991.
14. R.H. Thomas. A majority consensus approach to concurrency control for multiple copies databases. *ACM Trans. on Database Systems*, 4(2):180–209, 1979.

Observation of Software for Distributed Systems with RCL *

Alexander I. Tomlinson and Vijay K. Garg

email: {alext,vijay}@pine.ece.utexas.edu
homepage: http://maple.ece.utexas.edu/
Department of Electrical and Computer Engineering
The University of Texas at Austin, Austin, Texas 78712

Abstract. Program observation involves formulating a query about the behavior of a program and then observing the program as it executes in order to determine the result of the query. Observation is used in software development to track down bugs and clarify understanding of a program's behavior, and in software testing to ensure that a program behaves as expected for a given input set. RCL is a recursive logic built upon conjunctive global predicates. Computational structures of common paradigms such as butterfly synchronization and distributed consensus can be expressed easily in RCL. A nonintrusive decentralized algorithm for detecting RCL predicates is developed and proven correct.

1 Introduction

Posets have a recursive structure that has not been exploited much in research on observation predicates. This paper presents a poset predicate logic which exploits this recursive structure. The result, RCL, is a logic which is simple yet powerful. Recursive logics are intuitive because there are fewer constructs and rules to remember. They are powerful because the full power of the logic is available at each level of recursion. Boolean logic is an example of a recursive logic: it is simple, elegant and powerful.

RCL is based upon *conjunctive global predicates* (CGP). A CGP is defined to be a conjunction of local predicates. For example, let l_i be a predicate on the local state of process i. Then we can define a CGP g to be $l_1 \wedge l_2 \wedge l_3$. All CGPs, including g, are evaluated on global states. For example if c is a global state containing local states $\sigma_1, \sigma_2, \sigma_3$, then g is true in global state c if and only if l_i is true in local state l_i, for $1 \leq i \leq 3$.

An RCL formula takes a set of CGPs and specifies a pattern in which the individual CGPs must occur in a computation in order for the formula to be "true" in that computation. Some patterns which can be specified with RCL include butterfly synchronization, data collection, and distributed consensus. These examples are demonstrated in this paper.

* Research supported in part by NSF Grant CCR 9110605, TRW faculty assistantship award, GM faculty fellowship, a grant from IBM, and an MCD University Fellowship.

We begin with a review of related work, and then continue with a description of the computation model and notation. Then we define the syntax and semantics of RCL, and give examples of RCL formulas and how they can be applied in the observation of distributed programs. We present a distributed algorithm for online detection of RCL formulas and prove its correctness. The algorithm is based on existing algorithms for detecting CGPs, which are not considered in detail in this paper. Due to space constraints, the complexity of the algorithm is not discussed in detail. See [19] for a complete analysis of the algorithm's complexity.

2   Related Work

There has been much work in observing unstable global states of distributed computations. Babaoğlu and Marzullo [1] and Schwartz and Mattern [18] both survey recent work on detecting consistent global states in a distributed system. Recently, Chase and Garg [2] have shown that global predicate detection is an NP-complete problem. In that paper they define the property of *linearity* and show that there exists a polynomial detection algorithm for any linear predicate. CGP (discussed later) is an example of a linear predicate.

Cooper and Marzullo [4] present algorithms for online detection of three types of global predicates. The first type is "global predicate g was *possibly* true at some point in the past". The second type is "g was *definitely* true at some point in the past", and the third type is "g is *currently* true". The third type may require delaying certain processes of the execution.

Observation of general global predicates is very expensive. As a result, researchers have devised classes of global predicates which can be efficiently observed. One such class is *relational global predicates* as described by Tomlinson and Garg [20].

Another such class is *conjunctive global predicates* (CGP) as described by Garg and Waldecker [10]. Garg and Waldecker present strong and weak [11, 10, 12] forms of CGP which correspond to *possibly* and *definitely* of Cooper and Marzullo [4]. The *weak CGP* is true in a computation if there exists a global state in the computation which satisfies the CGP. The *strong CGP* is true in a computation if all runs consistent with the computation must enter a global state in which the CGP is true.

Some researchers have taken the idea of conjunctive global predicates (CGP) and extended them to form poset predicates. One can define an ordering relation on global states and then define a sequence of CGP. There have been several approaches to this that differ primarily in their definition of the ordering on global states.

Chiou and Korfage [3] define *event normal form* predicates which are sequences of CGP. In their sequencing relation, global state a precedes global state b if and only if each local state in a happens-before all local states in b.

Haban and Weigel [13] gave an early attempt to define poset predicates with recursive structure. They used local events (essentially the same as local predicates) as primitives and build global events from them with a set of binary relations that include alternation, conjunction, happens-before, and concurrency. All events (global and local) have vector timestamps which are used to determine if two events are related according to one of the four relations. The new global event inherits a timestamp from one of the constituent events. For example, consider alternation: if $e_1 \mid e_2$ is a global event which is said to occur whenever either e_1 or e_2 occurs. The event $e_1 \mid e_2$ inherits the vector time of whichever event actually occurred (i.e., either e_1 or e_2). Even though their definitions lead to ambiguities (resulting from timestamp inheritance) as demonstrated in [14], the work was noteworthy in that it was an early attempt to develop a recursive poset predicate logic.

The above systems are based on global predicates, but many systems have been designed around the local predicate too. One of the early works in this area was Miller and Choi's *sequence of local predicates* [17]. These are an ordered list of local predicates $p_1, \ldots p_k$. This predicate is true in an execution if and only if there exists a sequence of local states $\sigma_1, \ldots \sigma_k$ (sequenced by Lamport's happens-before relation) such that local predicate p_i is true in local state σ_i.

Hurfin, Plouzeau and Raynal [15] extended the sequence of local predicates to the *atomic sequence of local predicates*. In this class, occurrences of local predicates can be forbidden between adjacent predicates in a sequence of local predicates. The example given above for linked predicates could be expanded to include: "local predicate r_i never occurs in between local predicates p_i and p_{i+1}". Each local predicate can belong to a different process in the computation.

Fromentin, Raynal, Garg and Tomlinson [6] developed *regular patterns*, which are based upon regular expressions. A regular pattern is specified by a regular expression of local predicates. For example pq^*r is true in a computation if there exists a sequence of consecutive local states $(s_1, s_2, \ldots, s_n)$ such that p is true in s_1, q is true in $s_2, \ldots, s_{n-1}$, and r is true in s_n. Note that the states in the sequence need not belong to the same process – two states are consecutive if they are adjacent in the same process or one sends a message and the other receives it. In [9], the same authors extend regular patterns to allow patterns on directed acyclic graphs instead of just strings.

3 Model and Notation

We use the following notation for quantified expressions:

$$(\text{ op free\_var\_list : range\_of\_free\_vars : expr })$$

For example, $(\forall i : 0 \leq i \leq 10 : i^2 \leq 100)$ means that for all i such that $0 \leq i \leq 10$, we know that $i^2 \leq 100$. The operator "op" need not be restricted to universal or existential quantification. It can be any commutative associative operator (e.g., $min, \cup, +$). For example, if S_i is a finite set, then $(+u : u \in S_i : 1)$ equals the cardinality of S_i.

Any distributed computation can be modeled as a decomposed partially ordered set (deposet) of process states [5]. A deposet is a partially ordered set $(P, \leadsto)$ such that:

1. P is partitioned into N sets P_i, $1 \leq i \leq N$.
2. Each set P_i is a total order under some relation $\prec_{im}$.
3. $\prec_{im}$ does not relate two elements which are in different partitions.
4. Let $\rightarrow$ be the transitive closure of $\prec_{im} \cup \rightsquigarrow$. Then $(P, \rightarrow)$ is an irreflexive partial order.

An execution that consists of N processes can be modeled by a deposet where P_i is the set of local states at process i which are sequenced by $\prec_{im}$ the $\rightsquigarrow$ relation represents the ordering induced by messages; and $\rightarrow$ is Lamport's *happens before* relation[16]. For convenience, we use P_i to represent two quantities: the set of local states at process i (as it was defined), and the process i itself. Similarly, we use P to denote both the set of all local states and the set of all processes.

The concurrency relation on P is defined as $u \| v = (u \not\rightarrow v) \wedge (v \not\rightarrow u)$. $\preceq$ denotes the reflexive transitive closure of $\prec_{im}$. For convenience, $s.next = t$ and $t.prev = s$ whenever $s \prec_{im} t$.

A global state is a subset $c \subseteq P$ such that no two elements of c are ordered by $\rightarrow$. We define $\overline{P}$ to be the set of all global states in $(P, \rightarrow)$. We also use the terms "cut" and "antichain" to refer to an element of $\overline{P}$. A "chain" is a set of states which are totally ordered by $\rightarrow$. For example, each set P_i is a chain.

All formulas in RCL are evaluated on closed posets. Evaluating a formula on a poset which is not closed is not a defined operation. A poset P is closed if and only if every state which is ordered in between two elements of P is also in P. Another way of saying this is that P is closed if and only if its prefix-closure intersected with its suffix-closure equals P. Prefix and suffix closure of a poset A are denoted by $\overleftarrow{A}$ and $\overrightarrow{A}$.

$$\overleftarrow{A} \triangleq \{x \mid (\exists y : y \in A : x \rightarrow y \vee x = y)\}$$

$$\overrightarrow{A} \triangleq \{x \mid (\exists y : y \in A : y \rightarrow x \vee x = y)\}$$

$$closed(A) \triangleq A = (\overleftarrow{A} \cap \overrightarrow{A})$$

We define another closure operation which performs closure between any two subposets of a poset. For example, $[A..B]$ is the poset which includes posets A and B and all in between local states. Usually, A and B are cuts, but it is convenient to use the more general definition that they are subposets. This allows us to say, for example, that $[c..P]$ is the set of all states in or after cut c but still in poset P. We also define $(A..B)$ to be an open-ended version of $[A..B]$.

$$[A..B] \triangleq \overrightarrow{A} \cap \overleftarrow{B}$$

$$(A..B) \triangleq \overrightarrow{A} \cap \overleftarrow{B} - (A \cup B)$$

The *cutset* of a poset P and a formula f is the set of all cuts c of P such that $[P..c]$ satisfies f. The expression $\Psi(P, f)$ refers to this set and is defined as follows:

$$\Psi(P, f) \triangleq \{c \mid c \in \overline{P} \wedge [P..c] \models f\}$$

Cutsets will be used to prove the correctness of the RCL detection algorithm. It turns out that cutsets are lattices. The correctness proof shows that, given a computation P and a formula f, the detection algorithm returns the infimum of $\Psi(P, f)$, which is the unique first cut c of P such that $[P..c]$ satisfies f.

We also define two ordering relations on subposets: weak and strong precedes. The ordering relations are usually used on cuts, but the more general relation suffices. Subposet A weakly precedes subposet B if and only if B is entirely contained within the suffix closure of A and they have no elements in common.

$$A \prec B \triangleq B \subseteq \vec{A} \wedge A \cap B = \emptyset$$

Strong precedes requires not only that B must be contained in the suffix closure of A, but also that each element in A must happen-before every element in B. Clearly, this implies that A weakly precedes B as well. Strong precedes corresponds to barrier synchronization: there is a barrier synchronization between A and B if and only if A strongly precedes B. It is defined as follows:

$$A \prec\!\!\prec B \triangleq (\forall a, b : a \in A \wedge b \in B : a \rightarrow b)$$

4 Syntax and Semantics

A formula in RCL is evaluated on a poset. One can think of a formula as a boolean function whose argument is a poset. The rules for constructing well formed formulas are given by the syntactic definitions shown below:

$$f = S \mid f \triangle f$$
$$S = g \mid g\langle f \rangle S \mid g \langle\!\langle f \rangle\!\rangle S$$

The basic component of a formula is a conjunctive global predicate (CGP), which is represented by the terminal symbol g. The symbol S is a sequence of CGP formulas. The symbol f is a conjunction of these sequences, and the $\triangle$ operator is similar to boolean AND operator.

When S is fully expanded, it has the form $g\langle f \rangle g\langle f \rangle g \ldots g\langle f \rangle g$. When such a sequence is true on a poset, then each g corresponds to an antichain. The regions in between these antichains are subposets upon which the f's in the sequence are evaluated. This is explained in more detail in the section on semantics.

The symbol 'g' represents any global state based predicate which meets certain requirements. Mathematically, g is a set containing exactly those antichains upon which the predicate (that g represents) is true. One of these requirements is that this set forms a lattice.

Another requirement is that the antichains in g cannot contain any extraneous local states. For example, if g represents a conjunctive global predicate with components at process 1 and 2, then each antichain in g can only contain local states from these two processes. No others are needed to evaluate the predicate g. The reason for this requirement is to ensure the proper interpretation of sequences of g's – the ordering should be based on the necessary states only.

It is clear from the syntax that g is a valid RCL formula. The truth of such a formula is determined by the following rule:

$$P \models g \triangleq closed(P) \wedge (g \cap \overline{P} \neq \emptyset)$$

This rule states that g is true in P if and only if P is closed and some antichain in g is also in P. This is similar to saying that there exists a global state in P in which the global predicate g is true.

The $\triangle$ operator is essentially the same as the boolean AND operator. We use $\triangle$ in order to avoid confusion with its boolean counterpart. This is especially useful in proofs where the two operators frequently appear in the same expression.

$$P \models f_1 \triangle f_2 \triangleq (P \models f_1) \wedge (P \models f_2)$$

The last two rules are the heart of RCL. They show how to evaluate a recursive formula. The only difference between them is the ordering between the cuts.

$$P \models g\langle f \rangle S \triangleq (\exists a, b : a, b \in \overline{P} : a \prec b \wedge a \models g \wedge (a..b) \models f \wedge [b..P] \models S)$$

The strong-precedes counter part to the above formula is:

$$P \models g\langle\!\langle f \rangle\!\rangle S \triangleq (\exists a, b : a, b \in \overline{P} : a \prec\!\!\prec b \wedge a \models g \wedge (a..b) \models f \wedge [b..P] \models S)$$

Now consider the following formula:

$$g_1\langle f_2 \rangle g_2\langle f_3 \rangle g_3 \ldots \langle f_n \rangle g_n$$

This formula holds on a poset P if and only if there exist cuts a_i in P such that $a_{i-1} \prec a_i$ and $a_i \models g_i$ and $(a_{i-1}..a_i) \models f_i$. Figures 1 and 2 show some examples.

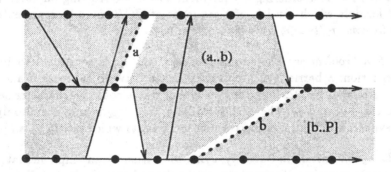

Fig. 1. Example of a poset structure which could satisfy $g\langle f \rangle g$. The cuts a and b divide the poset into regions as indicated by the shading. Notice that each region is closed.

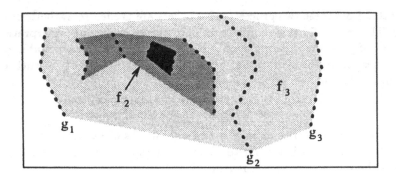

Fig. 2. Example of a structure which might satisfy $g_1 \langle f_2 \rangle g_3 \langle f_4 \rangle g_5$. The CGPs, g_i, need not be full width antichains, which is why they are shown in a "free-form" manner. Each f_i is another RCL formula. Two levels of recursion are shown for f_2, where f_2 has the structure $g \langle f \rangle g \langle g \langle f \rangle g \rangle g$.

5 Examples

This section gives examples of some useful RCL formulas. The first three examples show previous debugging logics which are special cases of RCL. In each example, p_i is a predicate on the state of a single process.

Sequence of local predicates: Consider the sequence of local predicates as defined in [17]: $(p_1, p_2, \ldots p_n)$. Each local predicate p_i is a special case of CGP. Therefore, this sequence of local predicates is equivalent to the RCL formula $p_1 \langle true \rangle p_2 \ldots \langle true \rangle p_n$.

Event Normal From (ENF): Chiou and Korfage [3] define *event normal form* predicates which are sequences of CGP. In their sequencing relation, global state a precedes global state b if and only if each local state in a happens-before all local states in b. This ordering is equivalent to the $\twoheadleftarrow$ ordering on cuts. Thus, an ENF formula which consists of a sequence of *CGP* could be represented in RCL as follows: $g_1 \langle\!\langle true \rangle\!\rangle g_2 \langle\!\langle true \rangle\!\rangle g_3 \ldots \langle\!\langle true \rangle\!\rangle g_n$.

Barrier Synchronization: The strong precedes relation is equivalent to barrier synchronization: a barrier synchronization exists between two cuts if and only if they are related by $\twoheadleftarrow$. Suppose two global states could be characterized by the predicates g_1 and g_2. The RCL formula $g_1 \langle\!\langle true \rangle\!\rangle g_2$ is true if and only if a barrier synchronization takes place between two cuts which satisfy g_1 and g_2.

Butterfly Synchronization: Butterfly synchronization is an implementation of barrier synchronization. Its structure can be defined recursively. Let X denote a set of process identifiers and let X_l and X_h be a partition of this set into upper and lower halves. $BF(X)$ will be defined to be an RCL formula which is true when there exists a butterfly synchronization between the processes named in X.

$BF(X)$ is the formula *true* if the size of X equals 1. Otherwise, $BF(X)$ equals the formula $g_X \langle\!\langle BF(X_l) \triangle BF(X_h) \rangle\!\rangle g_X$. Figure 3 shows an example.

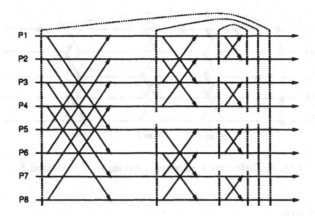

Fig. 3. Butterfly synchronization example. See section 5.

Distributed Consensus: Consider a fixed connection network. A phase consists of a message exchange on each edge. After phase i, each node has data from all nodes within distance i.

Let $g_{\{1,2\}}$ denote a CGP which is true on all antichains which contain exactly one state from each of process 1 and 2. Consider the example shown in figure 4. The communication structure of distributed consensus on the edge between nodes 1 and 2 that network can be captured by the following RCL formula:

$$g_{\{1,2,3,4,5\}} \langle\!\langle g_{\{1,2,3\}} \langle\!\langle g_{\{1,2\}} \langle\!\langle true \rangle\!\rangle g_{\{1\}} \rangle\!\rangle g_{\{1\}} \rangle\!\rangle g_{\{1\}}$$

The innermost form, $g_{\{1,2\}} \langle\!\langle true \rangle\!\rangle g_{\{1\}}$, is true after phase 1. The form that surrounds that becomes true after phase 2. The entire formula becomes true after phase 3 at which time consensus is complete since number of phases equals maximum distance between any two nodes.

Data Collection: It is common practice in distributed computing to scatter data among a set of computers, have each computer perform some operation on the data, and then collect the results. The collection phase of this operation can characterized with an RCL formula. Using the notation defined above, the formula is shown below.

$$g_{\{1,2,3,4,5,6,7,8\}} \langle\!\langle true \rangle\!\rangle g_{\{2,4,6,8\}} \langle\!\langle true \rangle\!\rangle g_{\{4,8\}} \langle\!\langle true \rangle\!\rangle g_{\{8\}}$$

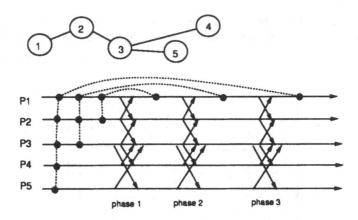

Fig. 4. Distributed consensus example. See section 5.

6 Algorithm

The algorithm is implemented by the function $fc()$. Given a cut a, and a formula f, the function $fc(a, f)$ finds the first cut b such that $[a..b]$ satisfies f. If there is no such cut, then $fc(a, f)$ returns $\top$.

The function $fc(a, f)$ calls three subroutines ($findCGP$, $advance$, and sup) as it parses an RCL formula. The pattern of subroutine calls depends on the syntactic structure of the formula.

Function $findCGP$: Function $findCGP$ finds the first cut in a poset that satisfies a given conjunctive global predicate (CGP). An efficient, decentralized, token-based algorithm for detecting CGP appears in [8]. The subroutine $findCGP$ is assumed to be a procedural interface to this algorithm. Thus, any process can call $findCGP$ to spawn the distributed token based algorithm described in [8]. The calling process is blocked until the underlying distributed algorithm completes at which point the result is returned by $findCGP$. Note that the blocked process is part of the RCL detection algorithm, not the underlying computation. In [8], it is shown that $findCGP$ has $O(NM)$ message, time and space complexity, where N is the number of processes, and M is the number of application messages.

Function sup: Function sup is takes one or more cuts as input and returns the supremum of those cuts. Cuts are represented as vector clock values, and the sup function takes the component-wise maximum of the vector clock values. The sup function can be implemented with computational complexity $O(NB)$, where B is the number of cuts input to sup, and N is the number of processes.

Function $advance(a, b)$: Function $advance(a, b)$ advances cut b until cut a strongly precedes it. It then returns the advanced cut. The $advance$ function can be implemented with $O(N^2)$ computational complexity.

The function $fc(a, f)$ takes a cut a and a formula f and returns the first cut b such that $[a..b]$ satisfies f. If there is no such cut, then $fc(a, f)$ returns $\top$. Four definitions of $\Psi(a, f)$ are shown below, one for each syntactic form of f.

$\underline{fc(a, g\langle f\rangle S)}$

$a_1 = fc(a, g)$;
$a_2 = fc(a_1, f)$;
$a_3 = fc(a_2, S)$;
return a_3;

$\underline{fc(a, g\langle\!\langle f\rangle\!\rangle S)}$

$a_1 = fc(a, g)$;
$a_2 = fc(a_1, f)$;
$a_3 = advance(a_1, a_2)$;
$a_4 = fc(a_3, S)$;
return a_4;

$\underline{fc(a, g)}$

return $findCGP(a, g)$;

$\underline{fc(P, f_1 \triangle f_2)}$

return $sup(fc(P, f_1), fc(P, f_2))$;

The algorithm is recursive and it mirrors the syntax structure of RCL. The recursion always bottoms out in the $fc(a, g)$ function. The poset which is being searched is a global read only structure and is not shown in the above descriptions. Only the subroutines $findCGP$ and $advance$ need access to the poset structure.

The correctness proof consists of showing that $fc(a, f) = \inf \Psi(P, f)$, where $a = \inf \overline{P}$. There are several properties of RCL which enable us to prove this. For example, RCL is monotonic with respect to set inclusion. Using this property of monotonicity, it can be shown that a cutset is a lattice. The lattice property allows us to implement $fc(a, g\langle f\rangle S)$ in a greedy fashion: first finding g, then f and finally S.

7 RCL Properties and Algorithm Proof

The logic is monotonic with respect to set inclusion over closed posets. If $P \subseteq R$, R is closed, and P satisfies some formula, then R also satisfies that formula. This definition of monotonicity is more encompassing than other commonly used definitions which use the "advancement of time" as the ordering relation instead of set inclusion. Using set inclusion has the benefit that if a formula is true for a given subcomputation, then it remains true not only when (local) states are added to the end of the subcomputation, but also when states are included from before the computation or even concurrent with it. That is, the poset which represents the computation can "grow" in any direction (as long as it remains closed). Lemma 1 proves that RCL is monotonic.

Lemma 1. *Monotonicity:* $P \subseteq R \wedge closed(R) \wedge P \models f \Rightarrow R \models f$

Proof: Appears in [19]. ∎

Given any poset P and formula f, the cutset of (P, f) forms a lattice. Cuts can be represented with vector clocks, and the infimum of two cuts is the component-wise minimum of their vector clocks. The supremum is the component-wise maximum. The same operators are used in cutset lattices.

Lemma 2. *Lattice:* $a, b \in \Psi(P, f) \Rightarrow \inf(a, b) \in \Psi(P, f) \wedge \sup(a, b) \in \Psi(P, f)$

Proof: Appears in [19]. ∎

The following is a statement that the algorithm is correct:

$$c = \inf \overline{P} \Rightarrow fc(c, f) = \inf \Psi(P, f)$$

This statement is proven by structural induction on f. First we show that it is correct for g, then we show that if it is correct for f_1 and f_2, then it is correct for $f_1 \triangle f_2$, and finally we show that if it is correct for g, f and S, then it is correct for $g\langle f \rangle S$ and $g\langle\!\langle f \rangle\!\rangle S$.

Lemma 3. $fc(c, g) = \inf \Psi(P, g)$, *where* $c = \inf \overline{P}$

Proof: Proof of a token based algorithm for the case when g is defined to be a CGP appears in [7]. ∎

Lemma 4. $fc(c, f_1 \triangle f_2) = \inf \Psi(P, f_1 \triangle f_2)$, *where* $c = \inf \overline{P}$

Proof:

$fc(c, f_1 \triangle f_2)$
$= \{$ from the algorithm $\}$
$\sup\{fc(c, f_1), fc(c, f_2)\}$
$= \{$ by induction, and since $c = \inf \overline{P}$ $\}$
$\sup\{\inf \Psi(P, f_1), \inf \Psi(P, f_2)\}$
$= \left\{ \begin{array}{l} \text{the sup of the first cut to satisfy } f_1 \text{ and the first cut to} \\ \text{satisfy } f_2 \text{ is equal to the first cut which satisfies both } f_1 \\ \text{and } f_2. \end{array} \right\}$
$\inf\{a \mid a \in \overline{P} \wedge [P..a] \models f_1 \wedge [P..a] \models f_2\}$
$= \{$ semantics $\}$
$\inf\{a \mid a \in \overline{P} \wedge [P..a] \models f_1 \triangle f_2\}$
$= \{$ defn $\Psi(\ )$ $\}$
$\inf \Psi(P, f_1 \triangle f_2)$

∎

The next lemma is used in the proof of $fc(P, g\langle f \rangle S)$ several times. Presenting it here greatly simplifies the presentation of the proof for $fc(P, g\langle f \rangle S)$. Figure 5 shows the structure of the posets in this lemma.

Lemma 5. $Z(a_1, a_2, b_1, b_2, f, P)$, *which is defined as:*
$$a_1, a_2, b_1, b_2 \in \overline{P} \wedge a_1 \preceq b_1 \wedge [b_1..b_2] \models f \wedge a_2 = \inf \Psi([a_1..P], f)$$
$$\Rightarrow$$
$$a_2 \preceq b_2 \wedge [a_1..a_2] \models f$$

Proof:

The following are assumed:
$$a_1, a_2, b_1, b_2 \in \overline{P}$$
$$a_1 \preceq b_1$$

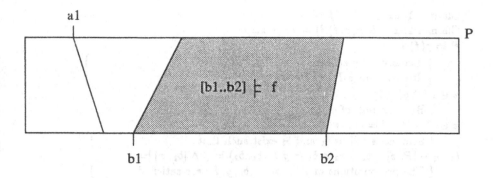

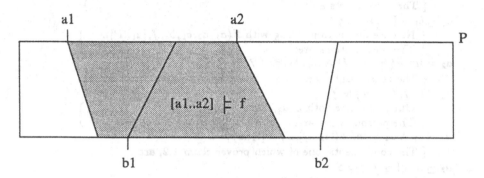

Fig. 5. Structure of posets in lemma 5.

$$[b_1..b_2] \models f$$
$$a_2 = inf\ \Psi([a_1..P], f)$$

Note that $[b_1..b_2] \models f$ implies that $b_1 \preceq b_2$, which in turn implies that $[b_1..b_2] \subseteq [a_1..b_2]$. Thus, by monotonicity, $[a_1..b_2] \models f$. This implies that $b_2 \in \Psi([a_1..P], f)$. And since $a_2 = inf\ \Psi([a_1..P], f)$, then $a_2 \preceq b_2$. ∎

Lemma 6. $fc(c, g\langle f\rangle S) = inf\ \Psi(P, g\langle f\rangle S)$, where $c = inf\ \overline{P}$

Proof: From the algorithm, it is clear that $fc(c, g\langle f\rangle S) = a_3$, where:
$$a_1 = fc(c, g)$$
$$a_2 = fc(a_1, f)$$
$$a_3 = fc(a_2, S)$$
By structural induction, we also know that:
$$a_1 = inf\ \Psi([c..P], g)$$
$$a_2 = inf\ \Psi([a_1..P], f)$$
$$a_3 = inf\ \Psi([a_2..P], S)$$

We must show that $a_3 = inf\ \Psi(P, g\langle f\rangle S)$. The proof is divided into two cases:
   Case 1: $P \models g\langle f\rangle S$
   Case 2: $P \not\models g\langle f\rangle S$
Case 1: $P \models g\langle f\rangle S$
In order to show $a_3 = inf\ \Psi(P, g\langle f\rangle S)$, it suffices to show:

Claim 1.1: $a_3 \in \Psi(P, g\langle f\rangle S)$, and

Claim 1.2: $x \in \Psi(P, g\langle f\rangle S) \Rightarrow a_3 \prec x$.

$P \models g\langle f\rangle S$

$\left\{\begin{array}{l} \text{Let } x \text{ be any element in } \Psi(P, g\langle f\rangle S). \\ \text{It exists since } P \models g\langle f\rangle S. \end{array}\right\}$

$x \in \Psi(P, g\langle f\rangle S)$

$\quad$ { By definition of cutsets: $\qquad\qquad\qquad\qquad\qquad\qquad$ }

$x \in P \wedge [P..x] \models g\langle f\rangle s$

$\quad$ { Semantics tell us b_1 and b_2 exist such that: $\qquad\qquad\qquad$ }

$b_1, b_2 \in \overline{[P..x]} \wedge b_1 \prec b_2 \wedge b_1 \models g \wedge (b_1..b_2) \models f \wedge [b_2..x] \models S$

$\quad$ { The preconditions of $Z(\bot, a_1, \bot, b_1, g, P)$ are satisfied: $\quad$ }

$a_1 = \inf \Psi(P, g) \wedge [\bot..b_1] \models g \wedge \bot \preceq \bot$

$\quad$ { The consequents are: $\qquad\qquad\qquad\qquad\qquad\qquad\quad$ }

$a_1 \preceq b_1 \wedge [\bot..a_1] \models g$

$\left\{\begin{array}{l} \text{Now we do the same thing with } Z(a_1, a_2, b_1, b_2, f, [a_1..P]). \\ \text{The preconditions are:} \end{array}\right\}$

$a_2 = \inf \Psi([a_1..P], f) \wedge (b_1..b_2) \models f \wedge a_1 \preceq b_1$

$\quad$ { The consequents are: $\qquad\qquad\qquad\qquad\qquad\qquad\quad$ }

$a_2 \preceq b_2 \wedge (a_1..a_2) \models f$

$\left\{\begin{array}{l} \text{One more time with } Z(a_2, a_3, b_2, x, S, [a_2..P]). \\ \text{The preconditions are:} \end{array}\right\}$

$a_2 \preceq b_2 \wedge a_3 = \inf \Psi([a_2..P], S) \wedge [b_2..x] \models S$

$\quad$ { The consequents, one of which proves claim 1.2, are: $\quad$ }

$a_3 \preceq x \wedge [a_2..a_3] \models S$

$\quad$ { We have collected the following true statements: $\qquad$ }

$a_1 \preceq a_2 \wedge a_1 \models g \wedge (a_1..a_2) \models f \wedge [a_2..a_3] \models S$

$\quad$ { And by semantics: $\qquad\qquad\qquad\qquad\qquad\qquad\quad$ }

$a_3 \in \overline{P} \wedge [P..a_3] \models g\langle f\rangle S$

$\quad$ { Which leads us to claim 1.1: $\qquad\qquad\qquad\qquad\quad$ }

$a_3 \in \Psi(P, g\langle f\rangle S)$

Case 2: $P \not\models g\langle f\rangle S$

$P \not\models g\langle f\rangle S$

$\quad$ { semantics $\qquad\qquad\qquad\qquad\qquad\qquad\qquad\qquad\quad$ }

$\neg(\exists c_1, c_2 :: c_1, c_2 \in \overline{P} \wedge c_1 \prec c_2 \wedge c_1 \models g \wedge (c_1..c_2) \models f \wedge [c_2..P] \models S)$

$\quad$ { Instantiate c_1, c_2 with a_1, a_2 and apply de Morgan's law: $\quad$ }

$a_1 \notin \overline{P} \vee a_2 \notin \overline{P} \vee a_1 \not\prec a_2 \vee a_1 \not\models g \vee (a_1..a_2) \not\models f \vee [a_2..P] \not\models S$

$\left\{\begin{array}{l} \text{Since } (a_1 \notin \overline{P} \Rightarrow a_1 \not\models g) \text{ and } (a_1 \not\prec a_2 \Rightarrow (a_1..a_2) \not\models f) \text{ and} \\ (a_2 \notin \overline{P} \Rightarrow [a_2..P] \not\models S), \text{ then:} \end{array}\right\}$

$a_1 \not\models g \vee (a_1..a_2) \not\models f \vee [a_2..P] \not\models S$

$\quad$ { By definition of a_1, a_2 and a_3: $\qquad\qquad\qquad\qquad\quad$ }

$a_1 = \top \vee a_2 = \top \vee a_3 = \top$

$\quad$ { By definition of a_3: $\qquad\qquad\qquad\qquad\qquad\qquad\quad$ }

$a_3 = \top$

$\blacksquare$

Lemma 7. $fc(c, g\langle\!\langle f\rangle\!\rangle S) = \inf \Psi(P, g\langle\!\langle f\rangle\!\rangle S)$, where $c = \inf \overline{P}$

Proof: Similar to the proof for weak sequences: $g\langle f\rangle S$

$\blacksquare$

8 Conclusions

An RCL formula is essentially a specification of a pattern of CGPs. If each CGP occurs in a computation, and they occur in the correct pattern, then the RCL formula is true in that computation.

RCL is a recursive logic, which means that the patterns can be recursive. logics are intuitive because there are fewer constructs and rules to remember. They are also powerful because the full power of the logic is available at each level of recursion. As a result of these properties of recursive logics, RCL is simple yet powerful. Computational structures of common paradigms such as butterfly synchronization and distributed consensus can be expressed easily in RCL.

A high level algorithm for detecting whether or not a computation satisfies a given RCL formula was presented. The complexity of the algorithm was not analyzed due to space constraints, but it is shown in [19] to be quite efficient – about the same as the complexity of detecting a single CGP.

9 Acknowledgments

We are grateful to Don Pazel at IBM for many fruitful discussions on the topic of breakpoint logic.

References

1. Ö. Babaoğlu and K. Marzullo. *Consistent global states of distributed systems: fundamental concepts and mechanisms, in Distributed Systems*, chapter 4. ACM Press, Frontier Series. (S. J. Mullender Ed.), 1993.
2. C. Chase and V. K. Garg. On techniques and their limitations for the global predicate detection problem. In *Proc. of the Workshop on Distributed Algorithms*, France, September 1995.
3. H. K. Chiou and W. Korfhage. Enf event predicate detection in distributed systems. In *Proc. of the Principles of Distributed Computing*, pages 91–100, Los Angeles, CA, 1994. ACM.
4. R. Cooper and K. Marzullo. Consistent detection of global predicates. In *Proc. of the Workshop on Parallel and Distributed Debugging*, pages 163–173, Santa Cruz, CA, May 1991. ACM/ONR.
5. C. J. Fidge. Partial orders for parallel debugging. *Proceedings of the ACM SIG-PLAN/SIGOPS Workshop on Parallel and Distributed Debugging, published in ACM SIGPLAN Notices*, 24(1):183–194, January 1989.
6. E. Fromentin, M. Raynal, V. K. Garg, and A. I. Tomlinson. On the fly testing of regular patterns in distributed computations. In *Proc. of the 23rd Intl. Conf. on Parallel Processing*, St. Charles, IL, August 1994.
7. V. K. Garg and C. Chase. Distributed algorithms for detecting conjunctive predicates. In *Proc. of the IEEE International Conference on Distributed Computing Systems*, pages 423–430, Vancouver, Canada, June 1995.
8. V. K. Garg, C. Chase, J. R. Mitchell, and R. Kilgore. Detecting conjunctive channel predicates in a distributed programming environment. In *Proc. of the 28th*

Hawaii International Conference on System Sciences, pages 232–241, Vol II, January 1995.

9. V. K. Garg, A. I. Tomlinson, E. Fromentin, and M. Raynal. Expressing and detecting general control flow properties of distributed computations. In *Proc. of the 7th IEEE Symposium on Parallel and Distributed Processing*, San Antonio, TX, October 1995.

10. V. K. Garg and B. Waldecker. Detection of unstable predicates in distributed programs. In *Proc. of 12th Conference on the Foundations of Software Technology & Theoretical Computer Science*, pages 253–264. Springer Verlag, December 1992. Lecture Notes in Computer Science 652.

11. V. K. Garg and B. Waldecker. Detection of weak unstable predicates in distributed programs. *IEEE Transactions on Parallel and Distributed Systems*, 5(3):299–307, March 1994.

12. V. K. Garg and B. Waldecker. Detection of strong unstable predicates in distributed programs. *IEEE Transactions on Parallel and Distributed Systems*, Submitted.

13. D. Haban and W. Weigel. Global events and global breakpoints in distributed systems. In *Proc. of the 21st International Conference on System Sciences*, volume 2, pages 166–175, January 1988.

14. G. Hoagland. A debugger for distributed programs. Master's thesis, University of Texas at Austin, Dept. of Electrical and Computer Engineering, Austin, TX, August 1991.

15. M. Hurfin, N. Plouzeau, and M. Raynal. Detecting atomic sequences of predicates in distributed computations. In *Proc. of the Workshop on Parallel and Distributed Debugging*, pages 32–42, San Diego, CA, May 1993. ACM/ONR. (Reprinted in SIGPLAN Notices, Dec. 1993).

16. L. Lamport. Time, clocks, and the ordering of events in a distributed system. *Communications of the ACM*, 21(7):558–565, July 1978.

17. B. P. Miller and J. Choi. Breakpoints and halting in distributed programs. In *Proc. of the 8th International Conference on Distributed Computing Systems*, pages 316–323, San Jose, CA, July 1988. IEEE.

18. R. Schwarz and F. Mattern. Detecting causal relationships in distributed computations: In search of the holy grail. *Distributed Computing*, 7(3):149–174, 1994.

19. A. I. Tomlinson. *Observation and Verification of Software for Distributed Systems*. PhD thesis, University of Texas at Austin, Dept. of Electrical and Computer Engineering, Austin, TX, August 1995.

20. A. I. Tomlinson and V. K. Garg. Detecting relational global predicates in distributed systems. In *Proc. of the Workshop on Parallel and Distributed Debugging*, pages 21–31, San Diego, CA, May 1993. (Reprinted in SIGPLAN Notices, Dec. 1993).

Partiality and Approximation Schemes for Local Consistency in Networks of Constraints*

Nick D. Dendris, Lefteris M. Kirousis, Yannis C. Stamatiou, and
Dimitris M. Thilikos

Department of Computer Engineering and Informatics,
Patras University, Rio, 265 00 Patras, Greece
and
Computer Technology Institute,
P.O. Box 1122, 261 10 Patras, Greece.
e-mail: {dendris, kirousis, stamatiu, sedthilk}@cti.gr

Abstract. A constraint network is arc consistent if any value of its variables is compatible with at least one value of any other variable. The Arc Consistency Problem (ACP) consists in filtering out values of the variables of a given network to obtain one that is arc consistent, without eliminating any solution, i.e. a total value assignment that satisfies all constraints. Enforcing arc consistency, or the more general k-consistency, in a constraint network is a widely used preprocessing step before identifying the set of solutions. ACP is known to be inherently sequential, so in this paper we examine the problem of achieving partial consistency, i.e. filtering out values from the variables so that each value in the new network is compatible with at least one value of not necessarily all, but a constant fraction of the other variables. We call such networks partially arc consistent. We give an algorithm that, for any constraint network, outputs a partially arc consistent subnetwork of it in sublinear ($O(\sqrt{n}\log n)$) parallel time using $O(n^2)$ processors. This is the first (to our knowledge) sublinear-time parallel algorithm with polynomially many processors that removes at least a constant fraction of the local inconsistencies of a general constraint network, without eliminating any solution. We also generalize the notion of partiality to the k-consistency problem: We give an algorithm that is faster than the previously known k-consistency algorithms, and which finds a partially k-consistent subnetwork of any given network. Finally, we propose several approximation schemes to a total solution of ACP, and show that they are P-complete. This indicates that the approach of partial solutions, rather than that of approximation schemes, is more promising for parallelism.

1 Introduction

A *constraint network* comprises of n variables, a *domain* of permissible values for each of the variables, and a set of *constraint relations*, each binding the

---

* This research was partially supported by the European Union ESPRIT Basic Research Projects ALCOM II (contract no. 7141), GEPPCOM (contract no. 9072) and Insight II (contract no. 6019).

values of a collection of the variables. The Constraint Satisfaction Problem (CSP) is the problem of determining the n-tuples of values from the domains that are compatible with all constraints. This is a fundamental problem in Artificial Intelligence (an interesting special case of it is scene labeling, a preprocessing stage of object recognition). For a review of recent results see [5, 17, 19]. CSP is, in general, intractable. Many NP-complete problems such as graph colouring, 3-SAT, can be directly expressed in terms of CSP. There is an extensive literature on special cases of CSP that can be solved efficiently sequentially, or even by fast parallel algorithms (see, e.g., [2, 7, 9, 15, 22]).

A constraint network is called arc consistent if any value from any domain is compatible with at least one value of any other variable. *Discrete relaxation* is the process of removing values from the domains of the variables until the network is transformed to an arc consistent one, but in a way that no n-tuple of values that satisfies all constraints is eliminated. In other words, discrete relaxation solves the Arc Consistency Problem (ACP). It removes local inconsistencies and so it reduces the search space. Therefore, a network that has been transformed to an arc consistent one is expected to be easier to handle from the point of view of obtaining a solution for the corresponding CSP.

There are many efficient implementations of discrete relaxation for sequential models of computation (see [18]; see also [11] for an overview and recent results and see [6] for interesting experimental results). Parallel algorithms for the arc consistency problem have been studied in [4, 20]. However, in general, the problem is inherently sequential, i.e. it is P-complete [13]. There are interesting results on identifying cases where an exponential parallel speedup is possible (see [14]).

In this paper, in order to overcome the inherent sequentiality of the problem, we introduce and investigate the parallel complexity of partial solutions for ACP. We justify this approach by defining several natural approximation schemes to the problem of arc consistency, and showing them to be P-complete. We also extend the notion of partiality to cover the case of the more general k-consistency.

An α-partial solution of ACP is one where every value of any variable is compatible with values of at least $\alpha(n-1)$ of the other variables ($0 < \alpha \leq 1$). Let Δ be the maximm degree in the graph of constraints, and q be the maximum number of values for a variable. Usually, q is a small constant. For example, in many applications of CSP in Computer Vision, q is no more than 3 or 4 (see [21]; see [8] for some recent results). Also, in the majority of cases, Δ is a constant factor of the number of variables. We show that, for any $\alpha < 1 - \frac{q}{q+1}\frac{\Delta}{n-1}$, there is a parallel algorithm that runs in time $O(\sqrt{nq}\log(nq))$ and produces an α-partial solution for ACP that contains the actual ACP solution.

Our algorithm operates as follows: For as long as there are more than $\sqrt{nq}$ values that are compatible with values of less than $n-1$ variables, the algorithm removes them in parallel steps. The target is to capture the steps of the original discrete relaxation that offer sufficiently high degree of parallelism. Afterwards, the algorithm removes values that are compatible with less than $\alpha(n-1)$ variables. A combinatorial lemma ensures that these steps will not be more than

$O(\sqrt{nq})$. The algorithm requires $O(n^2q^2)$ processors on the Concurrent-Read Exclusive-Write (CREW) shared-memory parallel machine model (PRAM).

Notice that a partial solution for ACP is, in itself, a removal of some local inconsistencies. To our knowledge, this is the first sublinear-time parallel algorithm that achieves a nontrivial degree of elimination of local inconsistencies, using polynomially many processors (see [4] for results with exponentially many processors).

We justify this approach for partial ACP solutions by defining several natural approximation schemes to the total solution of ACP, and proving them to be P-complete. Let D be the disjoint union of all domains of a network, D_{AC} be the disjoint union of all domains *after* the application of discrete relaxation, and $R_{AC} = D - D_{AC}$. Our first approximation scheme asks to determine, for a given $\epsilon \in (0,1]$, a set D' such that $D_{AC} \subseteq D' \subseteq D$ and $\epsilon|D'| \leq |D_{AC}|$. We prove that, unless P=NC, no such set D' can be found by a parallel algorithm in NC (for the class NC see [12]). We show the same for the approximation scheme that, given an $\alpha \in (0,1]$, asks for a set R' such that $R' \subseteq R_{AC}$ and $\alpha|R_{AC}| \leq |R'|$ (it is interesting that we had to apply completely different approaches to prove these two results).

The other two approximation schemes that we introduce are related to the degree of incompatibility that a value must exhibit in order to be removed by a relaxation-type algorithm. We define the support of a value d to be the maximum k such that, when we remove values compatible with less than k variables, d is not removed and no domain is emptied. We also define the elimination degree of d as the least k such that, when we remove values not supported by more than k variables, d is not removed and no domain is emptied (see next sections for formal definitions). The support and the elimination degree of a value have sum that is equal to $n - 1$. However, an algorithm approximating within constant approximation factor one of these parameters, does not yield a constant factor approximation of the other. We show that approximating any of these parameters is P-complete. Thus, looking for partial solutions, and not approximation schemes, seems to be the correct way to obtain sublinear-time parallel algorithms that approach the solution of ACP.

To this extent, in the last section of the paper, we generalize our partiality results to the case of the k-consistency, a generalized form of arc consistency, as well as for the case of strong k-consistency. A constraint network is k-consistent if for any $(k - 1)$-tuple of compatible values for $k - 1$ variables and for any kth variable, there exists a value for it that is compatible with the $(k - 1)$-tuple. If a network is i-consistent for all $i = 2, \ldots, k$, then we say that it is strongly k-consistent (see [5, 17]). k-consistency can be achieved in time $O(n^k q^k)$ [3].

For k-consistency, we define the notion of partiality as follows: a network is α-partially k-consistent if any $(k - 1)$-tuple of compatible values can be extended with a compatible value from the domain of not necessarily all, but at least $\alpha(n - k + 1)$ other variables. If a network has the α-partial k-consistency property for all $i = 2, \ldots, k$, then we call it α-partially strongly k-consistent. We give a parallel algorithm for the CREW PRAM that, in time $O((nq)^{\frac{k-1}{2}} \log(nq))$

and using $O(n^k q^k)$ processors, outputs an α-partially strongly k-consistent sub-network of any given constraint network that contains the solution of the strong k-consistency problem.

2 Definitions and Preliminaries

A *binary constraint network* consists of a set of variables $X_1, \ldots, X_n$, a set of variable domains $D_1, \ldots, D_n$, and a set of binary constraints $\mathcal{C}$. For $i = 1, \ldots, n$, D_i is the set of permissible values for variable X_i. A constraint $R_{ij} \in \mathcal{C}$ is a subset of $D_i \times D_j$ ($i \neq j$). A value d_i of X_i is *compatible* with a value d_j of X_j ($i \neq j$) iff there is no $R_{ij} \in \mathcal{C}$ such that $(d_i, d_j) \notin R_{ij}$. For a given constraint network $\mathcal{N}$ with variables $X_1, \ldots, X_n$, domains $D_1, \ldots, D_n$, and set of constraints $\mathcal{C}$, a *domain-reduced subnetwork* of $\mathcal{N}$ is any constraint network $\mathcal{N}'$ with the same variables, domains $D_1' \subseteq D_1, \ldots, D_n' \subseteq D_n$, and set of constraints $\mathcal{C}'$ containing exactly the restrictions of the elements of $\mathcal{C}$ to the corresponding domains.

The *Constraint Satisfaction Problem* (CSP), for a given constraint network, asks for all the n-tuples $(d_1, \ldots, d_n)$ of values such that $d_i \in D_i$, $i = 1, \ldots, n$, and for every $R_{ij} \in \mathcal{C}$, (d_i, d_j) is in R_{ij}. Such an n-tuple is called a *solution* of the CSP. The decision version of the CSP is to determine if any solution exists.

Given an instance Π of the CSP with n variables, its *constraint graph* (we denote this graph by G^Π or just G when no confusion may arise) has n vertices which correspond to the variables of Π and it contains an edge $\{v_i, v_j\}$ iff the corresponding variables are bound by a constraint.

In this paper we use another graph representation of a CSP instance Π, which we call the *compatibility graph*. Let q denote the maximum number of values that can be assigned to a variable, i.e. $q = \max\{|D_i|, i = 1, \ldots, n\}$. The compatibility graph C^Π (or just C, when no confusion may arise) of Π is an n-partite graph. The ith part of C^Π corresponds to variable X_i of Π and it has exactly $|D_i|$ vertices, one for each value of X_i. For simplicity, we use the same notation for the vertices of the compatibility graph and their corresponding values (domain-elements) in the network. In the compatibility graph C^Π, two vertices d_i and d_j are connected by an edge iff the corresponding values $d_i \in D_i$ and $d_j \in D_j$ are compatible. We denote by N the total number of vertices of C^Π, i.e. $N = |D|$ where $D = \bigcup_{i=1,\ldots,n} D_i$ and $N \leq nq$. In other words, N is equal to the total number of values in the constraint network (values in the domains of different variables are considered to be different). An n-tuple $(d_1, \ldots, d_n)$ is a solution of Π iff the set of vertices that correspond to the values in the n-tuple induces a subgraph of C^Π which is an n-clique. Therefore, the problem of finding all the solutions of a CSP instance, or of determining whether any solution at all exists, can be reduced to the problem of identifying *all* n-cliques of its compatibility graph, or of determining whether at least one such clique exists.

A constraint network $\mathcal{N}$ is *arc consistent* if the following holds: for any variable X_i, for any value $d_i \in D_i$, and for any other variable X_j, there exists at least one value assignment $d_j \in D_j$ such that d_i is compatible with d_j. The *Arc Consistency Problem* (ACP) is the problem of finding a maximal (with respect to the domains) arc consistent subnetwork $\mathcal{N}_{AC}$ of $\mathcal{N}$. Also, $\mathcal{N}_{AC}$ is unique, for if there was another maximal arc consistent subnetwork $\mathcal{N}_{AC}$ with domains $A'_1, \ldots, A'_n$, then the domain-reduced subnetwork of $\mathcal{N}$ with domains $A_1 \cup A'_1, \ldots, A_n \cup A'_n$ would also be arc consistent, a contradiction to the maximality of $\mathcal{N}_{AC}$ and $\mathcal{N}'_{AC}$. We call $\mathcal{N}_{AC}$ the *solution* of ACP.

In terms of the compatibility graph C, ACP is formulated as follows: Find the maximal (with respect to the set of vertices) induced subgraph C_{AC} of C such that any vertex in any of its parts is connected with vertices in $n-1$ other parts.

All values $d_i \in D_i$, $i = 1, \ldots, n$, which participate in some solution of the CSP belong in the domain A_i of the ACP solution. Since CSP is an NP-complete problem, a common approach for solving it involves a preprocessing step to make the network arc consistent and reduce the variable domains, before employing exhaustive search to determine the actual solutions. Arc consistency is achieved by the procedure known as *discrete relaxation*. The algorithm is the following:

Algorithm: discrete relaxation
Input: a constraint network $\mathcal{N}$
Output: $\mathcal{N}_{AC}$
begin
    while $\exists X_i$ and $d_i \in D_i : \exists j \neq i :$ d_i is incompatible with all labels in D_j
    do $D_i \leftarrow D_i - \{d_i\}$ **od**
end

The algorithm removes values that are incompatible with the values in the domain of at least one variable, until no such values exist. Discrete relaxation runs in optimal $O(eq^2)$time [1], where e is the number of constraints and q is the maximum number of values in the domain of a variable.

In [13] it was proved that the arc consistency problem is P-complete, in the sense that given a constraint network $\mathcal{N}$, it is P-complete to decide whether a given value for a variable belongs in the corresponding domain of $\mathcal{N}_{AC}$. Here, we state without proof the following stronger results:

Theorem 1. *It is P-complete, given a constraint network $\mathcal{N}$, to decide whether the discrete relaxation empties the domains of the variables in $\mathcal{N}$.*

Theorem 2. *It is P-complete to solve the ACP in constraint networks where the constraint graph is planar and has maximum degree 3.*

The proof for Theorem 2 uses a reduction from the Planar Circuit Value Problem (see [10]).

Several implementations of the discrete relaxation for the PRAM model of computation are described in [20]. The best of these algorithms, PAC-4, executes in $O(nq)$ time using $O(n^2q^2)$ processors.

3 Partial Arc Consistency

In this section we introduce the notion of partiality in arc consistency, and give a sublinear-time parallel algorithm that finds partial solutions to the ACP (for other relevant combinatorial problems that admit to partial solutions see [16]).

A constraint network $\mathcal{N}$ is α-*partially arc consistent* if no variable domain is empty and any value of a variable is compatible with values of at least a constant factor of the other variables, i.e. with values of $\alpha(n-1)$ other variables ($0 < \alpha \leq 1$). The α-*partial arc consistency problem* is the problem of finding a maximal (with respect to the domains) α-partially arc consistent subnetwork $\mathcal{N}_{AC,\alpha}$ of $\mathcal{N}$. When $\alpha = 1$, this problem is the same as the original ACP. In a way similar with the case of $\mathcal{N}_{AC}$, one can see that $\mathcal{N}_{AC,\alpha}$ is unique.

For any value d in a constraint network, the support-degree of d is the number of variables in the network that contain a value compatible with d. In the compatibility graph C, the support-degree of a vertex d in C is the number of parts in C that contain vertices adjacent to d.

In terms of the compatibility graph C, the α-partial arc consistency problem is formulated as follows: Find the maximal (with respect to the set of vertices) induced subgraph $C_{AC,\alpha}$ of C such that any vertex has support-degree at least $\alpha(n-1)$ and no part is empty. Such a graph is called α-partially arc consistent.

In this paper we introduce a modified version of the discrete relaxation algorithm, which we call α-*discrete relaxation*, that solves the α-partial arc consistency problem. We will present the algorithm in terms of the compatibility graph. Given a real value α, $0 < \alpha \leq 1$ and a compatibility graph C, the α-discrete relaxation is the following:

Algorithm: α-discrete relaxation
Input: a compatibility graph $C = (V, E)$
Output: $C_{AC,\alpha}$
begin
   while $\exists d \in V$: support-degree$(d) < \alpha(n-1)$
   do
      $V \leftarrow V - \{d\}$
      if an empty part appears in C
      then terminate and return the empty graph
   od
end

The algorithm outputs a subgraph $C_{AC,\alpha}$ of C that contains C_{AC} as a subgraph. Notice that, if during the α-discrete relaxation a part is emptied, then the ACP has the empty solution.

α-discrete relaxation has a parallel version which we call *parallel α-discrete relaxation*. In this version, on each step of the while loop, the identification and the elimination of the vertices to be removed is done in parallel in $O(\log(nq))$ time using $O(n^2 q^2)$ processors on the CREW PRAM. Therefore the α-partial arc consistency problem can be solved in $O(nq \log(nq))$ parallel time using $O(n^2 q^2)$ processors on the CREW PRAM.

Maximality of an α-partially arc consistent subgraph of C is not crucial, as long as this subgraph contains C_{AC}. It turns out that we can do better than computing $C_{AC,\alpha}$. That is, given a constraint network $\mathcal{N}$, its constraint graph C and its compatibility graph G, and for any $\alpha < 1 - \frac{q}{q+1}\frac{\Delta}{n-1}$, we can compute in sublinear parallel time an α-partially arc consistent subgraph of $C_{AC,\alpha}$ that contains C_{AC}. In the above, $\Delta = \Delta(G)$ is the maximum degree in the constraint graph G.

The algorithm is strongly based on the following technical lemma. This lemma ensures the existence of a partially consistent subgraph of C whenever C is sufficiently dense.

Lemma 3. *Let $\mathcal{N}$ be a constraint network. Let also G, C be its constraint and compatibility graph respectively. Assume that G is a graph with maximum degree Δ and let $l_\Delta = n - 1 - \Delta$. Let n be the number of parts in C, N the total number of values and q the largest cardinality of any part. Let $l_2 < l_1 \leq \Delta$ be two integers such that $l_2 < \frac{l_1}{1+q}$. If C contains $\leq m$ edges with support-degree$< l_\Delta + l_1$, then, if we apply $(\frac{l_\Delta + l_2}{n-1})$-discrete relaxation on C either no more than $\frac{m(l_1 - l_2)}{l_1 - l_2(1+q)}$ vertices are going to be removed or all the vertices of some part are going to be removed.*

Proof Let d be a vertex in some part D_i of C and v_i be the vertex of the constraint graph G corresponding to D_i. If two variables are not bound by any constraint, *all* pairs of their values are compatible. Thus, since there are at least l_Δ vertices in the constraint graph G not adjacent to v_i, any vertex in D_i is adjacent to all the vertices of at least l_Δ other parts of C. We define as var(d) the vertex of the constraint graph corresponding to the part of C that contains d. Also, for any vertex d in C:

(i) $U(d, C)$ is the collection of parts in C that contain a vertex adjacent to d and a vertex non adjacent to d.

(ii) $F(d, C)$ is the collection of parts in C where *all* the vertices are adjacent to d.

(iii) $W(d, C)$ is the collection of parts in C corresponding to vertices of the constraint graph G non adjacent to var(d).

(iv) $H(d, C)$ is the collection of parts that contain vertices adjacent to d in C and correspond to vertices adjacent to var(d) in G.

Notice that for a vertex d of C we have that support-degree$(d) = |U(d, C)| + |F(d, C)| = |W(d, C)| + |H(d, C)|$.

Let $\alpha = \frac{l_\Delta + l_2}{n-1}$, and assume that the α-discrete relaxation empties no part in C. Let R_α be the set of vertices in C that are removed by the α-discrete relaxation. Let also $T_\alpha \subseteq R_\alpha$ be the vertices in R_α that have support-degree$\geq l_\Delta + l_1$ in C. We denote by P the set of edges that connect a vertex $d \in T_\alpha$ with a vertex $d' \in R_\alpha$ where var(d), var(d') are adjacent verices in the constraint graph. Notice also that $|R_\alpha| \leq m + |T_\alpha|$. We observe the following:

(1) Let d_s be the vertex in T_α that is removed on the sth step of the α-discrete relaxation and C_s be the compatibility graph before step s. Then, support-degree$(d_s) < l_\Delta + l_2$ in C_s. Since during the α-elimination no part is emptied, we have that $F(d_s, C) = F(d_s, C_s)$. Also, $d_s \in T_\alpha$, so support-degree$(d_s) \geq l_\Delta + l_1$ in C. Recall that for any d, support-degree$(d) = |U(d, C)| + |F(d, C)|$. In this way we get that $|U(d_s, C) - U(d_s, C_s)| = |U(d_s, C)| - |U(d_s, C_s)| > l_1 - l_2$. Notice that var$(d_s)$ must be adjacent to all the vertices of G corresponding to the members of $U(d_s, C)$. Let E_{d_s} be the set of edges connecting d_s with the vertices in C that are contained in the members of $U(d_s, C) - U(d_s, C_s)$. Then $|E_{d_s}| > l_1 - l_2$. Notice also that for any $d_s, d_{s'} \in T_\alpha$, $E_{d_s} \cap E_{d_{s'}} = \emptyset$ and for any $d_s \in T_\alpha$, $E_{d_s} \subseteq P$. Thus $P \supseteq \bigcup_{d_s \in T_\alpha} E_{d_s}$, hence $|P| > (l_1 - l_2)|T_\alpha|$.

(2) For the vertex d_s that is removed at some step s, we have that $|W(d_s, C_s)| \geq l_\Delta$ (C_s is the compatibility graph before step s). Also, support-degree$(d_s) < l_\Delta + l_2$, in C_s. As support-degree$(d_s) = |W(d_s, C_s)| + |H(d_s, C_s)|$, we have that $H(d_s, C_s) < l_2$. Therefore the number of vertices of C_s that are adjacent to d_s and belong in members of $H(d_s, C_s)$, is less than ql_2. Using this fact, we have that $|P| < ql_2|R_\alpha| \leq ql_2(m + |T_\alpha|)$.

We conclude that:

$$(l_1 - l_2)|T_\alpha| < ql_2(m + |T_\alpha|) \Rightarrow$$
$$|T_\alpha|(l_1 - l_2(q+1)) < ql_1 m \Rightarrow$$
$$|T_\alpha| < \frac{kl_2 m}{l_1 - l_2(q+1)} \Rightarrow$$
$$|R_\alpha| < m + \frac{kl_2 m}{l_1 - l_2(q+1)} \Rightarrow$$
$$|R_\alpha| < m\frac{l_1 - l_2}{l_1 - l_2(q+1)}$$

and this completes the proof of the lemma. $\qquad \square$

Theorem 4. *Let $\mathcal{N}$ be a constraint network with n variables each one having at most q values. Let also G, C be its constraint and compatibility graph respectively. If G has n vertices and maximum degree Δ, then for any $\alpha < 1 - \frac{q}{q+1}\frac{\Delta}{n-1}$, there exists a parallel algorithm that returns a subgraph of $C_{AC,\alpha}$ which contains C_{AC} and where any vertex has support-degree at least $\alpha(n-1)$. This algorithm executes in $O(\sqrt{nq}\log(nq))$ time and uses $O(n^2 q^2)$ processors on the CREW PRAM.*

Proof We present the algorithm in terms of the compatibility graph:

Algorithm: fast α-discrete relaxation
Input: A compatibility graph $C = (V, E)$
Output: An α-partial arc consistent subgraph of $C_{AC,\alpha}$ that contains C_{AC}, where $\alpha < 1 - \frac{q}{q+1}\frac{\Delta}{n-1}$
begin
$\quad N \leftarrow |V|$

while $\exists V' \subseteq V : (|V'| \geq \sqrt{N}) \wedge (\forall d \in V' : \text{support-degree}(d) < n - 1)$
do
    for each $v \in V'$ **in parallel**
    do $V \leftarrow V - \{v\}$ **od**
    if an empty part appears in C
    then terminate and return the empty graph
od
    run parallel α-discrete relaxation
end

It is easy to see that each vertex in the subgraph the algorithm outputs has support-degree no less than $\alpha(n-1)$.

For the time and processor bounds, we observe that the identification of each set V' can be performed in $O(\log N)$ parallel time, using $O(N^2)$ processors on a CREW PRAM. In each such step, the algorithm removes (in constant time) at least $\sqrt{N}$ vertices inside the parallel for-loop. Therefore, the while-loop has at most $\sqrt{N}$ iterations. Also, after the removal of V', checking whether any part of C is empty requires $O(\log q)$ parallel time. Summing up, the algorithm exits the while-loop in time $O(\sqrt{N} \log N)$. Also, after the end of the while loop, there are less than $\sqrt{N}$ vertices in V that have support-degree strictly less than $n-1$. Then, by Lemma 3 and by seting $l_1 = \Delta$, $l_2 = \alpha(n-1) - l_\Delta$, we get that α-discrete relaxation will remove no more than

$$\sqrt{N} \frac{\Delta - \alpha(n-1) + l_\Delta}{\Delta - (q+1)(a(n-1) - l_\Delta)} =$$

$$\sqrt{N} \frac{n - 1 - \alpha(n-1)}{(n-1)(\frac{\Delta}{n-1} - (q+1)(\alpha - \frac{l_\Delta}{n-1}))} =$$

$$\sqrt{N} \frac{1 - \alpha}{\frac{\Delta}{n-1} + (q+1)\frac{l_\Delta}{n-1} - \alpha(q+1)} =$$

$$\sqrt{N} \frac{1 - \alpha}{1 + q\frac{n-1-\Delta}{n-1} - \alpha(q+1)} =$$

$$\sqrt{N} \frac{1 - \alpha}{(1 - \alpha)(1 + q) - q\frac{\Delta}{n-1}}$$

vertices. For any $\alpha < 1 - \frac{q}{q+1}\frac{\Delta}{n-1}$, this number is $O(\sqrt{N})$. Thus, the parallel α-discrete relaxation will execute in $O(\sqrt{N} \log N)$ parallel time, using $O(N^2)$ processors on the CREW PRAM. It follows that fast α-discrete relaxation executes in $O(\sqrt{N} \log N)$ time using $O(N^2)$ processors. Since $N \leq nq$, we have the required time and processor bounds for fast α-discrete realxation.   □

4 Parallel Intractability Results

In this section, we justify our choice for partial solutions to ACP by presenting several natural approximation schemes for it which are P-complete. The relevant proofs are too technical, and we omit them due to lack of space.

4.1 Domain Approximations

Let $\mathcal{N}$ be a constraint network with n variables and variable domains $D_1, \ldots, D_n$. Let also $A_1, \ldots, A_n$ denote the domains in $\mathcal{N}_{AC}$. If $D = \bigcup_i \{D_i, i = 1, \ldots, n\}$, $D_{AC} = \bigcup_i \{A_i, i = 1, \ldots, n\}$ and $R_{AC} = D - D_{AC}$, then the following hold:

Theorem 5. *For a given constraint network $\mathcal{N}$, and for any $\epsilon \in (0, 1]$, it is P-complete to find a set D' such that $D_{AC} \subseteq D' \subseteq D$ and $\epsilon |D'| \leq |D_{AC}|$.*

Theorem 6. *For a given constraint network $\mathcal{N}$, and for any $\epsilon \in (0, 1]$, it is P-complete to find a set R' such that $R' \subseteq R_{AC}$ and $\epsilon |R_{AC}| \leq |R'|$.*

4.2 Support-degree Approximations

Given a constraint network $\mathcal{N}$ and a value d of variable X_i, we define the *support* of d to be the maximum integer h such that, after the $(\frac{h}{n-1})$-relaxation, d is not removed and no domain is empty. Recall that an $(\frac{h}{n-1})$-relaxation removes values with support-degree less than h, until no such values exist.

The problem of computing support(d) is P-complete, since if we could efficiently compute the support of any value in $\mathcal{N}$, then, by comparing these values with $n - 1$, we would compute all the values in $\mathcal{N}_{AC}$.

Computing support(d) is a maximization problem, so we formulate the following approximation scheme for it, for any constant factor $\epsilon \in (0, 1]$: find an integer support(d)$_{\text{approx}}$ such that

$$\epsilon \cdot \text{support}(d) \leq \text{support}(d)_{\text{approx}} \leq \text{support}(d)$$

Theorem 7. *It is P-complete to approximate support(d) by a factor $\epsilon > 1/q$, where q is the maximum number of values in the domain of a variable.*

The proof of the above theorem uses a reduction from the Monotone Circuit Value Problem [10].

We also define the *elimination degree* of d to be the least integer h such that a $(\frac{n-1-h}{n-1})$-relaxation does not remove d and does not empty any variable domain. It holds that support(d) + elimin(d) = $n - 1$, so it is P-complete to find elimin(d). The computation of the elimination degree of d is a minimization problem, so we consider the following approximation for it, for some constant factor $\lambda \geq 1$: find an integer elimin(d)$_{\text{approx}}$ such that:

$$\lambda \cdot \text{elimin}(d) \geq \text{elimin}(d)_{\text{approx}} \geq \text{elimin}(d)$$

Theorem 8. *It is P-complete to approximate elimin(d) by any factor $\lambda \geq 1$.*

Theorem 8 is proved using reduction from the Propositional Horn Satisfiability Problem [13].

5 Extensions for Arbitrary Degree of Consistency

In this section we will extend our results to cover the case of k-consistency, both for CSPs with binary constraints, and for constraint networks where the relations are not necessarily binary. In the latter case, a *constraint network* consists of a set of variables $X_1, \ldots, X_n$, a set of variable domains $D_1, \ldots, D_n$, and a set of constraints $\mathcal{C}$. Let $I = \{1, \ldots, n\}$ and $I_r = \{i_1, \ldots, i_r\} \subseteq I$ such that $j < j' \Rightarrow i_j < i_{j'}$. A constraint $R_{I_r} \in \mathcal{C}$ of arity r is a subset of $D_{i_1} \times \ldots \times D_{i_r}$. A set of value assignments $\{d_{i_1}, \ldots, d_{i_r}\}$ where $d_{i_j} \in X_{i_j}, j = 1, \ldots, r$, is *compatible* iff there is no $R_{I_r} \in \mathcal{C}$ such that $(d_{i_1}, \ldots, d_{i_r}) \notin R_{I_r}$.

The *Constraint Satisfaction Problem* (CSP) for the general case, asks for all the n-tuples $(d_1, \ldots, d_n)$ of values such that $d_i \in D_i, i = 1, \ldots, n$, and for every $R_{I_r} \in \mathcal{C}, (d_{i_1}, \ldots, d_{i_r}) \in R_{I_r}$. Such an n-tuple is called a *solution* of the CSP.

Similarly with the notion of the constraint graph, for a CSP instance Π with n variables, its *constraint hypergraph* G^Π has n vertices corresponding to the variables. G^Π contains a hyperedge $\{d_{i_1}, \ldots, d_{i_r}\}, r \geq 2$, iff the corresponding variables are bound by a constraint. Also, its *compatibility hypergraph* C^Π has exactly one vertex for each value in the domain of each variable. A set of vertices $\{d_{i_1}, \ldots, d_{i_r}\}, d_{i_j} \in V_{i_j}, j = 1, \ldots, r$, is a hyperedge in C^Π iff the set of values that corresponds to $\{d_{i_1}, \ldots, d_{i_r}\}$ is compatible. An n-tuple $(d_1, \ldots, d_n)$ of values is a solution of Π iff any subset of the corresponding vertex set of C^Π that has cardinality ≥ 2 is a hyperedge.

A constraint network $\mathcal{N}$ is *k-consistent* if the following holds: for any set of value assignments $\{d_{i_1}, \ldots, d_{i_{k-1}}\}$ such that any subset of it is compatible, and for any $i \notin \{i_1, \ldots, i_{k-1}\}$, there exists a value assignment $d_i \in D_i$ such that every subset of $\{d_{i_1}, \ldots, d_{i_{k-1}}\} \cup \{d_i\}$ is compatible. The *k-Consistency Problem* is the problem of finding a maximal (with respect to the domains) k-consistent subnetwork $\mathcal{N}_k$ of $\mathcal{N}$. It is easy to see that $\mathcal{N}_k$ is unique. We call $\mathcal{N}_k$ the *solution* of k-consistency.

A constraint network $\mathcal{N}$ is *strongly k-consistent* if it is i-consistent for any $i, i = 2, \ldots, k$ and for any set of at most k compatible variables in it, all its subsets are also compatible.

The *strong k-consistency problem* is the problem of finding a maximal (with respect to the domains) strongly k-consistent subnetwork $\mathcal{N}_k^s$ of $\mathcal{N}$.

A hyperedge $\{d_{i_1}, \ldots, d_{i_r}\}$ of a compatibility hypergraph C is a *hyperclique* iff any subset of $\{d_{i_1}, \ldots, d_{i_r}\}, r \geq 2$, is a hyperedge. We call such a hyperclique an *r-hyperclique*; also, any vertex d in C is considered to be an 1-hyperclique. The *support-degree* of an r-hyperclique $\{d_{i_1}, \ldots, d_{i_r}\}$ is the number of parts not in $\{V_{i_1}, \ldots, V_{i_r}\}$ that contain a vertex d such that $\{d_{i_1}, \ldots, d_{i_r}, d\}$ is an $(r + 1)$-hyperclique.

In terms of the compatibility hypergraph, the k-consistency problem is formulated as follows: Find its maximal (with respect to the set of vertices) induced subhypergraph C_k such that any $(k - 1)$-hyperclique has support-degree equal to $n - k + 1$ (we call such a hypergraph k-consistent). The *strong k-Consistency Problem* is the problem of finding a maximal (with respect to the set of vertices) induced subhypergraph C_k^s that is i-consistent for all $i, i = 2, \ldots, k$, and any i-hyperedge is an i-hyperclique, for all $i, i = 1, \ldots, k$.

k-consistency is achieved by the procedure called k-*relaxation*. The optimal serial implementation of it executes in $O((nq)^k)$ time (see [3]). We will now describe a variation of it in terms of the compatibility hypergraph.

Algorithm: k-relaxation
Input: a compatibility hypergraph $C = (V, E)$
Output: C_k
begin
   while $\exists$ a hyperclique $e : (|e| = k - 1) \wedge (\text{support-degree}(e) < n - k + 1)$
   do
      $E' \leftarrow \{e' : e' \text{ is a hyperclique with } |e'| \leq k \text{ and } e \subseteq e'\}$
      $E \leftarrow E - E'$
   od
end

k-relaxation can be implemented to execute in $O((nq)^{k-1} \log(nq))$ parallel time using $O((nq)^k)$ processors on the CREW PRAM model.

A constraint network $\mathcal{N}$ is α-*partially k-consistent* if the following holds: for any set of values $\{d_{i_1}, \ldots, d_{i_{k-1}}\}$, such that every subset of it is compatible, there are at least $\alpha(n-k+1)$ domains not in $\{D_{i_1}, \ldots, D_{i_{k-1}}\}$ that contain at least one value d such that every subset of $\{d_{i_1}, \ldots, d_{i_{k-1}}\} \cup \{d\}$ is compatible, and each variable has at least one value compatible with other values in $\mathcal{N}$. The α-*partial k-consistency problem* is the problem of finding a maximal (with respect to the domains) α-partially k-consistent subnetwork $\mathcal{N}_{k,\alpha}$ of $\mathcal{N}$.

Let C be a compatibility hypergraph. A part of C whose vertices do not belong to any k-hyperclique is called a k-empty part. In terms of the compatibility hypergraph, the α-partial k-consistency problem is formulated as follows: Find the maximal (with respect to the set of vertices) induced subhypergraph $C_{k,\alpha}$ of C such that any $(k-1)$-hyperclique has support-degree $\geq \alpha(n-k+1)$ and at no part of it is k-empty. Such a hypergraph is called α-partially k-consistent. Also, the α-*partial strong k-consistency problem* is the problem of finding a maximal (with respect to the set of vertices) induced subhypergraph $C_{k,\alpha}^s$ of C that is α-partially i-consistent for any $i, i = 2, \ldots, k$, and any hyperedge e in $C_{k,\alpha}^s$ of size at most k is a hyperclique. Notice that by restricting out attention to $C_{k,\alpha}^s$, we do not lose any solution of the CSP.

In order to solve the α-partial strong k-consistency problem, we introduce a modified version of the k-relaxation algorithm, called *strong (α, k)-relaxation*. Given an integer $\alpha \leq 1$ and a hypergraph C as input, the algorithm is the following:

Algorithm: strong (α, k)-relaxation
Input: a compatibility hypergraph C
Output: $C_{k,\alpha}^s$
begin
   while $\exists$ a hyperedge $e : (|e| \leq k) \wedge (e \text{ is not a hyperclique})$

```
do E ← E - {e} od
while ∃ a hyperclique e : (|e| = k - 1) ∧ (support-degree(e) < α(n - k + 1))
do
    E' ← {e' : e' is a hyperclique with |e'| ≤ k and e ⊆ e'}
    E ← E - E'
    if a k-empty part appears in C
    then terminate and return the empty hypergraph
od
for i = k - 1 downto 2
do
    E' ← {e : e is an (i - 1)-hyperclique
          and support-degree(e) ≤ α(n - i + 1)}
    E ← E - E'
    if an i-empty part appears in C
    then terminate and return the empty hypergraph
od
end
```

The first while-loop is a preprocessing stage that ensures that all hyperedges are hypercliques. This property guarantees that, during the execution of the rest of the algorithm, no hyperedge which is not a hyperclique appears. Also, the second while-loop is a variant of the k-consistency algorithm that outputs an α-partially k-consistent hypergraph. This hypergraph is subsequently transformed by the for-loop into a strongly consistent one. It is interesting that in order to obtain strong consistency (i.e. i-consistency for $i < k$), the for-loop executes only the first parallel step of the variant of i-relaxation.

In fact, we prove the following (the proof is omitted due to lack of space):

Theorem 9. *α-partial strong k-relaxation solves the problem of α-partial strong k-consistency and can be implemented to execute in $O(n^k q^k)$ serial time and in parallel in $O((nq)^{k-1} \log(nq))$ time using $O(n^k q^k)$ processors on the CREW PRAM.*

We will now present a faster parallel algorithm that, given a compatibility hypergraph C and a positive value $\alpha < 1/(q + 1)$, outputs a subhypergraph of $C^s_{k,\alpha}$ that contains C^s_k and is α-partially strongly k-consistent.

Algorithm: fast strong (α, k)-relaxation
Input: a compatibility hypergraph C
Output: An α-partially strongly k-consistent subhypergraph of $C^s_{\alpha,k}$
              that contains C^s_k, where $\alpha < \frac{1}{q+1}$

```
begin
    H ← {e : e is a hyperclique of C and |e| = k - 1}
    while ∃H' ⊆ H : (|H'| > (nq)^(k-1/2)) ∧ (∀e ∈ H' : support-degree(e) < n - k + 1)
    do
        in parallel remove from C all hypercliques e ∈ H'
        H ← H - H'
```

      if a k-empty part appears
      then terminate and return the empty graph
   od
   run strong (α, k)-relaxation
end

Theorem 10. *Let C be a compatiblity hypergraph. For any $\alpha < 1/(1 + q)$, the above algorithm returns a strong α-partial k-consistent subhypergraph of $C^s_{\alpha,k}$ which contains C^s_k, in $O((N)^{\frac{k-1}{2}} \log(nq))$ time using $O((nq)^k)$ processors on the CREW PRAM.*

The proof of the above theorem follows the idea of the proof of Theorem 4, using a lemma similar to Lemma 3.

References

1. Bessiere C.: Arc-consistency and arc-consistency again. Artificial Intelligence **65** (1994) 179–190
2. Cohen D.A., Cooper M.C., and Jeavons P.G.: Characterizing tractable constraints. Artificial Intelligence **65** (1994) 347–361
3. Cooper M.C.: An optimal k-consistency algorithm. Artificial Intelligence **41** (1990) 89–95
4. Cooper P.R. and Swain M.J.: Arc consistency: parallelism and domain dependence. Artificial Intelligence **58** (1992) 207–235
5. Dechter R.: Constraint networks. In: S. Shapiro (ed.) Encyclopedia of Artificial Intelligence, Wiley, New York, 2nd ed. (1992) 276–285
6. Dechter R. and Meiri I.: Experimental results of preprocessing algorithms for constraint satisfaction problems. Artificial Intelligence **68** (1994) 211–241
7. Dechter R. and Pearl J.: Tree clustering for constraint networks. Artificial Intelligence **38** 353–366
8. Dendris N.D., Kalafatis I.A., and Kirousis L.M.: An efficient parallel algortihm for geometrically characterising drawings of a class of 3-D objects. Journal of Mathematical Imaging and Vision **4** (1994) 375–387
9. Freuder E.C.: Complexity of k-tree structured constraint satisfaction problems. Proceedings of the Eighth National Conference on Artificial Intelligence, Boston, Mass. (1990) 4–9
10. Goldschlager L.M.: The monotone and planar circuit value problems are log-space complete for P. SIGACT News **9(2)** (1977) 25–29
11. Van Hentenryck P., Deville Y., and Teng C.M.: A generic arc-consistency algorithm and its specializations. Artificial Intelligence **57** (1992) 291–321
12. Karp R.M. and Ramachandran V.: Parallel algorithms for shared-memory machines. In: J. van Leeuwen (ed.) Handbook of Theoretical Computer Science, Elsevier, Amsterdam (1990)
13. Kasif S.: On the parallel complexity of discrete relaxation in constraint satisfaction networks. Artificial Intelligence **45** (1990) 275–286
14. Kasif S. and Delcher A.L.: Local consistency in parallel constraint satisfaction networks. Artificial Intelligence **69** (1994) 307–327

15. Kirousis L.M.: Fast parallel constraint satisfaction. Artificial Intelligence **64** (1993) 147–160

16. Lieberherr K.J. and Specker E.: Complexity of partial satisfaction. J. of the ACM **28** (1981) 411–421

17. Mackworth A.K.: Constraint satisfaction. In: S. Shapiro (ed.) Encyclopedia of Artificial Intelligence, Wiley, New York, 2nd ed. (1992) 285–293

18. Mackworth A.K. and Freuder E.C.: The complexity of some polynomial network consistency algorithms for constraint satisfaction problems. Artificial Intelligence **25** (1985) 65–74

19. Mackworth A.K. and Freuder E.C.: The complexity of constraint satisfaction revisited. Artificial Intelligence **59** (1993) 57–62

20. Samal A. and Henderson T.C.: Parallel consistent labeling algorithms. International Journal of Parallel Programming **16(5)** (1987) 341–364

21. Waltz D.: Understanding line drawings of scenes with shadows. The Psychology of Computer Vision, McGraw-Hill, New York (1975) 19–91

22. Zhang Y. and Mackworth A.K.: Parallel and distributed algorithms for finite constraint satisfaction problems. Proceedings 3rd IEEE Symposium on Parallel and Distributed Processing, Dallas, TX (1991) 394–397

Maximal Extensions of Simplification Orderings

Deepak Kapur[1]* and G. Sivakumar[2]

[1] Computer Science Department
State University of New York
Albany, NY 12222
kapur@cs.albany.edu
[2] Computer Science Department
Indian Institute of Technology
Bombay, India
siva@cse.iitb.ernet.in

Abstract. Several well-founded syntactic orderings have been proposed in the literature for proving the termination of rewrite systems. *Recursive path orderings (RPO)* and their extensions are the most widely used in theorem proving systems such as *RRL, REVE, LP*. While these orderings can be total (up to equivalence) on ground terms, they are not **maximal.** That is, when used to compare non-ground terms, there can be terms such that for all ground substitutions, the first term is bigger than the second term, but these orderings declare the two terms as not comparable. A new family of orderings induced by precedence on function symbols, much like RPO, is developed in this paper. Terms are compared by comparing their paths. These ordering are shown to be maximal, and are hence called *maximal path orderings*. The maximal extension of RPO can be defined using symbolic constraint solving procedures. Such a decision procedure can check, given two terms s and t, whether there is a ground substitution σ that makes $\sigma(s)$ bigger than $\sigma(t)$ using RPO. A new decision procedure for the existential fragment of ordering constraints expressed using RPO is given based on the idea of **depth bounds**. It is shown that given two terms s and t, if there is a ground substitution σ which makes $\sigma(s)$ bigger than $\sigma(t)$ using RPO, then there is a ground substitution within depth $k * d + k$ which is also a solution, where k is the number of variables in s and t, and d is the maximum of the depths of s and t.

1   Introduction

Rewrite systems provide an interesting and useful model of computation based on the simple inference rule of "replacing equals by equals." Rewrite techniques have proved successful in many areas including theorem proving, specification and verification, and proof by induction.

The power of the rewriting approach stems from the ability to "orient" equality ($\leftrightarrow$), which is symmetric, into a directed "rewrite" relation ($\rightarrow$), which is

* Partially supported by NSF grants CCR-9404930 and INT-9416687.

anti-symmetric, using a "well-founded ordering." The rules are used for "simplifying" expressions by repeatedly replacing instances of left-hand sides by the corresponding right-hand sides. For example, the rules below express addition and multiplication over natural numbers.

$$
\begin{aligned}
0 + x &\rightarrow x \\
s(x) + y &\rightarrow s(x + y) \\
0 * x &\rightarrow 0 \\
s(x) * y &\rightarrow y + (x * y)
\end{aligned}
$$

A sample derivation chain is $s(0) * s(0) \rightarrow s(0) + (0 * s(0)) \rightarrow s(0) + 0 \rightarrow s(0 + 0) \rightarrow s(0)$.

Termination of such derivations is crucial for using rewriting in proofs and computations. Syntactic "path orderings" based on a precedence relation $\succ_f$ on function symbols have been developed to prove termination of a set of rewrite rules. A comprehensive survey is [3].

The Recursive Path Ordering (RPO) [2] is the most commonly used ordering. When $\succ_f$ is a total precedence, RPO is total (up to equivalence) on ground terms (terms without variables). That is, given two distinct ground terms, either they are equivalent under the ordering, or one of them is bigger.

A seemingly obvious way to extend RPO maximally to non-ground terms would be to say that s is bigger than t iff every ground instance of s is bigger than the corresponding ground instance of t. This maximal extension of RPO is hard to define directly however. When RPO is lifted in a direct way to non-ground terms, it is not maximal (examples in Section 4). The paths ordering (KNSS) [6] or the recursive decomposition ordering (IRDS) [8] are attempts to extend RPO. See [12] for a survey of these extensions. These orderings are the same as RPO when comparing ground terms if the precedence $\succ_f$ is total, but they strictly include RPO when applied to non-ground terms. However, even these extensions are also not maximal (examples in Section 4). In this paper, we define a new ordering based on paths that is also total on ground terms (up to equivalence) and prove that it is maximal on non-ground terms. However, it is not compatible with RPO and compares some ground terms differently.

Computing the maximal extension of RPO can be done using a decision procedure to check if given two terms s and t, there is a ground substitution σ that makes $\sigma(s)$ bigger than $\sigma(t)$ using RPO. Constraint solving procedures have been developed for this [1] [5]. This problem has also been proved NP-complete in [9]. In this paper, we develop a new method for computing the maximal extension of RPO using a bound on the depth of the substitution required.

The rest of this paper is organized as follows. In Section 2, we give the relevant definitions and background. In Section 3, we define a new ordering based on paths and prove it to be maximal. In Section 4, we show how to build a maximal extension of RPO using a new constraint solving procedure. We conclude in Section 5 with discussion of related work and suggestions for future work.

2   Rewrite Systems and Simplification Orderings

Let $T(F, X)$ be a set of terms constructed from a (finite) set F of function symbols and a (countable) set X of variables. We normally use the letters a through h for function symbols; s, t, and u through w for arbitrary terms; x, y, and z for variables. Each function symbol $f \in F$ has an *arity* $n \geq 0$; *constants* are function symbols of arity zero. Variable-free terms are called *ground* .

A term t in $T(F, X)$ may be viewed as a finite ordered tree. Internal nodes are labeled with function symbols (from F) of arity greater than 0. The outdegree of an internal node is the same as the arity of the label. Leaves are labeled with either variables (from X) or constants. We use $root(t)$ to denote the symbol labeled at the root of the tree corresponding to t. The *depth* of a term is the length of the longest root to leaf path in the tree. A subterm of t is called *proper* if it is distinct from t. By $t \mid_\pi$, we denote the *subterm* of t rooted at *position* π. Let u be a term and π a position in u. We use $u[\cdot]_\pi$ to denote the *context* for position π in u. Loosely speaking, the context is the tree obtained by deleting the subterm at position π leaving a "hole" in the term. We use $u[t]_\pi$ to denote a term that has t plugged in as a subterm at the "hole" in the context $u[\cdot]_\pi$.

A substitution σ is a mapping from variables to terms such that $x\sigma \neq x$ for a finite number of variables x's. The depth of a substitution σ is the maximum of the depths of the terms used in this mapping. A substitution can be extended to be a mapping from terms to terms. We use $t\sigma$ to denote the term obtained by applying a substitution σ to a term t. For example, the ground substitution $\sigma = \{x \mapsto k(a), y \mapsto d\}$ when applied to the term $t = f(x, y)$ gives the ground term $t\sigma = f(k(a), d)$.

A rewrite *rule* over a set $T(F, X)$ of terms is an ordered pair (l, r) of terms such that the variables in r also appear in l, and is written $l \to r$. A *rewrite system* (or *term rewriting system*) R is a set of such rules. Rules can be used to replace instances of l by corresponding instances of r.

One approach to proving termination of rewrite systems is to use simplification orderings [2]. A simplification ordering has the following properties.

1. **Subterm Property**: $u[t] \succ t$ for any term t and a non-empty context $u[\cdot]$.
2. **Monotonicity**: $s \succ t$ implies that $u[s]_\pi \succ u[t]_\pi$, for all contexts $u[\cdot]$, terms s and t, and positions π.

We have omitted the deletion property which is needed only if there are function symbols with varying arity. Any simplification ordering $\succ$ on terms is well-founded, that is, there is no infinite descending chain of terms $t_1 \succ t_2 \succ t_3 \dots$.

A simplification ordering that also has the following property

- **Stability**: for all terms s and t, $s \succ t$ implies that for all substitutions σ, $s\sigma \succ t\sigma$.

can be used for proving termination of rewrite systems. If for every rule $l \to r$ in R, $l \succ r$ in a simplification ordering $\succ$ which is stable under substitutions, then $\to$ is terminating.

2.1 Recursive Path Ordering

Let $\succ_f$ be a well-founded precedence relation on a set of function symbols F. The recursive path ordering $\succ_{rpo}$ extends $\succ_f$ to a well-founded ordering on terms. We use $\{\!\{t_1, \ldots, t_n\}\!\}$ to denote a multi-set of terms. We use $\succ_{mul}$ to denote the extension of $\succ$ to multisets. $S \succ_{mul} T$ if for every element t in $T - S$, we can find an element s in $S - T$ such that $s \succ t$.

Definition 1. [2] $t = f(t_1, \ldots, t_n) \underset{\sim rpo}{\succ} g(s_1, \ldots, s_m) = s$ iff one of the following holds.

1. $t_i \underset{\sim rpo}{\succ} s$ for some i, $1 \leq i \leq n$.
2. $f \succ_f g$, and $t \succ_{rpo} s_j$ for all j, $1 \leq j \leq m$.
3. $f \sim_f g$, and $\{\!\{t_1, \ldots, t_n\}\!\} \underset{\sim rpo}{\succ} \{\!\{s_1, \ldots, s_m\}\!\}$.

Example 1. Let $F = \{f, g, a, b\}$, with $f \succ_f g$ and $a \succ_f b$. Then,

$$t = f(g(a, b), b) \underset{rpo}{\succ} g(f(b, b), f(g(b, b), a)) = s$$

Since $f \succ_f g$, we have to compare each top-level subterm of s with t. This leads to the comparison of the multiset $\{\!\{g(a, b), b\}\!\}$ with $\{\!\{b, b\}\!\}$ in one case and $\{\!\{g(b, b), a\}\!\}$ in the other. In both cases we have the desired relation using RPO definition recursively.

It can be shown $\succ_{rpo}$ is a simplification ordering [2]. Also, two distinct ground terms are either equivalent or comparable under $\succ_{rpo}$.

In Case 3 of the definition of RPO, we can use sequence comparison instead of multisets by assigning a status (left-to-right or right-to-left) for each operator. This is useful to orient axioms like associativity. In this paper, we are assuming that all operators have only multi-set status. This is only to make notations easier and the relevant proofs can be carried out even if operators are assigned lexicographic status.

2.2 Maximal Extensions of Orderings

Let $\geq$ be a quasi-ordering that is total on ground terms. That is, given distinct ground terms s and t, one of the following holds – $s \succ t$ or $t \succ s$ or $s \sim t$. A quasi-ordering with the stability property cannot be total on general terms. For example, no quasi-ordering should be able to compare $s = f(x)$ with $t = f(y)$, since it is possible to construct two different ground substitutions σ_1 and σ_2 such that $\sigma_1(s) \succ \sigma_1(t)$ whereas $\sigma_2(t) \succ \sigma_2(s)$. Such terms have to be necessarily *incomparable*. An obvious question is whether it is possible to define an ordering in which only such terms s and t are incomparable.

Definition 2. Let $\geq$ be a total quasi-ordering on ground terms. The ordering $\geq$ is **maximal** if for any two terms t and s, $t \succ s$ iff for all ground substitutions σ, $t\sigma \succ s\sigma$

While orderings like RPO are total on ground terms, when $\succ_f$ is total, and satisfy the stability property (i.e. $s \succ t$ implies for all ground σ, $s\sigma \succ t\sigma$), they are not maximal. They do not possess the only-if property (i.e. for all ground σ, $s\sigma \succ t\sigma$ implies $s \succ t$) as we show in Section 4.

3 Maximal Path Ordering

In this section, we define a new simplification ordering $\succ_{mp}$ on terms. This ordering simply converts a term into a multiset of paths (strings) and compares these paths using the precedence relation $\succ_f$ on function symbols and constants in F. This idea is similar to the one used in the *Simple Path Ordering* [10] (see also [3]) except that path comparisons are done quite differently. The related *Path of Subterm Ordering* is shown in [12] to be same as RPO when comparing ground terms with total underlying precedence. Also it is properly included in KNSS or IRDS when comparing general terms, and hence is not maximal.

We assume that there is at least one constant in F. We show that $\succ_{mp}$ is total on ground terms when $\succ_f$ is total, and also prove that it is maximal.

Definition 3. Let t be a term. A full path P in t is defined as follows.

1. If $t = x$, a variable, $P = x$ is the only full path in t.
2. If $t = c$, a constant, $P = c$ is the only full path in t.
3. If $t = f(t_1, \ldots, t_n)$, $P = f \cdot P_1$ is a full path in t provided P_1 is a full path in some t_i.

For example, $f \cdot g \cdot a$ is a full path in the term $f(g(a, b), x)$. The multiset of paths $Mpaths(t)$ of a term t is the multiset of all full paths of t. For example, $Mpaths(f(g(a, b), x)) = \{\!\{ f \cdot g \cdot a, f \cdot g \cdot b, f \cdot x \}\!\}$. Note that any full path must end in either a variable or a constant. All other symbols in a full path are non-constant function symbols. Any prefix of a full path will be simply referred to as a path. Note that a proper prefix of a full path cannot end in a constant or a variable.

Definition 4. Two paths are equivalent $P_1 \sim_p P_2$, if one of the following holds.

1. Both are empty, or both are same variable.
2. $P_1 = f \cdot \mu_1$, $P_2 = g \cdot \mu_2$ and $f \sim_f g$ and $\mu_1 \sim_p \mu_2$.

We now define how to compare two paths P_1 and P_2 given a precedence relation $\succ_f$ on F.

Definition 5. $P_1 \succ_p P_2$ if one of the following holds.

1. P_2 is empty and P_1 is not.
2. Both paths end in the same variable, i.e. $P_1 = \gamma_1 \cdot x$, $P_2 = \gamma_2 \cdot x$, and, further, $\gamma_1 \succ_p \gamma_2$.

3. $P_1 = \gamma_1 \cdot x$ (ends in a variable), $P_2 = \gamma_2 \cdot c$ (ends in a constant) and $\gamma_1 \cdot d \succ_p P_2$ for every constant d that is minimal in $\succ_f$. (Note that if $\succ_f$ is total, there is only one smallest constant to try.)

4. Neither path ends in a variable, and for every symbol g in P_2 (that is $P_2 = \gamma_2 \cdot g \cdot \mu_2$), there is a symbol f in P_1 (that is, $P_1 = \gamma_1 \cdot f \cdot \mu_1$), such that one of the following holds.

   (a) $f \succ_f g$.
   (b) $f \sim_f g$ and $\mu_1 \succ_p \mu_2$ (greater suffix).
   (c) $f \sim_f g$ and $\mu_1 \sim_p \mu_2$ and $\gamma_1 \succ_p \gamma_2$ (greater prefix).

Note that two paths ending in different variables are incomparable. Also, a path ending in a constant cannot be greater than a path ending in a variable. Case 3 above is crucial and different from other orderings. Also, the Simple Path Ordering [10] does not keep information about the prefixes (Case 4c), which is useful in some cases.

For example, with $f \succ_f g \succ_f b \succ_f a \succ_f h$, we have

1. $f \cdot g \cdot b \succ_p f \cdot a$
2. $f \cdot f \cdot x \succ_p f \cdot g \cdot g \cdot x$
3. $h \cdot h \cdot x \succ_p h \cdot a$

Note the last example above is correct because we can replace x by a (the smallest constant) when doing comparisons. From now, we leave out the $\cdot$ symbol in paths, whenever obvious, and write a path simply as gfx.

First, we have to show that $\succ_p$ is an ordering. That is, it is irreflexive and transitive. Irreflexivity ($P \not\succ_p P$) is easy to prove by induction on the length of P, and considering the leftmost occurence of a maximal function symbol in P.

Lemma 6. $\succ_p$ *is transitive. That is, given paths P, Q and R, and $P \succ_p Q$, and $Q \succ_p R$, then $P \succ_p R$.*

Proof. By induction on the sum of the lengths P, Q and R using different cases of the definition. The interesting case is when Case 4 is used to show both $P \succ_p Q$ and $Q \succ_p R$, and is explained below.

Since $Q \succ_p R$, for every decomposition $R = \gamma_3 \cdot g \cdot \mu_3$, there is a decomposition $Q = \gamma_2 \cdot f \cdot \mu_2$, that "defeats" it. Note that for this we must have $f \geq g$. Now since $P \succ_p Q$, there is a decomposition $P = \gamma_1 \cdot h \cdot \mu_1$ that defeats $Q = \gamma_2 \cdot f \cdot \mu_2$. Note that for this we must have $h \geq f$.

We can show that the same decomposition of P also defeats $R = \gamma_3 \cdot g \cdot \mu_3$ by considering the various possible relations among f, g and h. If either $h \succ_f f$, or $f \succ_f g$, it is easy. In, the only remaining case ($h \sim_f f \sim_f g$), we have to compare the suffixes first and then, if necessary the prefixes. Since these paths are smaller in length, we can use induction to complete the proof.

This establishes that $\succ_p$ is indeed an ordering. Also, we can lift this to multiset of paths in the usual way. We use $\succ_{mul}$ to denote the extension of $\succ_p$ to multisets. $S \succ_{mul} T$ if for every element p in $T - S$, we can find an element q in $S - T$ such that $q \succ_p p$.

We now note some useful properties of path comparisons which can be easily proved by induction on length of the paths.

Lemma 7. *Any path is bigger than a proper suffix of itself. That is $\gamma \cdot P \succ_p P$ for any non-empty γ.*

Lemma 8. *If $P \cdot x \succ_p Q \cdot x$, then replacing x by any full path (from any term) preserves the ordering.*

Lemma 9. $f \cdot P \succ_p f \cdot Q$ *iff* $P \succ_p Q$.

Lemma 10. *Let $\succ_f$ be total. If P and Q are full paths both ending in constants, or both ending in the same variable, then $P \succ_p Q$ or $Q \succ_p P$ or $P \sim_p Q$.*

That is, two paths can be incomparable only if they end in different variables, or one ends in a variable and the other in a constant.

We lift the ordering on paths to an ordering on terms as follows.

Definition 11. $s \succ_{mp} t$ if $Mpaths(s) \succ_{mul} Mpaths(t)$,

Example 2. Let $F = f, g, d, c, b, a$ and $\succ_f$ is $f \succ g \succ d \succ c \succ b \succ a$.
Let $s = g(f(x,x), f(a, f(d,b)))$ and $t = f(x, f(c,b))$.
Now, $Mpaths(s) = \{gfx, gfx, gfa, gffd, gffb\}$
and $Mpaths(t) = \{fx, ffc, ffb\}$.
Since $gfx \succ_p fx$, $gffd \succ_p ffc$ and $gffd \succ_p ffb$,
we have $Mpaths(s) \succ_{mul} Mpaths(t)$. So, we have $s \succ_{mp} t$.

The following properties show that $\succ_{mp}$ is a well-founded ordering with the stability property.

Lemma 12. $\succ_{mp}$ *has the subterm property: $u[s] \succ_{mp} s$ for any non-empty context $u[\cdot]$.*

The proof of this follows from the property that every full path in s is a proper suffix of some full path in $u[s]$.

Lemma 13. $\succ_{mp}$ *has the replacement property: $s \succ_{mp} t$ implies $u[s] \succ_{mp} u[t]$.*

This is a consequence of Lemma 9.

Lemma 14. $\succ_{mp}$ *has the stability property: $s \succ_{mp} t$ implies for any substitution σ, $s\sigma \succ_{mp} t\sigma$.*

This is a consequence of Lemma 8.

Lemma 15. *Suppose $\succ_f$ is total, and s, t are ground terms. Then $s \succ_{mp} t$ or $t \succ_{mp} s$ or $s \sim_{mp} t$.*

This follows from Lemma 10.

We have shown that $\succ_{mp}$ is a well-founded ordering that is total on ground terms (up to equivalence) when $\succ_f$ is total. We now show that $\succ_{mp}$ is maximal. First, we consider an example and then give a proof.

Example 3. Let $F = f, g, d, c, b, a$ and $\succ_f$ is $f \succ g \succ d \succ c \succ b \succ a$.
Let $s = g(f(x,x), f(a, f(c,b)))$ and $t = f(x, f(d,b))$.
s and t are incomparable, because
$Mpaths(s) = \{\!\{gfx, gfx, gfa, gffc, gffb\}\!\}$
and $Mpaths(t) = \{\!\{fx, ffd, ffb\}\!\}$. There are no paths in $Mpaths(t)$ which can defeat gfx which is a path in s. By using this, we can construct a substitution that makes s bigger than t, by putting a suitable value in x (say $x \mapsto f(d,d)$). Similarly, there are no paths in $Mpaths(s)$ to beat ffd a path in t. This again implies that there is a substitution $x \mapsto a$ that makes fdd the biggest path, making t bigger than s.

Theorem 3.1 *Let $\succ_f$ be a total precedence relation and s, t be terms that are incomparable using $\succ_{mp}$. Then, there are ground substitutions σ_1 and σ_2 such that $t\sigma_1 \succ_{mp} s\sigma_1$ and $s\sigma_2 \succ_{mp} t\sigma_2$*

Proof. Without loss of generality, assume first that there are no common full paths in s and t. This is possible because these paths get dropped anyway when doing multiset comparisons. Since it is not the case that $Mpaths(s) \succ_{mul} Mpaths(t)$, there is at least one path P in $Mpaths(t)$, that cannot be beaten by any path Q in $Mpaths(s)$. We will now construct explicitly a σ_1 so that all paths in $s\sigma_1$ can be beaten by the paths in $t\sigma_1$ starting with P.

- Suppose $P = \gamma \cdot x$.
  Let a be the smallest constant in $\succ_f$. Let $\sigma = \{y \mapsto a\}$ for all variables y, and $\sigma_1 = \sigma \circ \{x \mapsto s\sigma\}$. That is in σ_1, all variables other than x get a, and x gets s with all variables in s replaced by a. We prove below that $t\sigma_1 \succ_{mp} s\sigma_1$. Since paths ending in the same variable are always comparable (Lemma 10), P must be bigger than all paths Q in s ending in x, since none of them could defeat P. By Lemma 8, after applying σ_1 to t and s, all paths obtained from Q in $s\sigma_1$ are smaller than the corresponding path in $t\sigma_1$ obtained from P.
  By the construction above, any other path in $s\sigma_1$ (not obtained from a path ending in x in s) is a proper suffix of a path obtained from P in $t\sigma_1$ and hence smaller. Thus $t\sigma_1 \succ_p s\sigma_1$.
- Otherwise, P must end in a constant.
  No path in $Mpaths(s)$ can beat P. This means that P is already bigger than all full paths in s that end in constants. And also that all paths ending in variables in $Mpaths(s)$ cannot beat P even when the variable is replaced by a the smallest constant in F. Thus, for the substitution σ where all variables get a, we have $t\sigma \succ_{mp} s\sigma$.

Similarly, one can construct a σ_2 for which s beats t.

The ordering $\succ_{mp}$ is well-founded, has the stability property, is total on ground terms when $\succ_f$ is total, as well as maximal. Using the dynamic programming technique proposed in [6] for showing the complexity of a related path ordering, it can be shown that given a precedence relation $\succ_f$, $\succ_{mp}$ can be checked in polynomial time We now point out a "limitation" of this ordering when handling axioms like distributivity. Suppose we compare $t = x * (y + z)$ with $s = (x * y) + (x * z)$. $Mpaths(t) = \{\!\{*x, * + y, * + z\}\!\}$, $Mpaths(s) = \{\!\{+ * x, + * x, + * y, + * z\}\!\}$. If $+ \succ_f *$, then $Mpaths(s) \succ_{mul} Mpaths(t)$ and $s \succ t$. However, if $* \succ_f +$, then the two terms become incomparable. Thus, there is no way to orient distributivity axioms as $x * (y + z) \rightarrow (x * y) + (x * z)$ using this ordering. We call this a "limitation" because this flexibility may be needed in some applications. Also, there is no simple way (like the lexicographic status used in LRPO) to handle axioms like associativity.

4   Maximal Extension of RPO

RPO and its extensions like KNSS [6] and IRDS [8] are the most widely used simplification ordering because they can be used to show the termination of most commonly encountered examples. These orderings can orient the distributivity axiom, for example, as $x * (y + z) \rightarrow (x * y) + (x * z)$ with proper choice of the precedence relation.

However, none of these extensions are maximal as revealed in the following example (with only one variable) where $F = f, g, d, c, b, a$ and $\succ_f$ is $f \succ g \succ d \succ c \succ b \succ a$.

Example 4. Let $s = g(f(x,x), f(a, f(d,b)))$ and $t = f(x, f(c,b))$.

In this example, s and t are incomparable under RPO (and the extensions proposed). But it is true that for all ground substitutions σ, $s\sigma > t\sigma$. The point is, that if the value substituted for x is not bigger (under RPO) than $f(c,b)$ then the second subterm of $s \succ t$. If it is bigger than $f(c,b)$, then the first subterm of s can beat t. So, any maximal extension of RPO should be able to conclude that $s \succ t$. As we saw in the previous section $\succ_{mp}$ can handle this example correctly.

Contrast this with the following slightly changed example, where the terms must remain incomparable.

Example 5. Let $s = g(f(x,x), f(a, f(c,b)))$ and $t = f(x, f(d,b))$.

Here, t is bigger for values of x less than $f(d,b)$, and s is bigger otherwise.

The maximal extension of RPO can be computed if the following problem can be solved. Given F, a total $\succ_f$ over F and two terms s and t from $T(F, X)$, is there a ground substitution σ such that $s\sigma \succ_{rpo} t\sigma$ or $s\sigma \sim_{rpo} t\sigma$? Constraint solving procedures have been developed to solve this problem in [1] and [5]. This problem has also been proved NP-complete in [9].

Since the satisfiability problem is decidable, we could give the following definition of a maximal extension $\succ_{mrpo}$ of RPO.

Definition 16. $s \succ_{mrpo} t$ iff there is no ground σ, $t\sigma \succ_{rpo} s\sigma$ or $t\sigma \sim_{rpo} s\sigma$.

This is clearly a proper extension of RPO (i.e. $s \succ_{rpo} t$ implies $s \succ_{mrpo} t$) since RPO has the stability property. It is also maximal by definition. We now develop a new technique using bounds on the depth of the substitution needed to solve the satisfiability problem for RPO.

Throughout this section we assume we are working over terms $T(F, X)$ with $\succ_f$ a total precedence on F. We also assume that there is at least one constant and at least one non-constant function symbol in F. Let a be the smallest constant and f the smallest non-constant symbol. For simplicity of notation in the proofs, we assume that f is binary (arity). Let $C = \{a_n > a_{n-1} > ... > a\}$ be the set of constants smaller than f if any exist, otherwise let $C = \{a\}$.

First we define the notion of a successor for any ground term, which we need later in the proofs. This notion is very similar to the one used in [1].

Definition 17. Let s be a ground term (i.e. in $T(F)$). The successor term $next(s)$ is defined as follows.

$$next(s) = \begin{cases} a_{i+1} & \text{if } s = a_i \in C \text{ with } i < n \\ f(a, a) & \text{if } s \text{ is biggest constant in C} \\ f(a, s) & s = g(s_1, \ldots, s_m) \text{ with } g \succ_f f \\ f(a, next(s_2)) & \text{if } s = f(s_1, s_2) \text{ and } s_1 \sim_{rpo} s_2 \\ & \text{Note: it can also be } f(next(s_2), a) \text{ in this case.} \\ f(s_1, next(s_2)) & \text{if } s = f(s_1, s_2) \text{ and } s_1 \succ_{rpo} s_2 \\ f(next(s_1), s_2) & \text{if } s = f(s_1, s_2) \text{ and } s_2 \succ_{rpo} s_1 \end{cases}$$

For example with $g \succ_f f \succ_f b \succ_f a$, we can build the following chains of terms using the next function.

$$\ldots f(a, f(a, a)) \underset{rpo}{\succ} f(b, b) \underset{rpo}{\succ} f(a, b) \underset{rpo}{\succ} f(a, a) \underset{rpo}{\succ} b \underset{rpo}{\succ} a$$

$$\ldots f(f(a, b), g(a)) \underset{rpo}{\succ} f(f(a, a), g(a)) \underset{rpo}{\succ} f(b, g(a)) \underset{rpo}{\succ} f(a, g(a)) \underset{rpo}{\succ} g(a)$$

Also, from the definition of $next$ it is clear that the depth of $next(t)$ is at most one more than depth of t.

The following result (proved by induction on depth of t) shows that there can be no term between $next(t)$ and t.

Lemma 18. Let s, t be two ground terms. $s \succ_{rpo} t$ iff $s \underset{\sim rpo}{\succ} next(t)$.

We call a ground substitution σ a *solution* to the constraint $s \succ t$ (similarly $s \sim t$), if $s\sigma \succ_{rpo} t\sigma$. A constraint $s \succ t$ (or $s \sim t$) is *feasible* if it has a solution. We now show that if $s \succ t$ has a solution, then there is a bound on the depth of a solution substitution σ needed. Since there are only finitely many ground substitutions for the variables in s, t up to a given depth, it is possible to enumerate them and thus decide if $s \succ t$ is satisfiable.

A ordering constraint satisfaction problem is the problem of determining whether a quantifier-free formula expressed using predicates $\sim$ and $\succ$ on terms can be satisfied using some substitution for variables appearing in the formula. It is easy to see that such a formula can be simplified to an equivalent formula that is a disjunction of conjunctions of atomic formulas of the form $s \sim t$ and $s \succ t$ in which at least one of s, t is a variable. Below we show that if a quantifier-free formula is satisfiable, there is a polynomial bound on the depth of a substitution showing its satisfiability.

Lemma 19. *Let $E = \{s_1 \sim t_1, ..., s_n \sim t_n\}$ be a set of equivalence (under RPO) constraints that is feasible (simultaneously). Let E have k variables $(x_1, ..., x_k)$ and let the maximum depth of s_i, t_j be d. Then, there is a solution γ for E, whose depth is at most $k * d$.*

Proof. This problem is really the same as unification (modulo commutativity), and it is known that there is is a a σ (sequential) that is linear in size (sum of the sizes of s_i and t_i) from which this result follows. We give a complete proof below, however, as some of the ideas here will also be used in the proof of the main result later.

The proof is by induction on size of E, defined as follows.
$Size(E) = \langle k, \{\!\{ < depth(s_i), depth(t_i) > \}\!\} \rangle$.
That is, the size is a pair whose first component is the number of variables in E and the second component is a multiset of depth pairs of each constraint in E. These pairs are compared lexicographically from left-to-right.

The base case is therefore when there are no variables in E. Since E is feasible, the empty substitution (size 0) is the answer. Otherwise, we enumerate the different possibilities below and show that in each case E can be replaced by a "smaller" set of constraints which are equivalent to E.

1. Some $s_j \sim t_j$ is non-variable on both sides. That is, $s_j = f(u_1, ..., u_n)$ and $t_j = g(v_1, ..., v_m)$. Since E is feasible it must be the case that $f \sim_f g$ and $m = n$ and we decompose this constraint. That is, we can replace $s_j \sim t_j$ with the multiset $\{\!\{ u_k \sim v_{p(k)} \}\!\}$ for some permutation p and get an equivalent set of constraints that is also feasible, but whose size measure is smaller. By induction, we are done.

2. Otherwise, all constraints are of the form $x_i \sim t_j$ with at least one side a variable. We will assume that the left-hand side of all constraints is a variable. Since $\sim$ is symmetric we can always re-arrange this way.

   (a) Some constraint is a variable on both sides, that is of the form $x_i \sim x_j$. Replace x_i by x_j everywhere getting an equivalent set of constraints with one less variable. Done by induction.

   (b) Otherwise, all constraints must be of the form $x_i \sim t_j$ with left-hand side variable and the right-hand side non-variable. If some variable occurs in two different constraints as left-hand sides. $x_i \sim t_j$ and $x_i \sim t_k$ say, where t_i and t_j are non-variables, replace these two constraints by $x_i \sim t_j$ and the decomposed form of $t_j \sim t_k$ to get a feasible set of constraints with smaller size measure.

   (c) Otherwise, we have that no variable occurs as left-hand side of two different constraints.

Suppose there is a variable x that does not occur on any left-hand side. Let $x \mapsto a$ where a is any constant. Replace x by a everywhere and we get a feasible set of constraints with one less variable.

Finally, each variable occurs exactly once in a left-hand side. If all the right-hand sides are non-ground terms, then E cannot be feasible. This is because each constraint $x_1 \sim t[x_2]$ implies that $x_1 \succ x_2$. Since each variable appears on the left-hand side it must be strictly bigger than at least one other variable which is impossible. (This is similar to the "occur-check" in unification.) So, there must be at least one constraint, $x_1 \sim t$ say, whose right-hand side is ground with depth of t at most d. Replacing x_1 by t in all right-hand sides to a similar problem with $k-1$ variables increasing the depth by at most d. We can eliminate each variable one by one this way with each step increasing the depth by at most d. Thus there is a solution within depth $k * d$.

We now consider both equivalence constraints $(s_i \sim t_i)$ and ordering constraints $(s_j \succ t_j)$ together and show that there is a depth bound within which a solution can be obtained.

Theorem 4.1 Let $E = \{u_i \sim v_i\}$ be set of equivalence constraints and $I = \{s_j \succ t_j\}$ be a set of ordering constraints. Let k be the number of variables in E and I, and d the maximum depth of any term in E or I. If there is a substitution σ that is a solution for both E and I, then there is a substitution γ within depth $k * d + k$ that is also a solution for E and I. Furthermore, if for any j, $s_j \sigma \neq next(t_j \sigma)$ then it is also the case that $s_j \gamma \neq next(t_j \gamma)$, i.e. "gaps" are preserved.

Proof. Let us define the size measure of $\langle E, I \rangle$ as follows. The size measure is used to do a proof by induction of the statement.

Let m be the number of variables in terms in I.

$Size(\langle E, I \rangle) = \langle m, \{\!\!\{ < depth(u_j), depth(v_j) > ... < depth(s_i), depth(t_i) > \}\!\!\} \rangle$.

That is, the size measure is a pair whose first component is the number of variables in I, ordering constraints, and the second component is the multiset of depth pairs of terms in each constraint (both E and I). Size measure are compared lexicographically from left to right.

As stated earlier, the proof is by induction on the size measure of $\langle E, I \rangle$. The base case is when there are no variables in I. That is, I is always true (only ground terms) and we have only equivalence constraints. By the previous Lemma, the result follows.

Otherwise, we enumerate the cases below.

1. There is a constraint $s_j \succ t_j$ in I with both sides non-variable. By using different cases in the definition of RPO, we decompose this constraint into smaller ones as below. Let $s_j = f(ss_1, ..., ss_m)$ and $t_j = g(tt_1, ...tt_n)$. One of the following cases in the definition of RPO must apply since the constraint is feasible.

(a) $ss_i \underset{\sim rpo}{\succ} t_j$ for some i, $1 \leq i \leq m$.

(b) $f \succ_f g$, and $s_j \succ_{rpo} tt_i$ for all i, $1 \leq i \leq n$.

(c) $f \sim_f g$, and $\{\!\{ ss_1, ..., ss_m \}\!\} \underset{\sim rpo}{\succ} \{\!\{ tt_1, ...tt_n \}\!\}$.

In each case, $\langle E, I \rangle$ can be replaced by by a smaller $\langle E', I' \rangle$ for which also σ is a solution. Note that in Case 1 we have to guess correctly which subterm to use, and in Case 3 we have to match the subterms correctly, but since $\langle E, I \rangle$ is feasible (using σ) this must be possible. By the induction hypothesis, a γ within the required depth for $\langle E', I' \rangle$ can be found, and this γ is also a solution to $\langle E, I \rangle$.

Similarly if there is a constraint in E which is a non-variable on both sides, we can decompose (as in previous Lemma) and get a simpler problem.

2. Otherwise, one side of each constraint in I and E is a variable.

There are two cases:

(i) σ assigns equivalent ground terms for distinct variables. In this case, there is a "smaller" constraint set $\langle E', I' \rangle$ with fewer variables (in which such variables are made equal), and for which σ is a solution; so we can construct the desired γ by induction.

(ii) substitutions assigned by σ can be used to totally order variables as: $x_k \succ x_{k-1} \succ ... \succ x_1$. That is, x_k gets the biggest value in σ. This case is considered below.

First note that in E we cannot have any constraint of the form $x_i \sim t[x_k]$ since x_k is the biggest variable under σ. So, the only constraint in E involving x_k can be of the form $x_k \sim v_i$. If x_k occurs more than once as left-hand side in E, then E can be simplified, as in previous lemma, and a γ can be constructed by induction. Otherwise, there can be only one constraint involving x_k in E, and it must be of the form $x_k \sim v_i$ (v_i cannot contain x_k).

Now consider all constraints in I involving x_k.

(a) There can be no constraint of the form $x_i \succ t[x_k]$ since this will not be feasible for σ.

(b) If there is any constraint in I of the form $t[x_k] \succ x_i$ where $t[x_k]$ is non-variable, we can replace it by $x_k \succ x_i$ and get a smaller constraint set that also has σ as a solution. By induction, we can construct the desired γ.

(c) There can only be **upper bounds** of the form $s_i \succ x_k$ (where s_i does not have x_k) and **lower bounds** of the form $x_k \succ t_j$ (where t_j does not have x_k). Note that for any such upper bound s_i and any lower bound t_j, we must have that $s_i \sigma \succ_{rpo} next(t_j \sigma)$ since we have $s_i \sigma \succ_{rpo} x_k \sigma \succ_{rpo} t_j \sigma$. That is, σ has a "gap" between these terms, which we will need preserved by γ when using the inductive hypothesis later.

If there is a constraint $x_k \sim v_i$ in E, simply replace x_k by v_i in I. As x_k is not occurring as a proper subterm of any term in I, we get a simpler problem $\langle E, I' \rangle$. with one variable less in I', and depths have not increased. So by induction, we are done.

Otherwise, x_k does not occur in E. If there are no upper bound constraints on x_k in I, then remove all the lower bound constraints on x_k in I. The resulting $\langle E, I' \rangle$ is smaller, so by the induction hypothesis, a substitution γ_1 within depth $(k-1)*d+k-1$ can be constructed. Applying this substitution on the lower bounds of x_k, we can extend γ_1 to be γ which includes a substitution of x_k to be $next(t_m \gamma_1)$, where $t_m \gamma_1$ is the greatest lower bound on x_k. Since $next$ can increase the depth at most by one, the depth $\gamma(x_k)$ is at most $d + (k-1)*d + (k-1) + 1 = k*d + k$. If there are no lower bounds on x_k in I, then remove all the upper bound constraints on x_k from I. The resulting $\langle E, I' \rangle$ is smaller, so by the induction hypothesis, a substitution γ_1 of the required depth can be constructed. Extend γ_1 to include x_k gets a, the smallest constant, as the substitution γ.

The last case is when there are both lower and upper bounds on x_k. For each lower bound t_j and each upper bound s_i on x_k, we add $s_i \succ t_j$, and then remove all lower and upper bound constraints on x_k. The resulting $\langle E, I' \rangle$ is of smaller measure as there is one less variable in I', and maximum depths are same. By induction hypothesis, we obtain a substitution γ_1 within depth $(k-1)*d$ that is a solution for $\langle E, I' \rangle$; further $s_i \gamma_1 \succ next(t_j \gamma_1)$ by induction hypothesis since $s_i \sigma \succ next(t_j \sigma)$. Like in the case of no upper bounds on x_k, we extend γ_1 to γ which includes a substitution for x_k obtained by $next(t_m)$, where t_m is the maximum of $t_1 \gamma_1, \cdots, t_j \gamma_1$, lower bounds on x_k. Again the depth of substitution for x_k is $(k-1)*d + (k-1) + d + 1 = k*d + k$. So we are done.

5 Conclusion

We have defined maximal extensions of simplification orderings to non-ground terms. We have proposed a new simplification ordering ($\succ_{mp}$) based on paths which is maximal. For a given precedence relation $\succ_f$, comparing two terms s and t using this ordering can be done in polynomial time (see [6] where the time complexity for a similar ordering is worked out). The proof of Theorem 3.1 also gives a construction which can be used to compute in case $s \not\succ_{mp} t$, a σ such that $t\sigma \succ_{mp} s\sigma$. In contrast, $\succ_{mrpo}$ check for the maximal extension of RPO is co-NP complete [9]. Thus, for applications in which a maximal simplification ordering is needed, $\succ_{mp}$ is much more efficient to implement than $\succ_{mrpo}$. Further, $\succ_{mp}$ has the nice property that along with the existential fragment of ordering constraints (expressed using $\succ_{mp}$ as well as $\sim_{mp}$), the universal fragment of ordering constraints is also decidable.

We have also given a new method for constructing the maximal extension of RPO by giving a bound on the depth within which a substitution must exist (if at all) that makes one term bigger than another under RPO. This bound gives a simple decision procedure for the problem: given s, t and a total precedence $\succ_f$, is there a ground σ such that $s\sigma \succ_{rpo} t\sigma$? We just enumerate all possible substitutions up to this depth bound. Further work is possible in developing optimizations when enumerating the substitutions as many of them may not need to be tried. It should be interesting to study the efficiency of this method compared to the constraint solving methods for related problems given in [1, 5, 9, 11, 4].

The definition of the maximal extension of RPO given in the paper uses the decision procedure and is hence, indirect. A challenging problem is to come up with a direct definition (like $\succ_{mp}$) which is compatible with RPO (or at least allows orienting distributivity in the desired direction) and is maximal. One promising approach is to define a new notion of $\succ_f$ which is not a precedence relation only on symbols in F, but also treats variables and ground terms as new constants for doing comparisons.

References

1. Comon, H. (1990). Solving Symbolic Ordering Constraints. *International Journal of Foundations of Computer Science*, 1, 4, 387-411.
2. Dershowitz, N. (1982). Orderings for term rewriting systems. *Theoretical Computer Science*, Vol. 17, No. 3, March 1982, 279-301.
3. Dershowitz, N. (1987). Termination of rewriting. *J. of Symbolic Computation*, 3, 69-116.
4. Johann, P., and Socher-Ambrosius R. (1994). Solving Simplification Ordering Constraints. *Constraints in Computational Logics*, In Springer-Verlag LNCS 845, 352-367.
5. Jouannaud, J.-P., and Okada, M. (1991). Satisfiability of Systems of Ordinal Notations with the Subterm Property is Decidable. *Proceedings of ICALP 1991* In Springer-Verlag LNCS 510, 455-468.
6. Kapur, D., Narendran, P., and Sivakumar, G. (1985). A path ordering for proving termination of term rewriting systems. Proc. *10th CAAP*, Berlin, LNCS 185, 173-187.
7. Knuth, D.E. and Bendix, P.B. (1970). Simple word problems in universal algebras. In: *Computational Problems in Abstract Algebras*. (ed. J. Leech), Pergamon Press, 263-297.
8. Lescanne, P., (1990). On the recursive decomposition ordering with lexicographical status and other related orderings, *J. Automated Reasoning*, 6, 1, 39-49.
9. Nieuwenhuis, R. (1993). Simple LPO constraint solving methods. *Information Processing Letters*, 47, 65-69.
10. Plaisted, D. A., (1978). Well-founded orderings for proving termination of systems of rewrite rules. Report UIUCDCS-R-78-943, University of Illinois.
11. Plaisted, D. A., (1993). Polynomial Time Termination and Constraint Satisfaction Tests. In Springer-Verlag LNCS 690, 405-420.
12. Steinbach, J. (1989). Extensions and comparison of simplification orderings. Proc. *3rd International Conf. on Rewriting Techniques and Applications (RTA-89)*, Chapel Hill, NC, 434-448.

Average Polynomial Time Is Hard for Exponential Time Under sn-Reductions

Rainer Schuler

Theoretische Informatik, Universität Ulm, D-89069 Ulm, Germany
email: rsc@informatik.uni-ulm.de

Abstract. Let $P_{P\text{-comp}}$ be the set of decision problems that are solvable in average polynomial time for every polynomial time computable distribution. In [BCGL92] Ben-David, Chor, Goldreich, and Luby show that if NP is contained in $P_{P\text{-comp}}$, then E=NE. Here we discuss the more general question whether NP is reducible to $P_{P\text{-comp}}$. As a first step we show that E is not truth-table reducible to $P_{P\text{-comp}}$. On the other hand, we are able to prove that E is strong nondeterministic Turing reducible to $P_{P\text{-comp}}$. In fact the reduction is deterministic using an advice of polynomial length. A consequence of this result is that it is unlikely that $P_{P\text{-comp}}$ is in P/poly, since this implies that E is in P/poly and a collapse of E and the polynomial time hierarchy to Σ_2^p. Therefore, the assumption that $P_{P\text{-comp}} \subseteq P/poly$ implies that $P \neq NP$.

1 Introduction

In [Lev84, Lev86] Levin introduced a notion of average polynomial time computability. This notion is motivated by the question whether all NP problems are efficiently solvable on average, which for practical applications might be of the same importance as the well known $P \overset{?}{=} NP$ question. Levin showed that there exist (reasonable) distributions and problems in NP that are complete in the following sense: If one assumes that one such problem is efficiently solvable on average for the fixed distribution, then all problems in NP are efficiently solvable for every reasonable distribution.

A partial solution to the question whether all of NP can be solved efficiently on average was given by Ben-David, Chor, Goldreich, and Luby [BCGL92]. They show that if every NP-problem is solvable in polynomial time on average for every polynomial time computable distribution, then E = NE. This indicates that this inclusion is unlikely but so far no stronger (collapse) consequences from this assumption have been shown.

In this paper we discuss the more general question whether NP is reducible to the set of problems that are solvable in polynomial time on average for every polynomial time computable distribution. Let $P_{P\text{-comp}}$ denote this class of decision problems [SY92]. Then the above question can be stated as whether $NP \subseteq P_{P\text{-comp}}$ holds or not and the result of Ben-David, Chor, Goldreich, and Luby [BCGL92] can be rephrased as follows: If $NP \subseteq P_{P\text{-comp}}$ then E = NE.

Concerning the question whether NP is reducible to $P_{\text{P-comp}}$, we give two surprising results. First, it can be shown that E is not truth-table reducible to $P_{\text{P-comp}}$. As a consequence it follows that NP is not truth-table reducible to $P_{\text{P-comp}}$ unless NP is properly contained in Exptime. It seems difficult to improve this upper bound from truth-table reductions to Turing reductions if we restrict our attention to polynomial time computable distributions. As shown in [SW95] Turing reductions (to sets in NP) transform polynomial time computable distributions to distributions that are polynomial time samplable with an NP oracle.

On the other hand, as our second result we are able to give a lower bound. Under strong nondeterministic Turing reductions (sn-reductions for short), $P_{\text{P-comp}}$ contains complete sets for Exptime. This is also true for non-uniform many-one reductions, where the reduction function is allowed to access an advice string of polynomial length which depends only on the length of the input. As an immediate consequence it follows that $P_{\text{P-comp}}$ is hard for the polynomial time hierarchy under sn-reductions, thus giving us a lower bound.

Furthermore, the assumption that $P_{\text{P-comp}} \subseteq P/\text{poly}$ implies that Exptime $\subseteq$ P/poly and thus a collapse of Exptime and the polynomial time hierarchy to the second level of the hierarchy [KL80]. Similarly, if we assume that $P_{\text{P-comp}} \subseteq$ NP/poly then we get a collapse to the third level [Yap83]. Since P is a proper subclass of Exptime both assumptions imply that $P \neq NP$.

2 Preliminaries

In this paper we follow the standard notation and definitions of computational complexity theory. Let $\Sigma = \{0,1\}$ be fixed. All sets considered are subsets of Σ^*, the set of all finite strings over Σ. We use $|x|$ to denote the length of a string x and $\leq$ to denote the lexicographic ordering of strings. For every set A, let A^n denote the set of all strings in A of length n and $||A||$ denote the cardinality of A. Let $z\Sigma^k$ denote the set of strings y such that $|y| = |z| + k$ and z is a prefix of y. For every transducer T, we identify the output of T with a dyadic rational number, i.e. if $T(x) = d_1 \ldots d_n$, then we identify $T(x)$ with the number $d_1 2^{-1} + \cdots + d_n 2^{-n}$.

A function $\mu : \Sigma^* \to [0,1]$ is a density function if $\sum_x \mu(x) = 1$. The distribution function μ^* of μ is defined by $\mu^*(x) = \sum_{y \leq x} \mu(y)$.

We will in particular discuss two types of polynomial time bounded reducibilities. A set A is strong nondeterministic Turing reducible [Lon82] to a set B, $A \leq_T^{sn} B$, if and only if $A \in \text{coNP}(B) \cap \text{NP}(B)$. A different characterization of strong nondeterministic Turing reducibility given in the following lemma is due to Selman [Sel78].

Lemma 1. $A \leq_T^{sn} B$ if and only if $\text{NP}(A) \subseteq \text{NP}(B)$.

A set A is P/poly-many-one reducible to a set B, $A \leq_m^{P/\text{poly}} B$, if and only if there exists a polynomial time bounded reduction function $f : \Sigma^* \to \Sigma^*$ and a function $h : \mathbb{N} \to \Sigma^*$ such that for some polynomial p, $|h(n)| = p(n)$ and for all x, $x \in A \Leftrightarrow f(x, h(|x|)) \in B$.

Note that if B is in P/poly (or NP/poly) and $A \leq_m^{P/poly} B$, then A is in P/poly (or NP/poly respectively). Let E denote $\bigcup_{c>0} \mathrm{Dtime}(2^{cn})$ (linear exponential time) and Exptime denote $\bigcup_{c>0} \mathrm{Dtime}(2^{n^c})$.

Preliminaries for average case complexity

For a detailed introduction to average case complexity the reader is referred to [Gur91]. The very foundation of the average-case complexity as considered here, is Levin's notion of polynomial time on average [Lev84, Lev86]. Let μ be a density function. A function $f : \Sigma^* \to \mathbb{N}$ is *polynomial on μ-average*, if there exists a constant $\epsilon > 0$ such that

$$\sum_{x \neq \lambda} \frac{f^\epsilon(x)}{|x|} \mu(x) < \infty.$$

A randomized (decision) problem is a pair consisting of a decision problem and a density function. A randomized decision problem (A, μ) is in AP (average polynomial time) if there exists a deterministic Turing machine M such that $A = L(M)$ and Time_M, the running time of M, is polynomial on μ-average.

In general, a decision problem is solvable in average polynomial time for some distributions but not for all. In [Mil91, LV92] it was shown that there exist distributions with the property that the average-case complexity of every algorithm is equal to its worst-case complexity. (For polynomial time algorithms such distributions can be computed in exponential time). Therefore it is necessary to restrict our attention to reasonable realistic distributions. Basically two classes of distribution functions have been considered to investigate the question whether NP (decision or search) problems are solvable efficiently on average. Here we will focus on the set of distribution functions that are computable in polynomial time. The class of polynomial time samplable distribution functions is more general, but in [IL90] it was shown that for randomized algorithms (and NP-search problems) it does not make any difference which class is considered.

A function $f : \Sigma^* \to \mathbb{R}^+$ is polynomial time computable [KF82], if there exists a polynomial-time bounded transducer T such that for all x and all $k > 1$

$$|f(x) - T(x, 1^k)| \leq 2^{-k}.$$

Let P-comp denote the set of density functions with polynomial time computable distribution functions. A decision problem D is in $P_{P\text{-comp}}$ if and only if for every polynomial time computable distribution function μ^* (i.e. for every $\mu \in$ P-comp) the randomized decision problem (D, μ) is in AP [SY92]. That is, for every density function μ in $P_{P\text{-comp}}$, D is solvable in time polynomial on μ-average.

It is not difficult to see that $P \subseteq P_{P\text{-comp}} \subseteq E$ [SY92]. Furthermore, if a function f is polynomial on μ average, then there exists a constant $\epsilon > 0$ such that for all (but finitely many) x, $f^\epsilon(x) \leq |x|/\mu(x)$.

Proposition 2. *For every set $A \in P_{P\text{-comp}}$, every density function $\mu \in P\text{-comp}$, and every polynomial p there exists a Turing machine M, a constant c, and a polynomial q such that for all x, $\text{Time}_M(x) \leq 2^{c|x|}$ and if $\mu(x) \geq 1/p(|x|)$, then $\text{Time}_M(x) \leq q(|x|)$.*

We now recall some properties of the class $P_{P\text{-comp}}$ which will be used in the next sections [Sch95, SY95].

Lemma 3. *Let h be defined inductively by $h(0) = 1$ and $h(k) = 2^{h(k-1)}$, $k > 0$. There exists a set $A \in \text{Dtime}(n^{\log n})$ such that*

- $\|A^n\| = 1$, *if $n = h(k)$ for some k, and $A^n = \emptyset$ otherwise.*
- $A \cap B \in P_{P\text{-comp}}$ *for every set $B \in E$.*

Lemma 4. *There exists a set $A \in \text{Dtime}(n^{\log n})$ such that*

- $A^n = z\Sigma^{m+1}$ *for a string z, if $n = 2m + 1$, and $A^n = \emptyset$ otherwise.*
- $A \cap B \in P_{P\text{-comp}}$ *for every set $B \in E$.*

The lemmas follow from [SY95] Lemma 6 (see also [Sch95]). We give here a different proof in terms of Kolmogorov complexity. For definitions and properties of Kolmogorov complexity see e.g. [LV93].

Proof. Let $KT[t, s \mid y]$ denote the set of all strings that can be computed by a universal machine in time t from y and a description of length less than or equal to s.

For every polynomial time computable function $k : \mathbb{N} \to \mathbb{N}$, $k(n) \leq n/\log n$, define a set Z of strings as follows. For every n and all $i \leq k(n)$ let

$$z_{i,n} = \min_{z, |z| = \log n} \{z \notin KT[\log n - 1, n^{\log i} \mid z_{1,n} \ldots z_{i-1,n}]\}.$$

Then for all i let $Z_{i,n} = z_{1,n} \ldots z_{i,n} \Sigma^{n-i \log n}$ and define $Z = \bigcup_{n>0} Z_{k(n),n}$. Note that $z_{i,n}$ is computable in time $n \cdot n^{\log i}$ and therefore $Z \in \text{Dtime}(n^{\log n})$.

Claim 1 *For every $\mu \in P\text{-comp}$ there exists constants t and n_0 such that for all $n \geq n_0$ and for all $i \leq k(n)$*

$$\mu(Z_{i,n}) \leq 2^{(t-i)(\log n)/2}$$

Proof. Choose t such that for some transducer $T \in \text{Dtime}(n^{\log t - 1})$ it holds that for all x, $|T(x) - \mu^*(x)| \leq 2^{-2|x|}$. Now choose n_0 large enough such that a (self-delimiting) description of T is smaller than $(\log n_0)/2 - 1$. Note that if μ^* (i.e. T) is computable in time $O(n^{\log t - 1})$, then for every string s, $|s| \leq n$, $\mu(s\Sigma^{n-|s|})$ is also computable in time $O(n^{\log t - 1})$.

For all $i \leq t$ it holds that $\mu(Z_{i,n}) \leq \mu(\Sigma^n) \leq 2^{(t-i)(\log n)/2}$.

Now assume that the claim does not hold and let i be the smallest number such that $\mu(Z_{i,n}) > 2^{(t-i)(\log n)/2}$. Then the number of strings z' such that $\mu(z_{1,n} \ldots z_{i-1,n} z' \Sigma^{n-i \log n}) > 2^{(t-i)(\log n)/2}$ is smaller than $2^{(\log n)/2}$.

Hence, given $z_{1,n} \ldots z_{i-1,n}$, it is possible to describe $z_{i,n}$ by a description of T and an index of length $(\log n)/2$. To compute $z_{i,n}$ it is necessary to simulate the transducer T at most $n^{1/2}$ times, which can be done in time $O(n^{\log t})$.

This implies that $z_{i,n} \in KT[\log n - 1, n^{\log i} \mid z_{1,n} \ldots z_{i-1,n}]$, a contradiction to the choice of $z_{i,n}$.

To prove Lemma 4 choose $k(n) = \lfloor n/(2\log n)\rfloor$ and let Z be defined as above. Then for all n, $Z^n = z_{1,n}\ldots z_{k(n),n}\Sigma^{n-k(n)\log n}$. Choose l such that $l + k(n)\cdot \log n = \lfloor n/2\rfloor$ and let $Z'^n = z_{1,n}\ldots z_{k(n),n}0^l\Sigma^{\lceil n/2\rceil}$. Now define $A^n = Z'^n$ if $n = 2m + 1$ for some m and $A^n = \emptyset$ otherwise.

Let μ be any distribution in P-comp and B be any set in E, i.e. $B \in \text{Dtime}(2^{dn})$ for some constant d. To compute $A \cap B$ it suffices to compute A on Σ^* and B on strings in A. Let f denote the time bound of the following algorithm computing $A \cap B$ in this way.

input x, $n = |x|$
    if n is even **then reject**
    for $i = 1$ **to** $k(n)$ **do**
        compute $z_{i,n} = \min_{z,|z|=\log n}\{z \notin \text{KT}[\log n - 1, n^{\log i} \mid z_{1,n}\ldots z_{i-1,n}]\}$
        if $z_{1,n}\ldots z_{i,n}$ is not a prefix of x **then reject**
    compute $l = (n-1)/2 - k(n)\cdot \log n$
    if $z_{1,n}\ldots z_{k(n),n}0^l$ is not a prefix of x **then reject**
    if $x \in B$ **then accept else reject**

The following estimation shows that f is polynomial on average. Let t and n_0 be the constants of Claim 1 and choose $\epsilon \leq 1/(8\max\{\log t, d\})$. Then for some constant N

$$\sum_{x\neq\lambda}\frac{f^\epsilon(x)}{|x|}\mu(x) \leq N + \sum_{n>n_0}\sum_{i\leq t}\frac{n^{\epsilon\log i}}{n}\mu(\Sigma^n) \; +$$

$$\sum_{n>n_0}\sum_{t<i\leq k(n)}n^{\epsilon\log i}\cdot 2^{(t-i)(\log n)/2} \; +$$

$$\sum_{n>n_0}2^{\epsilon dn}\cdot 2^{(t-\lfloor n/(2\log n)\rfloor)(\log n)/2} \; < \; \infty.$$

Similarly, Lemma 3 follows if we choose $k(n) = \lfloor n/\log n\rfloor$ and define $A^n = Z^n$ if $n = h(l)$ for some l and $A^n = \emptyset$ otherwise, where h is the function defined in Lemma 3.

3 Main results

First, in Theorem 5, we give an upper bound result and show that no hard set for E is truth-table reducible to $P_{\text{P-comp}}$. Therefore, a proof showing that NP is truth-table reducible to $P_{\text{P-comp}}$, implies that NP is different from Exptime.

Our second result gives a (surprising) lower bound on the complexity of the sets in $P_{\text{P-comp}}$. It turns out that $P_{\text{P-comp}}$ is hard for Exptime under strong nondeterministic Turing reductions. It is not difficult to see that $P_{\text{P-comp}}$ is properly included in E and, as shown in [Sch95], $P_{\text{P-comp}}$ contains P as a proper subclass. Intuitively, Theorem 7 shows that it is possible to pad exponential time sets (and hence hide the necessary exponential time computations) in such a way that no polynomial time computable distribution can give high probability to these instances.

As a further result, it turns out that proving that all sets in $P_{\text{P-comp}}$ can be computed by polynomial size circuits is equivalent to showing that Exptime has polynomial size circuits.

Theorem 5. *There exists a set in E that is truth-table reducible to no set in $P_{\text{P-comp}}$.*

Proof. Let A be the set defined in Lemma 3. Using diagonalization it is easy to define a set $A' \subseteq A$ such $A' \in E - \text{Dtime}(2^n)$. Let L be a test language for A'

$$L = \{0^n \mid \exists x, |x| = n \wedge x \in A'\}.$$

Obviously $L \in E$. Now assume that there exists a set B in $P_{\text{P-comp}}$ such that L is truth-table reducible to B. We will show that this implies that $A' \in \text{Dtime}(2^n)$, a contradiction. First note that for all x, x is in A' iff $x \in A$ and $0^{|x|} \in L$ and therefore if L is in $\text{Dtime}(2^n)$, then A' is in $\text{Dtime}(n^{\log n} + 2^n) = \text{Dtime}(2^n)$. Thus it remains to show that $L \in \text{Dtime}(2^n)$.

Let M be a non-adaptive oracle Turing machine such that $L = L(M, B)$ and for all x, $\text{Time}_M(x) \le p(|x|)$ for a polynomial p.

Since B is in $P_{\text{P-comp}}$, it follows from Proposition 2 that there is a Turing machine M_B and a constant c such that $B = L(M_B)$ and for all x, $\text{Time}_{M_B}(x) \le 2^{c|x|}$.

Let Q_k denote the set of strings y, such that y is queried by M with oracle B on input $0^{h(k)}$ and $2c|y| \ge h(k)$. Let $Q = \bigcup_k Q_k$. Consider a density function μ defined as follows.

$$\mu(y) = \begin{cases} 1/(\|Q_k\| \cdot k^2), & \text{if } y \in Q_k, \text{ where } k = \min_{i>0}\{|y| \le p(h(i))\} \\ 0, & \text{otherwise.} \end{cases}$$

Note that for all $y \in Q_k$ it holds that $2c|y| \ge h(k)$ and therefore the simulation of M on input $0^{h(k)}$, i.e. the computation of the set Q_k, is polynomial in $|y|$. Choose k_0 such that $h(k_0)/(2c) > p(h(k_0 - 1))$. Then for all $k > k_0$, if $x \in Q_k$ and $y \in Q_{k+1}$ then $x < y$. Hence

$$\mu^*(y) = \sum_{k:h(k) \le 2c|y|} \sum_{z:z \in Q_k \wedge z \le |y|} \mu(z)$$

is polynomial time computable. On the other hand, if y is queried by M on input $0^{h(k)}$ but y is not in Q_k, then $y \in B$ can be decided by simulating M_B on input y in time $2^{c|y|} \le 2^{h(k)/2}$.

Since $B \in P_{\text{P-comp}}$, $\mu \in \text{P-comp}$ and for all $y \in Q$, $\mu(y) \ge 1/q(|y|)$ for some polynomial q, it follows by Proposition 2 that there is a Turing machine M_B' and a polynomial r such that $B = L(M_B')$ and for all $y \in Q$, $\text{Time}_{M_B'}(y) \le r(|y|)$. Now we are ready to define a Turing machine N for L as follows.

> **input x**
> **if** $x = 0^{h(k)}$ for some k **then**
> (1)   simulate $M(0^{h(k)})$ and for every oracle query y
> (2)   **if** $2c|y| \ge h(k)$ and $y \in L(M_B')$ **or**
> (3) $2c|y| < h(k)$ and $y \in L(M_B)$ **then** continue with answer yes

(4)   **else** continue with answer no
      if $M(0^{h(k)})$ accepts **then accept**
   reject

The time bound of N can be calculated as follows. Consider the above procedure of the machine N. Line (1) is computable in time $p(n)$, line (2) is computable in time $r(p(n))$ and line (3) is computable in time $O(2^{n/2})$. Since lines (2) and (3) are repeated for at most $p(n)$ times, the machine is overall time bounded by $O(2^n)$.

Corollary 6. *If* NP *is truth-table reducible to* $P_{P\text{-comp}}$ *then* NP $\neq$ Exptime.

Theorem 7. *For every* $H \in$ Exptime *there exists a set* $L \in P_{P\text{-comp}}$ *such that*

$$H \leq_T^{sn} L \text{ and } H \leq_m^{P/poly} L.$$

Proof. Since Exptime is many one reducible to E and the reductions are transitive, it suffices to consider sets in E. Let H be any problem in E, then $H \in$ Dtime(2^{cn}) for some constant c. Let A be the set defined in Lemma 4 and define a coding L of H as follows

$$L = \{x \mid x \in A \wedge \forall z, y : (x = zy1 \wedge |y| = |z|) \to y \in H)\}.$$

Note that L is a subset of A and computable in E. Thus, from Lemma 4 it follows that $L \in P_{P\text{-comp}}$. Furthermore, for every length n, there is a string z, $|z| = n$, such that for all strings $y \in \Sigma^n$, $zy0 \in L$, and for every string $y \in \Sigma^n$, y is in H if and only if $zy1 \in L$. This shows that $H \leq_m^{P/n} L$. Since, by definition, for every length n there is exactly one string z, $|z| = n$ such that $zy0 \in L$ for any string $y \in \Sigma^n$, it follows that H and $\overline{H}$ are in NP(L). Therefore $H \leq_T^{sn} L$.

Corollary 8. *There exist sets in* $P_{P\text{-comp}}$ *which are complete for* Exptime *under strong nondeterministic Turing reductions.*

Corollary 8 gives us a lower bound on the complexity of $P_{P\text{-comp}}$. In particular it shows that $P_{P\text{-comp}}$ is $\leq_T^{sn}$-hard for the polynomial time hierarchy.

Recall that P/poly and NP/poly are closed under $\leq_m^{P/poly}$ reductions.

Corollary 9. *If* $P_{P\text{-comp}} \subseteq$ P/poly, *then* Exptime *collapses to* Σ_2^p.

This follows since NP $\subseteq$ P/poly implies that the polynomial time hierarchy collapses to its second level [KL80]. Similarly, if Exptime is contained in NP/poly then Exptime collapses to the third level of the polynomial time hierarchy [Yap83].

Corollary 10. *If* $P_{P\text{-comp}} \subseteq$ NP/poly, *then* Exptime *collapses to* Σ_3^p.

Note that both corollaries imply that P is different from NP, since P is a proper subset of Exptime. Furthermore since $P_{P\text{-comp}}$ is contained in E it follows that $P_{P\text{-comp}} \subseteq$ P/poly (or NP/poly) if and only if Exptime $\subseteq$ P/poly (NP/poly) respectively).

References

[BCGL92] S. BEN-DAVID, B. CHOR, O. GOLDREICH, AND M. LUBY. On the theory of average case complexity. *Journal of Computer and System Sciences* 44, 193–219 (1992).

[Gur91] Y. GUREVICH. Average case complexity. *Journal of Computer and System Sciences* 42(3), 346–398 (1991). A special issue on FOCS'87.

[IL90] R. IMPAGLIAZZO AND L.A. LEVIN. No better ways to generate hard NP-instances than picking uniformly at random. *Proc. 31st IEEE Symposium on Foundations of Computer Science* pages 812–821 (1990).

[KF82] K.I. KO AND H. FRIEDMAN. Computational complexity of real functions. *Theoretical Computer Science* 20, 323–352 (1982).

[KL80] R.M. KARP AND R.J. LIPTON. Some connections between nonuniform and uniform complexity classes. *Proc. 12th ACM Symposium on Theory of Computing* pages 302–309 (1980).

[Lev84] L. LEVIN. Problems, complete in "average" instance. *Proc. 16th ACM Symposium on Theory of Computing* page 465 (1984).

[Lev86] L. LEVIN. Average case complete problems. *SIAM Journal on Computing* 15, 285–286 (1986).

[Lon82] T. LONG. Strong nondeterministic polynomial-time reducibilities. *Theoretical Computer Science* 21, 1–25 (1982).

[LV92] M. LI AND P.M.B. VITÁNYI. Average case complexity under the universal distribution equals worst-case complexity. *Information Processing Letters* 42, 145–149 (1992).

[LV93] M. LI AND P. VITÁNYI. "An introduction to Kolmogorov complexity and its applications". Springer-Verlag, New York (1993).

[Mil91] P.B. MILTERSON. The complexity of malign ensembles. *Proc. 6th Structure in Complexity Theory Conference* pages 164–171 (1991).

[Sch95] R. SCHULER. Some properties of sets tractable under every polynomial-time computable distribution. *Information Processing Letters*. to appear.

[Sel78] A.L. SELMAN. Polynomial time enumeration reducibility. *Theoretical Computer Science* 14, 91–101 (1978).

[SW95] R. SCHULER AND O. WATANABE. Towards average-case complexity analysis of NP optimization problems. In "Proc. 10th Structure in Complexity Theory Conference", pages 148–159 (1995).

[SY92] R. SCHULER AND T. YAMAKAMI. Structural average case complexity. In "Proc. 12th Foundations of Software Technology and Theoretical Computer Science", LNCS 652, pages 128–139 (1992).

[SY95] R. SCHULER AND T. YAMAKAMI. Sets computable in polynomial time on average. In "Proc 1st International Computing and Combinatorics Conference", LNCS (1995).

[Yap83] C. YAP. Some consequences of non-uniform conditions on uniform classes. *Theoretical Computer Science* 26, 287–300 (1983).

On Self-Testing without the Generator Bottleneck

S. Ravikumar[1] * and D. Sivakumar[2] **

[1] Department of Computer Science, Cornell University, Ithaca, NY 14853.
[2] Department of Computer Science, State Univ. of New York at Buffalo, Buffalo, NY 14260.

Abstract. Suppose P is a program designed to compute a linear function f on a group G. The task of *self-testing* f, that is, testing if P computes f correctly on most inputs, usually involves checking if P computes f correctly on all the generators of G. Recently, F. Ergün presented self-testers that avoid this *generator bottleneck* for specific functions. In this paper, we generalize Ergün's results, and extend them to a much larger class of functions. Our results give efficient self-testers for polynomial differentiation, integration, arithmetic in large finite field extensions, and constant-degree polynomials over large rings.

1 Introduction

The notions of program result-checking, self-testing, and self-correcting [BK89, Lip91, BLR93] have proved to be very useful concepts. The theoretical developments in this area are at the heart of the recent breakthrough results on probabilistically checkable proofs and the subsequent results that show non-approximability of hard combinatorial problems. From a practical viewpoint, these methods offer realistic and efficient tools for program verification. See the survey articles by Blum and Wasserman [BW94] and by Madhu Sudan [Sud94] for interesting expositions and pointers.

Suppose we are given a program P designed to compute a function f. Informally, a *self-tester* for f enables us to estimate the fraction of the inputs on which P computes f correctly. Alternatively, the role of a self-tester is to either certify that P computes f correctly on most inputs, or reject P as largely erroneous. A *result-checker* for a function f takes a program P and an input q and outputs "CORRECT" if P always computes f correctly, and outputs "WRONG" if $P(q) \neq f(q)$. Given a program P that computes f correctly on most inputs, a *self-corrector* for f is a program that uses P as an oracle and computes f correctly on every input with high probability. The existence of a self-tester/corrector pair implies the existence of a result-checker. Blum and Kannan [BK89] and Blum, Luby, and Rubinfeld [BLR93] have laid out the minimal requirements for self-testers and result-checkers. These programs should not compute f directly or require the value of f at too many locations, that is, they should be "different" from the program being tested, and should be at least as efficient as the fastest algorithm for f.

Self-testers are usually built for a *class* $\mathcal{F}$ of functions by identifying certain properties of the class $\mathcal{F}$ that are *robust*. Informally, property Π is said to be a robust characterization of a function class $\mathcal{F}$ if the following two conditions hold: (1) every $f \in \mathcal{F}$ satisfies

---

* Research supported in part by ONR Young Investigator Award N00014-93-1-0590. Email: `ravi@cs.cornell.edu`

** Research supported in part by NSF grant CCR-9409104. Email: `sivak-d@cs.buffalo.edu`

Π, and (2) if P is a function (program) that satisfies Π on most inputs, then P must agree with some $g \in \mathcal{F}$ on most inputs. For example, consider the family $\mathcal{F}$ of linear functions on a group $(G, \circ)$. Here the linearity property $\Pi(f) \equiv (\forall p, q \in G)[f(p \circ q) = f(p) \circ f(q)]$ is robust: every $f \in \mathcal{F}$ satisfies Π, and moreover, if some program P satisfies Π for most $p, q \in G$, then there is a linear function $g \in \mathcal{F}$ with which P agrees most of the time. Moreover, it is often the case that given the faulty program P, a "self-corrected" version P_{sc} of P can be built that correctly computes (with high probability) the "ghost function" g on every input.

Various linear and low-degree polynomials, and functions described by functional equations have been shown to have self-testers and/or self-correctors [BLR93, Lip91, GLR+91, RS92, RS93, Rub94]. Using the notions of robust characterizations and self-correcting, the self-testing scheme for a program P that purports to compute a linear function f can be described as follows. First check if the program P satisfies the linearity property on most inputs $p, q \in G$. By the robustness of linearity, this implies that there is a linear function g on G that can be computed by the self-corrector P_{sc} of P. Using P_{sc}, verify that g agrees with f on the generators of the group G. If P passes these tests, it is then shown by induction that g agrees with f on all inputs. Since P agrees with g on most inputs, it follows that P computes f correctly on most inputs.

If the group G has many generators, for example, if the group is a vector space of large dimension, then testing if P_{sc} agrees with f on the generators adds to the overhead in the testing process. This is called the *generator bottleneck*. It is desirable to build self-testers that make only $O(1)$ calls to the program P being tested. Such self-testers are called *efficient* in [BLR93]. The problem of generator bottleneck was first addressed by Ergün [Erg95], who shows how to obtain efficient self-testers for specific functions like Fourier transforms, polynomial multiplication, etc.

We generalize and extend Ergün's idea to a larger class of functions. First we show that the technique can be applied in a more natural and general setting, namely that of vector spaces. We prove a general theorem that gives a condition for a linear function f to be efficiently self-testable on a large vector space. This approach has an extra benefit: it allows us to give conditions for efficient self-testing in terms of the matrices of linear operators. Applying this result, we obtain self-testers for many functions that include polynomial differentiation, integration, the polynomial "mod" function, etc., besides Fourier transforms. We then extend this to the case of multilinear functions, and derive the result for polynomial multiplication as a consequence. Another application we give is for large finite fields: we show that multilinear functions over finite field extensions of dimension n can be efficiently self-tested with $O(1)$ calls, independent of the dimension n.

The next step we take is to extend the result to the case of nonlinear functions. We give the first self-testers for exponentiation functions that avoid the generator bottleneck. Consider the function that computes the square of a polynomial over a finite field: $f(p) = p^2$. Here we do not have the nice property of linearity that was crucial in the proof for the linear functions. Instead, we use the fact that the interpolation identity for polynomials gives a robust characterization. We exhibit a self-tester for the function $f(p) = p^d$ that makes $O(d)$ calls to the program being tested.

The passage from constant degree exponentiation to constant degree polynomials is much harder. First we show a reduction from multiplication to the computation of low-degree polynomials. Using this reduction and the notion of result-checker, we construct a self-tester for degree d polynomials over finite field extensions of dimension n that make $O(2^d)$ calls to the program being tested.

2 Linear Functions over Vector Spaces

In this section, we generalize the idea of [Erg95] to the case of linear functions on a vector space. We begin with the basic definitions.

Definition 1. Let f be a function on a domain D. A (γ, ε)-*self-tester* for f is a probabilistic program T that, given a program P for f, and a parameter δ, satisfies the following conditions.

- $\Pr_{X \in D}[P(X) \neq f(X)] \leq \gamma \Rightarrow \Pr[T^P \text{ outputs "PASS" }] \geq 1 - \delta$.
- $\Pr_{X \in D}[P(X) \neq f(X)] \geq \varepsilon \Rightarrow \Pr[T^P \text{ outputs "FAIL" }] \geq 1 - \delta$.

A ε-*self-corrector* for P is a probabilistic program R that takes an input Y and a parameter δ, and satisfies the following condition:

- $\Pr_{X \in D}[P(X) \neq f(X)] \geq \varepsilon \Rightarrow \Pr[R^P(Y) = f(Y)] \geq 1 - \delta$.

A *checker* (or result-checker) for f is a probabilistic program C that, given a program P for f, an input Y, and a parameter δ, satisfies the following conditions:

- $\Pr_{X \in D}[P(X) = f(X)] = 1 \Rightarrow \Pr[C^P(Y) \text{ outputs "CORRECT" }] = 1$.
- $P(Y) \neq f(Y) \Rightarrow \Pr[C^P(Y) \text{ outputs "WRONG" }] \geq 1 - \delta$.

Let V be a vector space of finite dimension n over a field $\mathbb{K}$, and let f be a function from V into a ring R. We are interested in building a self-tester for the case where f is a *linear* function, that is, $f(c\alpha + \beta) = cf(\alpha) + f(\beta)$ for all $\alpha, \beta \in V$ and $c \in \mathbb{K}$. For simplicity, we will assume that the field $\mathbb{K}$ is finite; our testers work for infinite fields as well, but additional complications arise. For example, it is not clear how to choose a random element from the field of real numbers.

The self-tester we build follows the framework of [BLR93]. For $1 \leq i \leq n$, let e_i denote the vector that has a 1 in the i-th position and 0's in the other positions. The vectors $e_1, e_2, \ldots, e_n$ form a collection of basis vectors that span V. Viewed as an abelian group under vector addition, V is generated by $e_1, \ldots, e_n$. Suppose P is a program that purports to compute f. The framework of [BLR93] comprises the following two tests:

(1) *Linearity Test*: By checking if $P(\alpha + \beta) = P(\alpha) + P(\beta)$ on many randomly chosen inputs $\alpha, \beta \in V$, ensure that P satisfies the linearity property.

It is shown in [BLR93] that the linearity property is robust, that is, if the program P passes Test (1) sufficiently often, then there is a unique linear function g that agrees with P on most inputs. Moreover, as mentioned in the Introduction, there is a self-corrector P_{sc} that uses P as an oracle and computes g correctly on every input (with high probability). The second test ensures that g agrees with f on the generators.

(2) *Basis Tests*: For $1 \leq i \leq n$, verify that $g(e_i) = f(e_i)$.

Here $g(e_i)$ will be computed using the program P_{sc}. If P_{sc}, the self-corrector for P, passes Test (2), then one can show by induction, using the linearity of g, that g is identical to the function f.

There are two problems with this: one is that the self-tester is inefficient—if the inputs are n-element vectors, the self-tester makes $O(n)$ calls to the program, which is not desirable. The requirement in [BLR93] for an *efficient* self-tester is that it make $O(1)$ calls to the program being tested. Secondly, the self-tester needs to know the correct value of f on n different points, which is also undesirable. Our primary interest is to avoid this *generator bottleneck*. Ergün [Erg95] introduced an elegant trick that accomplishes this for specific functions. The key idea is to find an easy uniform way that "converts" one generator into the next generator. We illustrate this idea through the following example.

Let V denote the group of all degree n polynomials under addition. The elements $1, x, x^2, \ldots, x^n$ generate V. Given any generator x^k, multiplying by x gives the next generator x^{k+1}. For a polynomial $p \in V$ and a scalar $c \in \mathbb{K}$, let $E_c(p)$ denote the function that evaluates $p(c)$. Clearly E_c is linear and satisfies the simple relation $E_c(xp) = cE_c(p)$. Suppose P is a program that purports to compute E_c, and assume that P has passed Test (1) given above. Then we know that there is a "ghost linear function" g that agrees with P on most inputs. Now, rather than verify that $g(e) = f(e)$ for all generators e, we may instead verify that g satisfies the property $g(xp) = cg(p)$ everywhere. By an easy induction, this implies that g agrees with f at all the generators. By the linearity of g, it would follow that g agrees with f on all inputs.

We are now faced with the task of verifying $g(xp) = cg(p)$ for all $p \in V$, which is hopeless. Ergün [Erg95] shows that in order to verify this, it suffices to check if $g(xp) = cg(p)$ holds *almost* everywhere. That is, pick many random $p \in V$, ask the program P_{sc} to compute the values of $g(p)$ and $g(xp)$ in two calls, and cross-check that $g(xp) = cg(p)$ holds. In other words, the property "$(\forall p)[g(xp) = cg(p)]$" satisfies a *robustness* condition. Notice that the number of points on which the self-tester needs to know the value of f is just one, in contrast to knowing the values of f on all the n generators (as in the original approach of [BLR93]).

We note that this idea has a natural generalization to vector spaces. Let θ denote the linear operator on V that "rotates" the coordinate axes that span V; that is,

$$\theta = \left(e_2^T e_3^T \cdots e_n^T e_1^T \right)^T$$

The matrix θ defines a one-to-one correspondence from the set of basis vectors to itself: for every i, $\theta(e_i) = e_{(i+1) \bmod n}$. The computational payoff is achieved when there is a simple relation between $f(\alpha)$ and $f(\theta(\alpha))$ for all vectors $\alpha \in V$. More specifically, we show that the generator bottleneck can be avoided if there is an easily computable function h such that $f(\theta(\alpha)) = h(\theta, \alpha, f(\alpha))$ for all $\alpha \in V$. If the function f is linear, the linearity of the operator θ implies that h is always linear in its second argument in the following sense: $h(\theta, \alpha + \beta, f(\alpha + \beta)) = h(\theta, \alpha, f(\alpha)) + h(\theta, \beta, f(\beta))$. What is more important is that h be easy to compute, given just α and $f(\alpha)$.

Using this scheme, we show that besides the functions self-tested in [Erg95], many other natural functions f have a suitable candidate for h. Recall that if P passes Test (1) there is a unique linear function g that agrees with P on most inputs, and that can be computed by P_{sc} (using P as an oracle). The basis tests of [BLR93] can now be replaced by:

(2) Verify that $g(e_1) = f(e_1)$.
(3) For many randomly chosen α's, verify that $g(\theta(\alpha)) = h(\theta, \alpha, g(\alpha))$.

Theorem 2. *Suppose f is a linear function from the vector space V into a ring R, and suppose P is a program for f. Let $\varepsilon < 2/9$, and suppose P satisfies the following condition:*

(1) $\Pr_{\alpha, \beta \in V}[P(\alpha + \beta) \neq P(\alpha) + P(\beta)] \leq \varepsilon.$

Then the function g defined by $g(\alpha) = \text{majority}_{\beta \in V}\{P(\alpha + \beta) - P(\beta)\}$ is a linear function on V, and g agrees with P on at least $1 - 2\varepsilon$ fraction of the inputs. Furthermore, suppose g satisfies the following conditions:

(2) $g(e_1) = f(e_1)$.
(3) $\Pr_{\alpha \in V}[g(\theta(\alpha)) \neq h(\theta, \alpha, g(\alpha))] \leq \varepsilon.$

Then $g(\alpha) = f(\alpha)$ for all $\alpha \in V$.

Remarks. The above theorem only brings out two robust properties of linear functions. The fact that this yields a self-tester is presented in Theorem 4. Note that hypotheses (1), (2), and (3) above are *conditions* on P and g, not *tests* performed by a self-tester.

Proof. The proof that the function g is linear and self-corrects P is due to [BLR93] (the "2/9" in the theorem is the bound obtained there). For the rest of this proof, we will assume that g is linear and that it satisfies conditions (2) and (3) above.

We first argue that it suffices to prove that for every $\alpha \in V$, $g(\theta(\alpha)) = h(\theta, \alpha, g(\alpha))$. By condition (2), g agrees with f on the first basis vector. For $i > 1$, the basis vector e_i can be obtained by $\theta(e_{i-1})$. If g satisfies $g(\theta(\alpha)) = h(\theta, \alpha, g(\alpha))$ everywhere, it would follow that g computes f correctly on all the basis vectors. Finally, since g is linear, it computes f correctly on all of V, since the vectors in V are just linear combinations of the basis vectors.

Fix an arbitrary element $\alpha \in V$. We will show that the probability over a random $\beta \in V$ that $g(\theta(\alpha)) = h(\theta, \alpha, g(\alpha))$ is positive. Since the equality is independent of β and holds with non-zero probability, it must be true with probability 1. Now

$$
\begin{aligned}
\Pr_{\beta \in V}[g(\theta(\alpha)) &= g(\theta(\beta + \alpha - \beta)) \\
&= g(\theta(\beta) + \theta(\alpha - \beta)) \\
&= g(\theta(\beta)) + g(\theta(\alpha - \beta)) \\
&= h(\theta, \beta, g(\beta)) + h(\theta, \alpha - \beta, g(\alpha - \beta)) \\
&= h(\theta, \beta + \alpha - \beta, g(\beta) + g(\alpha - \beta)) \\
&= h(\theta, \alpha, g(\alpha))] \geq 1 - 2\varepsilon > 0.
\end{aligned}
$$

The first equality in the above is just rewriting. The second equality follows from the linearity of θ. The third equality follows from the fact that g is linear. If the r.v. β is distributed uniformly in V, the r.v.'s β and $\alpha - \beta$ are distributed identically and uniformly in V. Therefore, by the assumption that g satisfies condition (3), the fourth equality fails with probability at most 2ε. The fifth equality uses the fact that h is linear, and the last equality uses the fact that g is linear. $\qquad\qquad\square$

The foregoing theorem shows that if P (and g) satisfy certain conditions, then g, which can be computed using P, is identically equal to the function f. To build a self-tester, we will verify that each of these conditions holds. Since the vector space V could be very large, the self-tester cannot check these conditions exhaustively. However, by random sampling, the self-tester can *estimate* the required probabilities to a high degree of accuracy.

The self-tester takes two parameters as input: an *accuracy parameter* ε that specifies the conditions P is expected to meet, and a *confidence parameter* δ, which is an upper bound on the probability that the self-tester fails, and which also specifies the allowed error on the various estimates made by the self-tester. An efficient self-tester should only make $k^{O(1)}$ calls to P, where $k = \frac{1}{\varepsilon} + \frac{1}{\delta}$. The self-tester we describe performs three tests, and we will apportion the error allowance δ equally among the three tests, that is, each test will fail with probability at most $\delta/3$. To this end, we use the following form of the Chernoff bounds:

Theorem 3 (Chernoff Bounds). *Let $X_1, \ldots, X_\ell$ be independent Bernoulli random variables with $\Pr[X_i = 1] = p$ for all i, and let $X = \sum_i X_i$. Then*

$$\Pr\left[\left|\frac{X}{\ell} - p\right| \geq \lambda\right] \leq 2e^{-2\lambda^2\ell}.$$

The self-tester comprises the following tests:

(1) *Linearity Test*: Let $\ell = (1/2\delta^2)\ln(6/\delta)$. Pick uniformly and independently at random $\alpha_1, \beta_1, \alpha_2, \beta_2, \ldots, \alpha_\ell, \beta_\ell$ from V. For $1 \leq i \leq \ell$, let X_i be a Bernoulli random variable that is 1 iff $P(\alpha_i) \neq P(\alpha_i + \beta_i) - P(\beta_i)$. Let $X = \sum_i X_i$, and let $\tau = X/\ell$. If $\tau > \varepsilon + \delta$, then output "FAIL."

(2) *Basis Test*: Let $\ell = (1/2\delta^2)\ln(6/\delta)$. Pick uniformly and independently at random $\beta_1, \beta_2, \ldots, \beta_\ell$ from V, and let $u = \text{majority}_{\beta_i}\{P(e_1 + \beta_i) - P(\beta_i)\}$. If $u \neq f(e_1)$ then output "FAIL."

(3) *Inductive Test*: Let $\ell = (1/2\delta^2)\ln(12/\delta)$. Pick uniformly and independently at random $\alpha_1, \alpha_2, \ldots, \alpha_\ell$ from V. Let $k = (1/2\delta^2)\ln(12\ell/\delta)$. For each α_i, pick uniformly and independently at random $\beta_{i1}, \beta_{i2}, \ldots, \beta_{ik}$ from V, and let $u_i = \text{majority}_{\beta_{ij}}\{P(\alpha_i + \beta_{ij}) - P(\beta_{ij})\}$. Also for each α_i, pick uniformly and independently $\gamma_{i1}, \gamma_{i2}, \ldots, \gamma_{ik}$ from V, and let $v_i = \text{majority}_{\gamma_{ij}}\{P(\theta(\alpha_i) + \gamma_{ij}) - P(\gamma_{ij})\}$. For each i, let X_i be a Bernoulli random variable that is 1 iff $v_i \neq h(\theta, \alpha_i, u_i)$. Let $X = \sum_i X_i$, and let $\tau = X/\ell$. If $\tau > \varepsilon + \delta$, then output "FAIL."

If all the above tests were completed successfully, output "PASS."

Theorem 4. *For any* $\delta < 1$ *and* $\varepsilon < 2/9$, *the above three tests comprise a* $(0, 2\varepsilon)$-*self-tester for* f. *That is, if a program* P *computes* f *correctly on all inputs, then the self-tester outputs "PASS" with probability* 1, *and if* P *computes* f *incorrectly on more than* ε *fraction of the inputs, then the self-tester outputs "FAIL" with probability at least* $1 - \delta$.

Proof. Clearly if P always computes f correctly, the tester always outputs "PASS." For the other direction, we will show that if P passes the tests, then P computes f correctly on at least $1 - 2\varepsilon$ fraction of the inputs. By Theorem 2, we know that the function g equals P on at least $1 - 2\varepsilon$ fraction of the inputs, and that g is identical to f. It suffices to verify that conditions (1), (2) and (3) of the hypothesis of Theorem 2 are verified by the self-tester with high probability. By our choice of parameters and by the Chernoff bounds, with probability at least $1 - \delta$, all the probability estimates made by the self-tester are accurate to within δ, and the result follows. The only point worth mentioning is that in the Inductive test, the error allowance $\delta/3$ was further apportioned into two equal halves. With error at most $\delta/6$, the tester computes $u_i = g(\alpha_i)$ correctly for all the α_i's, and with error at most $\delta/6$, the tester estimates the probability that g fails the inductive test to within δ. $\square$

2.1 Applications

We present some applications of Theorem 4. We remind the reader that a linear function f on a vector space V is efficiently self-testable without the generator bottleneck if there is a (linear) function h that is easily computable and that satisfies $f(\theta(\alpha)) = h(\theta, \alpha, f(\alpha))$ for all $\alpha \in V$. In each of our applications f, we show that a suitable function h exists that satisfies the above conditions. Recall the example of the function $E_c(p) = p(c)$, where the identity $E_c(xp) = cE_c(p)$ holds; in the applications below, we will only establish similar "recurrences." Also, for the sake of simplicity, we do not give all the technical parameters required; these can be computed by routine calculations, following the proofs of the theorems in the last subsection. Details will be given in the full version of the paper.

Our first application concerns linear functions of polynomials. Besides obtaining self-testers for polynomial evaluation and Fourier Transforms as in [Erg95], we obtain new self-testers for polynomial differentiation, integration, and the "mod" function of polynomials. Moreover, the vector space setting lets us generalize some of these results in terms of the matrices that compute linear transforms of vector spaces.

Let $P_n \subseteq \mathbb{K}[x]$ denote the group of polynomials in x of degree $\leq n$ over a field $\mathbb{K}$. The group P_n forms a vector space under usual polynomial addition and scalar multiplication by elements from $\mathbb{K}$. The polynomials $1, x, x^2, \ldots, x^n$ span P_n, and a polynomial $p(x) = \sum_i p_i x^i$ has vector representation $(p_n, p_{n-1}, \ldots, p_1, p_0)$. The linear operator θ in this case is just multiplication by x, thus $\theta(p) = xp$. To handle the case that multiplying p by x results in a polynomial of degree $n + 1$, our testers will pad the polynomials with an extra zero coefficient (for x^{n+1}). For simplicity of exposition, we will suppress this detail; this does not cause any technical complication.

Polynomial Evaluation. For any $c \in \mathbb{K}$, let $E_c(p)$ denote, as described before, the function that returns the value $p(c)$. This function is linear. Moreover, the relation between

$E_c(xp)$ and $E_c(p)$ is simple and linear: $E_c(xp) = cE_c(p)$. To self-test a program that claims to compute E_c, the inductive test (test (3)) is simply to choose many random p's, and verify that $E_c(xp) = cE_c(p)$ holds.

Vandermonde Operators and the Discrete Fourier Transform. If $u_1, u_2, \ldots, u_{n+1}$ are $n+1$ distinct elements of $\mathbb{K}$, then one may wish to evaluate a polynomial $p \in P_n$ simultaneously on all $n+1$ points. The ideas for E_c extend easily to this case, for $E_u(xp) = uE_u(p)$ for any $u \in \mathbb{K}$, and these relations hold simultaneously.

Let ω be a principal $(n+1)$-st root of unity in $\mathbb{K}$. The operation of converting a polynomial from its coefficient representation to pointwise evaluation at the powers of ω is known as the Discrete Fourier Transform (DFT). DFT has many fundamental applications that include fast multiplication of integers and polynomials. With our notation, the DFT of a polynomial $p \in P_n$ is simply $F(p) = (E_{\omega^0}(p), E_{\omega^1}(p), \ldots, E_{\omega^n}(p))$. The DFT F is linear, and $F(xp) = (\omega^0 E_{\omega^0}(p), \ldots, \omega^n E_{\omega^n}(p))$. Notice that here the function "h" is really n functions h_{ω^i} for $0 \le i \le n$. The self-tester will simply choose p's randomly, request the program to compute $F(p)$ and $F(xp)$, and verify for each i, $0 \le i \le n$, that $(F(xp))[i] = \omega^i (F(p))[i]$ holds.

This suggests the following generalization (for the case of arbitrary vector spaces). Simultaneous evaluation of a polynomial at $d+1$ points $u_1, \ldots, u_{n+1}$ corresponds to multiplying the vector p by a Vandermonde matrix M, where $M_{ij} = u_i^{j-1}$. The ideas used to test simultaneous evaluation of polynomials and the DFT extend to give a self-tester for any linear transform that is represented by a Vandermonde matrix.

The matrix for the Discrete Fourier Transform can be written as a Vandermonde matrix F, where $F_{ij} = \omega^{ij}$. The inverse of the DFT, that is, converting a polynomial from pointwise representation to coefficient form, also has a Vandermonde matrix whose entries are given by $\widetilde{F}_{ij} = (1/\det F)\omega^{-ij}$. It follows that the inverse Fourier Transform can be self-tested efficiently. Another point worth mentioning here is that in carrying out the inductive test (test (3)), the self-tester does not have to compute $\det F$. All it needs to do is verify that for many randomly chosen p's, the identity $(\widetilde{F}(xp))[i] = \omega^{-i}(\widetilde{F}(p))[i]$ holds.

Operators in elementary Jordan canonical form. A linear operator M is said to be in elementary Jordan canonical form if for some $c \in \mathbb{K}$, all the diagonal entries of M are c's, and all the elements to the left of the main diagonal (the first non-principal diagonal in the lower triangle of M) are 1's. It is easy to verify that $M\theta = \theta M + M'$, where M' is a matrix that has a -1 in the top left corner and a 1 in the bottom right corner and zeroes elsewhere. Therefore, for every $v = (v_n, v_{n-1}, \ldots, v_2, v_1)^T$ in the vector space, $M(\theta(v)) = \theta(M(v)) + (-v_n, 0, \ldots, 0, v_1)^T$. This gives an easy way to implement the inductive test in the self-tester.

An attempt to extend this to matrices in Jordan canonical form, or even to diagonal matrices, seems not to work. However, if a diagonal or "shifted" diagonal matrix has a special structure, then we can obtain self-testers that avoid the generator problem. For example, the matrix corresponding to the differentiation of polynomials has a special structure: it contains the entries $n, n-1, \ldots, 1$ on the diagonal below the main diagonal. We exploit this to obtain the following:

Differentiation. Differentiation of polynomials is a linear function $D : P_n \to P_{n-1}$. We have the explicit form for h: $D(xp) = p + xD(p)$. Integration of polynomials is a linear function $I : P_n \rightarrow P_{n+1}$. The explicit form for h is $I(xp) = xI(p) - I(I(p))$. Hence we can avoid the generator problem for these functions.

Higher Order Differentiation. Let D^k denote the k-th differential operator. It is easy to write a recurrence-like identity for D^k in terms of D^j, $j < k$. This gives us a self-tester in the "library setting" described in [BLR93, Rub90], where one assumes that there are programs to compute all these differential operators. However, if we wish to self-test a program that only computes D^k, this assumption is not valid. To remedy this, we will use the following lemma.

Lemma 5. *If p is a polynomial in x that is differentiable k times, then*

$$\sum_{i=0}^{k} \binom{k}{i} (-x)^{k-i} D^k(x^i p) = k! p$$

Proof. By induction on k. The base case $(k = 0)$ is obviously true. Let $k > 0$, we have $S_k = \sum_{i=0}^{k+1} \binom{k+1}{i} (-x)^{k+1-i} D^{k+1}(x^i p)$. Since $D^{k+1}(x^i p) = D^k(ix^{i-1}p + x^i D(p))$ and since differentiation is linear, we have

$$S_k = \left[\sum_{i=0}^{k+1} \binom{k+1}{i} (-x)^{k+1-i} D^k(ix^{i-1}p) \right] + \left[\sum_{i=0}^{k+1} \binom{k+1}{i} (-x)^{k+1-i} D^k(x^i D(p)) \right].$$

Since the first term $(i = 0)$ in the first sum vanishes, and since $i\binom{k+1}{i} = (k+1)\binom{k}{i-1}$, the first sum evaluates to $(k+1) \sum_{i=1}^{k} \binom{k}{i-1} (-x)^{k+1-i} D^k(ix^{i-1}p)$, which equals $(k+1)!p$ by the inductive hypothesis. Hence it suffices to show that the second sum evaluates to 0. This summation can be written as $\sum_{i=0}^{k} \binom{k+1}{i} (-x)^{k-i+1} D^k(x^i D(p)) + D^k(x^{k+1} D(p))$. By Pascal's identity, this sum can be split into the terms $\sum_{i=0}^{k} \binom{k}{i-1} (-x)^{k-i+1} D^k(x^i D(p))$, $\sum_{i=0}^{k} \binom{k}{i} (-x)^{k-i+1} D^k(x^i D(p))$, and $D^k(x^k \cdot xD(p))$. Regrouping terms and applying the inductive hypothesis, this equals $(-x)k!D(p) + k!(xD(p)) = 0$. $\square$

By the "robustness" of the inductive test, testing if a program P satisfies this identity for most p suffices to ensure that it satisfies this identity everywhere. This implies that if P computes $D^k(p)$ correctly, then it computes $D^k(x^k p)$ correctly. To carry the induction through, we need to test k base cases: $P(1), P(x), P(x^2), \ldots, P(x^{k-1})$.

Mod Function. Let $q \in \mathbb{K}[x]$ be a monic irreducible polynomial. Let $M_q(p)$ denote the mod function with respect to q, that is, $M_q(p) = p \bmod q$. This is a linear function when the addition is interpreted as mod q addition. Since q is monic, the degree of $M_q(p)$ is always less than $\deg(q)$. If $c \in \mathbb{K}$ is the coefficient of the highest degree term in p, we have

$$M_q(xp) = \begin{cases} xM_q(p) & \text{if } \deg(xM_q(p)) < \deg(q) \\ xM_q(p) - cq & \text{if } \deg(xM_q(p)) = \deg(q) \end{cases}$$

As before, in testing if a program P computes the function M_q, step (3) of the self-tester is to choose many p's at random, compute $P(p)$ and $P(xp)$, and verify that one of the identities $P(xp) = xP(p)$ or $P(xp) = xP(p) - cq$ holds (depending on the degree of p).

3 Multilinear Functions

A function $f(x_1, \ldots, x_k)$ is *multilinear* if it is linear in each of its variables when the other variables are fixed. Following [Erg95], our main motivating example is polynomial multiplication: suppose we wish to test a program that purports to multiply polynomials of degree $\leq n$ over a finite field $\mathbb{K}$. The naive approach would require testing the program at n^2 pairs of generators. Blum, Luby, and Rubinfeld [BLR93] give a "bootstrap" self-tester that makes $O(\log n)$ calls to the program. Ergün's work presents an efficient self-tester that avoids the generator bottleneck. As in the previous section, we generalize this to arbitrary multilinear functions over large groups. For k-variate functions, we need to test the program at n^k different k-tuples of generators, where n is the number of generators of the group. In this section, we give a scheme to efficiently self-test multilinear functions, where the self-tester is required to know the correct value of f at only one point. For easy visualization, we request the reader to consider polynomial multiplication as an archetypal example.

We proceed in the same manner as before. Let $\langle \alpha \rangle$ denote $(\alpha_1, \ldots, \alpha_k) \in V^k$, and let $\langle \alpha, i \rangle$ denote $(\alpha_1, \ldots, \alpha_{i-1}, \theta(\alpha_i), \alpha_{i+1}, \ldots, \alpha_k) \in V^k$. If $f(\langle \alpha, i \rangle) = h_i(\theta, \langle \alpha \rangle, g(\langle \alpha \rangle))$, then h_i is multilinear in $\langle \alpha \rangle$. Moreover if f is symmetric, then all the h_i's are identical. The self-tester for f consists of the following tests:

(1) *Multilinearity Test*: Use the self-tester for multilinearity from [BFL91, FGL$^+$91], which also works in our vector space setting, to verify that P agrees with some multilinear function g on all but an ε fraction of the inputs. (Note, however, that these tests impose a limit $\varepsilon \leq \varepsilon(k) = (1/k)^{O(1)}$. We will focus, for the moment, on the case where $1/2^k \ll \varepsilon(k)$; more details will be given in the full paper.) If the program P passes this test, perform the next two tests using P_{sc}, the self-corrector for P, to compute g.

(2) *Basis Test*: Verify that $g(\langle e_1 \rangle) = f(\langle e_1 \rangle)$.

(3) *Inductive Test*: For $1 \leq i \leq k$, verify that $\Pr_{\alpha \in V^k}[g(\langle \alpha, i \rangle) \neq h_i(\theta, \langle \alpha \rangle, g(\langle \alpha \rangle))] \leq \varepsilon + \delta$.

With a total of $(1/\delta)^{O(1)}$ calls to P, we can arrange to have the probability that all three tests succeed to be at least $1 - \delta$. Output "PASS" iff P passes all these tests.

It is not hard to see that multilinear functions over groups (for example, multiplication of polynomials over a finite field $\mathbb{K}$) are randomly self-reducible, and hence self-correctable. For the rest of the proof, we assume that g is multilinear. Since Test (2) ensures that g agrees with f on $\langle e_1 \rangle$, it suffices to prove that for every $i = 1, \ldots, k$ and every $\langle \alpha \rangle \in V^k$, $g(\theta(\langle \alpha, i \rangle)) = h_i(\langle \alpha \rangle, \theta, g(\langle \alpha, \rangle))$. By induction, we can then conclude that g computes f at all generators and hence everywhere.

For any $\langle \alpha \rangle \in V^k$, we prove that $g(\theta(\langle \alpha, i \rangle)) = h_i(\theta, \langle \alpha \rangle, g(\langle \theta \rangle))$. For simplicity, we give the proof for $k = 2$ and $i = 2$.

$$\Pr_{\beta_1, \beta_2 \in V^2}[g(\alpha_1, \theta(\alpha_2))$$
$$= g(\beta_1 + \alpha_1 - \beta_1, \theta(\beta_2 + \alpha_2 - \beta_2))$$
$$= g(\beta_1, \theta(\beta_2 + \alpha_2 - \beta_2)) + g(\alpha_1 - \beta_1, \theta(\beta_2 + \alpha_2 - \beta_2))$$
$$= g(\beta_1, \theta(\beta_2)) + g(\beta_1, \theta(\alpha_2 - \beta_2)) + g(\alpha_1 - \beta_1, \theta(\beta_2)) + g(\alpha_1 - \beta_1, \theta(\alpha_2 - \beta_2))$$

$$
\begin{aligned}
&= h_2(\theta, \beta_1, \beta_2, g(\beta_1, \beta_2)) + h_2(\theta, \beta_1, \alpha_2 - \beta_2, g(\beta_1, \alpha_2 - \beta_2)) + \\
&\quad h_2(\theta, \alpha_1 - \beta_1, \beta_2, g(\alpha_1 - \beta_1, \beta_2)) + h_2(\theta, \alpha_1 - \beta_1, \alpha_2 - \beta_2, g(\alpha_1 - \beta_1, \alpha_2 - \beta_2)) \\
&= h_2(\theta, \beta_1, \alpha_2, g(\beta_1, \alpha_2)) + h_2(\theta, \alpha_1 - \beta_1, \alpha_2, g(\alpha_1 - \beta_1, \alpha_2)) \\
&= h_2(\theta, \alpha_1, \alpha_2, g(\alpha_1, \alpha_2))] \geq 1 - 4\varepsilon > 0.
\end{aligned}
$$

The first equality is rewriting. Multilinearity of g implies the second and third equalities. If the true probability $\Pr[g(\langle \alpha, i \rangle) \neq h_i(\theta, \langle \alpha \rangle, g(\langle \alpha \rangle))] < \varepsilon$, then the fourth equality fails with probability less than 4ε. The rest of the equality follows from the multilinearity of h_2 and g. If $\varepsilon < 1/4$, this probability is non-zero. The fact that the probability is independent of β_1, β_2 yields the desired result.

It is easy to see that the above proof extends to an arbitrary k so long as $\varepsilon < 1/2^k$. Thus, we obtain the following theorem whose proof mirrors the proof of Theorem 4.

Theorem 6. *If f is a k-variate linear function, then for any $\delta < 1$ and $\varepsilon < 1/2^k$, the above three tests comprise a $(0, 2^k \varepsilon)$-self-tester for f that succeeds with probability at least $1 - \delta$.*

3.1 Applications

The function $M(p_1, p_2)$ that multiplies two polynomials is symmetric and linear in each variable. Moreover, since $M(x p_1, p_2) = x M(p_1, p_2)$, polynomial multiplication has an efficient self-tester.

An interesting application of polynomial multiplication, together with the "mod" function described in Section 2.1, is the following. It is well-known that a degree n (finite) extension $\mathbb{K}$ of a finite field $\mathbb{F}$ is isomorphic to the field $\mathbb{F}[x]/(\alpha)$, where α is an irreducible polynomial of degree n over $\mathbb{F}$. Under this isomorphism, each element of $\mathbb{K}$ is viewed as a polynomial of degree $\leq n$ over $\mathbb{F}$, addition of two elements $p, q \in \mathbb{K}$ is just their sum $p + q$ as polynomials, and multiplication of $p, q \in \mathbb{K}$ is given by $pq \bmod \alpha$. It follows that linear functions over a finite extension of a finite field can be self-tested without the generator bottleneck, that is, the number of calls made to the program being tested is independent of the degree of field extension!

4 Nonlinear Functions

In this section, we consider nonlinear functions. Specifically, we deal with exponentiation and constant degree polynomials over groups with n generators, and show how to self-test them without the generator bottleneck. Since a clear notion of multiplication is required, and since we need the Lagrange interpolation formula (for random self-reducibility of polynomials), we will restrict ourselves to the case of rings. An interesting example is the ring of polynomials over finite fields, and for simplicity, we restrict our discussion below to this ring. It is obvious that exponentiation and constant degree polynomials are clearly defined over this ring.

We first consider the function $f(p) = p^d$ for some d (that is, raising a polynomial to the d-th power). Suppose a program P claims to perform this exponentiation for all degree n polynomials $p \in P_n \subseteq \mathbb{K}[x]$, the ring of all polynomials over the field $\mathbb{K}$. Using the "low-degree" test of Rubinfeld and Sudan [RS93] we can first test if the function computed by P is a degree d polynomial. As before, we can also verify that g, the self-corrected version of P, satisfies $g(e_1) = f(e_1)$.

The induction identity $f(xp) = x^d f(p)$ also applies, so it remains to show how to verify that $g(xp) = x^d g(p)$ for all $p \in P_n$. This is where the difficulty comes. The obvious idea would be to emulate the case of linear functions and verify that $g(xp) = x^d g(p)$ for most randomly chosen p. It turns out, however, that proving that this property is robust, that is, proving that random verification suffices to ensure that this identity holds everywhere, is not so easy as before. Recall that in the proof for the linear case, we expressed p as $p - q + q$ for some random q, and proved using the linearity of g and θ that $g(\theta p) = h(\theta, p, g(p))$ holds with positive probability, and hence with probability one. There we made essential use of the fact that $g(x(p-q)+xq) = g(x(p-q))+g(xq)$. When g is nonlinear, this identity is no longer true.

To circumvent this problem, we follow the ideas of Rubinfeld and Sudan [RS93], who extended the [BLR93] result from linear functions to low-degree polynomials. They used the fact that the Lagrange interpolation identity for polynomials is robust. Here our task is to verify that $g(xp) = x^d g(p)$ for all p. We will show that if this equality holds on a large fraction of the p's, then it holds for all p. Before proceeding to do so, we pause to state the following elementary fact concerning the Lagrange interpolation identity.

Fact 7 *Let g be a degree d polynomial. For any $p \in P_n$, if $q_1, q_2, \ldots, q_{d+1}$ are distinct elements of P_n,*

$$g(p) = \sum_{i=1}^{d+1} g(q_i) \prod_{j \neq i} \frac{p - q_j}{q_i - q_j}; \qquad \text{and also} \qquad g(xp) = \sum_{i=1}^{d+1} g(xq_i) \prod_{j \neq i} \frac{p - q_j}{q_i - q_j}.$$

The self-tester for $f(x) = x^d$ comprises the following tests:

(1) *Degree Test*: Use the low-degree test of [RS93] to verify that P computes a degree d polynomial.
(2) *Basis Test*: Using P_{sc} to compute g, verify that $g(e_1) = f(e_1)$.
(3) *Inductive Test*: Using P_{sc} to compute g, verify that $\Pr_\alpha[g(x\alpha) \neq x^d(g(\alpha))] \leq \varepsilon + \delta$.

As before, the all the above tests can be made to succeed with probability $1 - \delta$, using $(1/\delta)^{O(1)}$ calls to P. Let β denote the probability that the $d + 1$ random choices from the domain produce distinct elements. Usually, the domain is large enough so that β is close to 1. Below we sketch the proof that if $\varepsilon < \beta/(d+1)$ and P_{sc} passes Tests (1) and (3), then g satisfies $g(xp) = x^d(g(p))$ everywhere.

$$\Pr_{q_1, \ldots, q_{d+1}} [g(xp) = \sum_{i=1}^{d+1} a_i g(xq_i)$$

$$= \sum_{i=1}^{d+1} a_i x^d g(q_i)$$

$$= x^d \sum_{i=1}^{d+1} a_i g(q_i)$$

$$= x^d g(p)] \geq \beta - (d+1)\varepsilon > 0.$$

Here $a_i = \prod_{j \neq i}(p - q_j)/(q_i - q_j)$. The first equality is Fact 7, and applies since g has been verified to be a degree d polynomial. Since the q_i's are uniformly and identically distributed, by Test (3), the second equality fails with probability $< (d+1)\varepsilon$. The third equality is just rewriting, and the fourth equality is due to Fact 7. Since the equality $g(xp) = x^d g(p)$ holds independent of q_i's, if $\varepsilon < \beta/(d+1)$, it holds always.

Theorem 8. *The function $f(p) = p^d$ has an efficient self-tester that avoids the generator bottleneck.*

Next we consider extending the above for arbitrary degree d polynomials. Clearly the low-degree test and the basis test work as before. The interpolation identity is valid, too. The missing ingredient is the availability of nice identities like "$f(xp) = x^d f(p)$." We get around this difficulty using the notion of *program result-checkers* [BK89]. Recall that a checker for a function f is a probabilistic program that, given a program P purporting to compute f and an input p, outputs "CORRECT" if $P(q) = f(q)$ for all q, outputs "WRONG" if $P(p) \neq f(p)$. The checker may call the program P on inputs other than p, but must access P only as a black-box; it does not have access to the code of P. If a function has a self-tester and a self-corrector, then it has a checker.

Suppose $f(p)$ is a degree d polynomial in p, and suppose P purports to compute f. The most obvious way to perform self-testing would be to randomly choose many inputs p, compute $P(p)$ and check if $P(p) = f(p)$. This way, one can easily estimate the probability that $P(p) = f(p)$. The difficulty, however, is that the self-tester needs to know the correct value of $f(p)$ on many randomly chosen inputs. In other words, the self-tester is not "different" from the function being self-tested. To get around this, we employ a suggestion due to Ronitt Rubinfeld. We show how to use the (possibly faulty) program P to compute the correct value of $f(p)$ in a "certifiable" manner—this is where the checker comes in.

We first prove a technical lemma that helps us express d-ary multiplication in terms of f—that is, we establish a reduction from the multilinear function $\prod_{i=1}^{d} p_i$ to the nonlinear function f. This reduction is a generalization to degree d of the elementary identity $pq = ((p+q)^2 - (p-q)^2)/4$, but stronger in that it works for arbitrary polynomials of degree d, not just degree-d exponentiation. Given this reduction, we can build a program P' that performs c-ary multiplication for any $c \leq d$ (if $c < d$, we can simply multiply by extra 1's). The program P' can then be self-tested without the generator bottleneck, using the result of Section 3. We can then build a checker for multiplication, and using this we can package P' into a multiplication program Q that *certifies* its output as "CORRECT" or "WRONG." The checker itself, and hence Q, may fail with probability β for some $\beta > 0$. We can then build a program Q' that computes f using the reliable multiplication program Q. Finally, we pick random p's, and check if $P(p) = Q'(p)$, and estimate the probability that P computes f correctly.

Lemma 9. *Let $p_1, \ldots, p_d$ be distinct variables. Let $0 \leq m < 2^d$. Denote by m_i the i-th least significant bit of m, that is, $m = m_d m_{d-1} \ldots m_1$. Then*

$$\sum_{m=0}^{2^d - 1} \left(\prod_{i=1}^{d} (-1)^{m_i} \right) \left(\sum_i (-1)^{m_i} p_i \right)^c = \begin{cases} 0 & \text{if } c < d \\ 2^d d! \prod_{i=1}^{d} p_i & \text{if } c = d \end{cases}$$

Proof.

$$\sum_{m=0}^{2^d-1} \left(\prod_{i=1}^{d} (-1)^{m_i} \right) \left(\sum_i (-1)^{m_i} p_i \right)^c$$

$$= \sum_{m=0}^{2^d-1} \left(\prod_{i=1}^{d} (-1)^{m_i} \right) \sum_{\substack{n_1+\cdots+n_d=c \\ 0 \le n_1,\ldots,n_d \le c}} \binom{c}{n_1,\ldots,n_d} \prod_{i=1}^{d} (-1)^{m_i n_i} p_i^{n_i}$$

$$= \sum_{\substack{n_1+\cdots+n_d=c \\ 0 \le n_1,\ldots,n_d \le c}} \binom{c}{n_1,\ldots,n_d} \prod_{i=1}^{d} p_i^{n_i} \sum_{m=0}^{2^d-1} \prod_{i=1}^{d} (-1)^{m_i+m_i n_i}$$

$$= \sum_{\substack{n_1+\cdots+n_d=c \\ 0 \le n_1,\ldots,n_d \le c}} \alpha \left[\sum_{m=0}^{2^d-1} \prod_{i=1}^{d} (-1)^{(m_i+m_i n_i) \bmod 2} \right],$$

where α depends only on the p_i's and the n_i's. Fix $n_1,\ldots,n_d$. We claim that the inner summation is non-zero if and only if $c = d$ and $n_1 = \cdots = n_d = 1$. When $c = d$ and $n_1 = \cdots = n_d = 1$,

$$\sum_{m=0}^{2^d-1} \prod_{i=1}^{d} (-1)^{(m_i+m_i n_i) \bmod 2} = 2^d; \qquad \text{and} \qquad \alpha = d! \prod_{i=1}^{d} p_i.$$

For every i such that n_i is odd, $(-1)^{(m_i+m_i n_i) \bmod 2} = 1$. If $c < d$ or if all the n_i's are not 1, there must be at least one n_i which is zero (since they sum to c). Therefore, there must be at least one n_i which is even. Arbitrarily fix some j such that n_j is even; now,

$$(-1)^{(m_j+m_j n_j) \bmod 2} = (-1)^{m_j}.$$

For any setting of the bits of m other than m_j, the two terms that correspond to $m_j = 0$ and to $m_j = 1$ cancel each other, and the sum is zero. $\qquad\square$

Corollary 10. *For any polynomial $G(x) = \sum_i p_i x^i$ of degree d,*

$$\sum_{m=0}^{2^d-1} \left(\prod_{i=1}^{d} (-1)^{m_i} \right) G\left(\sum_{i=1}^{d} (-1)^{m_i} x_i \right) = p_d 2^d d! \prod_{i=1}^{d} x_i.$$

By Corollary 10, it is easy to see that if P computes f correctly on all inputs, the self-tester will output "PASS" with probability one. If $\varepsilon < 1/2^d$, the self-corrector for multiplication will work correctly, since a random call to d-ary multiplication results in 2^d uniformly and identically distributed calls to P. Therefore, the certified program Q computes multiplication with probability arbitrarily close to $1 - \beta$, and the certified program Q' computes $f(p)$ correctly for every p with probability arbitrarily close to $1 - d\beta$. If $P(p) \ne f(p)$ with probability $> \varepsilon$, then the probability that the self-tester will catch an error in one trial is $\varepsilon(1 - d\beta)$; by repeated trials, this can be boosted to any required success probability $1 - \delta$. The dependence on the running time of our self-tester on d is rather poor. However, if the degree n of the polynomials is large and if d is constant, this scheme is attractive. In particular, it helps us efficiently self-test constant degree polynomials on finite field extensions of large dimension.

Acknowledgments

We are grateful to Professor Ronitt Rubinfeld for valuable suggestions and guidance. We thank Professor Manuel Blum, Funda Ergün, and Mandar Mitra for useful discussions, and the anonymous referees for helpful comments.

References

[BFL91]   L. Babai, L. Fortnow, and C. Lund. Non-deterministic exponential time has two-prover interactive protocols. *Computational Complexity*, 1:3–40, 1991.

[BK89]   M. Blum and S. Kannan. Designing programs that check their work. In *Proc. 21st Annual ACM Symposium on the Theory of Computing*, pages 86–97, 1989.

[BLR93]   M. Blum, M. Luby, and R. Rubinfeld. Self-testing/correcting with applications to numerical problems. *J. Comp. Sys. Sci.*, 47(3):549–595, 1993. An earlier version appeared in STOC 1990.

[BW94]   M. Blum and H. Wasserman. Program result-checking: A theory of testing meets a test of theory. In *Proc. 35th Annual IEEE Symposium on Foundations of Computer Science*, pages 382–392, 1994.

[Erg95]   F. Ergün. Testing multivariate linear functions: Overcoming the generator bottleneck. In *Proc. 27th Annual ACM Symposium on the Theory of Computing*, pages 407–416, 1995.

[FGL+91]   U. Feige, S. Goldwasser, L. Lovasz, S. Safra, and M. Szegedy. Approximating clique is almost NP-complete. In *Proc. 32nd Annual IEEE Symposium on Foundations of Computer Science*, pages 2–12, 1991.

[GLR+91]   P. Gemmell, R. Lipton, R. Rubinfeld, M. Sudan, and A. Wigderson. Self-testing/correcting for polynomials and for approximate functions. In *Proc. 23rd Annual ACM Symposium on the Theory of Computing*, pages 32–42, 1991.

[Lip91]   R. Lipton. New directions in testing. In *Proc. of DIMACS Workshop on Distributed Computing and Cryptography*, pages 191–202, 1991.

[RS92]   R. Rubinfeld and M. Sudan. Testing polynomial functions efficiently and over rational domains. In *Proc. 3rd Annual ACM-SIAM Symposium on Discrete Algorithms*, pages 23–43, 1992.

[RS93]   R. Rubinfeld and M. Sudan. Robust characterizations of polynomials and their applications to program testing. TR 93-1387, Dept. of Computer Science, Cornell University, 1993. To appear in *SIAM Journal of Computing*.

[Rub90]   R. Rubinfeld. *A Mathematical Theory of Self-Checking, Self-Testing, and Self-Correcting Programs*. PhD thesis, University of California at Berkeley, 1990.

[Rub94]   R. Rubinfeld. Robust functional equations with applications to self-testing/correcting. In *Proc. 35th Annual IEEE Symposium on Foundations of Computer Science*, pages 288–299, 1994.

[Sud94]   Madhu Sudan. On the role of algebra in the efficient verification of proofs. In M. Agrawal, V. Arvind, and M. Mahajan, editors, *Proc. Workshop on Algebraic Methods in Complexity Theory (AMCOT)*, pages 58–68, 1994. Technical Report IMSc 94/51, The Institute of Mathematical Sciences, Madras, India.

Observing Behaviour Categorically*

Mogens Nielsen and Allan Cheng

BRICS, Department of Computer Science, University of Aarhus, Denmark
e-mail: {acheng,mnielsen}@daimi.aau.dk

Abstract. In an attempt to understand the relationships and differences between the extensive amount of research within the field of bisimulation equivalences, Joyal, Nielsen, and Winskel recently proposed an abstract category-theoretic definition of bisimulation. They identify spans of morphisms satisfying certain "path lifting" properties, so-called open maps, as an abstract definition of bisimilarity. Furthermore, it was shown how to capture Milner's bisimulation and a variant of history-preserving bisimulations for event structures. In this paper we review the theory of open maps and show that the theory, in fact, captures not only bisimulations but many other behavioural equivalences. We also briefly present presheaf models as an abstract model of computation.

1 Introduction

As a response to the numerous models for concurrency proposed in the literature Winskel and Nielsen have used category theory as an attempt to understand the relationship between models like event structures, Petri nets, trace languages, and asynchronous transition systems [WN94]. From the algebraic point of view many of the operators of CCS like process algebras have been recasted using category-theoretic concepts such as products, co-products. However, a similar convincing way of adjoining abstract equivalences to a category of models had been missing until Joyal, Nielsen, and Winskel proposed the notion of *span of open maps* [JNW93] as an abstract generalization of Park and Milner's strong bisimulation.

In this paper we briefly review the theory of open maps. Then, by examples, we show how not only bisimulations but, in fact, many other behavioural equivalences can be captured in a natural way using the theory, including trace equivalence, weak bisimulation, probabilistic bisimulation, and pomset based equivalences. We also briefly present presheaves as an abstract model for computation. Most proofs can be found in [CN95, JNW93, NW95].

---

* This work has been supported by The Danish Research Councils, The Danish Research Academy, and **BRICS**, Basic Research in Computer Science, Centre of the Danish National Research Foundation

Definition 8. Let $\mathcal{W}_1$ be the subcategory of $\mathcal{LTS}_1$ whose objects (observations) are of the form

$$i \xrightarrow{\alpha_1} r_1 \xrightarrow{\alpha_2} \cdots \xrightarrow{\alpha_n} r_n \ , \tag{11}$$

where all states are distinct, and whose morphisms are the identity morphisms and morphisms whose domains are observations having only one state (the empty word).

Notice that we know automatically that the associated $\mathcal{W}_1$-bisimulation is an equivalence relation. This follows from the fact that $\mathcal{LTS}$ has pullbacks (so one automatically gets an equivalence for arbitrary choice of observation (sub)category!).

And now the main theorem of this section.

Theorem 9. *Given two lts's T_1 and T_2. Then:*

T_1 and T_2 are trace equivalent if and only if they are $\mathcal{W}_1$-bisimilar.

Having identified trace equivalence we now continue by exploring other possibilities. In the next section we try to take "invisible" or "silent" actions into account.

3.3 Weak Bisimulation

Weak bisimulation according to Milner differs from strong bisimulation in at least two respects. First, a special "invisible" action, usually denoted τ, is required to be a member of the set of labels. Second, an α labelled transition in one labelled transition system is no longer required to be simulated exactly by an α labelled transition in the other system. It may be preceded and succeeded by several τ transitions. We write $r \xrightarrow{t} r'$ if $t = \alpha_1\alpha_2\cdots\alpha_n$ and $r \xrightarrow{\tau^*} r_1 \xrightarrow{\alpha_1} r'_1 \xrightarrow{\tau^*} \cdots \xrightarrow{\tau^*} r_n \xrightarrow{\alpha_n} r'_n \xrightarrow{\tau^*} r'$ for some $r_1, \cdots, r'_n$. Furthermore, a τ transition needn't be simulated by any transitions at all.

We start by defining a category $\mathcal{LTS}_\tau$, *labelled transition systems with τ-moves*, and a subcategory of observations, $\mathcal{W}_\tau$, in $\mathcal{LTS}_\tau$. Then, we show that $\mathcal{W}_\tau$-bisimilarity corresponds to Milner's weak bisimulation.

The objects of $\mathcal{LTS}_\tau$ are the same as those from $\mathcal{LTS}$. However, we assume that the set of actions Act contains the special "invisible" action τ. On such objects, we give Milner's definition of weak bisimulation [Mil89], here adapted to the case where we consider initial states of *lts*'s.

Definition 10. Given two *lts*'s T_1 and T_2. A relation $R \subseteq S_1 \times S_2$ is said to be a *weak bisimulation* over T_1 and T_2 if

$$(i_1, i_2) \in R \ , \tag{12}$$

$$((r, s) \in R \wedge r \xrightarrow{\alpha} r') \Rightarrow \text{ for some } s', (s \xRightarrow{\hat{\alpha}} s' \wedge (r', s') \in R) \ , \tag{13}$$

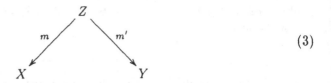

$$(3)$$

Notice that if $\mathcal{M}$ has pullbacks, it can be shown that $\sim_{\mathcal{P}}$ is always an equivalence relation. The important observation is that pullbacks of open maps are themselves open maps. For more details, the reader is referred to [JNW93].

In the rest of this paper, we shall present quite a few concrete examples of following the above presented steps.

3 Transition Systems

As a preliminary example of a category of models of computation $\mathcal{M}$ we present *labelled transition systems*.

Definition 3. A *labelled transition system (lts)* over Act is a tuple

$$(S, i, Act, \longrightarrow) , \qquad (4)$$

where S is a set of states with *initial state i*, Act is a set of actions ranged over by α, β, ... , and $\longrightarrow \subseteq S \times Act \times S$ is the transition relation. For the sake of readability we introduce the following notation. Whenever (s_0, α_1, s_1), (s_1, α_2, s_2), ..., $(s_{n-1}, \alpha_n, s_n) \in \longrightarrow$ we denote this as $s_0 \xrightarrow{\alpha_1} s_1 \xrightarrow{\alpha_2} \cdots \xrightarrow{\alpha_n} s_n$ or $s_0 \xrightarrow{v} s_n$ where $v = \alpha_1 \alpha_2 \cdots \alpha_n \in Act^*$. Also, we assume that all states $s \in S$ are reachable from i, i.e., there exists a $v \in Act^*$ such that $i \xrightarrow{v} s$.

3.1 Strong Bisimulation

Let us briefly recall Park and Milner's definition of strong bisimulation.

Definition 4. Let $T_1 = (S_1, i_1, Act, \longrightarrow_1)$ and $T_2 = (S_2, i_2, Act, \longrightarrow_2)$. A *strong bisimulation between T_1 and T_2* is a relation $R \subseteq S_1 \times S_2$ such that

$$(i_1, i_2) \in R , \qquad (5)$$

$$((r, s) \in R \land r \xrightarrow{\alpha}_1 r') \Rightarrow \text{ for some } s', (s \xrightarrow{\alpha}_2 s' \land (r', s') \in R) , \qquad (6)$$

$$((r, s) \in R \land s \xrightarrow{\alpha}_2 s') \Rightarrow \text{ for some } r', (r \xrightarrow{\alpha}_1 r' \land (r', s') \in R) . \qquad (7)$$

T_1 and T_2 are said to be *strongly bisimilar* if there exists a strong bisimulation between them.

Henceforth, whenever no confusion is possible we drop the indexing subscripts on the transition relations and write $\longrightarrow$ instead.

Labelled transition systems are equipped with simulation morphisms as follows, and hence turned into a category of models for computation, $\mathcal{LTS}$.

Definition 5. Let $T_1 = (S_1, i_1, Act, \longrightarrow_1)$ and $T_2 = (S_2, i_2, Act, \longrightarrow_2)$. A morphism $m : T_1 \longrightarrow T_2$ is a function $m : S_1 \longrightarrow S_2$ such that

$$m(i_1) = i_2 , \tag{8}$$

$$s \xrightarrow{\alpha}_1 s' \Rightarrow m(s) \xrightarrow{\alpha}_2 m(s') . \tag{9}$$

Composition of morphisms is defined as the usual composition of functions. The intuition behind this specific choice of morphism is that an α labelled transition in T_1 must be simulated by an α labelled transition in T_2. Isomorphisms are "equality up to identity of states".

In [JNW93], the authors choose as observation the *full* subcategory $\mathcal{W}$ of $\mathcal{LTS}$ whose objects are finite synchronisation trees with at most one maximal branch, i.e. labelled transition systems of the form

$$i \xrightarrow{\alpha_1} s_1 \xrightarrow{\alpha_2} \cdots \xrightarrow{\alpha_n} s_n$$

$\mathcal{W}$ is essentially "words over alphabet Act with prefix as extension". With this choice, $\mathcal{W}$-bisimilarity corresponds to Park and Milner's strong bisimulation, as shown in [JNW93].

Theorem 6. *Given two lts's T_1 and T_2. Then:*

T_1 *and* T_2 *are strongly bisimilar if and only if they are $\mathcal{W}$-bisimilar.*

In the following we shall "rediscover" other well-known behavioural equivalences by varying $\mathcal{M}$ and $\mathcal{P}$.

3.2 Trace Equivalence

Trace equivalence is perhaps the first, simplest (and oldest) behavioural equivalence between labelled transition systems that one can think of. It is defined for labeled transition systems.

Definition 7. Given an *lts* $T = (S, i, Act, \longrightarrow)$. The *traces/language of T*, denoted $L(T)$, is defined as

$$L(T) = \{v \in Act^* \mid i \xrightarrow{v} r \text{ for some } r \in S\} . \tag{10}$$

Two *lts's*, T_1 and T_2, are said to be *trace equivalent* iff $L(T_1) = L(T_2)$.

Our model of computations will be $\mathcal{LTS}$ as defined above. The only thing we vary relative to the characterization of strong bisimulation, is our choice of observation category—actually we only slightly restrict our choice of observation extension:

2 Open Maps, an Introduction

In this section we briefly recall the basic definitions from [JNW93].

As presented there, the general setting requires several steps. First, a category which represents a *model of computation* has to be identified. We denote this category $\mathcal{M}$. A morphism $m : X \longrightarrow Y$ in $\mathcal{M}$ should intuitively be thought of as a simulation of X in Y. Then, within $\mathcal{M}$, we choose a subcategory of "observation objects" and "observation extension" morphisms between these objects. We denote this *category of observations* by $\mathcal{P}$. Given an observation (object) P in $\mathcal{P}$ and a model X in $\mathcal{M}$, then P is said to be an *observable behaviour* of X if there exists a morphism $p : P \longrightarrow X$ in $\mathcal{M}$, representing a particular way of realizing P in X.

Next, we identify morphisms $m : X \longrightarrow Y$ which have the property that whenever an observable behaviour of X can be extended via f in Y then that extension can be matched by an extension of the observable behaviour in X.

Definition 1. Open Maps
A morphism $m : X \longrightarrow Y$ in $\mathcal{M}$ is said to be $\mathcal{P}$-*open* if whenever $f : O_1 \longrightarrow O_2$ in $\mathcal{P}$, $p : O_1 \longrightarrow X$, $q : O_2 \longrightarrow Y$ in $\mathcal{M}$, and the diagram

$$
\begin{array}{ccc}
O_1 & \xrightarrow{\ p\ } & X \\
{\scriptstyle f}\downarrow & & \downarrow{\scriptstyle m} \\
O_2 & \xrightarrow{\ q\ } & Y
\end{array}
\qquad (1)
$$

commutes, i.e., $m \circ p = q \circ f$, there exists a morphism $h : O_2 \longrightarrow X$ in $\mathcal{M}$ such that the two triangles in the diagram

$$
\begin{array}{ccc}
O_1 & \xrightarrow{\ p\ } & X \\
{\scriptstyle f}\downarrow & {\scriptstyle h}\nearrow & \downarrow{\scriptstyle m} \\
O_2 & \xrightarrow{\ q\ } & Y
\end{array}
\qquad (2)
$$

commute, i.e., $p = h \circ f$ and $q = m \circ h$. When no confusion is possible, we refer to $\mathcal{P}$-open morphisms as *open maps*.

The abstract definition of bisimilarity is as follows.

Definition 2. $\mathcal{P}$-bisimilarity
Two models X and Y in $\mathcal{M}$ are said to be $\mathcal{P}$-*bisimilar*, written $X \sim_{\mathcal{P}} Y$, if there exists a *span of open maps* from a common object Z:

$$((r, s) \in R \land s \xrightarrow{\alpha} s') \Rightarrow \text{ for some } r', (r \xRightarrow{\hat{\alpha}} r' \land (r', s') \in R) . \qquad (14)$$

The function $\hat{\ }: Act^* \longrightarrow Act^*$ removes all τ's from its argument [Mil90].

T_1 and T_2 are said to be *weakly bisimilar* if there exists a weak bisimulation as defined above.

Guided by Milner's intuitive understanding of how an action may be simulated, we modify the morphisms between two *lts*'s as follows.

Definition 11. Given two *lts*'s, $T_j = (S_j, i_j, Act, \longrightarrow_j), j = 1, 2$. A morphism between T_1 and T_2, $m : T_1 \longrightarrow T_2$, is a function m from S_1 to S_2, such that

$$m(i_1) = i_2 , \qquad (15)$$

$$r \xrightarrow{\alpha} r' \text{ implies } m(r) \xRightarrow{\hat{\alpha}} m(r') . \qquad (16)$$

Composition of morphisms is defined as the usual composition of functions. This defines the category $\mathcal{LTS}_\tau$. $\mathcal{W}_\tau$, the category of observations, is defined as follows.

Definition 12. Let $\mathcal{W}_\tau$ be the subcategory of $\mathcal{LTS}_\tau$ whose objects are of the form

$$i \xrightarrow{\alpha_1} r_1 \xrightarrow{\alpha_2} \cdots \xrightarrow{\alpha_n} r_n , \qquad (17)$$

where all the states are distinct. Moreover, there will be a morphism f from an observation

$$i \xrightarrow{\alpha_1} r_1 \xrightarrow{\alpha_2} \cdots \xrightarrow{\alpha_n} r_n \qquad (18)$$

to another observation

$$i' \xrightarrow{\alpha_1} r'_1 \xrightarrow{\alpha_2} \cdots \xrightarrow{\alpha_n} r'_n \xrightarrow{\alpha_{n+1}} \cdots \xrightarrow{\alpha_{n+k}} r'_{n+k} , \qquad (19)$$

if $f(i) = i'$, $f(r_j) = r'_j$ for $1 \le j \le n$, and $k \ge 0$.

Notice that morphisms between observations are required to simulate actions in the "strong" sense—i.e., no additional τ's may be added. As a matter of fact, $\mathcal{W}_\tau$ is exactly the category of observations from the characterization of strong bisimulation, just adapted to the assumption of possible τ-moves.

Having defined $\mathcal{M}$ as $\mathcal{LTS}_\tau$ and $\mathcal{P}$ as $\mathcal{W}_\tau$ we now characterize the open maps.

Lemma 13. *A morphism* $m : T_1 \longrightarrow T_2$ *is* $\mathcal{W}_\tau$-*open if and only if it satisfies the following "zig-zag" property:*

If $m(r) \xrightarrow{\alpha} s$ *then there exists an* r' *such that* $r \xRightarrow{\hat{\alpha}} r'$ *and* $m(r') = s$.

Based on this characterization it is not difficult to see that $\mathcal{W}_\tau$-bisimilarity coincides with weak bisimulation.

Theorem 14. *Given two lts's T_1 and T_2. Then,*

T_1 and T_2 are weakly bisimilar if and only if T_1 and T_2 are W_τ-bisimilar.

Having seen now how to characterize trace equivalence and weak bisimulation, it should be an easy exercise for the reader to work out a characterization of "weak trace equivalence". As mentioned, our main purpose with this section was to indicate the generality of the open map approach relative to behavioural equivalences for labelled transition systems. In the following sections we apply the approach to other models for computation, probabilistic and non-interleaving.

4   Probabilistic Transition Systems

In this section we show that the *probabilistic bisimulation* of Larsen and Skou [LS91] can be characterized using the general setting in Sect. 2. We will however apply the theory in a slightly different way. Until now, we have tried to characterize $\mathcal{P}$-bisimilarity between objects of $\mathcal{M}$, for the specific choices of $\mathcal{P}$ and $\mathcal{M}$. In this section we will focus on $\mathcal{P}$-bisimilarity between objects of a subcategory of $\mathcal{M}$. This application of the theory of open maps still turns out "successful".

Intuitively, Larsen and Skou's *probabilistic bisimulation* differs from strong bisimulation in at least two respects. First, to each labelled transition there is associated a real number from the interval $[0;1]$ which is to be understood as the probability with which the transition can be performed. Second, it is no longer single labelled transitions between two states that have to be matched but a set of identically labelled transitions into an equivalence class of probabilistic bisimilar states.

Based on [LS91] we start by defining a category $\mathcal{PPTS}$, *partial probabilistic transition systems*, corresponding to $\mathcal{M}$, and a subcategory of observations, $\mathcal{P}$, in $\mathcal{PPTS}$. Then, we show that $\mathcal{P}$-bisimilarity in the full subcategory of *probabilistic transition systems*, $\mathcal{PTS}$, in $\mathcal{PPTS}$, corresponds to Larsen and Skou's probabilistic bisimulation. Contrary to Larsen and Skou we do not assume lower limit on the probability of transitions. Because we wish to allow arbitrary small probabilities and for technical reasons, we consider $\mathbb{R}^*$, the field of hyperreal numbers, instead of $\mathbb{R}$, the field of real numbers. $\mathbb{R}^*$ is the proper ordered extension of $\mathbb{R}$ containing infinitesimals. An element $\epsilon \in \mathbb{R}^*$ is infinitesimal if $0 < |\epsilon| < r$ for all positive real numbers r. We reserve the symbol ϵ to denote infinitesimals. For a thorough presentation the reader is referred to [Kei76].

Definition 15. A *partial probabilistic transition system (ppts)* is a tuple

$$T = (Pr, i, Act, Can, \mu) , \qquad (20)$$

where Pr is a set of *processes* (or states), i is the *initial state*, Act is the set of *observable actions* that processes may perform, Can is an Act-indexed family of sets of processes, and μ is a family of *partial probability distributions* indexed by states and actions. For any action $a \in Act$ and any process $r \in Pr$, $r \in Can_a$ indicates that the process r can perform an a-action, in which case $\mu_{r,a} : Pr \to [0;1] \subseteq \mathbb{R}^*$ is a function such that $\sum_{r'} \mu_{r,a}(r') \leq 1$.

In general, we do not require the sum to be equal to 1; hence the name *partial* probability distribution. If all $\mu_{r,a}$ are probability distributions, i.e., $\mu_{r,a}$ maps into the real numbers and $\sum_{r'} \mu_{r,a}(r') = 1$, we leave out the term *partial* and refer to T as a *probabilistic transition system (pts)*. $\mu_{r,a}(r') = \mu$ can intuitively be read as "r can perform the action a and with probability μ become r' ".

Given a *ppts* T. We shall use the following notations:

$r \xrightarrow{a}_\mu r'$ whenever $r \in Can_a$ and $\mu_{r,a}(r') = \mu$

$r \xrightarrow{a} r'$ whenever $r \xrightarrow{a}_\mu r'$ for some $\mu > 0$

$r \xrightarrow{a}$ whenever $r \in Can_a$

$r \xrightarrow{a}\!\!\!\!/$ whenever $r \notin Can_a$

$r \xrightarrow{a}_\mu S$ whenever S is *any* set of processes,
$$r \in Can_a \text{ and } \mu = \sum_{r' \in S} \mu_{r,a}(r').$$

We assume the set Act to be fixed and that all processes in Pr are reachable from the initial state via transitions having non-zero probabilities. Finally, two *ppts* will be said to be *distinct* if their sets of processes are disjoint.

Next, we define morphisms between *ppts's*.

Definition 16. A *ppts-morphism* f between two *ppts's*, $T_j = (Pr_j, i_j, Act, Can_j, \mu_j)$, $j = 1, 2$, is a function between Pr_1 and Pr_2 such that

$$f(i_1) = i_2 , \tag{21}$$

$$f(r) \xrightarrow{a} f(r') \text{ whenever } r \xrightarrow{a} r' , \tag{22}$$

$$\text{If } r \xrightarrow{a} r' \text{ and } f(r) \xrightarrow{a}_{\mu'} f(r') \text{ then } \sum_{r \xrightarrow{a}_\mu r'' f(r'')=f(r')} \mu \le \mu' . \tag{23}$$

The intuition behind (23) is that all transitions from r in T_1 which are simulated by a transition from $f(r)$ can occur with a probability which is no higher than the probability of the simulating transition from $f(r)$.

Let $\mathcal{PPTS}$ denote the category of partial probabilistic transition systems, whose objects are *ppts's* and morphisms are *ppts-morphisms*, with composition of morphisms defined as the usual composition of functions. Let $\mathcal{PTS}$ denote the full subcategory of $\mathcal{PPTS}$ whose objects are *pts's*.

In our model of computation, $\mathcal{PPTS}$, we identify the following subcategory $\mathcal{P}$ of observations.

Definition 17. Let $\mathcal{P}$ be the full subcategory of $\mathcal{PPTS}$ whose objects are *ppts's* of the following form

$$i \xrightarrow{a_1}_{\epsilon_1} r_1 \xrightarrow{a_2}_{\epsilon_2} \cdots \xrightarrow{a_n}_{\epsilon_n} r_n , \tag{24}$$

for some natural number n, distinct states, and actions $a_1, \ldots, a_n \in Act$. Notice that all the probabilities are infinitesimals.

The intuition behind using only infinitesimals on the transitions is that we will only be interested in whether or not a transition can occur rather the probability with which is occurs. This is only true because PTS are the models which we consider.

This time, we postpone the investigation of the existence of pullbacks in $PPTS$. Instead, we now try to characterize the P-open maps in $PPTS$ between any two pts's.

Lemma 18. *A morphism* $m : T_1 \longrightarrow T_2$ *between two* pts*'s is* P-*open if and only if it is "zig-zag" in the following sense:*

If $m(r) \xrightarrow{a} s$ *then there exists an* r' *such that* $r \xrightarrow{a} r'$ *and* $m(r') = s$.

Going through the proof in [CN95] one realises why only infinitesimal probabilities are allowed on the observations from P. Allowing arbitrary probabilities would imply that two pts's which are related by an open map m would be locally isomorphic in the following sense: if $m(r) \xrightarrow{a}_\mu s'$, then there exists an $r \xrightarrow{a}_\mu r'$ such that $m(r') = s'$.

¿From the definition of the morphisms in $PPTS$ one observes the following facts:

- If $m : T_1 \longrightarrow T_2$ and T_1 is a pts then T_2 must also be a pts.
- $PPTS$ does not have pullbacks, neither does PTS. Consider the following example which illustrates three pts's.

Let T, T_1, and T_2 denote the pts's from left to right. Clearly there are uniquely determined morphisms from T_1 to T and from T_2 to T. Together, they form a diagram which does not have a pullback.

However, for P-bisimilarity to be an equivalence relation (a transitive relation, to be more precise) it is in general not necessary for the category $\mathcal{M}$ to have pullbacks. The following weaker result suffices.

Theorem 19. *Given two* P-*open morphisms between* pts*'s,* $m_1 : T_1 \longrightarrow T_0$ *and* $m_2 : T_2 \longrightarrow T_0$. *There exists a* pts T *and* P-*open morphisms* π_1 *and* π_2 *such that*

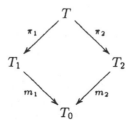

is a commuting square.

Noticing that the composition of two open maps is itself an open map [JNW93], we obtain the following corollary.

Corollary 20. *P-bisimilarity between* pts's *is an equivalence relation.*

If, as done by Larsen and Skou in [LS91], we had assumed a lower limit γ on the probability of transitions (*minimal probability assumption*) and only considered the field of real numbers it would have been hard to obtain a result as the above, at least for us. The problem is that "pullback like" constructions of T involves expressions of the type $\frac{\mu_1\mu_2}{\mu}$ which may denote values smaller than γ.

Next, we recall the definition of probabilistic bisimulation from [LS91]. We have adapted it to the case where the probabilistic transition systems have initial states.

Definition 21. Let $T_j = (Pr_j, i_j, Act, Can_j, \mu_j)$, where $j = 1, 2$, be two distinct pts's. A *probabilistic bisimulation* between T_1 and T_2 is an equivalence $\equiv$ on $Pr = Pr_1 \cup Pr_2$ such that $i_1 \equiv i_2$ and whenever $r \equiv s$, then the following holds:

$$\forall a \in Act. \forall S \in Pr/_{\equiv} . r \xrightarrow{a}_\mu S \Leftrightarrow s \xrightarrow{a}_\mu S , \tag{25}$$

where the notation $r \xrightarrow{a}_\mu S$ was defined after Definition 15.

Now to the main result of this section.

Theorem 22. *Given two* pts's, T_1 *and* T_2. *Then :*

T_1 *is probabilistic bisimilar to* T_2 *if and only if* T_1 *is* P-*bisimilar to* T_2.

5 Non-interleaving Models

We start by recalling a well-known model of "true concurrency", event structures [Win80, Win86], and the characterization of the non-interleaving bisimulation from [JNW93] and [NW95]. We then proceed to look for a non-interleaving generalization of our characterization of trace equivalence.

Definition 23. Define a *(labelled prime) event structure* to be a structure $(E, \leq, \#, l)$ consisting of a set E, of *events* which are partially ordered by $\leq$, the *causal dependency relation*, and a binary, symmetric, irreflexive relation $\# \subseteq E \times E$, the *conflict relation*, and a labelling function $l : E \longrightarrow Act$ assigning an action-label from Act to each event in E, which satisfy

$$c \{e' \mid e' \leq e\} \text{ is finite,}$$
$$e \# e' \leq e'' \Rightarrow e \# e''$$

for all $e, e', e'' \in E$.

Say two events $e, e' \in E$ are *concurrent* iff $\neg(e \leq e'$ or $e' \leq e$ or $e \# e')$.

The finiteness assumption restricts attention to discrete processes where an event occurrence depends only on finitely many previous occurrences. The axiom on the conflict relation expresses that if two events causally depend on events in conflict then they too are in conflict.

Guided by our interpretation we can formulate a notion of computation state of an event structure $(E, \leq, \#, l)$. Taking a computation state of a process to be represented by the set x of events which have occurred in the computation, we expect that

$$e' \in x \ \& \ e \leq e' \Rightarrow e \in x$$

—if an event has occurred then all events on which it causally depends have occurred too—and also that

$$\forall e, e' \in x. \neg(e \# e')$$

—no two conflicting events can occur together in the same computation.

Definition 24. Let $(E, \leq, \#, l)$ be an event structure. Define its *configurations*, $\mathcal{D}(E, \leq, \#, l)$, to consist of those finite subsets $x \subseteq E$ which are

- *conflict-free:* $\forall e, e' \in x. \ \neg(e \# e')$ and
- *downwards-closed:* $\forall e, e'. \ e' \leq e \in x \Rightarrow e' \in x$.

Events manifest themselves as atomic jumps from one configuration to another, in a way which allows one to regard event structures as transistion systems with a notion of concurrency. Notice that the event structurs presented here is not just an arbitrary choice from the world of so-called non-interleaving models.It is well known, see e.g. [WN94], that these structures bear a strong relationship to many other models like net systems [Thi87] and asynchrounous transition systems [Shi85], [Bed88]—a fact which we shall explore later.

Definition 25. Let $(E, \leq, \#, l)$ be an event structure. Let x, x' be configurations. Write

$$x \xrightarrow{e} x' \iff e \notin x \ \& \ x' = x \cup \{e\}.$$

Proposition 26. *Two events e_0, e_1 of an event structure are concurrent iff there exist configurations x, x_0, x_1, x' such that*

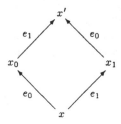

Morphisms on event structures are defined as follows:

Definition 27. Let $ES = (E, \leq, \#, l)$ and $ES' = (E', \leq', \#', l')$ be event structures over the same action set Act. A *morphism* from ES to ES' consists of a total function $\eta : E \longrightarrow E'$ on events which satisfies $l' \circ \eta = l$ and

$$x \in \mathcal{D}(ES) \Rightarrow \eta x \in \mathcal{D}(ES') \ \&$$
$$\forall e_0, e_1 \in x . \eta(e_0) = \eta(e_1) \Rightarrow e_0 = e_1.$$

A morphism $\eta : ES \longrightarrow ES'$ between event structures expresses how behaviour in ES determines behaviour in ES'. The function η expresses how the occurrence of events in ES are simulated by events in ES'.

We write $\mathcal{E}$ for the category of event structures; composition is the usual componentwise composition of functions.

partial identity on L.

Choosing $\mathcal{M}$ as $\mathcal{E}$ leaves open the choice of an observation category $\mathcal{P}$. However, Pratt's *pomsets* are a natural "concurrent" counterpart of the sequential observations from Sec. 2, as suggested in [JNW93]. Formally, we may identify pomsets with the full subcategory, $\mathcal{PO}$, of $\mathcal{E}$ whose objects consist of event structures, in which $\#$ is empty (no conflict), i.e. structures of the form $P = (E, \leq, \emptyset, l)$, representing individual non-sequential "runs" of systems. Notice that we may "dually" identify Milners's synchronization trees with event structures with empty concurrency relation.

It turns out that with this choice of pomsets $\mathcal{PO}$-bisimilarity is a strengthening of a previously studied equivalence.

Definition 28. (Rabinovitch-Trakhtenbrot [RT88], van Glabeek-Goltz [GG89])
A *history-preserving bisimulation* between two labelled event structures ES_1 and ES_2 over a common labelling set Act consists of a set H of triples (x_1, f, x_2) where x_1 is a configuration of ES_1, x_2 a configuration of ES_2 and f is a isomorphism between them (regarded as pomsets), such that $(\emptyset, \emptyset, \emptyset) \in H$ and, whenever $(x_1, f, x_2) \in H$

(i) if $x_1 \xrightarrow{e} x_1'$ in ES_1 then there exists an e' such that $x_2 \xrightarrow{e'} x_2'$ in ES_2 and $l_1(e) = l_2(e')$ and $(x_1', f', x_2') \in H$ with $f \subseteq f'$, for some x_2' and f'.

(ii) if $x_2 \xrightarrow{e'} x_2'$ in ES_2 then there exists an e such that $x_1 \xrightarrow{e} x_1'$ in ES_1 and $l_1(e) = l_2(e')$ and $(x_1', f', x_2') \in H$ with $f \subseteq f'$, for some x_1' and f'

A history-preserving bisimulation H is *strong* when it further satisfies

(I) $(x, f, y) \in H$ & $x' \subseteq x$, for a configuration x' of ES_1 implies $(x', f', y') \in H$, for some $f' \subseteq f$ and $y' \subseteq y$.

(II) $(x, f, y) \in H$ & $y' \subseteq y$, for a configuration y' of ES_2, implies $(x', f', y') \in H$, for some $f' \subseteq f$ and $x' \subseteq x$.

In [NW95] the following theorem was shown.

Theorem 29. $\mathcal{PO}$-*bisimilarity between labelled event structures coincides with strong history-preserving bisimilarity.*

One interesting observation made in [NW95] is that the characterization of strong history-preserving bisimilarity is very robust with respect to the choice of observations. Formally, the characterization holds also for a very restricted full subcategory of $\mathcal{PO}$ containing only objects corresponding to sequential words, "sticks", and "lollipops" like

This raises the question of robustness of the characterization also with respect to the choice of observation extension—to be addressed in the next section.

5.1 Pomset Equivalences

Let us start by recasting in the terminology of Pratt the above choice of observation extension defined formally in terms of event structure morphisms on pomsets. The following terminology is from [Pra86].

Definition 30. Given a pomset $P = (E, \leq, \emptyset, l)$, any $(E, \leq', \emptyset, l)$ for which $\leq \subseteq \leq'$ is called an *augment* of P. Any restriction of $(E, \leq, \emptyset, l)$ to a downwards closed subset of E is called a *prefix* of P.

On pomsets, a morphism η from P to P' amounts to "P being an augment of a prefix of P'"—a very generous notion of extension. An alternative would be the restricted notion of extension corresponding to the one from the characterization of trace equivalence for transition systems.

Definition 31. Let $\mathcal{PO}_1$ denote the subcategory of $\mathcal{PO}$ whose morphisms are the identities and morphisms with the empty pomset as domain.

With this choice, we would expect to get some kind of equivalence in terms of "pomset-languages" of event structures.

Definition 32. Let $ES = (E, \leq, \#, l)$ be an event structure. The *pomset language* of ES, $\mathcal{P}(ES)$, is defined as the set of restrictions of ES to $\mathcal{D}(E)$. Following Pratt, $\alpha\mathcal{P}(E)$ denotes the augmentation closure of $\mathcal{P}(E)$.

And the $\mathcal{PO}_1$-bisimilation may now be characterized as follows.

Theorem 33. *Two event structures ES and ES' are $\mathcal{PO}_1$-bisimilar iff $\alpha\mathcal{P}(ES) = \alpha\mathcal{P}(ES')$.*

Here, we have presented our results for a particular choice of $\mathcal{M}$, event structures. However, it is well-known that event structurs form a coreflective subcategory of many other models studied in the literature, notably the net systems of [Thi87] and the asynchronous transition systems of [Shi85] and [Bed88]. And one of the advantages of working in this abstract setting is that our results carry over immediately to these other models via general theorems like the following from [NW95].

Theorem 34. *Let M be a coreflective subcategory of N with R right adjoint to the inclusion functor from M to N and P a subcategory of M. Then:*

(i) M_1, M_2 are P-bisimilar in M iff M_1, M_2 are P-bisimilar in N.
(ii) N_1, N_2 are P-bisimilar in N iff $R(N_1), R(N_2)$ are P-bisimilar in M.

As a corollary of this, we get imediate characterizations of PO- and PO_1-bisimulations on nets and asynchronous transition systems, which in turn may be easily formulated directly in terms of e.g. the bisimulations and pomset equivalences for one-safe nets studied in [JM93], just to mention a few examples.

However, again our purpose has been mainly to illustrate the generality of the open map approach. Obviously, many alternative choices exist, e.g. with respect to the choice of extension morphisms for pomsets, but we leave it to the interested reader to try some of them out.

6 Presheaf Models

The approach of open maps as presented above may be seen as an attempt to provide a meta-theory for the numerous suggested behavioural equivalences for models of computation. The basic idea is to parameterize such equivalences in the general theory by the choice of observations and their extensions. However, as suggested in [JNW93] one may take this approach one step further and also define the models of computation themselves as derived from the notion of observations.

Consider a category of models M and a choice of observation category forming a subcategory P of M. Consider the *canonical functor* from the category of models M to the topos of presheaves $[P^{op}, Set]$. The functor

$$M \longrightarrow [P^{op}, Set]$$

takes an object X of M to the presheaf $M(-, X)$, and a morphism $f : X \longrightarrow Y$ in M to the natural transformation

$$M(-, f) : M(-, X) \longrightarrow M(-, Y)$$

whose component at an object P of P is the function $M(P, X) \longrightarrow M(P, Y)$ taking g to $f \circ g$.

In general, this canonical functor will not be full and faithful, i.e. a full embedding. However, it folows from [JNW93] that this indeed the case of M being event structures and P pomsets, PO, – and also in the case of M being Milner's synchronization trees and P words, W. In these cases the embeddings extend the Yoneda embedding of $P \longrightarrow [P^{op}, Set]$. Now, if we regard presheaves as the model M' and the image of P under the Yoneda embedding as its observation category P', we can apply the general set-up above, to obtain the class of P'-open morphisms of the presheaf category. They form a category of *open maps* of the topos $[P^{op}, Set]$, in the sense of Joyal and Moerdijk [JM]. As a matter of fact, the two notions of P-open and open map agree formally in the following sense [JNW93] (a subcategory P of M is dense if every object of M is the colimit of objects from P):

Proposition 35. *Let $\mathcal{P}$ be a dense, full subcategory of $\mathcal{M}$. A morphism f : $X \longrightarrow Y$ of $\mathcal{M}$ is $\mathcal{P}$-open iff the morphism $\mathcal{M}(-, f) : \mathcal{M}(-, X) \longrightarrow \mathcal{M}(-, Y)$ is an open map (in the sense of [JM]).*

In [JNW93] the authors argue for presheaves as a model of parallel computation. In general there are many more objects in the presheaf categories than in the original models. In the case of W as observations, the presheaf catergory is essentially a generalization from synchronization trees to synchronization forrests, and in the case of W_1 the presheaf category amounts to a category of multisets of words. In the case of pomsets as observations, the extra objects are much more difficult to explain, but also of potentially much more significance. In [JNW93] indications of this are indicated in terms of the search for suitable models for unfolding of general Petri Nets and a treatment of the concept of refinement.

In some ways, from a computer-science viewpoint, presheaf models are less concrete and harder to motivate than traditional models like event structures. In another way they give a more uniform representation of processes as coherent collections of possible "computational observations". For more details we refer to [JNW93].

7   Conclusion

In this paper we have reviewed Joyal, Nielsen, and Winskel's theory of open maps which was proposed as an abstract definition of equivalences in categories of models of computation. Furthermore, we have presented different applications of the theory, illustrating how many new and existing behavioural equivalences may be captured by the theory. And finally we briefly introduced presheaf models as a unifying framework for models of computation.

Many interesting developments had to be left out of our presentation. One example is the game theoretic characterizations of bisimulations as presented in [NC95], including games for strong history-preserving bisimulation. Another is recent developments towards meta-theorems for behavioural equivalences defined by open maps, e.g. parameterized proofs of congruence properties with respect to process algebraic operators.

Acknowledgements

We acknowledge much of the presented material here to Andre Joyal and Glynn Winskel. We are grateful to Edmund Robinson for pointing out the elegant way of capturing trace equivalence, which was a definite improvement on a previous approach of ours.

References

[Bed88] M.A. Bednarczyk. *Categories of asynchronous systems*, PhD thesis in Computer Science, University of Sussex, report no.1/88, 1988.

[CN95] . A. Cheng and M. Nielsen. Open maps (at) work. Research Series RS-95-23, BRICS, Department of Computer Science, University of Aarhus, April 1995.

[GG89] R. J. Van Glabeek and U. Goltz. Equivalence notions for concurrent systems and refinement of actions. In *Proceedings of MFCS'89*. Springer-Verlag (*LNCS* 379), 1989.

[JM93] L. Jategaonkar and A. Meyer Deciding True Concurrency Equivalences on Finite Safe Nets In *Proceedings of ICALP'93*. Springer-Verlag (*LNCS* 700), 409–417, 1993.

[JM] A. Joyal, and I. Moerdijk. A Completeness Theorem for Open Maps. To appear.

[JNW93] A. Joyal, M. Nielsen, and G. Winskel. Bisimulation and open maps. In *Proc. LICS'93, Eighth Annual Symposium on Logic in Computer Science*, pages 418–427, 1993.

[Kei76] H. Jerome Keisler. *Foundations of Infinitesimal Calculus*. Prindle, Weber & Schmidt, Incorporated, 1976.

[LS91] K. G. Larsen and A. Skou. Bisimulation through Probabilistic Testing. *Information and Computation*, 94:1–28, 1991.

[Mil89] R. Milner. *Communication and Concurrency*. Prentice Hall International Series In Computer Science, C. A. R. Hoare series editor, 1989.

[Mil90] R. Milner. *Operational and Algebraic Semantics of Concurrent Processes*. In Handbook of Theoretical Computer Science, editor J. van Leeuwen, chapter 19. Elsevier Science Publishers, 1990.

[NC95] M. Nielsen and C. Clausen. Games and Logics for a Noninterleaving Bisimulation. *Nordic Journal of Computing*, 2:221–249, 1995.

[NW95] M. Nielsen and G. Winskel. Petri Nets and Bisimulation Research Series RS-95-4, BRICS, Department of Computer Science, University of Aarhus, January 1995. To appear in *Theoreticlal Computer Science*.

[Pra86] V. Pratt. Modelling Concurrency with Partial Orders. *International Journal of Parallel Programming*, 15(1):33–71, 1986.

[RT88] A. Rabinovitch and B. Trakhtenbrot. Behaviour Structures and Nets. *Fundamenta Informatica*, 11(4):357–404, 1988.

[Shi85] M. W. Shields. Concurrent machines. *Computer Journal*, 28, 449–465, 1985.

[Thi87] P. S. Thiagarajan. Elementary Net Systems. In *Petri Nets*, 26–59. Springer-Verlag (*LNCS* 254), 1986.

[Win80] G. Winskel. *Events in Computation*. PhD thesis, University of Edinburgh, 1980.

[Win86] G. Winskel. Event structures. In *Petri Nets: Applications and Relationships to Other Models of Concurrency*, 325–390. Springer-Verlag (*LNCS* 255), 1986.

[WN94] G. Winskel and M. Nielsen. *Models for Concurrency*. In Handbook of Logic in Computer Science, vol. 4, eds. S. Abramsky, D. V. Gabbay, T. S. E. Maibaum. Oxford University Press, 1995.

An Algorithm for Reducing Binary Branchings*

Paul Caspi[†]   Jean-Claude Fernandez[†]   Alain Girault[‡]

Abstract

In this paper we propose an algorithm suppressing useless boolean tests in object code, for programs translatable into deterministic labeled transition systems. This algorithm is based on the notion of test equivalence, a variant of the classical observational equivalence: a test is useless iff each branch leads to equivalent states, the test labels being considered as invisible actions.

Key Words

object code optimization, observational equivalence, "on the fly" bisimulation

1   Introduction

The aim of the paper is to apply a variant of observational equivalence to suppress useless boolean tests in object code. This will allow to replace "**if** *expr* **then** *A* **else** *A*" statements by "*A*". This situation arises in the field of automatic distribution of programs [2, 4], where the replication of code and the suppression of useless actions lead to programs non minimal with respect to boolean tests. We will represent programs under consideration by finite Deterministic Labeled Transition Systems (DLTS, for short). The DLTS represents the control flow structure of a program and each transition is labeled by a finite sequence of basic actions, e.g., assignments, read/write operations ...

In [8, 9], the concept of *bisimulation* is introduced, providing an appropriate notion of equivalence for labeled transition systems. In fact, various notions of bisimulation may be defined, depending on which actions are considered as observable. For instance, in the strong bisimulation, each action is observable. In the weak bisimulation, or observational equivalence, the internal communications, labeled by τ, are invisible. Here we shall use a notion of bisimulation where the branches of a test are invisible. Then we can say that a test is useless if the two branches lead to equivalent states. Such a test can then be eliminated.

*This work has been partially supported by GRECO Automatique Action C2A, Ministère de l'Enseignement Supérieur et de la Recherche and SCHNEIDER ELECTRIC.

†VERIMAG, Miniparc - ZIRST, rue Lavoisier, 38330 Montbonnot, FRANCE, tel: (33) 76.90.96.33, email: {Paul.Caspi,Jean-Claude.Fernandez}@imag.fr, VERIMAG is a joint laboratory of CNRS, INPG, UJF and VERILOG S.A. associated with IMAG

‡INRIA MEIJE project, Centre de Mathématiques Appliquées, BP 207, 06904 Sophia Antipolis cedex, FRANCE, tel: (33) 93.95.74.93, email: Alain.Girault@cma.cma.fr

Programs under consideration are particular in the following sense: each state has either only one successor by a visible action or two successors by a boolean test (binary branching). The behavior of a state q is an action tree, starting from q and representing all possible executions of q. The behavior of q will be noted $[\![q]\!]$.

The purpose of the algorithm we propose is to reduce such programs by suppressing useless branchings, that is branchings leading to equivalent states. This algorithm has to call for a second algorithm testing the equivalence of states. For instance, consider the following DLTS:

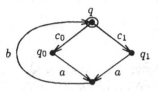

Fig. 1: A DLTS with one test

This DLTS has one test, labeled by c_0 and c_1. The behavior of state q_0 is $[\![q_0]\!] = a.b.[\![q]\!]$. Concerning state q_1, we have $[\![q_1]\!] = a.b.[\![q]\!]$. States q_0 and q_1 are thus equivalent and we can suppress the test.

The paper is organized as follows. In section 2, we give a syntactic characterization of DLTSs. In section 3, we give a variant of the classical notion of bisimulation to our DLTSs and we give a formal system defining it. In section 4, we give the implementation of the equivalence and the reduction algorithms. First we derive the implementation of the test equivalence algorithm from its formal system; in doing this, we use "on the fly" techniques that have proved their efficiency in the equivalence verification field [5, 7]. Second we describe the branching reduction algorithm. For each algorithm, we give the corresponding time and memory complexity. Finally we present an example of computation of the branching reduction and we conclude about this research work.

2 Labeled Transition Systems

We consider a finite alphabet of actions A. We split A into the set of visible actions V and the set of branchings C. We thus have $V \cup C = A$ and $V \cap C = \emptyset$. The set C is partitioned in two subsets C_0 and C_1 : for each $c_0 \in C_0$ there is a unique element, called its complement, in C_1 denoted by c_1. We shall let α, β range over A, a, b range over V and c_0, c_1, c_i range over C; c_i denotes either c_0 or c_1.

We consider finite DLTSs of the form $\{Q, q_0, A, T\}$, where Q is the set of states, q_0 is the initial state, A is the alphabet of actions, T is the transition function, with the notation $p \xrightarrow{\alpha}_T q$ for $(p, \alpha, q) \in T$ and where the only branchings are deterministic and binary ones. More precisely:

$$(p \xrightarrow{\alpha}_T p' \text{ and } p \xrightarrow{\beta}_T p'') \Rightarrow (\alpha = \beta \wedge p' = p'' \text{ or } \exists c_i \in C : \alpha = c_i \wedge \beta = c_{1-i})$$

In order to give syntactic constraints over states of DLTSs, we choose to formally represent them by CCS terms [8].

Definition 2.1 (regular deterministic term)
A regular deterministic binary term is $q ::= nil \mid x \mid a.q \mid c_0.q_0 + c_1.q_1 \mid rec\ x.q$, where x belongs to a finite set of variables.

Transitions are taken by such a term in the following way:

Definition 2.2 (transition relation)

- $c_0.q_0 + c_1.q_1 \xrightarrow{c_0} q_0$ • $c_0.q_0 + c_1.q_1 \xrightarrow{c_1} q_1$

- $a.q \xrightarrow{a} q$ • if $q \xrightarrow{a} q'$ then $rec\ x.q \xrightarrow{a} q'[rec\ x.q/x]$

Definition 2.3 (language transition [1])
For any language λ over A:

$$p \xrightarrow{\lambda} q \Leftrightarrow \exists a_1.a_2.\dots.a_n \in \lambda \wedge \exists q_1, q_2, \dots, q_{n-1} \in Q :$$
$$p \xrightarrow{a_1} q_1 \wedge q_1 \xrightarrow{a_2} q_2 \wedge \dots \wedge q_{n-1} \xrightarrow{a_n} q$$

We also define the function act giving the set of actions which can be performed in a term: $act\ (p) = \{\alpha \mid \exists q.\ p \xrightarrow{\alpha} q\}$.

Note that a finite binary DLTS q is a closed (each variable is bounded by a rec) and a well-guarded (each variable occurs within some sub-term $a.q'$ of q, i.e., there is no empty loop as $rec\ x.x$) regular deterministic term.

For instance, the DLTS shown in figure 1 is: $q = rec\ x.(c_0.a.b.x + c_1.a.b.x)$.

Finally we define the invisible terms that only take transitions labeled by actions of C:

Definition 2.4 (regular invisible term)
A regular invisible binary term is $s ::= nil \mid x \mid c_0.s_0 + c_1.s_1 \mid rec\ x.s$, where x belongs to a finite set of variables.

Now an Invisible Labeled Transition System (ILTS) is a closed and well-guarded regular invisible term. Note that an ILTS is also a DLTS. In the sequel, p and q will denote DLTSs, while s and t will denote ILTSs.

3   Test Equivalence of Behavior

In this section we address the problem of testing for the equivalence of two DLTSs. At a given time, a program state is entirely defined, on one hand by the set of actions it can compute at this time, and on the other hand by the states that will be reached after computing these actions.

Two programs are in equivalent states iff their possible behaviors starting from these states are identical: for each evolution of one of those programs by a given execution sequence, there exists an evolution of the other one by the same execution sequence, such that at each time the behaviors of the reached states are again identical.

Following this notion, we first define a special kind of equivalence relation well suited to our problem. Then we give a formal system of axioms and inference rules computing our equivalence relation. Finally we quickly show how the soundness of our solution can be proven.

3.1 Test Equivalence

In the sequel we only consider two DLTSs having the same set of states, alphabet and transition function, but distinct initial states. We formally define the test bisimulation as:

Definition 3.1 (test bisimulation)
A binary relation R over Q is a test bisimulation iff it is symmetric and satisfies $R \subseteq \mathcal{F}(R)$, where $\mathcal{F}$ is:

$$\mathcal{F}(R) = \{(p,q), \ \forall c_0, c_1, \forall p_0, p_1, p \xrightarrow{c_0} p_0 \wedge p \xrightarrow{c_1} p_1 \Rightarrow$$
$$\begin{cases} (p_0, q) \in R \wedge (p_1, q) \in R \ \textbf{or} \\ \exists q_0, q_1, q \xrightarrow{c_0} q_0 \wedge q \xrightarrow{c_1} q_1 \wedge (p_0, q_0) \in R \wedge (p_1, q_1) \in R \end{cases}$$
$$\forall a, \forall p_1, p \xrightarrow{a} p_1 \Rightarrow$$
$$\exists q_1, q_2, q \xrightarrow{C^{\bullet}} q_1 \xrightarrow{a} q_2 \wedge (p, q_1) \in R \wedge (p_1, q_2) \in R \}$$

We then define the test equivalence as the largest fix point of $\mathcal{F}$. We will note the test equivalence $\sim_t$.

Proposition 3.1
The test equivalence $\sim_t$ is an equivalence relation over DLTSs.

We now give the following formal system of axioms and inference rules that computes the test bisimulation. Here p and q are two DLTSs, s and t are two ILTSs, H is a set of couples of terms with the meaning that $H \overset{i}{\vdash} (p,q)$ or $H \overset{v}{\vdash} (p,q)$ iff p and q are test bisimilar provided that each element of H is itself a couple of test bisimilar terms. This notion of hypothesis, along with the meaning of the i and v labels will be explained afterwards:

Inference system 3.1 (test bisimulation)

- axiom 1: $H \overset{v}{\vdash} (p,p)$ • axiom 2: $H \overset{v}{\vdash} (s,t)$ • axiom 3: $H \cup (p,q) \overset{i}{\vdash} (p,q)$

- rule 1: $\dfrac{H \overset{v}{\vdash} (p,q)}{H \overset{i}{\vdash} (p,q)}$ • rule 2: $\forall a, \dfrac{H \overset{i}{\vdash} (p,q)}{H \overset{v}{\vdash} (a.p, a.q)}$

- rule 3: $\forall c_0, c_1, \dfrac{H \overset{i}{\vdash} (p_0, q_0), \ H \overset{i}{\vdash} (p_1, q_1)}{H \overset{v}{\vdash} (c_0.p_0 + c_1.p_1, c_0.q_0 + c_1.q_1)}$

- rule 4: $\forall c_0.c_1.$ $\dfrac{H \overset{v}{-} (p_0.q) \;.\; H \overset{i}{-} (p_1,q)}{H \overset{v}{-} (c_0.p_0 + c_1.p_1,q)}$ and the symmetrical rule

- rule 5: $\forall c_0,c_1.$ $\dfrac{H \overset{i}{-} (p_0.q) \;.\; H \overset{i}{-} (p_1.q)}{H \overset{i}{-} (c_0.p_0 + c_1.p_1,q)}$ and the symmetrical rule

- rule 6: $\dfrac{H \cup (rec\ x.p.q) \overset{v}{-} (p[rec\ x.p/x].q)}{H \overset{v}{-} (rec\ x.p.q)}$ and the symmetrical rule

Starting from a couple of terms (p,q), this inference system allows to build a proof tree whose nodes are the rules and whose leaves are the axioms. If we can build a proof tree with root $\emptyset \overset{v}{-} (p,q)$, then p and q are test bisimilar.

Axiom 1 draws advantage of the fact that both transition systems have the same transition function and only differ by their initial state.

Axiom 2 proves that two ILTSs are always test bisimilar.

Axiom 3 and rule 6 allow to compute the couples (p,q) at most once, even in the presence of loops in the DLTSs. Indeed, loops in the DLTSs could produce infinite sequences in the proof tree of the test bisimulation. Instead, rule 6 adds the roots of such sequences in the set H. These roots are either of the form $(rec\ x.p.q)$ or $(p.rec\ x.q)$. Then axiom 3 avoids computing infinitely such sequences. By construction. $H \overset{v}{-} (p.q)$ means that p and q are test bisimilar under the hypotheses that all elements of H are couples of test bisimilar terms. For instance. consider the two following DLTSs:

Fig. 2: Two DLTSs producing an infinite sequence

The two corresponding terms are $p = a.nil$ and $q = rec\ x.(c_0.x + c_1.a.nil)$. We apply successively rule 6 and rule 4 to have the following proof tree:

$\emptyset \overset{v}{-} (a.nil, rec\ x.(c_0.x + c_1.a.nil))$
rule 6
$\{(a.nil, rec\ x.(c_0.x + c_1.a.nil))\} \overset{v}{-} (a.nil, c_0.rec\ x.(c_0.x + c_1.a.nil) + c_1.a.nil)$
rule 4
$\{(a.nil. rec\ x.(c_0.x + c_1.a.nil))\} \overset{i}{-} (a.nil, rec\ x.(c_0.x + c_1.a.nil))$
$\{(a.nil. rec\ x.(c_0.x + c_1.a.nil))\} \overset{v}{-} (a.nil, a.nil)$
axiom 3 and axiom 1

The first branch of this proof tree is ended thanks to axiom 3 since the couple is in H, and the second one thanks to axiom 1 since we have $a.nil$ twice.

Now the detection of loops in the proof tree raises the problem of comparing a DLTS with an ILTS. For instance, let $p = a.nil$ and $q = rec\ x.(c_0.x + c_1.x)$. We have the following proof tree:

$$\emptyset \overset{v}{\vdash} (a.nil, rec\ x.(c_0.x + c_1.x))$$

$$\star\star\ rule\ 6\ \star\star$$

$$\{(a.nil, rec\ x.(c_0.x + c_1.x))\} \overset{i}{\vdash} (a.nil, c_0.rec\ x.(c_0.x + c_1.x) + c_1.rec\ x.(c_0.x + c_1.x))$$

$$rule\ 5$$

$$\{(a.nil, rec\ x.(c_0.x + c_1.x))\} \overset{i}{\vdash} (a.nil, rec\ x.(c_0.x + c_1.x))$$

$$\{(a.nil, rec\ x.(c_0.x + c_1.x))\} \overset{i}{\vdash} (a.nil, rec\ x.(c_0.x + c_1.x))$$

$$axiom\ 3\ twice$$

Both branches of the proof tree are ended with axiom 3. However we cannot conclude $\emptyset \overset{v}{\vdash} (p, q)$ since in order to apply the inference rule 6, there must be a $\overset{v}{\vdash}$ in the lower part. This is correct since, according to definition 3.1, p and q are not test bisimilar.

The labels i and v allow to accurately compare ILTSs with DLTSs:

- $H \overset{i}{\vdash} (p, q)$ is the root of a proof sub-tree whose nodes are exclusively rules 5 and whose leaves are exclusively axioms 3; this means that both sub-terms only perform branchings;

- $H \overset{v}{\vdash} (p, q)$ is the root of a proof sub-tree with at least one node being a rule 2, 3, 4 or 6, or one leaf being an axiom 1 or 2; this means that both sub-terms at least perform one identical action, either visible or invisible.

In the previous example, the $\overset{i}{\vdash}$ in axiom 3 prevents from concluding $\emptyset \overset{v}{\vdash} (p, q)$.

3.2 Soundness of the Inference System 3.1

This section is devoted to the proof of our algorithm. The following proposition proves the soundness of the inference system 3.1:

Proposition 3.2
$$\emptyset \overset{v}{\vdash} (p, q) \Rightarrow p \sim_t q.$$

The idea is to give an interpretation, and to prove to be consistent for the inference system 3.1. Then we will prove proposition 3.2 thanks to this interpretation.

We consider the lattice of binary relations on Q. We start with the test bisimulation operator $\mathcal{F}$ (see definition 3.1). The set $\mathcal{F}(R)$ includes only couples of DLTSs that are test bisimilar. We also define the monotonic operator $\mathcal{G}$ such that $\mathcal{G}(R)$ includes only couples of DLTSs that are test bisimilar and only perform tests:

Definition 3.2 (operator $\mathcal{G}$)

$$
\begin{aligned}
\mathcal{G}(R) \;=\; & \{(c_0.p_0 + c_1.p_1.q).(p_0,q) \in R \wedge (p_1,q) \in R\} \\
& \cup\, \{(p.c_0.q_0 + c_1.q_1).(p,q_0) \in R \wedge (p,q_1) \in R\} \\
& \cup\, \{(c_0.p_0 + c_1.p_1.c_0.q_0 + c_1.q_1).(p_0,q_0) \in R \wedge (p_1,q_1) \in R\}
\end{aligned}
$$

Then we define. by monotonicity of $\mathcal{F}$ and $\mathcal{G}$, the following operators:

Definition 3.3 (operators $\mathcal{G}^*$, $\mathcal{F}^i$ and $\mathcal{F}^v$)

- $\mathcal{G}^*$ is the greatest fix-point of operator $\lambda f.\mathcal{I} \cup (\mathcal{G} \circ f)$, $\mathcal{I}$ being the identity.

- $\mathcal{F}^i$ is the greatest fix-point of operator $\lambda f.\mathcal{G}^* \cup (\mathcal{F} \circ f)$.

- $\mathcal{F}^v = \mathcal{F} \circ \mathcal{F}^i$.

We then note R_H the greatest fix-point of operator $\lambda r.\mathcal{F}^v(r \cup H)$, and we define the following interpretation rules:

Definition 3.4 (interpretation rules)

- $I(H \overset{i}{\vdash} (p.q)) = ((p.q) \in \mathcal{F}^i(R_H \cup H))$

- $I(H \overset{v}{\vdash} (p.q)) = ((p.q) \in \mathcal{F}^v(R_H \cup H))$

Proposition 3.3
The interpretation rules are consistent for the inference system 3.1.

According to these interpretation rules, the sketch of the proof of proposition 3.2 is: $\emptyset \overset{v}{\vdash} (p.q) \Rightarrow \mathcal{F}^v(\emptyset) \Rightarrow p \sim_t q$. For the sake of concision, we omit the proofs of propositions 3.1. 3.2 and 3.3. but they can be found in [3].

4 Implementation

We first implement the test bisimulation and we give the time and memory requirements of the corresponding *equivalence* function. Then we describe the algorithm in charge of minimizing DLTSs thanks to the *equivalence* function: it is the *reduction* function. Once again we give the corresponding complexity. Finally we give the global complexity and we present an example of computation for the two algorithms.

4.1 Test Equivalence Algorithm

We want to implement the inference system 3.1. We have to perform a depth first traversal of the proof tree. The root will be given by the function *reduction*. Each node is of the form $H \vdash (p, q)$, with either i or v as a label. Each leaf is one of the three axioms of the inference system. However, rather than CCS terms, we work on DLTSs represented by their transition functions which are more suited to the algorithmic point of view. We thus have to perform a depth first traversal of the product of the two DLTSs that are given by the *reduction* function, according to the inference system 3.1. This product is a transition system whose states are of the form (p, q) and whose transitions are labeled by i or v. This transition system has two special states *success* and *fail*. The state *success* represents the certainly bisimilar couples, i.e., couples (p, p) or (s, t). The state *fail* represents the certainly not bisimilar states, i.e., couples $(a.p, b.q)$. Finally it has loops. The roots are either of the form $(rec\ x.p, q)$ or $(p, rec\ x.q)$. Note that the states precedings the roots in the loops were formerly the axiom 3 leaves of the proof tree.

In order to avoid an exponential computation, we can mark the previously visited states. However, this would need to build entirely the product, which would lead to memory overflow for large scale programs. The solution consists in building the product "on the fly", and in memorizing information about states into separate sets. The same method has already been applied by J-C.Fernandez and L.Mounier in [5] for the computing of the synchronous product of two labeled transition systems.

We use the following data structures:

- (q_{01}, q_{02}) is the product initial state.

- The stack St_state contains the current execution sequence, with the respective depth of its states.

- The stack St_trans contains the transitions remaining to be explored from the corresponding state of St_state, along with a label i or v telling whether these transitions are invisible or visible.

- $Equiv\_states$ is the set of bisimilar states: those are the states leading through an elementary sequence to the state *success*.

- Non_equiv_states is the set of non bisimilar states: those are the states leading to the state *fail*. By construction, $Equiv\_states \cap Non\_equiv\_states = \emptyset$.

- The flag non_equiv tells that the current branch of the product leads to *fail*.

- $Root$ is the set of loop roots of the product. Those are the states (p, q) belonging to the stack St_state and such that the current state of the sequence has (p, q) as one of its successors. This set is initialized to $\emptyset$ at each call of the *equivalence* function.

- $Visited$ is the set of previously visited states, which are neither in $Equiv\_states$ nor in Non_equiv_states. This set is initialized to $\emptyset$ at each call of the $equivalence$ function.

- To each state $(p.q)$ of $Visited$ corresponds a list of hypotheses H such that $H \overset{i}{-} (p.q)$ or $H \overset{v}{-} (p.q)$. These lists. initially empty, will be accessed and modified with the $hypo$ function.

These structures are manipulated with the usual primitives ($push$, top ...). In order to reach the successors of a given product state. we define the $successor$ function. For a given state $(p.q)$. it returns the list of its direct successors as well as a label i or v. according to the inference system 3.1:

```
function successor (p, q) :
begin
  if (p = q) then return ((v. {success})) :
  else if (p = a.p₁ and q = a.q₁) then return ((v, {(p₁, q₁)})) :
  else if (p = c₀.p₀ - c₁.p₁ and q = c₀.q₀ - c₁.q₁) then return ((v, {(p₀, q₀), (p₁, q₁)})) :
  else if (p = c₀.p₀ + c₁.p₁) then return ((i, {(p₀, q), (p₁, q)})) :
  else if (q = c₀.q₀ - c₁.q₁) then return ((i. {(p, q₀), (p, q₁)})) :
  else return ((v. {fail})) :
  endif
end
```

For all the complexity computations, we introduce the following notations:

- For a given function f. $T(f)$ and $\mathcal{M}(f)$ are respectively its time and memory requirements.

- For a given DLTS. n_s. n_l and n_t are respectively its number of states, loops and tests.

Concerning the $successor$ function. its time and memory requirements are $T(successor) = \mathcal{M}(successor) = O(1)$.

In order to detect elementary sequences, we need to keep track of when was performed the last test by the product. Then. by comparing the depth in the stack of this last test with the depth of a given loop root, we can tell whether there is a test in the current loop or not:

- if the last test depth is lesser than the root depth, then there is no test in the loop:

- otherwise the last test is inside the loop.

So as to obtain the root depth. we define the function $belongs\,(s, St\_state, d)$ which tests whether s belongs to St_state. and if so returns **true** and sets d to its depth in St_state.

So. the test equivalence algorithm is an "on the fly" depth first traversal of the product:

```
function equivalence ((q01, q02));
begin
  (* step 1: initialize the stacks, sets and flags *)
  (* step 2: do some verifications about (q01, q02) *)
  (f0, L0) := successor ((q01, q02));
  (* step 3: do some verifications about L0 *)
  push ((1, (q01, q02)), St_state);
  push ((f0, L0), St_trans);
  while (St_state ≠ ∅) do
    (d, (q1, q2)) := top (St_States);
    (f, L) := top (St_Trans);
    if (f = ι) then last_test := d; endif
    if (L ≠ ∅) then (* some successor of (q1, q2) remains *)
      (* step 4: choose one of these successors and treat it *)
    else (* top (St_trans) = (f, ∅): all successors of (q1, q2) have been checked *)
      (* step 5: conclude about (q1, q2), update Equiv_states and Non_equiv_states,
      and propagate the result *)
    endif
  endwhile
  (* step 6: return the result *)
end
```

Now the algorithm steps are:

1. The stacks, *Root* and *Visited* are initialized to ∅, *non_equiv* to **false** and *last_test* to 0.

2. The verifications about (q_{01}, q_{02}) consist in checking if it belongs to *Equiv_states* (resp. *Non_equiv_states*), in which case the algorithm returns **true** (resp. **false**).

3. The verifications about L_0 consist in checking if L_0 equals {*success*} (resp. {*fail*}), in which case we add (q_{01}, q_{02}) to *Equiv_states* (resp. *Non_equiv_states*) and the algorithm returns **true** (resp. **false**).

4. Once a state (q'_1, q'_2) is chosen in L, we check whether it belongs to *Equiv_states*, *Non_equiv_states*, *Visited*, *St_state*, or none of these:

   - If $(q'_1.q'_2) \in$ *Equiv_states* we do nothing.

   - Else if $(q'_1, q'_2) \in$ *Non_equiv_states* we set *non_equiv* to **true** and *top (St_trans)* to $(f, \emptyset)$. This prevents from exploring the remaining successors of the top state (q_1, q_2).

   - Else if $(q'_1, q'_2) \in$ *St_state*, then we face a loop in the product. We insert (q'_1, q'_2) in *Root* and check the flag *last_test*:
     - If there is no branching in the concerned loop, then $q'_1 \sim_t q'_2$: we thus insert (q'_1, q'_2) in *Equiv_states*.
     - Else we add (q'_1, q'_2) to the already attached hypotheses of (q_1, q_2).

   - Else if $(q'_1, q'_2) \in$ *Visited*, then (q'_1, q'_2) has been already visited during the current computation of *equivalence*. We extract the list of hypotheses attached to (q'_1, q'_2), and we explore it with the *explore* function:

- If all the hypotheses belong to $Equiv\_states$ we insert (q_1', q_2') in $Equiv\_states$.
- Else. if one of the hypotheses belongs to Non_equiv_states we insert (q_1', q_2') in Non_equiv_states, set non_equiv to **true** and $top\,(St\_trans)$ to $(f, \emptyset)$.
- Otherwise. we add the hypotheses attached to (q_1', q_2') to the already attached hypotheses of (q_1, q_2).

• Else we have to explore (q_1', q_2') successors. We call the $successor$ function to compute the list (f', L') of (q_1', q_2') successors and check L':

- If $L' = \{success\}$ we insert (q_1', q_2') in $Equiv\_states$.
- Else if $L' = \{fail\}$ we insert (q_1', q_2') in Non_equiv_states, set non_equiv to **true** and $top\,(St\_trans)$ to $(f, \emptyset)$.
- Otherwise we push (q_1', q_2') along with its depth (i.e., $d + 1$) into St_state. and push (f', L') into St_trans.

The algorithm of the $explore$ function is:

```
function explore (L):
begin
  while (L ≠ ∅) do
    choose and remove (q₁, q₂) in L :
    if ((q₁, q₂) ∈ Non_equiv_states) then return nonequiv ;
    else if ((q₁, q₂) ∉ Equiv_states) then return add;
    endif
  endwhile
  return equiv :
end
```

5. When all the successors of the current state have been checked, we pop St_state and St_trans. insert the current state in $Visited$, remove it from its hypotheses. and check the non_equiv flag:

• If $non_equiv = $ **true**. then the current state is not bisimilar: we insert it in Non_equiv_states and set $top\,(St\_trans)$ to $(f, \emptyset)$.

• Else we check the belonging to $Root$ and the hypotheses list:

- If the current state belongs to $Root$, we check the label f of the transition starting from itself: if the label is i then, we test if both terms are ILTSs. If they are, then according to axiom 2 the current state is bisimilar; we thus insert it in $Equiv\_states$. If they are not, then according to the rule 6 the current state is not bisimilar: we thus insert it in Non_equiv_states, set non_equiv to **true** and $top\,(St\_trans)$ to $(f, \emptyset)$.
- Else if the list of hypotheses is empty. then this state is bisimilar: we insert it in $Equiv\_states$.
- Otherwise, we check the label f off the transitions starting from q_1, q_2). If $f = v$ then we replace the label of the transition starting from the new top state by v, according to the rule 4 of inference system 3.1.

In order to check whether a DLTS is an ILTS or not, we give below the algorithm of the *invisible* function. Basically it is a depth first search of any visible action in the DLTS given in parameter, under an hypotheses list also given in parameter:

```
function invisible (q, L) ;
begin
  if (q ∈ L or q = nil) then return (true) ;
  else if (q = a.q₁) then return (false) ;
  else if (q = c₀.q₀ + c₁.q₀) then return (invisible (q₀, L ∪ {q})) ;
  else if (q = c₀.q₀ + c₁.q₁) then
    if (invisible (q₀, L ∪ {q})) then return (invisible (q₁, L ∪ {q})) ; else return (false) ; endif
  endif
end
```

6. In order to return the result, we check the flag *non_equiv*: if it is **true**, we return **false**, otherwise we return **true**. Indeed, the list of hypotheses attached to the initial state (q_{01}, q_{02}) is by construction empty, and thus we can conclude that $q_{01} \sim_t q_{02}$.

We now focus on the time and memory requirements of this algorithm. We keep the same notations as introduced for the *successor* function. The number of states of the product is n_s^2. The implementation consists in memorizing each state of the product in a single data structure called *States*. To each state (p, q) of *States*, we associate the following data:

- The sets to which (p, q) belongs. These sets are *St_state*, *Equiv_states*, *Non_equiv_states*, *Root* and *Visited*.

- For each state of the current execution sequence (i.e., belonging to *St_state*), the list of its remaining successors.

- For each state already computed (i.e., belonging to *Visited*), the list of its attached hypotheses.

A linkage between the states allows to represent the stack *St_state*. *States* is implemented by an open hash table [6]. Such a data structure allows to push and pop states in $O(1)$. The search and the insertion into a given set is in $O(n/h)$, where n is the number of states already inserted in the hash table, and h the hash size.

Each state is inserted in *Visited* once it has been computed, and is never computed anymore. Moreover, in each computation, the only non linear operations are the handling of the hypotheses (which is in $O(n_l)$) and the insertions in *Equiv_states*, *Non_equiv_states* ... (which are in $O(n/h)$). Hence, the time requirement is $\mathcal{T}(equivalence) = O(n_s^2 \times max(n_l, n/h))$.

The memory requirement is the cost of the *States* data structure. The size needed for a given state is a constant bit-string for the belonging in *St_state*, *Equiv_states* ..., and $O(n_l)$ for the list of hypotheses. Hence, $\mathcal{M}(equivalence) = O(n_s^2 \times n_l)$.

4.2 Branching Reduction Algorithm

We now want to implement the reduction of branching in DLTSs. This algorithm consists in calling the function *reduction* on the initial state of the program that we want to minimize. The *reduction* function is in charge of calling the *equivalence* function on each branching of the program. As described in subsection 4.1. this function answers **true** iff the corresponding branching must be reduced. Otherwise it answers **false**.

The first step is basically a depth first search of all the branchings in the DLTS of the program. This DLTS has of course loops. In order to compute its states only once. we keep track of the previously computed states by marking them with the *visited* flag. For any state p. this flag can be tested and set thanks to the *visited* function.

Now if a given branching $c_0.q_0 + c_1.q_1$ has to be reduced, then we must choose one of the two branches for continuing the reduction algorithm. In doing this, we must take care of branching loops. i.e.. those containing no visible actions. We thus use the *invisible* function defined in section 4.1.

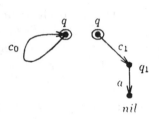

For instance, consider the DLTS q of figure 2: $q = rec\ x.(c_0.a.nil + c_1.x)$. According to the definition 3.1, $q \sim_t q_1$ and the branching must be reduced. Then we must choose one of the two branches. If we choose the left branch, the resulting DLTS will be $q = rec\ x.c_0.x$. If we choose the right branch. the resulting DLTS will be $q = c_1.a.nil$.

Fig. 3: The two resulting DLTSs

In this case. we must choose the right branch. Indeed, the right branch q_1 contains the visible action a. However. the left branch q is a branching whose right branch contains the visible action a. Thus the left branch contains also a visible action. But this left branch is precisely an invisible loop. In order to avoid such invisible loops. we initiate the *invisible* function with an hypotheses list containing the branching itself:

```
function reduction (q) :
begin
    if (visited (q)) thenreturn : endif
    visited (q) := true :
    if (q = a.q₁) then reduction (q₁) : return :
    else if (q = c₀.q₀ - c₁.q₀) then q := q₀ : reduction (q₀) ;
    else if (q = c₀.q₀ - c₁.q₁) then
        if (equivalence ((q₀.q₁))) then
            if (invisible (q₀, {q})) then q := q₁ : reduction (q₁) : else q := q₀ : reduction (q₀) ; endif
        endif
        return :
    endif
end
```

For the time and memory requirements, we keep the same notations as in subsection 4.1. The *visited* flag allows to compute each state of the DLTS only once. Thus $T(reduction) = O(n_s + n_t \times (T(equivalence) + T(invisible)))$.

Concerning the *invisible* function, since all states of the DLTS are computed at most once, the time requirement is $T(invisible) = O(n_s)$.

To each state of the DLTS, we attach one flag. Moreover, the memory requirement for the *invisible* function is the cost of the list L, which is at most $M(invisible) = O(n_s)$. Thus the memory cost added is $M(reduction) = O(n_s + n_s + n_t \times M(equivalence))$.

As a result, the final time and memory requirements are:

- $\boxed{T(reduction) = O(n_t \times n_s^2 \times max(n_l, n/h))}$

- $\boxed{M(reduction) = O(n_t \times n_s^2 \times n_l)}$

4.3 A Non-Trivial Example

In order to carefully understand the mechanism of the branching reduction, we study in this section a non trivial example of a DLTS (figure 4).

We now represent vertically the execution sequences of functions *reduction* and *equivalence*. A state between braces leads to two execution sequences, thanks to rules 3,4 or 5 of derivation system 3.1. "V", "E" or "N" respectively tell that the considered state is in the set *Visited*, *Equiv_states* or *Non_equiv_states_states*. Finally, "*fail*" or "*succ*" tell that the considered state directly leads to the *fail* or *success* state.

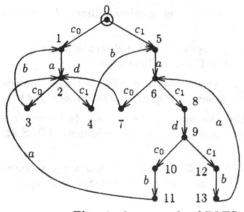

0				
(1,5)	1			5
[2,6]	2			V
(3,7)	(3,4)	3	4	
fail	(1,5)	1	5	
	N	V	6	
		(7,8)		7
		[2,9]		2
		(3,10)	(4,12)	V
		(1,11)	(5,13)	
		succ	succ	

Fig. 4: An example of DLTS with the execution sequences

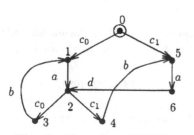

Fig. 5: The reduced DLTS

- $c_0.1 + c_1.5$ is not reduced since $(1,5) \rightsquigarrow fail$.

- $c_0.3 + c_1.4$ is not reduced since $(3,4) \rightsquigarrow (1,5) \in Non\_equiv\_states$.

- $c_0.7 + c_1.8$ is reduced since $(7,8) \rightsquigarrow success$.

- $c_0.10 + c_1.12$ is not computed since $invisible\ (7, \{6\}) = \textbf{false}$.

5 Conclusion

We have studied in this paper an algorithm for suppressing useless boolean tests in imperative sequential programs. In consists in applying a test equivalence procedure to each test of the program, and to suppress a test whenever both branches offer equivalent behaviors. In order to achieve this, we have represented our programs by means of finite binary deterministic labeled transition systems. Then the main idea was to base the test equivalence algorithm on axioms and inference rules over labeled transition systems.

This definition of our algorithm allows to implement it as a depth first traversal of the proof tree built by our formal system of axioms and rules, and starting with the left and right branch of the initial test. The difficulty was then, first to avoid exponential computation, and second to avoid building entirely the proof tree. We have thus implemented our algorithm as an "on the fly" depth first traversal with a marking of the previously visited states.

Finally, an important issue is that the definition of our algorithm by a inference system has allowed to write an elegant formal proof. This proof consists in showing the soundness of this inference system with respect to well chosen interpretation rules [3].

References

[1] D. Austry and G. Boudol. Algèbre de processus et synchronisation. *Theoretical Computer Science*, 30:91–131, April 1984.

[2] D. Callahan and K. Kennedy. Compiling programs for distributed memory multiprocessors. *Journal of Supercomputing*, 2(2):151–169, June 1988.

[3] P. Caspi, J.C. Fernandez, and A. Girault. An algorithm for reducing binary branchings: implementation and formal proof. Research Report INRIA, France, 1994.

[4] P. Caspi, A. Girault, and D. Pilaud. Distributing reactive systems. In *Seventh International Conference on Parallel and Distributed Computing Systems, PDCS'94*, Las Vegas, USA, October 1994. ISCA.

[5] J.C. Fernandez and L. Mounier. "on the fly" verification of behavioural equivalences and preorders. In K.G. Larsen, editor, *Proceedings of the 3rd workshop on computer-aided verification, CAV'91*, July 1991.

[6] D.E. Knuth. *The Art of Computer Programming*, volume III : Sorting and Searching of *Computer Science and Information Processing*. Addison-Wesley, Reading, Massachussets, 1973.

[7] K.G. Larsen. Efficient local correctness checking. In G.V. Bochmann and D.K. Probst, editors, *Proceedings of the 4th workshop on computer-aided verification, CAV'92*, July 1992.

[8] R. Milner. A calculus of communicating systems. *LNCS*, 92, 1980.

[9] D. Park. Concurrency and automata on infinite sequences. In *5th GI-Conference on Theorical Computer Science*. Springer Verlag, 1981. LNCS 104.

On the Complexity of Bisimilarity
for Value-Passing Processes
(Extended Abstract)

Michele Boreale[1] and Luca Trevisan[2]

[1] Istituto per l'Elaborazione dell'Informazione, Consiglio Nazionale delle Ricerche,
Via S. Maria 46, 56126 Pisa, Email: michele@dsi.uniroma1.it
[2] Dipartimento di Scienze dell'Informazione, Università di Roma "La Sapienza", Via
Salaria 113, 00198 Roma, Email: trevisan@dsi.uniroma1.it

Abstract. We study the complexity of deciding bisimilarity between
non-deterministic processes with explicit primitives for manipulating
data values. In particular, we consider a language with value-passing
(input/output of data) and parametric definitions of processes. We distinguish the case in which data cannot be tested and the case in which
a simple equality test over data is permitted.
In the first case, our main result shows that the problem is PSPACE-
hard for the full calculus. In the second case, we first show that the
problem is coNP-complete in the fragment with value-passing and no
parametric definitions. We then define a compositional polynomial-time
translation of the full calculus to the fragment with parametric definitions
but no value-passing. The translation preserves bisimilarity. This fact
establishes the decidability of the full calculus and the PSPACE-hardness
of the fragment without value-passing. In other words, once parametric
definitions and equality test are allowed, the adding of value-passing does
not increase neither the expressive nor the computational power.

1 Introduction

Recently, there has been a renewed interest in process calculi with explicit primitives to manipulate data values. In particular, several enriched versions of Milner's CCS [Mil80, Mil89, JP93, HL95, HL93] have been studied. In *pure*, i.e.
data-less, process calculi such as CCS, beside standard operators for describing
behaviours of processes (such as non-determinism + and parallel composition
|) only pure-synchronization actions (also called "pure" actions) are provided.
By contrast, process calculi with explicit treatment of data contain primitives
for expressing transmission and receipt of values at communication ports: this
feature is known as *value-passing*. Using the notation of [Mil80], output of v at
port a is written $\bar{a}v.$, while input at a is written $a(x).$; here the variable x acts as
a formal parameter. Besides being exchanged, usually data values can be used
as *parameters* in recursively defined processes and tested by means of predicates
to control the execution flow. Languages with explicit manipulation of values
permit a natural description of realistic systems. As an example, the recursively

defined process $C(x)$:

$$C(x) \quad \Leftarrow \quad [x < o](a(y).C(y) + \overline{b}x.C(x)) + [x \geq o]Error(x)$$

specifies a memory cell whose initial content is a number x; as long as this content is less than an overflow value o, the cell can either receive a new value at a, or transmit its content at b; as soon as the value x equals or exceeds o, a recovery process $Error$ is called.

A very peculiar kind of value-passing language is Milner, Parrow and Walker's π-calculus [MPW92], where the values being exchanged among processes are communication ports themselves (name-passing). This permits the description of systems with dynamical communication linkage.

When analysing concurrent systems, a central problem is to be able to decide whether two given descriptions (usually regarded as a specification and as an implementation) are equivalent or not, according to a chosen notion of equivalence (verification). The algebraic aspects of this problem are becoming now well-understood, also for value-passing processes [HL93, PS93, BD94]. On the contrary, a lot of questions concerning the decidability and the computational complexity of verification remain unanswered. A basic problem is to determine meaningful fragments of the calculi with values over which the verification problem is decidable. Then, a fundamental issue is to determine the abstract computational complexity of each of these fragments w.r.t. verification. Answering such questions would improve our understanding of the mathematical nature of processes. In practical cases, it could provide us with useful information to locate sources of inefficiency. In the present work, we will try to address some of these issues. We will restrict our attention to one of the most widely studied equivalences, Milner's bisimulation equivalence (or simply "bisimilarity"), written $\sim$ and described e.g. in [Mil89].

For processes manipulating values, a non-trivial aspect of the problem is that they have usually an operational description in terms of an infinite state-transition graph (they are infinite state), at least if the domain of data values is infinite. This is due to the fact that each input action $a(x)$. gives rise to infinitely many actual transitions, one for each different value. In [JP93], Jonsson and Parrow concentrate on a particular class of processes with values, the data independent ones, which cannot test data nor perform any kind of operation over them. They prove that the bisimilarity problem for such processes can be transformed into a bisimilarity problem for finite-state processes. For the latter, decision algorithms exist [PT87, KS90], which are polynomial in the sizes of the involved graphs (that can be however much larger than the syntactical size of the processes). A detailed comparison of our work with [JP93] is contained in Section 6.

In the present paper, we consider a calculus for describing non-deterministic processes. It should be naturally embedded in every "reasonable" language with explicit data manipulation. More precisely, besides permitting the execution of pure actions, we allow data values to be exchanged, used as parameters in recursive definitions and tested for equality. The latter is done via the matching predicate $[a = b]$, also considered in the π-calculus [MPW92]. This is perhaps the most elementary form of test one would admit on data. Not even negative tests, to decide inequality of data, are permitted.

Our goal is to classify and separate the computational complexity of the two basic operations for manipulating data, value-passing and parametric recursive definitions. This will be done both for the data-independent case (where matching is excluded) and for the data-dependent one (where matching is included).

In each of the two cases, we consider separately three (sub-)languages, obtained from the calculus with pure actions by adding either or both of value-passing and recursive definitions. Then we asses the decidability and the difference in complexity of these languages. In this analysis, we refer to the complexity classes P,NP, coNP and PSPACE. Recall that the latter contains both NP and coNP and that it is believed that this containment is strict (see e.g. [BC93]).

In the data-independent regime, we first note that the bisimilarity problem is solvable in polynomial time for the calculi allowing either, but not both, of recursive definitions or value-passing. Then we prove that the problem is PSPACE-hard for the full language, i.e.: every problem in the class PSPACE is not more difficult than the bisimilarity problem over the matching-free calculus. This improves on a NP-hardness result due to Jonsson and Parrow and is, to the best of authors' knowledge, the highest known lower-bound to the complexity of a decidable bisimilarity over a meaningful language.

In the data-dependent regime, we first show that, in the sublanguage with value-passing but no recursive definitions, the bisimilarity problem is decidable and is as difficult as the most difficult problems in coNP, i.e. it is coNP-complete. Then we analyze the complexity of the full language, with both value-passing and recursive definitions. We show that the language can be compositionally translated down to the fragment without value-passing, in a way that preserves bisimulation equivalence. The translation can be carried out in a time polynomial in the sizes of the processes. The result is interesting for two reasons. First, it gives us a procedure for deciding the bisimilarity problem in the full language, since the problem is easily seen to be decidable in the fragment without value-passing. Second, it ensures that the problem for the fragment without value-passing is just as complex as for the full language (which is of course PSPACE-hard). It is important to point out that the matching predicate plays a crucial role in the definition of the translation.

To sum up, in the absence of matching, value-passing and recursive definitions are separately tractable, but if we join them together the bisimilarity problem becomes very complex (PSPACE-hard). If matching is allowed, the presence of value-passing itself makes the problem coNP-complete. By contrast, the presence of recursive definitions themselves makes the problem PSPACE-hard; then, the adding of value-passing does not increase neither the expressive nor the computational power. These results are also summarized in Table 1.

The rest of the paper is organized as follows. In Section 2, syntax and semantics of the considered language are presented, and a few notions from complexity theory are recalled. Section 3 deals with the complexity of data-independent pro-

Language	Complexity	Reduces to ...
VP ($\mathcal{IL}_v$)	P	all
RD ($\mathcal{IL}_r$)	P	all
VP, RD ($\mathcal{IL}_{v,r}$)	PSPACE-hard	$\mathcal{L}_r$, $\mathcal{L}_{v,r}$
M, VP ($\mathcal{L}_v$)	coNP-complete	$\mathcal{IL}_{v,r}$, $\mathcal{L}_r$, $\mathcal{L}_{v,r}$
M, RD ($\mathcal{L}_r$)	PSPACE-hard	$\mathcal{L}_{v,r}$
M, VP, RD ($\mathcal{L}_{v,r}$)	PSPACE-hard	$\mathcal{L}_r$

VP = value-passing, RD = recursive definitions, M = matching.

Table 1. The complexity results of the paper.

cesses. As to data-dependent processes, the treatment of value-passing is contained in Section 4, while the relationship between the full language and the fragment without value-passing is investigated in Section 5. Comparison with related work and conclusive remarks are contained in Section 6.

2 Preliminaries

2.1 The Language

Below, we present first the syntax and then operational and bisimulation semantics of the language. The notation we use is that of value-passing CCS [Mil80, Mil89] and of π-calculus [MPW92]. We assume the following sets:

- a countable set *Act* of *pure actions* or *communications ports*, ranged over by $a, a', \ldots$;
- a countable set *Var* of *variables*, ranged over by $x, y, \ldots$;
- a set *Val* of *values*, ranged over by $v, v', \ldots$, containing at least two distinct elements;
- a countable set *Ide* of *identifiers* each having a non-negative *arity*. *Ide* is ranged over by *Id* and capital letters and is disjoint from the previous sets.

A *value expression* is either a variable or a value. Value expressions are ranged over by $e, e', \ldots$. We also consider the set $\overline{Act} = \{\overline{a} \mid a \in Act\}$ of *co-actions*, which represent output synchronizations. The set $Act \cup \overline{Act}$ will be ranged over by c.

The set of *terms* of our language, ranged over by $P, Q, \ldots$, is given by the operators of *pure synchronization prefix*, *input prefix*, *output prefix*, *non-determinism*, *matching* and *identifier*, according to the following grammar:

$$P ::= c.P \mid a(x).P \mid \overline{a}e.P \mid \sum_{i \in I} P_i \mid [e_1 = e_2]P \mid Id(e_1, \ldots, e_k)$$

where k is the arity of *Id*. We always assume that the index set I in $\sum_{i \in I} P_i$ is finite and sometimes write $P_1 + \cdots + P_n$ for $\sum_{i \in \{1, \ldots, n\}} P_i$. When I is empty, we use the symbol 0: $0 \stackrel{def}{=} \sum_{i \in \emptyset} P_i$. When no confusion may arise, we write c for $c.0$.

An occurrence of a variable x in a term P is said to be *bound* if it is within the scope of an input prefix $a(x)$; otherwise it is said a *free* occurrence. The set of variables which have a bound occurrence in P is denoted by $bvar(P)$, while the set of variables which have a free occurrence in P is denoted by $fvar(P)$; $var(P)$ is $bvar(P) \cup fvar(P)$. We define $val(P)$ as the set of values occurring in P. The *size* of a term P, indicated by $|P|$, is the number of symbols appearing in it; e.g., if $P = a(x).\bar{a}x.a'.0 + Id(x)$ then $|P| = 9$.

Substitution of the distinct variables $x_1, \ldots, x_n$ with the values $v_1, \ldots, v_n$, indicated by $\{v_1/x_1, \ldots, v_n/x_n\} = \{\tilde{v}/\tilde{x}\}$ and *composition* of two substitutions σ and σ', denoted by $\sigma\sigma'$, are defined as expected. We let $\sigma, \ldots$ range over substitutions. The function val is extended to substitutions in the obvious way and such notations as $val(P, Q, \sigma)$ will mean $val(P) \cup val(Q) \cup val(\sigma)$.

We presuppose an arbitrarily fixed *finite* set Eq of *identifiers definitions*, each of the form

$$Id(x_1, \ldots, x_k) \Leftarrow P$$

where $k \geq 0$ is the arity of Id. We require that the x_i are pairwise distinct and that $fvar(P) \subseteq \{x_1, \ldots, x_k\}$. In Eq, each identifier has a single definition. The requirement for the set Eq to be finite is motivated by the fact that we are only interested in syntactically finite processes.

Note that we have not made any assumption on whether the sets Var, Val and Act are pairwise disjoint or not. There are two particularly interesting cases: (i) if Act, Var and Val are pairwise disjoint we get a proper sublanguage of value-passing CCS [Mil80, Mil89]; this case will be referred to as the *simple value-passing* case; (ii) if $Act = Var = Val$, we get a proper sublanguage of the π-calculus [MPW92]; this case will be referred to as the *name-passing* case.

Most of our results will not depend on a particular such assumption. Also, they will not depend on whether Val is finite or infinite (though, of course, if the name-passing assumption is made, Val must be infinite, since Act is).

A process term P is said to be *closed* if $fvar(P) - Val = \emptyset$; in this case, P is said to be a *process*. According to this definition, all terms are processes in a name-passing setting. Processes are the terms we are most interested in. As we shall see, bisimulation semantics will be defined only over the set of processes.

Since we are interested in determining the contributions of different operators to the complexity of deciding bisimilarity, it is convenient to single different (sub)languages out of the syntax defined above. The *data-independent* languages $\mathcal{IL}_v$, $\mathcal{IL}_r$ and $\mathcal{IL}_{v,r}$ are defined as follows:

- $\mathcal{IL}_v$ contains all operators, but identifiers and matching;
- $\mathcal{IL}_r$ contains all operators, but input and output prefixes and matching;
- $\mathcal{IL}_{v,r}$ contains all operators, but matching.

The *data-dependent* languages $\mathcal{L}_v$, $\mathcal{L}_r$ and $\mathcal{L}_{v,r}$ are defined similarly, but with matching in addition. In particular, $\mathcal{L}_{v,r}$ is the full language.

The operational behaviour of our processes is defined by means of a transition relation. Its elements are triples (P, μ, P') written as $P \xrightarrow{\mu} P'$. Here, μ can be of three different forms: c, $\bar{a}v$ or $a(v)$. A *pure action* c represents a synchronization through the port c, without passing of data involved. An *output action* $\bar{a}v$ means transmission of the datum v through the port a. An *input action* $a(v)$ represents receipt of the datum v through the port a. We let μ range over actions. The transition relation is defined by the inference rules in Table 2.1. Note that $\xrightarrow{\mu}$ leads processes into processes.

$$(Sync) \; c.P \xrightarrow{c} P$$

$$(Inp) \; a(x).P \xrightarrow{a(v)} P\{v/x\}, \; v \in Val \qquad (Out) \; \bar{a}v.P \xrightarrow{\bar{a}v} P$$

$$(Match)\dfrac{P \xrightarrow{\mu} P'}{[v = v]P \xrightarrow{\mu} P'} \qquad (Sum)\dfrac{P_j \xrightarrow{\mu} P'}{\sum_{i \in I} P_i \xrightarrow{\mu} P'} \; j \in I$$

$$(Ide)\dfrac{P\{\tilde{v}/\tilde{x}\} \xrightarrow{\mu} P'}{Id(\tilde{v}) \xrightarrow{\mu} P'} \quad \text{if } Id(\tilde{x}) \Leftarrow P \text{ is in } Eq$$

Table 2. Inference rules for the transition relation $\xrightarrow{\mu}$.

On the top of the transition relation $\xrightarrow{\mu}$, we define *strong bisimulation equivalence* $\sim$, [Mil89, MPW92, PS93] as usual:

Definition 1 (Strong bisimulation equivalence). A binary symmetric relation $\mathcal{R}$ over processes in $\mathcal{L}_{v,r}$ is a *bisimulation* if, whenever $P \mathcal{R} Q$ and $P \xrightarrow{\mu} P'$, there exists Q' s.t. $Q \xrightarrow{\mu} Q'$ and $P' \mathcal{R} Q'$. We let $P \sim Q$, and say that P *is bisimilar to* Q, if and only if $P \mathcal{R} Q$, for some bisimulation $\mathcal{R}$.

From now on, we will omit the adjective "strong". A drawback of the above definition is that it requires considering the whole (possibly infinite) set of transitions of the two processes being compared. We will rely on an alternative "finitary" definition of bisimulation. It differs from the standard one in that, on the input action clause, case-analysis on just a *finite* set of values is required. In the sequel, we say that a value v is *fresh* if v does not occur in any previously mentioned process, nor in the set Eq.

Definition 2 (F-bisimulation). Let $\mathcal{R}$ be a symmetric relation over processes. We say that $\mathcal{R}$ is a *F-bisimulation* if, whenever $P \mathcal{R} Q$, then:

- $P \xrightarrow{\mu} P'$, with μ not an input action, implies $Q \xrightarrow{\mu} Q'$ for some Q' s.t. $P' \mathcal{R} Q'$, and

- for some fresh v_0, $P \xrightarrow{a(v)} P'$, with $v \in val(P, Q, Eq) \cup \{v_0\}$, implies $Q \xrightarrow{a(v)} Q'$ for some Q' s.t. $P' \mathcal{R} Q'$.

Define $P \sim_F Q$ if and only if $P \mathcal{R} Q$ for some F-bisimulation $\mathcal{R}$.

Intuitively, doing case-analysis on input actions by considering just one fresh value suffices, because, under certain conditions, bisimulation is preserved by replacements of values with fresh values. Indeed, we have:

Theorem 3. $P \sim Q$ *if and only if* $P \sim_F Q$.

2.2 Complexity Classes and Hard Problems

In the paper, we will measure the complexity of deciding bisimilarity for P and Q with a set of identifier definitions Eq, in function of the sum of the syntactical sizes of P, Q and of the processes occurring in Eq.

We will deal with the complexity classes P, NP, coNP and PSPACE, and with the notions of *polynomial reducibility*, *hardness* and *completeness*. It is known that P $\subseteq$ NP, coNP $\subseteq$ PSPACE, and it is strongly conjectured that all these classes are distinct. A problem is hard for a class $\mathcal{C}$ if every problem in $\mathcal{C}$ is polynomial-time reducible to it; a $\mathcal{C}$-hard problem is said to be $\mathcal{C}$-complete if it belongs to $\mathcal{C}$. Formal definitions can be found in any textbook of computational complexity theory, such as [BC93, Pap94].

3 Data-Independent Calculi

In this section we will deal with the complexity of the bisimulation problem in the three data-independent calculi. We will first restrict ourselves to the simple value-passing case (i.e. we assume that Var, Val and Act are pairwise disjoint) and we will argue how the achieved results apply to the name-passing case.

Theorem 4. *The bisimilarity problem for* $\mathcal{IL}_v$ *and* $\mathcal{IL}_r$ *is in* P.

Let us now consider the complexity of the bisimilarity problem in $\mathcal{IL}_{v,r}$. Jonsson and Parrow proved that such a problem is decidable and NP-hard [JP93]. We will improve on this result and we will show that the problem is indeed PSPACE-hard, by reducing to it a well-known PSPACE-complete problem. We first need some preliminary definitions in order to introduce *quantified boolean formulas*.

Let $U = \{x_1, \ldots, x_n\}$ be a set of boolean variables. If x is a variable in U, then x and $\neg x$ are said to be *literals*. A *(conjunctive) 3-clause* over U is the conjunction of three literals. A *formula in 3DNF* (disjunctive normal form) is the disjunction of a set of 3-clauses, e.g. $\phi \stackrel{def}{=} (x_1 \wedge \neg x_3 \wedge \neg x_4) \vee (x_2 \wedge \neg x_3 \wedge x_4)$. A *truth assignment* for U is a function $t : U \to \{\textbf{true}, \textbf{false}\}$. Given an assignment t, we associate in the usual way a truth value to literals, clauses and formulas. With a slight abuse of notation, we will admit that a literal may also be a member of the set $\{\texttt{true}, \texttt{false}, \neg\texttt{true}, \neg\texttt{false}\}$. Each assignment t will map these special literals to the expected truth values. Moreover, let $\phi\{b_1/x_1, \ldots, b_k/x_k\}$, where $b_i \in \{\texttt{true}, \texttt{false}\}$, be the formula obtained from ϕ by substituting b_i to x_i for $i = 1, \ldots, k$.

Definition 5. A *quantified boolean formula* (in short, QBF) is a formula $\phi = Q_1 x_1.Q_2 x_2 \ldots Q_n x_n.\phi'$, where ϕ' is a formula in 3DNF, $\{x_1, \ldots, x_n\}$ is the set of variables occurring in ϕ' and, for any $i = 1, \ldots, n$, $Q_i \in \{\exists, \forall\}$ is a *quantifier*.

Definition 6. A *quantified boolean formula* $\phi = Q_1 x_1.Q_2 x_2 \ldots.Q_n x_n.\phi'$ is *valid* if one of the following conditions holds:

1. $n = 0$ and in ϕ' there is a true clause c;
2. $Q_1 = \exists$, and $Q_2 x_2 \ldots Q_n x_n.\phi'\{\texttt{true}/x_1\}$ or $Q_2 x_2 \ldots Q_n x_n.\phi'\{\texttt{false}/x_1\}$ is valid;
3. $Q_1 = \forall$, and $Q_2 x_2 \ldots Q_n x_n.\phi'\{\texttt{true}/x_1\}$ and $Q_2 x_2 \ldots Q_n x_n.\phi'\{\texttt{false}/x_1\}$ are both valid.

Given a QBF ϕ, the QBF problem consists of deciding whether ϕ is valid: this is a PSPACE-complete problem [SM73], and it is easy to see that it remains PSPACE-complete even when restricted to formulas $\phi = Q_1 x_1 Q_2 x_2 \ldots, Q_n x_n.\phi'$ such that n is even and $Q_i = \exists$ if and only if i is odd. Let us call RQBF this restricted problem. We now come to describing the actual reduction.

Let $\phi = \exists x_1.\forall x_2.\ldots.\forall x_n.\phi'$ be an instance of RQBF, where $\phi' = c_1 \vee \ldots \vee c_m$, and $c_i = l_i^1 \wedge l_i^2 \wedge l_i^3$ for $i = 1, \ldots, m$. Let us define the processes $B_0, \ldots, B_n$, $T_0, \ldots, T_n$, $E_0, \ldots, E_n$ as shown in Table 3. There, in the clause for B_n, i_j is the index of the variable x_{i_j} occurring in literal l_i^j, and $w_{i_j} = y_{i_j}$ if $l_i^j = x_{i_j}$, while $w_{i_j} = z_{i_j}$ if $l_i^j = \neg x_{i_j}$.

We will prove that $B_0 \sim T_0$ if and only if ϕ is valid. The proof is split into three technical lemmas.

Lemma 7. *For any i, $0 \leq i \leq n$ and for any $(v_1, \ldots, v_i) \in \{\texttt{true}, \texttt{false}\}^i$, $B_i(v_1, \ldots, v_i, \neg v_1, \ldots, \neg v_i)$ is either bisimilar to E_i or bisimilar to T_i.*

Lemma 8. *For any i, $0 \leq i \leq n$, T_i is not bisimilar to E_i.*

Lemma 9. *Consider any i, $0 \leq i \leq n$ and any $(v_1, \ldots, v_i) \in \{\texttt{true}, \texttt{false}\}^i$. Let $\psi' \stackrel{def}{=} \phi'\{v_1/x_1, \ldots, v_i/x_i\}$, and define $\psi \stackrel{def}{=} Q_{i+1} x_{i+1} \ldots \forall x_n.\psi'$, where Q_{i+1} is the $(i + 1)$-th quantifier in ϕ. Then, ψ is valid if and only if $B_i(v_1, \ldots, v_i, \neg v_1, \ldots, \neg v_i) \sim T_i$.*

Proof. (Sketch) We proceed by induction on $n - i$. If $i = n$, then the proof is trivial. Now, fix any $(v_1, \ldots, v_i) \in \{\texttt{true}, \texttt{false}\}^i$ and let ψ' and ψ be as described in the hypothesis. We can assume by inductive hypothesis that for any $v_{i+1} \in \{\texttt{true}, \texttt{false}\}$, $B_{i+1}(v_1, \ldots, v_{i+1}, \neg v_1, \ldots, \neg v_{i+1}) \sim T_{i+1}$ if

$$E_n \Leftarrow \sum_{\substack{(v_1,v_2,v_3)\in\{\mathbf{true},\mathbf{false}\}^3 \\ (v_1,v_2,v_3)\neq(\mathbf{true},\mathbf{true},\mathbf{true})}} \bar{a}v_1.\bar{a}v_2.\bar{a}v_3$$

$$T_n \Leftarrow \sum_{(v_1,v_2,v_3)\in\{\mathbf{true},\mathbf{false}\}^3} \bar{a}v_1.\bar{a}v_2.\bar{a}v_3$$

$$B_n(y_1,\ldots,y_n,z_1,\ldots,z_n) \Leftarrow \sum_{i=1}^{m} \bar{a}w_{i_1}.\bar{a}w_{i_2}.\bar{a}w_{i_3} + \sum_{\substack{(v_1,v_2,v_3)\in\{\mathbf{true},\mathbf{false}\}^3 \\ (v_1,v_2,v_3)\neq(\mathbf{true},\mathbf{true},\mathbf{true})}} \bar{a}v_1.\bar{a}v_2.\bar{a}v_3$$

For any even i, $0 \leq i \leq n-2$:

$$B_i(y_1,\ldots,y_i,z_1,\ldots,z_i) \Leftarrow a.B_{i+1}(y_1,\ldots,y_i,\mathbf{true},z_1,\ldots,z_i,\mathbf{false}) + $$
$$a.B_{i+1}(y_1,\ldots,y_i,\mathbf{false},z_1,\ldots,z_i,\mathbf{true}) + a.E_{i+1}$$
$$T_i \Leftarrow a.T_{i+1} + a.E_{i+1}$$
$$E_i \Leftarrow a.E_{i+1}$$

For any odd i, $1 \leq i \leq n-1$:

$$B_i(y_1,\ldots,y_i,z_1,\ldots,z_i) \Leftarrow a.B_{i+1}(y_1,\ldots,y_i,\mathbf{true},z_1,\ldots,z_i,\mathbf{false}) + $$
$$a.B_{i+1}(y_1,\ldots,y_i,\mathbf{false},z_1,\ldots,z_i,\mathbf{true}) + a.T_{i+1}$$
$$T_i \Leftarrow a.T_{i+1}$$
$$E_i \Leftarrow a.E_{i+1} + a.T_{i+1}$$

Table 3. The reduction from QBF to bisimilarity in $\mathcal{IL}_{v,r}$.

and only if $Q_{i+2}x_{i+2},\ldots,\forall x_n.\psi'\{v_{i+1}/x_{i+1}\}$ is valid. We have to distinguish two cases, depending on whether i is odd or even. Let i be even (the other case is similar), then $Q_{i+1} = \exists$. Due to lemmas 7 and 8, we have that $B_i(v_1,\ldots,v_i,\neg v_1,\ldots,\neg v_i) \sim T_i$ if and only if a value $v_{i+1} \in \{\mathbf{true},\mathbf{false}\}$ exists such that $B_{i+1}(v_1,\ldots,v_{i+1},\neg v_1,\ldots,\neg v_{i+1}) \sim T_{i+1}$. By inductive hypothesis, the latter holds if and only if either $\forall x_{i+2}\ldots\forall x_n.\psi'\{\mathbf{false}/x_{i+1}\}$ is valid or $\forall x_{i+2}\ldots\forall x_n.\psi'\{\mathbf{false}/x_{i+1}\}$ is valid, that is, if and only if $\psi = \exists x_{i+1}\ldots\forall x_n.\psi'$ is valid. $\square$

The following corollary is just a special case ($i = 0$) of the previous lemma.

Corollary 10. $B_0 \sim T_0$ if and only if ϕ is valid.

The definition of the identifiers can be easily constructed in polynomial time, thus it immediately follows the main result of this section.

Theorem 11. *The bisimilarity problem in $\mathcal{IL}_{v,r}$ is PSPACE-hard*

Let us now consider the name-passing case arising when $Act = Val = Var$. The results regarding $\mathcal{IL}_v$ and $\mathcal{IL}_{v,r}$ still apply, while bisimilarity in $\mathcal{IL}_r$ can be shown to be PSPACE-hard by using the reduction of Table 3, provided that we replace the output action $\bar{a}v$ with the simple action v.

4 Data-Dependent Value-Passing

In this section we will show that the bisimilarity problem for the calculus $\mathcal{L}_v$ is coNP-complete. We will first present a reduction from the coNP-complete problem 3-TAUTOLOGY, thus establishing the coNP-hardness of the bisimilarity problem. Then we will show that it belongs to the class coNP.

The 3-TAUTOLOGY problem consists in testing whether a given formula ϕ in 3DNF (see the preceding section) is a tautology or not. From the results of [Coo71] it follows that any problem in coNP is polynomial-time reducible to 3-TAUTOLOGY, that is, the 3-TAUTOLOGY problem is coNP-hard.

Theorem 12. *The bisimilarity problem in $\mathcal{L}_v$ is coNP-hard.*

Proof. (Sketch) It is sufficient to prove that the 3-TAUTOLOGY problem is polynomial-time reducible to the bisimilarity problem in $\mathcal{L}_v$. Let $\phi = c_1 \vee \ldots \vee c_m$ be an instance of 3-TAUTOLOGY over the set of variables $\{x_1, \ldots, x_n\}$, let $c_i = l_i^1 \wedge l_i^2 \wedge l_i^3$ for $i = 1, \ldots, m$, and let x_{i_j} be the variable occurring in literal l_j^i. Let also b_{i_j} stand for **true** if $l_i^j = x_{i_j}$, and for **false** otherwise. Consider the processes $P(\phi)$, Q, P', Q' as defined in Table 4.

$$P(\phi) \stackrel{def}{=} a(y_1)\ldots a(y_n).P'$$
$$Q \stackrel{def}{=} a(y_1)\ldots a(y_n).Q'$$
$$P' \stackrel{def}{=} a.a + \sum_{i=1}^{n} a.([y_i = \text{true}]a + [y_i = \text{false}]a)$$
$$\qquad + \sum_{i=1}^{m} [y_{i_1} = b_{i_1}][y_{i_2} = b_{i_2}][y_{i_3} = b_{i_3}]a$$
$$Q' \stackrel{def}{=} a + a.a$$

Table 4. The reduction from 3-TAUTOLOGY to bisimilarity in $\mathcal{L}_v$.

It is possible to prove that ϕ is a tautology if and only if, for any $(v_1, \ldots, v_n) \in Val^n$, $P'\{v_1/y_1, \ldots, v_n/y_n\} \sim Q'$. This implies that $Q \sim P(\phi)$ if and only if ϕ is a tautology. By observing that Q and $P(\phi)$ are computable in polynomial time in the size of ϕ, the theorem follows. $\qquad\square$

Theorem 13. *The bisimilarity problem in $\mathcal{L}_v$ is in coNP.*

Proof. (Sketch) We prove that the *inequivalence* problem (given P, Q in $\mathcal{L}_v$, decide whether $P \not\sim Q$) is in NP. To this aim, it is sufficient to consider the nondeterministic algorithm in Figure 1 and to show that the following properties hold:

1. the algorithm runs in polynomial time (in the sizes of the terms);
2. if $P \sim Q$ all computations of the algorithm lead to rejection;

3. if $P \not\sim Q$ there exists a computation of the algorithm leading to acceptance.

Due to lack of space, details are omitted. The only subtle point is the third one, where also Theorem 3 is exploited. □

Algorithm **Non-equiv**
Input: P, Q
begin
  if not $B(P,Q)$ then **accept**
  else **reject**
end

$B(P,Q)$
begin
  Fix v_0 fresh; **guess** $v \in val(P,Q) \cup \{v_0\}$;
  $I := \{(\mu, P') \mid P \xrightarrow{\mu} P'$ and if μ is an input action then $\mu = a(v)$, for some $a \}$;
  $J := \{(\mu, Q') \mid Q \xrightarrow{\mu} Q'$ and if μ is an input action then $\mu = a(v)$, for some $a \}$;
  for each $\big((\mu, P'), (\mu, Q')\big) \in I \times J$ **do** $b(\mu, P', Q') := B(P', Q')$;
  return $\big(\forall(\mu, P') \in I. \exists(\mu, Q') \in J : b(\mu, P', Q') \wedge$
        $\forall(\mu, Q') \in J. \exists(\mu, P') \in I : b(\mu, P', Q') \big)$
end

Fig. 1. A nondeterministic algorithm for detecting inequivalence of processes in $\mathcal{L}_v$.

Corollary 14. *The bisimilarity problem in $\mathcal{L}_v$ is* coNP-*complete.*

5 Reducing Value-passing to Identifiers and Matching

We will exhibit a polynomial-time reduction of $\mathcal{L}_{v,r}$ to $\mathcal{L}_r$. It is convenient here to separate the case of simple value-passing (Val, Var and Act disjoint) and the case of name-passing ($Var = Val = Act$). We first deal with simple value-passing, and then indicate the necessary modifications to accommodate name-passing.

We will first give an informal account of the translation. As a first approximation, the idea is to express each input process $a(x).P$ as a nondeterministic sum $\sum_{v \in V} av.P\{v/x\}$. Here, each av is a pure action uniquely associated with the channel a and the value v; V is a set of values, which is finite, but large enough to represent all "relevant" input actual parameters. The idea stems from Definition 2 and from Milner's translation of CCS with values into pure CCS (with infinite summation [Mil89]). However, in the presence of nested input actions, this solution would give rise to an exponential explosion of the size of translated term. To overcome this drawback, we exploit the ability of identifiers of handling parameters. The idea is to translate $a(x).P$ as $\sum_{v \in V} av.A(v)$, where A is an auxiliary identifier defined by $A(x) \Leftarrow T$ and T is the translation of the subterm P.

We assume an arbitrarily large supply of auxiliary identifiers of arity j, A_1, A_2, A_3, ..., for any $j \geq 0$, each distinct from the identifiers defined in Eq. These auxiliary identifiers will be ranged over by the letter A.

The actual translation consists of two parts: for each term P in $\mathcal{L}_{v,r}$, we have to specify, in $\mathcal{L}_r$, a term $[\![P]\!]$, and a set of identifiers definitions, $\mathcal{D}(P)$, which defines the auxiliary identifiers occurring in $[\![P]\!]$. The definitions of $[\![P]\!]$ and $\mathcal{D}(P)$ are reported in Table 5. The definition of $[\![P]\!]$ is parametric with a chosen non-empty set $V_0 \subseteq_{fin} Val$ of values, appearing in the clauses for input and output prefixes. Thus we should have written $[\![P]\!]_{V_0}$ in place of $[\![P]\!]$; we have omitted the subscript V_0 as no confusion can arise. Note that the definition of $[\![P]\!]$ does not depend on that of $\mathcal{D}(P)$, while the latter does depend on the former.

$[\![P]\!]$ is defined as:

$$[\![c.P]\!] = c.[\![P]\!]$$

$$[\![\bar{a}v.P]\!] = \bar{a}v.[\![P]\!]$$

$$[\![\bar{a}x.P]\!] = \sum_{v \in V_0}[x = v]\bar{a}v.A(\tilde{y})$$
$$\text{where } \tilde{y} = fvar([\![P]\!])$$

$$[\![a(x).P]\!] = \sum_{v \in V_0} av.A(\tilde{y}, v)$$
$$\text{where } \tilde{y} = fvar([\![P]\!]) - \{x\}$$

$$[\![[e_1 = e_2]P]\!] = [e_1 = e_2][\![P]\!]$$

$$[\![\sum_{i \in I} P_i]\!] = \sum_{i \in I}[\![P_i]\!]$$

$$[\![Id(\tilde{e})]\!] = Id(\tilde{e})$$

$\mathcal{D}(P)$ is defined as:

$$\mathcal{D}(c.P) = \mathcal{D}(P)$$

$$\mathcal{D}(\bar{a}v.P) = \mathcal{D}(P)$$

$$\mathcal{D}(\bar{a}x.P) = \{A(\tilde{y}) \Leftarrow [\![P]\!]\} \cup \mathcal{D}(P)$$

$$\mathcal{D}(a(x).P) = \{A(\tilde{y}, x) \Leftarrow [\![P]\!]\} \cup \mathcal{D}(P)$$

$$\mathcal{D}([e_1 = e_2]P) = \mathcal{D}(P)$$

$$\mathcal{D}(\sum_{i \in I} P_i) = \bigcup_{i \in I} \mathcal{D}(P_i)$$

$$\mathcal{D}(Id(\tilde{e})) = \emptyset$$

Table 5. The reduction of $\mathcal{L}_{v,r}$ to $\mathcal{L}_r$.

The translation has to be applied to the set of identifiers definitions, Eq, as follows:

Definition 15. Let us define $\mathcal{D}(Eq)$ as

$$\bigcup_{Id(\tilde{y}) \Leftarrow P \in Eq} (\{Id(\tilde{y}) \Leftarrow [\![P]\!]\} \cup \mathcal{D}(P)).$$

The reduction proof can be split in two parts: *completeness* (if $P \sim Q$ in $\mathcal{L}_{v,r}$, then their translations are bisimilar in $\mathcal{L}_r$) and *correctness* (if the translations of P and Q are bisimilar in $\mathcal{L}_r$, then P and Q are bisimilar in $\mathcal{L}_{v,r}$).

Theorem 16 (Completeness). *For any two processes P_0 and Q_0 in $\mathcal{L}_{v,r}$, if $P_0 \sim Q_0$ then $[\![P_0]\!] \sim [\![Q_0]\!]$ in $\mathcal{L}_r$ equipped with the set of identifier definitions $\mathcal{D}(P_0) \cup \mathcal{D}(Q_0) \cup \mathcal{D}(Eq)$.*

Proof. (Sketch) For any two terms P and Q, define $P \prec Q$ if P is a *subterm* of Q, where the standard definition of subterm is extended by the axiom: $R\{\tilde{v}/\tilde{x}\} \prec Id(\tilde{v})$ if $Id(\tilde{x}) \Leftarrow R$ is in Eq.

Let $\mathcal{L}_r$ be equipped with the identifiers definitions $\mathcal{D}(P_0) \cup \mathcal{D}(Q_0) \cup \mathcal{D}(Eq)$. Over $\mathcal{L}_r$, consider the relation $\mathcal{R}$ defined thus:

$$\{([\![P]\!]\sigma_1, [\![Q]\!]\sigma_2) \mid [\![P]\!]\sigma_1 \text{ and } [\![Q]\!]\sigma_2 \text{ are closed}, P \prec P_0, \qquad (1)$$
$$Q \prec Q_0 \text{ and } P\sigma_1 \sim Q\sigma_2 \text{ in } \mathcal{L}_{v,r} \}. \qquad (2)$$

Then it is not difficult to prove that $\mathcal{R}$ is a bisimulation *up to* $\sim$ [Mil89] and, hence, $\mathcal{R} \subseteq \sim$. □

We now come to the correctness part. This is slightly more difficult, because we have to choose appropriately the parameter parameter V_0 of the translation. The choice depends also on whether or not Val is infinite. In the next theorem, we assume that Val is infinite; the case when Val is finite will be easily accommodated afterward. Intuitively, V_0 must contain all "relevant" values, i.e. all values appearing in the two processes being compared and in their subterms, *plus* a reserve of fresh values.

Theorem 17 (Correctness). *For any two processes P_0 and Q_0 in $\mathcal{L}_{v,r}$, if $[\![P_0]\!] \sim [\![Q_0]\!]$ in $\mathcal{L}_r$ equipped with $\mathcal{D}(P_0) \cup \mathcal{D}(Q_0) \cup \mathcal{D}(Eq)$, then $P_0 \sim Q_0$.*

Proof. (Sketch) Let the parameter V_0 of the translation be set as $V_0 = val(P_0, Q_0, Eq) \cup V$, for some $V \subseteq_{fin} Val$ s.t. $V \cap val(P_0, Q_0, Eq) = \emptyset$ and $|V| = |P_0| + |Q_0| + \sum_{Id(\tilde{x}) \Leftarrow R \in Eq} |R|$. Over $\mathcal{L}_{v,r}$, define the relation $\mathcal{R}$ as follows:

$$\{(P\sigma_1, Q\sigma_2) \mid P\sigma_1 \text{ and } Q\sigma_2 \text{ are closed}, P \prec P_0, Q \prec Q_0,$$
$$val(\sigma_1, \sigma_2) \subseteq V_0 \text{ and } [\![P]\!]\sigma_1 \sim [\![Q]\!]\sigma_2 \text{ in } \mathcal{L}_r\}.$$

It is not hard to show that $\mathcal{R}$ is an F-bisimulation, and hence (by Theorem 3) $P_0 \sim Q_0$. □

If Val is finite, then the above theorem can be proven by just letting $V_0 = Val$.

It is easily seen that the translation can be carried out in polynomial-time with the size of the problem. Thus, putting together Theorems 16 and 17, we get:

Theorem 18. *The equivalence problem in $\mathcal{L}_{v,r}$ is polynomial-time reducible to the equivalence problem in $\mathcal{L}_r$. Consequently, the equivalence problem in $\mathcal{L}_r$ is* PSPACE-*hard.*

We indicate now the necessary changes to accommodate the name-passing case ($Act = Var = Val$). In a name-passing input action $a(x).$, not only the formal parameter x, but also the channel a is subject to be possibly instantiated. It then suffices to replace the output (both $\bar{a}v.$ and $\bar{a}x.$) clauses and the input clause of the definition of $[\![\,.\,]\!]$ in Table 5 with the following two:

$$[\![a(x).P]\!] = \sum_{v \in V_0}[a = v]\sum_{w \in V_0} vw.A(\tilde{y}, w) \qquad \text{where } \tilde{y} = fvar([\![P]\!]) - \{x\}$$

$$[\![\bar{a}y.P]\!] = \sum_{v \in V_0}[a = v]\sum_{w \in V_0}[y = w]\overline{vw}.A(\tilde{y}) \quad \text{where } \tilde{y} = fvar([\![P]\!]).$$

The remaining clauses and the definition of $\mathcal{D}$ are left unchanged. It is easy to see that the translation is still polynomial and that the reduction proofs carry over essentially without modifications.

6 Conclusions

In this paper we have studied the decidability and the complexity of bisimilarity in fragments of CCS with values and of the π-calculus. We considered both a data-independent setting, in which processes are allowed to send and receive data, but cannot do any test on them, and a simple data-dependent one, in which processes can only perform equality tests. In the literature, some variant form of bisimulation have been proposed, such as *late* bisimilarity [MPW92, PS93] and *open* bisimilarity [San93]. Most of the results presented in previous sections extend to these equivalences. In particular, both late and open bisimilarity are PSPACE-hard over the data-independent processes, because the three equivalences coincide in this case (see e.g. [PS93]).

Our paper is mainly related to [JP93]. There, Jonsson and Parrow prove that bisimilarity is decidable in the data-independent language $\mathcal{IL}_{v,r}$, by showing that the infinitely many transitions due to an input action can be reduced to a single, suitably chosen, *schematic* action [JP93]. The latter is characterized as the receipt of the least value (w.r.t. to a fixed ordering of values) not "used" in the considered process. This approach yields the polynomial-time tractability of some restricted cases. On the other hand, the technique cannot be used in a data-dependent setting, mainly because in the presence of the equality test, determining the set of "used" values of a process becomes very complex (perhaps undecidable). In this paper, we have taken a less radical approach to deal with the infinite-state problem: instead of substituting infinitely many actions with a single one, we replace them with a "moderate" number of actions (the ones corresponding to the set V_0). Jonsson and Parrow also show that $\mathcal{IL}_{v,r}$ is NP-hard, by means of a quite involved reduction from the CLIQUE problem. Here, we have for the same language a stronger result with an easier technique.

In [HL95, HL93, San93, BD94, EL95], notions of *symbolic* bisimulation are investigated for both CCS with value-passing and π-calculus, aiming at a more efficient representation of bisimilarity. Our results show that, even for very simple fragments, it is very unlikely that efficient algorithms exist. It remains to be seen whether symbolic techniques give some benefits on the average.

A question that is left open by the present work is the exact complexity of bisimilarity in $\mathcal{IL}_{v,r}$ and $\mathcal{L}_{v,r}$. Moreover, other interesting fragments of CCS with values should be considered from a complexity point of view. For example, the parallel composition operator | has been considered in the case of *trace equivalence* [MS94], but nothing is known regarding bisimilarity.

References

[BC93]   D.P. Bovet and P. Crescenzi. *Introduction to the Theory of Complexity.* Prentice Hall, 1993.

[BD94]   M. Boreale and R. De Nicola. A symbolic semantics for the π-calculus. In *Proc. of CONCUR '94, LNCS 836*, pages 299–314. Springer-Verlag, 1994.

[Coo71]   S.A. Cook. The complexity of theorem proving procedures. In *Proc. of STOC '71*, pages 151–158, 1971.

[EL95]   U.H. Engberg and K.S. Larsen. Efficient simplification of bisimulation formulas. In *Proc. of TACAS '95*, pages 89–103, 1995.

[HL93]   M. Hennessy and H. Lin. Proof systems for message-passing process algebras. In *Proc. of CONCUR '93, LNCS 715*. Springer-Verlag, 1993.

[HL95]   M. Hennessy and H. Lin. Symbolic bisimulations. *Theoretical Computer Science*, 138:353–389, 1995.

[JP93]   B. Jonsson and J. Parrow. Deciding bisimulation equivalences for a class of non-finite state programs. *Information and Computation*, 107:272–302, 1993.

[KS90]   P.C. Kannellakis and S.A. Smolka. CCS expressions, finite state processes, and three problems of equivalence. *Information and Computation*, 86:43–68, 1990.

[Mil80]   R. Milner. *A Calculus of Communicating Systems.* LNCS, 92. Springer-Verlag, 1980.

[Mil89]   R. Milner. *Communication and Concurrency.* Prentice-Hall, 1989.

[MPW92]   R. Milner, J. Parrow, and D. Walker. A calculus of mobile processes, part I and II. *Information and Computation*, 100:1–41 and 42–78, 1992.

[MS94]   A.J. Mayer and L.J. Stockmeyer. The complexity of word problems – this time with interleaving. *Information and Computation*, 115:293–311, 1994.

[Pap94]   C.H. Papadimitriou. *Computational Complexity.* Addison-Wesley, 1994.

[PS93]   J. Parrow and D. Sangiorgi. Algebraic theories for name-passing calculi. Technical Report ECS-LFCS-93-262, University of Edinburgh, Department of Computer Science, 1993. To appear in *Information and Computation*.

[PT87]   R. Paige and R.E. Tarjan. Three partition refinement algorithms. *SIAM Journal on Computing*, 16(6):973–989, 1987.

[San93]   D. Sangiorgi. A theory of bisimulation for the π-calculus. In *Proc. of CONCUR '93, LNCS 715*. Springer-Verlag, 1993. To appear in *Acta Informatica*.

[SM73]   L. Stockmeyer and A. Meyer. Word problems requiring exponential time. In *Proc. of STOC '73*, pages 1–9, 1973.

On the Expressive Power of CCS

Ashvin Dsouza and Bard Bloom

Department of Computer Science
Cornell University
Ithaca, NY 14853, USA

Abstract. In the context of structured operational semantics, a useful
measure of the expressive power of a process algebra $\mathcal{P}$ is the class of
operator specifications definable in $\mathcal{P}$ up to a given equivalence. Our
goal in this study is to characterize this class of operators for CCS up to
strong and branching bisimulation. We use our results to motivate modifications to CCS to eliminate awkward constraints on expressiveness. In
particular, we present a strong case for adding a checkpointing operator.

1 Introduction

Process algebras are increasingly being used to formally specify and verify concurrent systems. Since there are many different settings for concurrency, *e.g.*
broadcast versus point-to-point communication and synchrony versus asynchrony,
a host of process algebras has emerged, including CCS/SCCS[Mil89], CSP[Hoa85],
ACP[BW91], LOTOS[BB87] and MEIJE[Bou85]. However, these process algebras have much in common, which leads to duplication of effort when studying them independently. This variety also makes it difficult to choose one of
these formalisms when modelling a system. These problems can be alleviated
by investigating the metatheory of process algebras to explore their underlying
generalities and relationships.

The behavior of most process algebras is defined using structured operational
semantics (SOS), whereby the behavior of a composite process is determined
from the behavior of its component processes. Such rules are easy to read, and
lend themselves readily to proof techniques such as structural induction. In this
setting, the metatheory of process algebras we use places syntactic constraints
on the rules that may be used to define process operators. By showing these
constraints to be sufficient for a particular property to hold, *e.g.* bisimulation
equivalence as a congruence, we establish a class of operators associated with
that property[dS85, BIM88]. This is analogous to the way in which constraints on
grammars give rise to regular, context-free, context sensitive and *r.e.* languages,
and the resulting benefits are similar.

Example 1. In [BIM88], the GSOS operators are proposed, which are specified
by finitely many rules of the following format:

$$\frac{\bigcup_{i=1}^{l}\left\{x_i \overset{a_{i,j}}{\to} y_i \mid 1 \leq j \leq m_i\right\} \cup \bigcup_{i=1}^{l}\left\{x_i \overset{b_{i,j}}{\nrightarrow} \mid 1 \leq j \leq n_i\right\}}{f(\vec{x}) \overset{a}{\to} C[\vec{x}, \vec{y}]}$$

where C, the *target* of the rule, is a term consisting of GSOS operators. A language designer who draws on this class of operators is guaranteed a well behaved operational semantics and bisimulation as a congruence. Another rule format with this property is is *tyft/tyxt*[GV89].

A natural question to ask of a class of operators $\mathcal{O}$ is whether there is a subclass $\mathcal{L}$ of $\mathcal{O}$ such that, for a given equivalence R on $\mathcal{O}$-terms, any operator in $\mathcal{O}$ can be simulated by a $\mathcal{L}$-term up to R. Hence, $\mathcal{L}$ contains all the expressive power of $\mathcal{O}$. [DB93] considers this for restrictions of GSOS.

The question can be turned around: Given $\mathcal{L}$, what is the class of operators $\mathcal{O}$ that can be realized using $\mathcal{L}$? In [dS85], MEIJE is used to program any operator specification of the following format up to strong bisimulation:

$$\frac{\left\{ x_i \xrightarrow{a_i} y_i \right\}_{i \in I}, \quad (\vec{a}, a) \in R}{f(\vec{x}) \xrightarrow{a} t[\vec{z}]} \tag{1}$$

where t is a term consisting of operators from this class, z_i is y_i if $i \in I$ and x_i otherwise, and R is a recursively enumerable relation. In [Mil89], Milner comments on the need for similar work for the asynchronous system CCS, which we address here. The benefits of carrying out such a study of expressiveness inoclude:

Language Insights: Determining the class of operator specifications definable in $\mathcal{L}$ requires examining the way the terms of $\mathcal{L}$ compute; this increases our understanding of $\mathcal{L}$.

Language Extensions: When writing a specification using a process algebra one can freely draw on the class of definable operators to model concepts from the problem domain at a high level[Blo95]. This improves the readability of specifications, and the new operators can be compiled into the basis language to be processed by basis-specific tools.

Language Comparison Criteria: Two languages can be compared using the classes of operator specifications definable in each of them. Alternatively, one could simply establish the interdefinability of their operators – *e.g.* [Bou85] does this for MEIJE and SCCS. It follows directly from our results that broadcasting is not definable in CCS up to strong or branching bisimulation, unless we extend CCS with checkpointing.

Language Design Hints: If minor changes to a language greatly increase the class of definable operators, or significantly simplify its characterization, then incorporating these changes should also make programming in the language more effective. This is similar to the extension of PCF with parallel-or[Plo77], where the new operator was motivated by a semantic argument.

1.1 Contributions of this Study

In this paper we investigate the operator specifications definable in CCS up to both strong and branching bisimulation. The operator class we encode in

CCS up to strong bisimulation includes all the CCS operators, and permits many useful specifications, *e.g.* a two-process mutual exclusion operator. A useful auxiliary result is a standard form for the subset of CCS consisting of the static CCS operators – parallel composition, restriction and renaming – augmented with finite state machines. We also suggest "splitting" the CCS complement and allowing finer restriction to eliminate some of the constraints on the definable operators.

Unfortunately, strong bisimulation is too fine an equivalence, given the asynchronous nature of CCS, and we are unable to get a cleanly characterized class of definable operators. When we switch to branching bisimulation we are able to obtain a simpler characterization, but we are forced to introduce a *trigger closure* restriction. This says that for every rule ρ for an operator f, and for each subsequence of antecedents of ρ, there must also be a rule η for f with this subsequence as its antecedents.

The reason we require trigger closure is that since CCS is asynchronous, the CCS encoding of f can only test the antecedents of the rules for f one at a time and will deadlock if it successfully tests some antecedents and then a subsequent one fails. Trigger closure guarantees that some rule is applicable at every stage during sequential antecedent testing.

We eliminate the trigger closure requirement without compromising the asynchronous nature of CCS by extending CCS with a checkpointing operator. This takes a running process and a stored process as arguments, reserving actions c and r to store the running process and to run the stored process respectively:

$$\frac{x \xrightarrow{\alpha} x', \ \alpha \in \{\tau\} \cup \mathsf{Act}\backslash\{r,c\}}{x \text{ was } y \xrightarrow{\alpha} x' \text{ was } y} \qquad \frac{}{x \text{ was } y \xrightarrow{c} x \text{ was } x} \qquad \frac{}{x \text{ was } y \xrightarrow{r} y \text{ was } y} \quad (2)$$

Interestingly, Milner has commented on CCS that "there is nothing canonical about [his] choice of basic combinators"[Mil89], and actually uses the same checkpointing operator as an example of a useful operator he did not include.

Related Work: Operators defined by rules of the format (1) are programmed up to strong bisimulation using MEIJE in [dS85]. A similar encoding of the format (1) using the guarded calculus PC is described in [Vaa93]. However, both settings are synchronous and consider strong bisimulation only.

In [Par90], a static asynchronous version of the format (1) is programmed up to weak bisimulation using a data flow network based algebra. This setting is somewhat different from CCS, and less intuitive – properties such as the representability of finite state transition systems are not immediately obvious. Also, the definable rules can't introduce nesting or a change of arity.

A notion of expressiveness that we do not consider here is the class of transition systems represented by the terms of a process algebra via its operational semantics. This is discussed in [Vaa93].

1.2 Organization

In Section 2, we describe the version of CCS that we use in this paper. Then, in Section 3, we establish a rule format that can be programmed in CCS up to strong bisimulation. In Section 4, we change the setting to branching bisimulation and describe how to encode in CCS a subset of the "simply BB-cool" operators for which branching bisimulation is a congruence. We also show how a checkpointing operator can eliminate trigger closure. All construction details can be found in [DB95].

2 CCS

We use CCS with a finite action set Act closed under complement, guarded recursion, and without infinite summation. By convention, variables $a, b, c, \ldots$ range over Act, while $\alpha, \beta, \gamma, \ldots$ range over Act $\cup \{\tau\}$. We assume a set $\mathcal{P}$ of process variables, and a set $\mathcal{X}$ of recursion variables that is disjoint from $\mathcal{P}$. The CCS terms are generated by the grammar:

$$t ::= p \mid X \mid 0 \mid \alpha.t \mid t + t \mid t|t \mid t\backslash B \mid t[f] \mid \mu X.t$$

where $p \in \mathcal{P}$, $X \in \mathcal{X}$ and can only occur in a term of the form $\mu X.t$, $B \subseteq$ Act, and f is a mapping from Act to Act respecting complements and preserving τ. The operational semantics is the standard one for CCS:

$$\frac{}{\alpha.p \xrightarrow{\alpha} p} \qquad \frac{p \xrightarrow{\alpha} p'}{p + q \xrightarrow{\alpha} p'} \qquad \frac{q \xrightarrow{\alpha} q', \; f(\alpha) = \beta}{q[f] \xrightarrow{\beta} q'[f]} \qquad \frac{p \xrightarrow{\alpha} p', \; \alpha, \overline{\alpha} \notin A}{p\backslash A \xrightarrow{\alpha} p'\backslash A}$$

$$\frac{p \xrightarrow{\alpha} p'}{p|q \xrightarrow{\alpha} p'|q} \qquad \frac{p \xrightarrow{a} p', \; q \xrightarrow{\overline{a}} q'}{p|q \xrightarrow{\tau} p'|q'} \qquad \frac{t[X := \mu X.t] \xrightarrow{\alpha} t'}{\mu X.t \xrightarrow{\alpha} t'}$$

where we omit the symmetric rules for $|$ and $+$.

In this paper, we restrict our attention to terms that do not create copies of their process variables. This is an assumption common to most expressiveness results of this kind, because a static structure in the underlying language is used to encode the definable operators. The only exception we are aware of is [DB93].

Definition 1. A CCS term t is *linear* if no process variable occurs more than once in t, and no process variable occurs in the scope of a recursion variable, *i.e.* in a subterm of the form $\mu X.t$.

We will extend Act where necessary with a finite number of tagged copies of it for coding purposes, following [dS85]. All morphisms only specified over Act will be taken to be the identity on this extension. In addition, we assume that process variables may only be instantiated with terms that produce actions from Act. Figure 1 summarizes the names of the operator classes we introduce in this paper and their underlying terms.

2.1 CCS Definability

A naïve approach to determining a class of CCS definable operators is to simply enumerate the terms of CCS, and assign an operator symbol to each of them, with rules corresponding to the transitions of the terms. However, this does not provide a decision procedure for whether an operator belongs to this class.

Given an equivalence relation R on CCS terms, our approach to determining the definable operators up to R consists of 3 steps:

Name	Linear Terms	$\subseteq$ CCS?	Operators	Equivalence
stCCS	using static CCS operators and finite state machines	yes	SSF	$\sim$
stCCS$^+$	stCCS and limited use of $+$	yes	SSF$^+$	$\sim$
CCS^l	CCS with trigger-closure	yes	tcSB^l	$\approx_b$
wasCCS^l	CCS with checkpointing	no	SB^l	$\approx_b$

Fig. 1. Definable operators and represented terms.

1. Obtain the most general rule format the operators could take, $\mathcal{F}_0$, by looking at how CCS terms compute.
2. Determine the constraints necessary to guarantee that R will be a congruence on $\mathcal{F}_0$. This is to ensure that the new operators can be used freely with the existing operators.
3. Encode as much of the result of step 2 as possible. We proceed by establishing a *standard form* for terms and using it for the encoding.

3 CCS-definability up to Strong Bisimulation

Strong bisimulation[Par81] is a branching equivalence that relates two processes if they can make the same decisions at the same times. It is the finest generally accepted notion of process equivalence.

Definition 2. A symmetric binary relation R between processes is a *strong bisimulation relation* iff, whenever we have $(p, p') \in R$ then $p \xrightarrow{a} q$ for some action a and process q implies that $p' \xrightarrow{a} q'$ for some process q' such that $(q, q') \in R$ holds. Processes p and p' are strongly bisimilar, $p \sim p'$, iff there is some strong bisimulation relation R such that $(p, p') \in R$.

Following the procedure outlined in Section 2.1, we proceed by proving a *ruloid lemma* to describe the behavior of CCS terms. A ruloid [BIM88] is similar to a rule but *describes* when a term can make a transition, rather than *specifying* when an operator can make a transition, in terms of process variable transitions. For example, the ruloids for the term $t = ((x_1|(a.0)) + x_2)$ are:

$$\frac{x_1 \xrightarrow{\alpha} y_1}{t \xrightarrow{\alpha} y_1|(a.0)} \qquad \frac{x_1 \xrightarrow{\bar{a}} y_1}{t \xrightarrow{\tau} y_1|0} \qquad \frac{}{t \xrightarrow{a} x_1|0} \qquad \frac{x_2 \xrightarrow{\alpha} y_2}{t \xrightarrow{\alpha} y_2}$$

Given an action α and term t, the following ruloid lemma describes precisely when t can make an α-transition.

Lemma 3. *If* $t[\vec{x}]$ *is a CCS term then there is a finite set of ruloids of the forms:*

$$\frac{}{t \xrightarrow{\alpha} t_1} \qquad \frac{x_i \xrightarrow{\beta} y_i}{t \xrightarrow{\alpha} t_2} \qquad \frac{x_i \xrightarrow{a} y_i, \; x_j \xrightarrow{b} y_j}{t \xrightarrow{\tau} t_3}$$

where the $\vec{x}$, $\vec{y}$ *are distinct,* $\alpha, \beta \in \text{Act} \cup \{\tau\}$ *and* $a, b \in \text{Act}$, *that describes exactly when* $t[\vec{x}]$ *can make a transition, where* y_i *must occur exactly once in* t_2, *and both* y_i *and* y_j *must occur exactly once in* t_3.

It follows from Lemma 3 that an encoding scheme involving nested control such as that used in [dS85] is not possible with CCS up to strong bisimulation, *i.e.* the target of a definable rule cannot have any nesting of operators. This is because the CCS term encoding the outermost operator of a nested term has to test the moves of the nested subterms. This testing will generate τ actions that are are visible to strong bisimulation but not permitted by the rule specifications. Also taking into account the restriction to linear CCS terms (Definition 1), a rule format for operator specifications definable in CCS up to strong bisimulation cannot be more general than the following subset of the format of Lemma 3:

$$\frac{}{f(\vec{x}) \xrightarrow{\alpha} g(\vec{x})} \qquad \frac{x_i \xrightarrow{\beta} y_i}{f(\vec{x}) \xrightarrow{\alpha} g(\vec{z})} \qquad \frac{x_i \xrightarrow{a} y_i, \; x_j \xrightarrow{b} y_j}{f(\vec{x}) \xrightarrow{\tau} g(\vec{z})} \qquad (3)$$

where $|\vec{z}| \leq |\vec{x}|$, the $\vec{z}$ are distinct, and each z_k is either some x_l which is not in the antecedents, or some y_l. Since format (3) is a restriction of GSOS, it follows immediately from [BIM88] that strong bisimulation will be a congruence for operators of this format.

3.1 Static CCS

For simplicity, we initially consider only the subset of CCS generated by the static operators, *i.e.* parallel composition, restriction, and renaming, and finite state machines, which are definable in CCS using recursion. This subset, stCCS, contains the operators that give *structure* or *shape* to CCS terms: the parallel composition operator is like a bus linking the components, the renaming operator encodes signals entering a bus, and the restriction operator controls access to a bus. Including finite state machines – easily programmed in CCS using prefixing, choice and recursion – provides a control component.

More formally, stCCS comprises the linear CCS terms generated by the following grammar:

$$t ::= p \mid \Omega \mid t[\varphi] \mid t \backslash A \mid t|t$$

where $p \in \mathcal{P}$ and Ω ranges over finite state machines A nice property of stCCS terms is that they have a standard form, which is significantly more general than the *standard concurrent form* described in [Mil89]:

Lemma 4. *If* t *is a stCCS term, with process variables* $\vec{P}$, *then* t *is strongly bisimilar to a stCCS term of the form:*

$$(p_1[\varphi_1] | \ldots | p_n[\varphi_n] | \Omega_t) \backslash B[\psi]$$

where ψ *satisfies* $\psi(\alpha) = \tau \Leftrightarrow \alpha = \tau$, *and the* φ_i *map* Act *to coding actions outside* Act. *We refer to such terms as being in* static standard form.

Since we need a uniform translation from operator specifications to stCCS terms, and we know that the target of the translation can be expressed in static standard form, we can analyze this to derive the constraints to add to the rule format (3) to characterize the operators definable in stCCS. Consider the following term t in static standard form:

$$(x_1[\varphi_1]|\ldots|x_n[\varphi_n]|\Omega_t)\setminus B[\psi] \tag{4}$$

The finite state machine (fsm), Ω_t, is a control element that can synchronize with the x_i. Hence, a change of state of the fsm Ω_t corresponds to a change of operator. This leads to the following two rule forms:

$$\frac{}{f(\vec{x}) \xrightarrow{\alpha} g(\vec{x})} \qquad \frac{x_i \xrightarrow{\alpha} y_i}{f(\vec{x}) \xrightarrow{\tau} g(\vec{z})}$$

The first rule, abbreviated $(f : \alpha \nearrow g)$, can be encoded as an autonomous transition of Ω_t. The second rule, abbreviated $(f : i; \alpha \nearrow g)$, can be encoded as a synchronization between Ω_t and some x_i in term (4).

Actions emitted by the process arguments that are not in the restriction set B of term (4) get renamed before leaving the term; there is no change of state in the finite automaton, and hence the operator does not change. This gives us the rule format:

$$\frac{x_i \xrightarrow{a} y_i}{f(\vec{x}) \xrightarrow{b} f(\vec{z})}$$

which we abbreviate as $(f : i; a \triangleright b)$.

Finally, we can control the communication between the x_i using the morphisms φ_i of term (4), but without changing the state of Ω_t, hence the rule format:

$$\frac{x_i \xrightarrow{a} y_i, \ x_j \xrightarrow{b} y_j}{f(\vec{x}) \xrightarrow{\tau} f(\vec{z})}$$

which we abbreviate as $(f : i\langle a|b\rangle j)$.

However, the CCS semantics also imposes further restrictions. Since complements in CCS are undirectional, the predicates $(f : i\langle \cdot | \cdot \rangle j)$ and $(f : i; \cdot \triangleright \cdot)$ must be closed under complementary actions, though $(f : i; \cdot \nearrow g)$ need not be. Also, the τ action is invisible to CCS restriction, so $(f : i; \tau \triangleright \tau)$ holds for for all x_i.

The remaining constraints are motivated by the bus-and-components analogy for the static standard form, shown in Figure 2. If some x_i emits a signal onto the bus, then there is only one way that the signal can be decoded if it gets through the filter, unless it synchronizes with a signal from Ω_t to produce a τ signal; hence we have:

REN: If $(f : i; a \triangleright \alpha)$ and $(f : i; a \triangleright \beta)$ then

$$\alpha = \beta \ \text{xor} \ \exists g.[[(f : \overline{\alpha} \nearrow g) \ \text{and} \ \beta = \tau] \ \text{or} \ [(f : \overline{\beta} \nearrow g) \ \text{and} \ \alpha = \tau]]$$

If two processes synchronize via the bus, then the encodings of the signals they synchronize must be CCS complements, and hence CCS requires that the decodings of these signals be complementary:

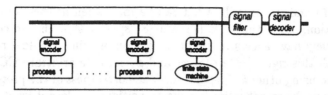

Fig. 2. Concrete view of stCCS standard form.

COMM: If $(f : i \langle a|b \rangle j)$ then $(f : i; a \triangleright \alpha) \Leftrightarrow (f : j; b \triangleright \overline{\alpha})$

We use constraint TRANS to express the transitivity of CCS communication, since only complementary signals can synchronize.

TRANS: If $(f : i \langle a|b \rangle j)$ and $(f : j \langle b|c \rangle k)$ then $(f : i \langle a|\overline{c} \rangle k)$

Finally, MOVE reflects the encoding of a change of operator as a state change of the fsm Ω_t that preserves the configuration of the processes on the bus.

MOVE: If operator f has a rule with consequent $f(\vec{x}) \xrightarrow{\alpha} g(\vec{z})$ then
$$(f : i; a \triangleright b) \iff (g : i; a \triangleright b)$$
$$(f : i \langle a|b \rangle j) \iff (g : i \langle a|b \rangle j)$$

We name the class of operators obeying these constraints SSF. It corresponds exactly to stCCS up to strong bisimulation:

Theorem 5. *If f is a SSF operator, then there is a stCCS term t such that $f(\vec{x}) \sim t$. Conversely, if t is a stCCS term, then there is a SSF operator f such that $t \sim f(\vec{x})$.*

Though the SSF rule format does not have a very clean characterization, it does permit some useful specifications:

Example 2. A mutual exclusion protocol for two processes can be described using the following operator specifications, where A is a set of actions not containing request, use or release:

$$\frac{x_1 \xrightarrow{\text{request}} y_1}{N(x_1, x_2) \xrightarrow{\tau} C_1(y_1, x_2)} \qquad \frac{x_2 \xrightarrow{\text{request}} y_2}{N(x_1, x_2) \xrightarrow{\tau} C_2(x_1, y_2)} \qquad (5)$$

$$\frac{x_1 \xrightarrow{\alpha} y_1, \ \alpha \in A}{N(x_1, x_2) \xrightarrow{\alpha} N(y_1, x_2)} \qquad \frac{x_2 \xrightarrow{\alpha} y_2, \ \alpha \in A}{N(x_1, x_2) \xrightarrow{\alpha} N(x_1, y_2)} \qquad (6)$$

$$\frac{x_1 \xrightarrow{\alpha} y_1, \ \alpha \in A \cup \{\text{use}\}}{C_1(x_1, x_2) \xrightarrow{\alpha} C_1(y_1, x_2)} \qquad \frac{x_2 \xrightarrow{\alpha} y_2, \ \alpha \in A}{C_1(x_1, x_2) \xrightarrow{\alpha} C_1(x_1, y_2)} \qquad \frac{x_1 \xrightarrow{\text{release}} y_1}{C_1(x_1, x_2) \xrightarrow{\tau} N(y_1, x_2)}$$
$$(7)$$

$$\frac{x_2 \xrightarrow{\alpha} y_2, \ \alpha \in A \cup \{\text{use}\}}{C_2(x_1, x_2) \xrightarrow{\alpha} C_2(x_1, y_2)} \qquad \frac{x_1 \xrightarrow{\alpha} y_1, \ \alpha \in A}{C_2(x_1, x_2) \xrightarrow{\alpha} C_2(y_1, x_2)} \qquad \frac{x_2 \xrightarrow{\text{release}} y_2}{C_2(x_1, x_2) \xrightarrow{\tau} N(x_1, y_2)}$$
$$(8)$$

The rules (5) and (6) describe what can happen when neither process is in the critical section, while the rules (7) and (8) specify that the processes x_1 and x_2 respectively have exclusive use of the action use till they do a release action to relinquish this right. Note that x_1 and x_2 may be arbitrary CCS processes, possibly also using other SSF operators. The correctness of the protocol is easily observed from the specification, as the transitions we do not want to allow are simply omitted.

Modifying CCS: We consider each of the constraints TRANS, REN, COMM and MOVE in turn, to see if they can be eliminated by a slight modification of CCS. We can remove TRANS by generalizing the notion of action complement in CCS to a set. Consider processes $p_1, \ldots, p_4$ communicating as shown:

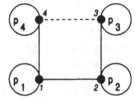

The solid lines indicate intended synchronization between the processes, while synchronization along the dotted line is enforced by CCS but not by the modified language. To see this, suppose port 1 were labelled a. In CCS, we would also have to have ports 2 and 4 labelled $\bar{a}$. Hence, port 3 would have to be labelled a, resulting in the communication link between ports 3 and 4. However, ports 2 and 4 need not be the same members of the generalized complement of a, and so port 3 need not occur in the generalized complement of port 4.

We can eliminate the constraint COMM by generalizing the CCS restriction operator, so that we can selectively restrict either an action or its complement. In fact, this reflects many real-world situations, *e.g.* interfaces on hardware devices. The constraints REN and MOVE ensure consistency and must remain. We are investigating the impact of these changes on the theory of CCS.

3.2   Representing Dynamic CCS Operators

A limitation of the SSF operator class is that it excludes the CCS prefixing and summation operators[1]. By extending stCCS to include limited dynamic behavior, we get a definable operator class that includes them.

Definition 6. A linear CCS term is a stCCS$^+$ term if it is either a stCCS term or of the form $\sum_{i \in I} a_i.t_i + \sum_{j \in J} t_j$, where the t_i are stCCS$^+$ terms and the t_j are stCCS terms.

The class of SSF$^+$-definable operators directly reflects the manner in which we added terms to stCCS to obtain stCCS$^+$:

---

[1] Since SSF is intended as an extension of CCS, this is not crippling in practice.

Definition 7. An operator f is an SSF$^+$ operator if it is a SSF operator, or every rule in its specification is of one of the following forms:

$$\overline{f(\vec{x}) \xrightarrow{a} g(\vec{x})} \tag{9}$$

$$\frac{x_i \xrightarrow{a} y_i}{f(\vec{x}) \xrightarrow{b} h(\vec{z})} \qquad \frac{x_i \xrightarrow{a} y_i,\ x_j \xrightarrow{b} y_j}{f(\vec{x}) \xrightarrow{c} h(\vec{z})} \tag{10}$$

where g is an SSF$^+$ operator and h is a SSF operator; in (10), we require $|\vec{z}| \leq |\vec{x}|$, and the $\vec{z}$ must be distinct and drawn from $\vec{x}$ and $\vec{y}$ as constrained by (3).

Theorem 8. *If f is a SSF$^+$ operator, then there is a stCCS$^+$ term t such that $f(\vec{x}) \sim t$. Conversely, if t is a stCCS$^+$ term, then there is a SSF$^+$ operator f such that $t \sim f(\vec{x})$.*

While the class of SSF$^+$ operators includes all the CCS operators, it remains open to characterize a class of operator specifications that represents all CCS terms up to strong bisimulation.

4   CCS-definability up to Branching Bisimulation

The ability of CCS to define new operators up to strong bisimulation is very constrained because strong bisimulation is sensitive to internal (τ) computation, and CCS is asynchronous. By not observing τ-actions we get around this limitation.

Definition 9. Given linear CCS terms t and t', and $s \in \text{ActSeq} = (\text{Act} \bigcup \{\tau\})^*$ such that $s = \alpha_1 \ldots \alpha_n$, $t \xrightarrow{s} t'$ if there exist $t_1, \ldots, t_n$ such that $t = t_1 \xrightarrow{\alpha_1} t_2 \xrightarrow{\alpha_2} \ldots \xrightarrow{\alpha_n} t_n = t'$. We write $t \xrightarrow{\tau^i} t'$ if, for some $i \in \mathbf{N}$, we have $t \xrightarrow{\tau^i} t'$. Given a CCS terms t and t', and action a, we have $t \xRightarrow{a} t'$ if and only if $t \xrightarrow{\tau^* a \tau^*} t'$; in addition, we have $t \Rightarrow t'$ if and only if $t \xrightarrow{\tau^*} t'$. We also extend $\Rightarrow$ to labels from ActSeq the same way as for $\rightarrow$.

We call ruloids for the $\Rightarrow$ relation *weruloids*. The following lemma describes the weruloids for linear CCS terms:

Lemma 10. *Given a linear CCS term t, and $s \in \text{ActSeq}$, there is a set $\mathcal{T}_{t,s}$ of weruloids of the form*

$$\frac{\left\{ x_i \xRightarrow{s_i} y_i \mid i \in I \right\}}{t \xRightarrow{s} t'[\vec{z}]}$$

where s and the s_i are in ActSeq, t' is a linear CCS term, and z_i is y_i if $i \in I$ and x_i otherwise, that describes exactly when t can make a $\xRightarrow{s}$ transition.

This suggests that we could encode something similar to the format (1), *i.e.* de Simone's, up to weak or branching bisimulation, subject to branching bisimulation being a congruence, and finiteness of operator specifications. We also introduce here a closure condition on sets of weruloids:

Definition 11. Given weruloid $\eta = \dfrac{\left\{ x_i \overset{s_i}{\Rightarrow} y_i \mid i \in \{i_1, \ldots, i_n\} \right\}}{t \overset{s}{\Rightarrow} t'}$ for a linear CCS term t, the *trigger* of η is $\langle (i_1, s_1), \ldots, (i_n, s_n) \rangle$, where $i_1 < i_2 < \ldots < i_n$. A set of weruloids for t is *trigger-closed* if, for each weruloid with trigger $\langle (i_1, s_1), \ldots, (i_n, s_n) \rangle$, $n > 1$, there is a weruloid with trigger $\langle (i_1, s_1), \ldots, (i_{n-1}, s_{n-1}) \rangle$.

4.1 Branching Bisimulation as a Congruence

We chose branching bisimulation over weak bisimulation because it has more intuitive appeal as a notion of branching equivalence[vG93] and, being finer, it is a congruence for more operators.

Definition 12. A symmetric relation R is a branching bisimulation relation if, for all actions $\beta \in \mathsf{Act} \cup \{\tau\}$, and for all $(p, p') \in R$, we have: if $p \overset{\beta}{\to} r$, then either

1. $\beta = \tau$ and $(r, p') \in R$, or
2. there are q', r' such that $p' \Rightarrow q' \overset{\beta}{\to} r'$, $(p, q') \in R$ and $(r, r') \in R$.

Processes p and p' are branching bisimilar, $p \approx_b p'$, iff there is some branching bisimulation relation R such that $R(p, p')$ holds.

The problem of determining a class of operators for which branching bisimulation is a congruence is examined in depth in [Blo93]:

Definition 13 [Blo93]. An operator f is *simply BB-cool* if

1. All rules ρ for f have the form $\dfrac{\left\{ x_i \overset{a_i}{\to} y_i \mid i \in I(\rho) \right\}}{f(\vec{x}) \overset{b}{\to} t}$, where $I(\rho)$ is a set of numbers telling which arguments of f take actions under rule ρ.
2. For all rules ρ for f, and all $i \in I(\rho)$, f has a *patience* rule:
$$\frac{x_i \overset{\tau}{\to} y_i}{f(\vec{x}) \overset{\tau}{\to} f(x_1, \ldots, x_{i-1}, y_i, x_{i+1}, \ldots, x_n)}$$
3. Only the patience rules of clause 2. may have hypotheses using τ's.

The only CCS operator that is not simply BB-cool is $+$, which also does not respect weak or branching bisimulation. For example, $a.0 \approx_b \tau.a.0$, but $a.0 + c.0 \not\approx_b \tau.a.0 + c.0$, for $c \neq a$.

Theorem 14 [Blo93]. *If process algebra $\mathcal{L}$ consists only of simply BB-cool operators, then branching bisimulation is a congruence for $\mathcal{L}$.*

4.2 CCS^l

We now present our CCS definability result for branching bisimulation.

Definition 15. A CCS term t is a CCS^l term if

1. t is generated by the grammar:
$$t ::= u \mid p \mid t + t \mid t|t \mid \alpha.t \mid t\backslash B \mid t[\varphi]$$
   where $p \in \mathcal{P}$ and u is a CCS term without process variables, and
2. t is linear, and the set of all its weruloids is trigger-closed.

This is analogous to the architectural expressions of [dS85], and there is also a analogous standard form for CCS^l terms, described by Lemma 16. However, we work in an asynchronous setting, using different primitives and a different equivalence.

Lemma 16. *If t is a CCS^l term then t is branching bisimilar to a CCS^l term of the form*
$$(x_1[\varphi_1]|\ldots|x_n[\varphi_n]|\mathsf{ctrl}_t)\backslash B \tag{11}$$
where the φ_i map Act *to coding actions and*

1. ctrl_t *is a closed CCS term, i.e. it has no process variables,*
2. *if* $\mathsf{ctrl}_t \xrightarrow{s} \mathsf{ctrl}'_t$, *$s$ contains some $a \in$ Act between any two occurrences of any $b_i \in \varphi_i(\mathsf{Act})$, and*
3. *if $i \neq j$ then $\varphi_i(\mathsf{Act}) \bigcap \varphi_j(\mathsf{Act}) = \{\}$*

Before defining the class of operators we can encode using CCS terms of the form (11), we introduce trigger closure for operator specifications, which will enable their CCS^l encodings to avoid deadlock.

Definition 17. Given a rule ξ for an operator f of the form
$$\frac{\left\{ x_i \xrightarrow{a_i} y_i \mid i \in \{i_1, \ldots, i_n\} \right\}}{f(\vec{x}) \xrightarrow{a} t}$$

the *trigger* of ξ is $\langle (i_1, a_1), \ldots, (i_n, a_n) \rangle$, where $i_1 < i_2 < \ldots < i_n$. Operator f is *trigger-closed* if, for every rule η for f with trigger $\langle (i_1, a_1), \ldots, (i_n, a_n) \rangle$, $n > 1$, there is also a rule with trigger $\langle (i_1, a_1), \ldots, (i_{n-1}, a_{n-1}) \rangle$. Given rule η, the subset of the rules for f that is required for f to be trigger closed is the *trigger-closure* of η.

All the CCS and Meije operators are trigger closed, but the CSP parallel composition operator $\|$ and the SCCS product operator $\times$ are not.

Definition 18. An operator f is a tcSB^l operator if the following three conditions hold:

1. f is a simply BB-cool operator, but with patience rules for each source variable, not just those occurring in the antecedent.

2. For each defining rule for f, the target must be a linear term, and if $x_i \xrightarrow{a} y_i$ is an antecedent of the rule, then x_i cannot occur in the target.
3. The specification of f is trigger closed.

The first constraint ensures that branching bisimulation is a congruence, and also accounts for CCS not being able to restrict τ-actions. The second constraint reflects the static embedding of the arguments in the encoding. Finally, trigger closure ensures that the completed antecedent tests always enable some rule to fire.

Example 3. The operator $\|_a$ forces its arguments to use action a only for local synchronization, and is not trigger closed:

$$\frac{x_1 \xrightarrow{a} y_1, \quad x_2 \xrightarrow{a} y_2}{x_1 \|_a x_2 \xrightarrow{a} y_1 \|_a y_2} \qquad \frac{x_1 \xrightarrow{\alpha} y_1, \quad \alpha \neq a}{x_1 \|_a x_2 \xrightarrow{\alpha} y_1 \|_a x_2} \qquad \frac{x_2 \xrightarrow{\alpha} y_2, \quad \alpha \neq a}{x_1 \|_a x_2 \xrightarrow{\alpha} x_1 \|_a y_2}$$

Suppose we were to encode this using a CCS^l term $t[x_1, x_2]$. Now, $a.0 \|_a b.0$ can only do a b-action via the third rule and stop. However, when the term $t[a.0, b.0]$ tries to test if the first rule can fire, it will successfully test the first argument for an a transition, but then this will fail for the second argument. Not being able to undo the test it has completed, t can only deadlock and hence cannot be branching bisimilar to $x_1 \|_a x_2$.

Theorem 19. *The* tcSB^l *operators are realizable in* CCS^l *up to branching bisimulation.*

Example 4. The following tcSB^l operator specification is not encodable in CCS up to strong bisimulation:

$$\frac{x \xrightarrow{a} y}{f(x) \xrightarrow{a} b.y} \qquad \frac{x \xrightarrow{\tau} y}{f(x) \xrightarrow{\tau} f(y)}$$

It can be encoded up to branching bisimulation in format (11):

$$(x[a^{\bullet}/a] \mid ctrl_f) \setminus \{a^{\bullet} \mid a \in Act\}$$

where

$$ctrl_f = \overline{a^{\bullet}}.a.b. \left(\mu X. \sum_{a \in Act} \overline{a^{\bullet}}.a.X \right)$$

Corollary 20. *Every* CCS^l *term represents a* tcSB^l *operator up to branching bisimulation.*

4.3 · was · needed for CCS

In this section, we add a checkpointing operator to CCS to work around the trigger closure restriction. The class of definable operators then becomes a subclass of the finitely specified de Simone operators (format (1)) with branching bisimulation as a congruence.

Definition 21. A wasCCS^l term t is a linear term generated by the following grammar:

$$t ::= u \mid p \mid t + t \mid t|t \mid \alpha.t \mid t\backslash B \mid t[\varphi] \mid t \text{ was } t$$
$$u ::= 0 \mid X \mid \alpha.u \mid u + u \mid u|u \mid u\backslash B \mid u[\varphi] \mid \mu X.u \mid u \text{ was } u$$

where was is specified by the rules (2), and the rest are as for CCS.

Lemma 22. *If t is a wasCCS^l term, then t is branching bisimilar to a term of the form*

$$((p_1 \text{ was } p_1)[\varphi_1]| \dots |(p_n \text{ was } p_n)[\varphi_n]|\text{ctrl}_t) \backslash B \tag{12}$$

where ctrl$_t$ is a closed wasCCS^l term.

The class of definable operators differs from Definition 18 only in that it omits the third clause (trigger closure):

Definition 23. An operator f is a SB^l operator if the following two conditions hold

1. f is a simply BB-cool operator, but with patience rules for every source variable, not just those occurring in the antecedent
2. For each defining rule ρ for f, the target must be a linear term, and if $x_i \xrightarrow{a} y_i$ is an antecedent of ρ, then x_i cannot occur in the target.

Example 5. The operator $\cdot \|_a \cdot$ described in Example 4.13 is a SB^l operator. It can be encoded in wasCCS^l, but as it is not a tcSB^l operator, it cannot be encoded in CCS^l.

By enclosing process arguments in the was operator, the wasCCS^l encoding of a SB^l operator f testing the antecedents of a rule for f can undo the effects of completed tests by issuing r signals to the arguments that were successfully tested, when faced with a test that can't succeed. If all the antecedents are successfully tested, a c signal is issued to all the arguments to cement the changes.

Theorem 24. *If f is a SB^l operator, then there is a wasCCS^l term t such that $f(\vec{x}) \approx_b t$. Conversely, if t is a wasCCS^l term, then there is a SB^l operator f such that $t \approx_b f(\vec{x})$.*

5 Conclusions

We have described preliminary work on a program to use expressiveness to better understand and compare process algebras. We have identified subsets of CCS that admit standard forms, and used these to characterize formats for operators definable up to strong and branching bisimulation. We also showed how to modify CCS to increase its expressiveness. We are currently applying the techniques we developed here to analyzing other process algebras, particularly LOTOS and CSP, in order to compare their expressiveness to that of CCS.

References

[BB87]   Tommaso Bolognesi and Ed Brinksma. Introduction to the ISO specification language LOTOS. *Computer Networks and ISDN Systems*, 14:25–59, 1987.

[BIM88] Bard Bloom, Sorin Istrail, and Albert R. Meyer. Bisimulation can't be traced (preliminary report). In *Proceedings of POPL 1988*, pages 229–239. IEEE Press, 1988.

[Blo93] Bard Bloom. Structural operational semantics for weak bisimulations. Technical Report TR 93-1373, Cornell, August 1993. To appear in Theoretical Computer Science.

[Blo95] Bard Bloom. Structural operational semantics considered as a specification language. In *Proceedings of POPL 1995*, pages 107–117. IEEE Press, 1995.

[Bou85] G. Boudol. Notes on algebraic calculi of processes. In K. Apt, editor, *Logics and Models of Concurrent Systems*, pages 261–303. Springer-Verlag, 1985. NATO ASI Series F13.

[BW91]   W. Baeten and P. Weijland. *Process Algebra*. Cambridge University Press, 1991.

[DB93]   Ashvin Dsouza and Bard Bloom. Towards a basis for GSOS. Unpublished manuscript, 1993.

[DB95]   Ashvin Dsouza and Bard Bloom. On the expressive power of ccs, 1995. http: //www.cs.cornell.edu/Info/People/dsouza/ccs-expr.ps.

[dS85]   R. de Simone. Higher-level synchronizing devices in MEIJE-SCCS. *Theoretical Computer Science*, 37(3):245–267, 1985.

[GV89]   Jan Friso Groote and Frits Vaandrager. Structured operational semantics and bisimulation as a congruence (extended abstract). In G. Ausiello, M. Dezani-Ciancaglini, and S. Ronchi Della Rocca, editors, *Automata, Languages and Programming: 16th International Colloquium*, volume 372 of *Lect. Notes in Computer Sci.*, pages 423–438. Springer-Verlag, 1989.

[Hoa85] C. A. Hoare. *Communicating Sequential Processes*. Prenctice Hall, 1985.

[Mil89] Robin Milner. *Communication and Concurrency*. Prentice Hall, 1989.

[Par81] D. Park. Concurrency and automata on infinite sequences. In P. Deussen, editor, *Theoretical Computer Science*, page 261. Springer-Verlag, 1981.

[Par90] J. Parrow. The expressive power of parallelism. *Future Generation Computer Systems*, 6:271–285, 1990.

[Plo77] G. D. Plotkin. LCF considered as a programming language. *Theoretical Computer Sci.*, 5(3):223–255, 1977.

[Vaa93] Frits Vaandrager. Expressiveness results for process algebras. In de Bakker, de Roever, and Rozenberg, editors, *Semantics: Foundations and Applications*, pages 609–620. Springer-Verlag, 1993. LNCS 666.

[vG93]   Rob van Glabbeek. A complete axiomatization for branching bisimulation congruence of finite-state behaviours. In *Symposium on Mathematical Foundations of Computer Science '93*. Springer Verlag, 1993. LNCS 711.

Polarized Name Passing

Martin Odersky

Department of Computer Science
University of Karlsruhe
76128 Karlsruhe, Germany
odersky@ira.uka.de

Abstract. We study a refinement of name passing in a process calculus, where names have input and output polarities. Building on a simple asynchronous reduction semantics, we develop a notion of polarized bisimulation and show that it is a congruence. We then give an encoding of Moggi's computational lambda calculus in polarized π which preserves all of Moggi's observational equivalences except the η-value rule.

1 Introduction

Modern process calculi are centered around name passing as the fundamental principle of interaction [10]. Name passing, first proposed by Engberg and Nielsen [3], was studied in depth in the Chemical Abstract Machine [1] and particularly in the π-calculus [11, 8]. It is at the same time simple and expressive enough for representing a wide range of computation and interaction patterns. Consequently, there has also been much work in using a name passing process calculus as the target of encodings of traditional functional and imperative languages, which pass structures as arguments [9, 20, 6]. However, it has turned out to be surprisingly difficult to capture observational equalities of those languages as process equivalences. For instance, one of Milner's original encodings of the call-by-value λ-calculus [7] did not validate the β-value reduction rule. The reason for some of these difficulties becomes clear if we look at the operations that can be performed on a name.

One can do three things with a name: Read from it, write to it or test two names for equality. The last operation is sometimes missing from the term language, but it is certainly a capability of an observing environment. Now consider encodings of functions. The standard call-by-value encoding of an abstraction $\lambda x.M$ introduces a fresh anonymous *handle* name, say a, and composes it with a reader agent $*a?x.M$ (where the $*$ signifies that the reader may be used arbitrarily often). A function call then involves writing an argument name to the handle a. But a process context may manipulate the handle arbitrarily; it could equally well install a competing reader, say $*a?y.N$, or it could test the handle for equality to itself or to some other handle. These added capabilities can invalidate common laws of functional programming, since the duplication of lambda abstractions in substitutions is not observable while the duplication of handles is.

Indeed, this discrepancy points to a general problem with unrestricted name passing: it fails to validate many seemingly reasonable equivalences which involve duplication of subterms. As an example, consider a worker agent, which reads data items from channel a and processes them.

$$WORK(a) \stackrel{\text{def}}{=} *a?x.HANDLE(x)$$

(Here, * means replication and $a?x....$ is an input term). Intuitively, we should be able to have several workers instead of one without changing the semantics of a program. That is, we would like the following equivalence to hold, for any process P:

$$P \mid WORK(a) \mid WORK(b) = P[a := c, b := c] \mid WORK(c).$$

However, such an equivalence would hold only for "well-behaved" clients P, which use the communication channels a and b only for writing. If P would contain a comparison of the channels a and b, say, then the two sides would clearly be not equivalent. Even if comparison of channel names were not available as a primitive, a process P could still distinguish the two sides by sending to channel a and reading from channel b. On the left-hand side, the read would surely be blocked, but on the right-hand side it might succeed.

This paper is motivated by the observation that unrestricted reading, writing, and comparison of names might be too powerful a capability (meaning too many laws are broken). We therefore discriminate between the three operations by associating *polarities* with names. A name with *output* polarity can used for writing but not for reading. Conversely, a name with *input* polarity can used for reading but not for writing. Finally, an equality test succeeds only if both polarities of a name are supplied. The rest of this paper formalizes this idea and studies its ramifications. In particular:

- We define a reduction system for a process calculus with polarized names. The *polarized π-calculus* builds on a simple asynchronous process calculus, similar to systems developed independently by Boudol [2] and by Honda [5], with the addition of polarities for names.
- We define *polarized bisimulation*, a new notion of process equivalence for polarized π-terms. We show that polarized bisimulation is a congruence, that it is reduction-closed and that it preserves output events.
- We give a new encoding of the call-by-value λ-calculus with let's into the π-calculus. We show that this encoding satisfies all equalities of Moggi's computational λ-calculus [13, 14], except the η-value equality. We also discuss why it is unreasonable to expect η-value to hold in a process calculus. Computational λ-calculus is an interesting test case for an encoding, since it embodies a complete set of laws for reasoning about functions in a sequential environment.

Polarities in message exchanges are also found in the PICT programming language [17] which is based π-calculus. They have been studied in detail by Pierce and Sangiorgi [16], albeit with a slightly different motivation. Pierce and Sangiorgi consider only sorted systems, while we concentrate on the unsorted version of polarized π. They add a neutral polarity, which allows unrestricted use of a name. We do not, since such a capability can always be modeled by passing both polarities of a name. Omitting the neutral polarity makes polarized π quite a bit simpler than their system, both in the structural and reduction rules and in the type system, where no subtyping is required. We can thus study process equivalence in considerably more detail, coming up with a tractable definition of bisimulation congruence for processes with polarized names. By contrast, Pierce and Sangiorgi use barbed bisimulation and barbed congruence [12] as process equivalences. Barbed congruence is more general, but also considerably harder to use than polarized bisimulation since it requires an argument over program contexts that comprise a parallel composition with terms of arbitrary complexity instead of just parallel compositions with atomic outputs $a!b$.

Sangiorgi [19] has also explored an alternative treatment of the problem of preserving equivalences from functional programming. His technique relies on translating λ to π

Constants	a, b, c	$\in$	C_a	
Variables	x, y, z	$\in$	$\mathcal{V}$	
Names	u, v, w	$=$	$a \mid x$	
Terms	P, Q, R	$=$	uv	Application (Output)
			$\mid ux.P$	Abstraction (Input)
			$\mid *ux.P$	(Eager) Replication
			$\mid \nu a.P$	Restriction
			$\mid P \mid Q$	Parallel Composition
			$\mid 0$	Identity
			$\mid [u \doteq v]P$	Match

Fig. 1. Names and terms in asynchronous π.

in such a way that sharing can never be observed outside of a translated term. This is achieved by "padding" the translation with forwarding terms that transmit arguments in only one direction. Thus, all translations interact with their environment only via fresh anonymous names. One disadvantage of this technique is the additional bulk introduced by the translation. Another problem is that, due to the complete elimination of sharing, constants in a functional language cannot be translated to names in the π-calculus — if they were, all comparisons between constants would show inequality. Instead, his translation maps constants to terms of a particular form called "triggers" [18].

The rest of this paper is organized as follows. Section 2 reviews Honda's formulation of asynchronous π-calculus. Section 3 discusses what changes when polarities are added to that calculus. Section 4 defines polarized bisimulation and shows that it is a congruence. Section 5 studies an encoding of the computational λ-calculus in polarized π. Section 6 concludes.

2   Asynchronous π-calculus

Asynchronous π-calculus is a general concept rather than a fixed set of rules. One instance was developed by Boudol [2], another, with a slightly different set of rules, by Honda and Tokoro [5, 4]. Honda's version, also called the ν-calculus, distinguishes between input-bound and ν-bound identifiers, while Boudol's version doesn't. In this section, we review and place in context transition relation and bisimulation theories of asynchronous π-calculus. We use in general Honda's formulation, but stick sometimes to the more conventional syntax of Boudol.

Figure 1 illustrates the term language of asynchronous π-calculus. We distinguish between an alphabet for *constants*, ranged over by a, b, c and an alphabet for *variables*, ranged over by x, y, z. Names are either constants or variables; they are ranged over by u, v, w. Term forms include application uv, abstraction $ux.P$, parallel composition $P \mid Q$, the inert term 0 and name restriction $\nu a.P$. Replication $*ux.P$ is restricted to abstraction terms. Replication is eager; that is, $*ux.P$ is equivalent to arbitrarily many copies of $ux.P$ executed in parallel. This is a minor deviation from the treatment of Honda and Yoshida, who consider lazy replication. As another minor change, we include a name matching construct $[u \doteq v]P$. Honda and Yoshida also study this construct, but introduce it as an extension of ν-calculus proper.

The free names $\text{fn}(P)$, and, analogously, the free constants $\text{fc}(P)$ and free variables $\text{fv}(P)$ of a term are defined as usual. We assume the following structural equivalences.

1. α-renaming:

   $$\nu a.P \equiv \nu b.P[a := b] \quad \text{if } b \notin \text{fc}(P) \qquad u x.P \equiv u y.P[x := y] \quad \text{if } y \notin \text{fv}(P)$$

2. Abelian monoid laws for parallel composition:

   $$P \mid Q \equiv Q \mid P, \qquad (P \mid Q) \mid R \equiv P \mid (Q \mid R), \qquad P \mid 0 \equiv P$$

3. Scope extrusion:

   $$P \mid \nu a.Q \equiv \nu a.(P \mid Q) \quad \text{if } a \notin \text{fc}(P)$$

4. (Eager) input-replication:

   $$*ux.P \equiv ux.P \mid *ux.P$$

5. The name matching law:

   $$[a \doteq a]P \equiv P$$

Reduction involves passing a name u from an output term au to an input term $a?x.P$. Reduction is asynchronous, *i.e.* , processes do not block on output:

   (COM) $\qquad ax.P \mid au \to P[x := u].$

Reduction is considered modulo syntactic equivalence. Reduction can be applied anywhere in a term except under an abstraction or a matching:

   (STR) $\qquad \dfrac{P' \equiv P \quad P \to Q \quad Q \equiv Q'}{P' \to Q'},$

   (NU) $\qquad \dfrac{P \to Q}{\nu a.P \to \nu a.Q},$

   (PAR) $\qquad \dfrac{P \to Q}{P \mid R \to Q \mid R}.$

Let $\twoheadrightarrow$ be the reflexive, transitive closure of reduction.

Starting with the definition of an observable action, we now adapt Honda's formulation of *asynchronous bisimulation* to the reduction-based context that we follow here.

Definition 2.1 An *output action* α is one of ab, $a(b)$ where a and b are names. A labeled transition relation $\overset{\alpha}{\Longrightarrow}$ is given by the rules:

(a) $P \overset{ab}{\Longrightarrow} P'$ if $P \twoheadrightarrow ab \mid P'$,

(b) $P \overset{a(b)}{\Longrightarrow} P'$ if $P \twoheadrightarrow \nu b.(ab \mid P')$.

Definition 2.2 A symmetric relation $\sim$ over closed terms is a (weak) *asynchronous bisimulation* if

   $P \sim Q$ and $P \overset{\alpha}{\Longrightarrow} P'$ implies $Q \overset{\alpha}{\Longrightarrow} Q'$ for some term Q' such that $P' \sim Q'$,

and

$P \sim Q$ implies $P \mid ab \sim Q \mid ab$, for all names a, b.

Let asynchronous *bisimilarity* $\approx_a$ be the maximal asynchronous bisimulation relation (which is shown to exist by a standard argument). Bisimilarity is extended to open terms by defining $P \approx_a Q$ if $P\sigma \approx_a Q\sigma$, for all closing substitutions σ.

Theorem 2.3 (Honda) $\approx_a$ is a congruence.

In fact, asynchronous bisimulation can be characterized as the largest congruence that satisfies the following two properties, see [4] for details.

(a) It is *reduction closed*: If $P \approx_a Q$ and $P \twoheadrightarrow P'$ then there is a Q' such that $Q \twoheadrightarrow Q'$ and $P' \approx_a Q'$.
(b) It is *sound*: If P and Q are both insensitive (that is, neither can interact with its environment), then $P \approx_a Q$.

3 From asynchronous to polarized π.

We add polarities to the asynchronous π-calculus by introducing for every (unpolarized) constant $a \in C_a$ an *input* constant $a?$ and an *output* constant $a!$. Let $C?$ be the set of all input constants and let $C!$ be the set of all output constants. Then the set of all polarized constants C is $C? \cup C!$.

In a slight misuse of notation, we also let letters a, b, c denote arbitrary (input or output) constants. Let $a^\perp$ denote the complement of a constant a, that is, $(a!)^\perp = a?$ and $(a?)^\perp = a!$.

Names and terms are defined as in asynchronous π, with the clarification that in the term $\nu a.P$, a is understood to be an unpolarized constant.

Alpha renaming is adapted to polarities as follows:

$$\nu a.P \equiv \nu b.P[a? := b?, a! := b!] \quad \text{if } b!, b? \notin \mathrm{fc}(P)$$
$$ux.P \equiv uy.P[x := y] \quad \text{if } y \notin \mathrm{fv}(P).$$

The other structural equivalences are as in asynchronous π, with the exception of name matching, which now requires both polarities of a name to be present:

$$[a! \doteq a?]P \equiv P.$$

Reduction is sharpened by requiring the proper polarities in the terms that take part in a reduction:

(COM) $\qquad a?x.P \mid a!u \rightarrow P[x := u].$

The resulting calculus is summarized in Figure 2.

Examples.

1. A forwarder
$$FWD(a, b) \overset{\text{def}}{=} *a?x.b!x.$$

This term transmits all messages received from channel a to channel b.

2. An input term with wrong polarity $a!x.0$. This term is inactive; it can never react with its environment.

Syntactic Domains

Unpolarized Constants	a, b, c	$\in \mathcal{C}_a$
Variables	x, y, z	$\in \mathcal{V}$
Names	u, v, w	$= a? \mid a! \mid x$
Terms	P, Q, R	$= uv \mid ux.P \mid *ux.P \mid \nu a.P \mid P\mid Q \mid 0 \mid [u \doteq v]P$

Structural Equivalences

$$(\alpha) \qquad \nu a.P \equiv \nu b.P[a? := b?, a! := b!] \qquad \text{if } b!, b? \notin \text{fc}(P)$$
$$ux.P \equiv uy.P[x := y] \qquad\qquad\quad \text{if } y \notin \text{fv}(P)$$

$$(\mid) \qquad P \mid Q \equiv Q \mid P$$
$$(P \mid Q) \mid R \equiv P \mid (Q \mid R)$$
$$P \mid 0 \equiv P$$

$$(\nu) \qquad P \mid \nu a.Q \equiv \nu a.P \mid Q \qquad \text{if } a \notin \text{fc}(P)$$

$$(*) \qquad *ux.P \equiv ux.P \mid * ux.P$$

$$(\doteq) \qquad [a! \doteq a?]P \equiv P$$

Reduction

$$(\text{COM}) \qquad a?x.P \mid a!u \;\rightarrow\; P[x := u]$$

Fig. 2. The polarized π calculus

3. An inactive output term $a?b$, again with wrong polarity.

4. An identity tester

$$TEST(x, y).P \stackrel{\text{def}}{=} \nu c.(x(c!) \mid yz.P).$$

When applied to a constant and its negation — as in $TEST(a!, a?).P$ — this construct enables execution of the term P, provided there are no other terms in the environment reading or writing a. The name matching construct $[x \doteq y]P$ has the same effect, but without the restrictions on the environment.

5. A bidirectional forwarder

$$EQ(a, b) \stackrel{\text{def}}{=} *a?x.b!x \mid * b?x.a!x,$$

which, when present, would "fool" the identity tests $TEST(a!, b?).P$ or $TEST(b!, a?).P$, but would not fool the matchings $[a! \doteq b?]P$ or $[b! \doteq a?]P$.

4 Polarized Bisimulation

In this section, we develop a definition of process equivalence in polarized π. The equivalence is based on a refined notion of bisimulation. Asynchronous bisimulation is too discriminating for polarized π, since it assumes that the identity of all names passed out of a tested process can be determined. We do not want to make the environment more powerful than what can be expressed in a process context. Hence we insist that the identity of a name a be known to the environment only if the environment has knowledge of both polarities of a.

In the absence of full knowledge about the identity of names, the environment needs to carry along a hypothesis that describes what it knows about the names that take part in interactions. We formalize this idea as follows.

Definition 4.1 A *correspondence hypothesis* (or just *correspondence*) H is a finite relation over $(C! \times C!) \cup (C? \times C?)$ that satisfies

(a) if $a! \, H \, b!$ and $a? \, H \, c?$ then $b = c$,
(b) if $b! \, H \, a!$ and $c? \, H \, a?$ then $b = c$.

Let $\mathcal{H}$ be the set of all correspondence hypotheses. Let $\text{dom}(H)$ be the set of polarized constants that are related on either side by H.

Essentially, a correspondence hypothesis records what an experimenter learns about anonymous names passed from the tested processes. In a bisimulation between P and Q, if process P passes out a constant a while process Q passes out b, then the correspondence (a, b) is added to the hypothesis. A hypothesis H is consistent if and only if every constant that exists in both polarities on one side of H corresponds precisely to another such constant on the other side. Consistency is enforced by the two conditions in the definition of correspondence hypotheses above.

Note the duality of correspondences to *distinctions* [11]: A correspondence records which names are surely equal, whereas a distinction records which names are surely different.

We now duplicate our development from Section 2 in working toward a definition of polarized bisimulation.

Definition 4.2 An *output action* α is one of ab, $a(b)$, where a and b are constants. Given a hypotheses H, a labeled transition relation $\overset{\alpha}{\Longrightarrow}_H$ is defined as follows.

(a) $P \overset{ab}{\Longrightarrow}_H P'$ if $P \rightarrow a!b \mid P'$.
(b) $P \overset{a(b)}{\Longrightarrow}_H P'$ if $P \rightarrow \nu b.(a!b \mid P')$ and $b, b^{\perp} \notin \text{dom}(H)$.

Definition 4.3 Two output actions α_1, α_2 are *consistent* with a correspondence H, in symbols $\alpha_1 \leftrightarrow_H \alpha_2$, if one of the following cases applies

$$a_1 b_1 \quad \leftrightarrow_H \quad a_2 b_2$$
$$a_1(b_1) \leftrightarrow_H a_2 b_2$$
$$a_1 b_1 \quad \leftrightarrow_H a_2(b_2)$$
$$a_1(b_1) \leftrightarrow_H a_2(b_2)$$

where $a_1! \, H \, a_2!$ and $H \cup (b_1, b_2)$ is a correspondence. The *refinement* $H.(\alpha_1, \alpha_2)$ of H by actions α_1 and α_2 is in each case $H \cup (b_1, b_2)$.

Definition 4.4 A family of relations $\{\sim_H \mid H \in \mathcal{H}\}$ over closed terms is a *polarized simulation* if

$P \sim_H Q$ and $P \overset{\alpha}{\Longrightarrow}_H P'$ implies $Q \overset{\alpha'}{\Longrightarrow}_H Q'$, for some action α' such that $\alpha \leftrightarrow_H \alpha'$ and some term Q' such that $P' \sim_{H.(\alpha, \alpha')} Q'$,

and

$P \sim_H Q$ implies $P \mid a_1!b_1 \sim_{H \cup (b_1, b_2)} Q \mid a_2!b_2$, for all names a_i, b_i such that $a_1 H a_2$ and $b_1 (H \cup =) b_2$.

$\{\sim_H \mid H \in \mathcal{H}\}$ is a *bisimulation* iff $\{\sim_H \mid H \in \mathcal{H}\}$ and $\{\sim_H^{-1} \mid H \in \mathcal{H}\}$ are simulations.

We usually write just $\sim$ for the family $\{\sim_H \mid H \in \mathcal{H}\}$.

Definition 4.5 Two closed terms P, Q are *bisimilar under a correspondence* H, written $P \approx_H Q$, if there is a stable polarized bisimulation $\sim$ such that $P \sim_H Q$. Two open terms P, Q are bisimilar under H iff $P\sigma_1 \approx_H Q\sigma_2$ for all closing substitutions σ_1, σ_2 such that $(x\sigma_1) H (x\sigma_2)$, for all variables x.

Given a finite set of unpolarized constants $A \subseteq C_a$, let δ_A be the correspondence $\{(a, a) \mid a \in A! \cup A?\}$. The *full* correspondence $H(P_1, \ldots, P_n)$ for terms $P_1, \ldots, P_n$ is given by

$$H(P_1, ..., P_n) = \delta_{fc(P_1, ..., P_n)}.$$

Definition 4.6 Two closed terms P, Q are *bisimilar*, written $P \approx Q$, if $P \approx_{H(P,Q)} Q$. Two open terms P, Q are bisimilar iff $P\sigma \approx Q\sigma$ for all closing substitutions σ.

To distinguish between the versions of $\approx$ with and without a hypothesis index we will use the notation $\approx_-$ for the former. We now show some important properties of polarized bisimulations.

Lemma 4.7 Let a be a polarized constant such that $a, a^\perp \notin fc(P, Q) \cup \text{dom}(H)$. Then

$$P \approx_H Q \quad \Leftrightarrow \quad P \approx_{H \cup \{(a, a)\}} Q.$$

Proposition 4.8 The indexed relation $\approx_-$ is a polarized bisimulation.

Proof: Assume that $P \approx_H Q$ and $P \xRightarrow{\alpha}_H P'$. Then there exists a bisimulation $\sim$ such that $P \sim_H Q$. It follows that $Q \xRightarrow{\alpha'}_H Q'$, for some action α' such that $\alpha \leftrightarrow_H \alpha'$ and some term Q' such that $P' \sim_{H.(\alpha, \alpha')} Q'$. But then also $P' \approx_{H.(\alpha, \alpha')} Q'$. $\square$

Proposition 4.9 $\approx$ is consistent: There are terms P, Q such that $P \not\approx Q$.

Proof: Pick $P \equiv 0, Q \equiv a!(a!)$. Then $a!(a!) \xRightarrow{a!(a!)}_{H(0, a!(a!))} 0$ but $0 \xslashedRightarrow{a!(a!)}_{H(0, a!(a!))}$. Hence there is no polarized bisimulation $\sim$ such that $P \sim_{H(0, a!(a!))} Q$. $\square$

The usefulness of $\approx$ for reasoning about process equivalences largely stems from the next result.

Theorem 4.10 The binary relation $\approx$ is a congruence.

Proof sketch: See appendix A. $\square$

Like asynchronous bisimulation and barbed bisimulation, polarized bisimulation is reduction-closed:

Proposition 4.11 Let $\{\sim_H \mid H \in \mathcal{H}\}$ be a polarized bisimulation. If $P \sim_H Q$ and $P \twoheadrightarrow P'$ then there exists a term Q' such that $Q \twoheadrightarrow Q'$ and $P' \sim_H Q'$.

Proof: Let $\sim$ be a polarized simulation and let P, Q, H be such that $P \sim_H Q$ and $P \twoheadrightarrow P'$. Pick a constant a such that $a?, a! \notin fc(P, Q)$. Then $P \mid a!(a!) \twoheadrightarrow P' \mid a!(a!)$, hence $P \mid a!(a!) \xRightarrow{a(a!)} P'$. By the definition of polarized bisimulation, $Q \mid a!(a!) \xRightarrow{\alpha'} Q'$, for some action α' such that $a(a!) \leftrightarrow_H \alpha'$, and term Q' such that $P' \sim_{H.(a(a!),\alpha')} Q'$. The only such action is again $a!(a!)$. That is, $Q \mid a!(a!) \twoheadrightarrow Q' \mid a!(a!)$. This implies with $a! \notin fc(Q)$ that $Q \twoheadrightarrow Q'$. $\square$

Corollary 4.12 $\approx$ is reduction-closed: if $P \approx Q$ and $P \twoheadrightarrow P'$ then there exists a term Q' such that $Q \twoheadrightarrow Q'$ and $P' \approx Q'$.

Proof: Assume $P \approx Q$ and $P \twoheadrightarrow P'$. By definition of $\approx$, $P \approx_{H(P,Q)} Q$. Since $\approx_-$ is a bisimulation, it follows with Proposition 4.11 that there is a term Q' with $Q \twoheadrightarrow Q'$ and $P' \approx_{H(P,Q)} Q'$. Since $fc(P',Q') \subseteq fc(P,Q)$ the proposition then follows with Lemma 4.7. $\square$

Definition 4.13 A term P can be observed at a constant $a \in fc(P)$, written $P \Downarrow_a$, if $P \twoheadrightarrow (a!u \mid P')$ or $P \twoheadrightarrow \nu u.(a!u \mid P')$, for some name u and term P'. A term is *insensitive*, if it can be observed at no constant a.

Proposition 4.14 Let a be a constant. If $P \Downarrow_a$ and $P \approx Q$ then $Q \Downarrow_a$.

Proof: Immediate from the definition of polarized bisimulation. $\square$

Corollary 4.15 [12] Let P be an insensitive process. Then

$$P \approx Q \iff Q \text{ is insensitive.}$$

Hence, polarized bisimulation is a sound equivalence, in the sense of Honda and Yoshida. Proposition 4.14 also gives us a link to the other most important notion of process equivalence for polarized calculi, *barbed bisimulation* [12].

Definition 4.16 A symmetric relation $\sim_b$ between polarized π terms is a (weak) *barbed bisimulation* if $\sim_b$ is reduction-closed and $P \sim_b Q$ and $P \Downarrow_a$ implies $Q \Downarrow_a$, for all terms P, Q and constants a. Let $\dot{\approx}_b$ be the largest barbed bisimulation.

Unlike polarized bisimulation, barbed bisimulation is not a congruence. One therefore defines *barbed congruence* $\approx_b$ to be the largest congruence contained in $\dot{\approx}_b$. The next proposition states that polarized bisimulation is a barbed bisimulation, and therefore is finer than barbed congruence.

Proposition 4.17 $\approx$ is a barbed bisimulation.

Proof: By Corollary 4.12, $\approx$ is reduction-closed, and by Proposition 4.17 it preserves output events. $\square$

In summary, we have arrived at a tractable definition of a congruence for processes that captures the notion of polarity in name accesses — "tractable" in that no quantification over all possible contexts is needed in proofs. In this sense, $\approx$ is the analogue of Honda and Yoshida's asynchronous bisimulation [4], with which it also shares the properties of soundness and reduction-closedness.

A currently open question is whether, analogously to asynchronous bisimulation, polarized bisimulation is maximal among all consistent, sound and reduction-closed congruences for polarized π, or equivalently, whether polarized bisimulation and barbed bisimulation coincide.

5 Encoding of Computational λ-Calculus

In this section, we give an encoding of a sequential program calculus into polarized π, and study what happens to the program equalities in the translation. As the source of our translation, we use Moggi's computational λ-calculus, λ_c. We concentrate on λ_c's laws for function application and local definitions, which hold for a wide class of sequential programming languages, including languages with side-effects or non-determinism.

Syntactic Domains

Terms	L, M, N	$::= V \mid E$	
Values	V	$::= x \mid \lambda x.M$	
Non-Values	E	$::= M\,N \mid \text{let } x = M \text{ in } N$	

Equivalences

(β_V)	$(\lambda x.M)V$	$= M[x := V]$	
(η_V)	$\lambda x.Vx$	$= V$	$(x \notin \text{fv}(V))$
(let_V)	let $x = V$ in M	$= M[x := V]$	
(id)	let $x = M$ in x	$= M$	
$(comp)$	let $y =$ let $x = L$ in M in N	$=$ let $x = L$ in let $y = M$ in N	
$(let.1)$	$E\,M$	$=$ let $x = E$ in xM	$(x \notin \text{fv}(M))$
$(let.2)$	$V\,E$	$=$ let $x = E$ in Vx	$(x \notin \text{fv}(V))$

Encoding in π-Calculus

$[x]k$	$= kx$	
$[\lambda x.M]k$	$= \nu a.(k(a!) \mid * a?(x, l).[M]l)$	$(a \text{ fresh})$
$[M\,N]k$	$= \nu a.([M](a!) \mid * a?x.\nu b.([N](b!) \mid * b?y.x(y, k)))$	$(a, b \text{ fresh})$
$[\text{let } x = M \text{ in } N]k$	$= \nu a.([M](a!) \mid * a?x.[N]k)$	

Fig. 3. Moggi's computational lambda calculus, λ_c.

By translating λ_c to polarized π we hope to find out which of these properties continue to hold in a process calculus. This stands in contrast to encodings such as Milner's [9], Boudol's [2], or Sangiorgi's [19], which have worked with pure λ-calculi.

Figure 3 describes our source calculus and the encoding function. We work with the untyped version of λ_c, which is presented in Moggi's technical report [13] but is omitted from his conference paper [14].

The encoding $[\cdot]$ of λ_c in polarized π takes an additional variable k as parameter, which determines where the result of a λ_c-term should be sent to. It differs from previous work [9, 19] in that the function and argument part of an application are evaluated in sequence. We will show below that this is necessary for preserving the $(let.1)$ equivalence. The target language of the encoding is a polyadic calculus, where messages can be tuples of names. Tuples can be expanded out using the translations below.

$$a!(u, v) = \nu b.(a!b \mid b?x\,.\,(x!u \mid b?x\,.\,x!v))$$
$$a?(x, y).M = a?z.\nu c.(z!c \mid c?x\,.\,(z!c \mid c?y.M))$$

This translation of pairs is a slight variation of Honda and Tokoro's "zip-lock" technique[1]

---

[1] Parallel compositions in the input term correspond to input prefixes in the output term and vice versa.

[5]. We add an additional message exchange at the beginning of the communication, in order to keep the direction in which messages are sent over the common channel, a (in Honda and Tokoro's original proposal this direction is reversed compared to single-name message exchanges).

Our first task will be to prove that the encoding maps the β-value reduction relation to a process equivalence. An important step on this way is the following *contraction lemma*.

Lemma 5.1 For all terms P, Q and constants a, b, c, d such that $a?, b?, c!, c? \notin \mathrm{fn}(P, Q)$,

$$(i) \qquad \nu a.\nu b.(P \mid * a?x.Q \mid * b?x.Q) \approx \nu c.(P[a, b := c] \mid * c?x.Q[a, b := c])$$
$$(ii) \qquad \nu a.(* d?y.P \mid * a?x.Q) \approx * d?y.\nu a.(P \mid * a?x.Q).$$

Proof sketch: (i) Abbreviate the sides of the equation by defining $R = P \mid * a?x.Q \mid * b?x.Q$ and $R' = P[a, b := c] \mid * c?x.Q[a, b := c]$. Let

$$\sim_H \stackrel{\mathrm{def}}{=} \begin{cases} \{(\nu a.\nu b.R, \nu c.R') \mid a?, b?, c!, c? \notin \mathrm{fn}(P, Q)\} & \text{if } a! \not{H} c!, b! \not{H} c! \\ \{(\nu a.R, R') \mid a?, b?, c!, c? \notin \mathrm{fn}(P, Q)\} & \text{if } a! \not{H} c!, b! H c! \\ \{(\nu b.R, R') \mid a?, b?, c!, c? \notin \mathrm{fn}(P, Q)\} & \text{if } a! H c!, b! \not{H} c! \\ \{(R, R') \mid a?, b?, c!, c? \notin \mathrm{fn}(P, Q)\} & \text{if } a! H c!, b! H c! \end{cases}$$

An analysis of possible reductions shows that $\{\sim_H \mid H \in \mathcal{H}\}$ is a polarized bisimulation. This implies $\nu a.\nu b.R \approx \nu c.R'$. An analogous proof shows (ii). □

Definition 5.2 Let $=_c$ be the smallest congruence over λ_c-terms that contains the equalities (β_V), (let_V), (id), $(comp)$, $(let.1)$, $(let.2)$.

Theorem 5.3 For any λ_c terms M, N and any constant a, if $M =_c N$ then $[M](a!) \approx [N](a!)$.

Proof sketch: One shows that each of the generating equalities of $=_c$ is a bisimulation. The proof for (β_V) makes use of the following equality, which relates substitutions in λ_c and polarized π:

$$[M[x := \lambda y.N]](a!) \approx \nu b.([M](a!)[x := b!] \mid * b?(y, k).[N]k),$$

This equality is shown by a structural induction on M, using Lemma 5.1 for the steps that deal with parallel composition and replicated input. □

Note that the $(let.1)$ equality requires a sequential left-to-right encoding of function application. If function-application was encoded as in [9, 19], where both function and argument evaluate in parallel, then $E M$ and let $x = E$ in xM would not have equivalent encodings. (To see why, substitute a non-terminating computation for E).

Note also that Theorem 5.3 does not extend to the η_V equality. A counterexample is:

$$[\lambda x.yx](a!) \approx \nu b.(a!(b!) \mid * b?(x, k).y(x, k)) \not\approx a!y = [y](a!)$$

The inequality can be observed by placing both sides in the context

$$c!(d!) \mid c?y.([\,] \mid a?z.[z \doteq d?]e!(e!)).$$

Why does η_V fail to hold? Essentially η_V is a completeness property; it states that every value is either a function or a variable that can be substituted only with functions. But a process context might substitute arbitrary names, including ones which can be tested for identity (and therefore cannot model a function).

6 Conclusion

We have proposed a refinement of name passing in the π-calculus where names carry polarities. We have shown that this refinement has a useful notion of bisimulation congruence which validates all observational equivalences of computational λ-calculus except η-value. An interesting topic for future work are encodings of other sequential constructs, such as those discussed in [20, 15]. In particular, we hope to find out more about which observational equivalences translate to polarized bisimulations between encoded terms.

Another open question concerns the relationship of polarized bisimulation with other notions of process equivalences. For the case without polarities, Honda and Yoshida have shown that asynchronous bisimulation includes all other consistent, sound and reduction-closed congruences. It remains to be seen whether a similar characterization holds for polarized bisimulation.

Acknowledgments

Thanks to John Maraist, David N. Turner and the anonymous referees for detailed and helpful comments.

References

1. Gérard Berry and Gérard Boudol. The chemical abstract machine. In *Proc. 17th ACM Symposium on Principles of Programming Languages*, pages 81–94, January 1990.
2. Gérard Boudol. Asynchrony and the pi-calculus. Research Report 1702, INRIA, May 1992.
3. U. Engberg and M. Nielsen. A calculus of communicating systems with label-passing. Report DAIMI PB-208, Computer Science Department, Aarhus University, 1986.
4. Keiho Honda and Nobuko Yoshida. On reduction-based process semantics. In *Proc. 13th Conf. on Foundations of Softawre Technology and Theoretical Computer Science*, pages 373–387, December 1993.
5. Kohei Honda and Mario Tokoro. An object calculus for asynchronous communication. In *Proc. 5th European Conference on Object-Oriented Programming*, pages 133–147, July 1991. Springer LNCS 512.
6. C.B. Jones. Process-algebraic foundations for an object-based design notation. Technical Report UMCS-93-10-1, University of Manchester, 1993.
7. Robin Milner. Functions as processes. Rapport de Recherche 1154, INRIA Sophia-Antipolis, February 1990.
8. Robin Milner. The polyadic π-calculus: A tutorial. Report ECS-LFCS-91-180, Laboratory for Foundations of Computer Science, Edinburgh University, October 1991.
9. Robin Milner. Functions as processes. *Mathematical Structures in Computer Science*, 2(2):119–141, 1992.
10. Robin Milner. Elements of interaction. *Communications of the ACM*, 36(1):78–89, January 1993. Turing Award lecture.
11. Robin Milner, Joachim Parrow, and David Walker. A calculus of mobile processes, part I and II. Report ECS-LFCS-89-85/86, Laboratory for Foundations of Computer Science, Edinburgh University, 1989.
12. Robin Milner and Davide Sangiorgi. Barbed bisimulation. In *Automata, Languages, and Programming, 19th International Colloquium*, 1992. Lecture Notes in Computer Science 623.
13. Eugenio Moggi. Computational lambda-calculus and monads. LFCS Report ECS-LFCS-88-66, University of Edinburgh, October 1988.

14. Eugenio Moggi. Computational lambda-calculus and monads. In *Proceedings 1989 IEEE Symposium on Logic in Computer Science*, pages 14–23. IEEE, June 1989.
15. Martin Odersky. Applying π: Towards a basis for concurrent imperative programming. In *Proc. 2nd ACM SIGPLAN Workshop on State in Programming Languages*, January 1995.
16. Benjamin Pierce and Davide Sangiorgi. Typing and subtyping for mobile processes. In *IEEE Symposium on Logic in Computer Science*, June 1993.
17. Benjamin C. Pierce, Didier Rémy, and David N. Turner. A typed higher-order programming language based on the Pi-calculus. Draft report; available in the PICT distribution, July 1993.
18. Davide Sangiorgi. *Expressing Mobility in Process Algebras: First-Order and Higher-Order Paradigms*. PhD thesis, Department of Computer Science, University of Edinburgh, 1992.
19. Davide Sangiorgi. An investigation into functions as processes. In *Proc. 9th International Conference on the Mathematical Foundation of Programming Semantics, New Orleans, Louisana*, pages 143–159, April 1993.
20. David Walker. π-calculus semantics of object-oriented programming languages. In Takayasu Ito and Albert R. Meyer, editors, *Proc. Theoretical Aspects of Computer Software*, pages 532–547. Springer-Verlag, September 1991. LNCS 526.

A Polarized Bisimulation is a Congruence

We sketch a proof of Theorem 4.10, which states that $\approx$ is a congruence. The proof is largely, but not completely, along the lines of [11].

We start with showing that $\approx$ is an equivalence relation. Given two bisimulations $\sim^1$, $\sim^2$, define their *composition* by

$$\sim^1 \cdot \sim^2 = \{\{(P,Q) \mid \exists H_1, H_2, R. \ H = H_1 \cdot H_2, \ P \sim^1_{H_1} R \sim^2_{H_2} Q\} \mid H \in \mathcal{H}\}$$

Proposition A.1 The composition $\sim^1 \cdot \sim^2$ of two polarized bisimulations $\sim^1$ and $\sim^2$ is a polarized bisimulation.

Proof sketch: A straightforward verification of closure properties. $\square$

Proposition A.2 The binary relation $\approx$ is an equivalence relation.

Proof sketch: Only transitivity is nontrivial. Assume that $P \approx_{H(P,R)} R$ and $R \approx_{H(R,Q)} Q$. Let $H' = H(P,Q,R)$. By Lemma 4.7, $P \approx_{H'} R$ and $R \approx_{H'} Q$. Then by the definition of composition for bisimulations, $P (\approx \cdot \approx)_{H'} Q$. By the other direction in Lemma 4.7, $P (\approx \cdot \approx)_{H(P,Q)} Q$. Since $\approx \cdot \approx$ is a bisimulation, it follows that $P \approx Q$. $\square$

Proposition A.3 If $P \approx Q$ then

(a) $\quad \nu a.P \approx \nu a.Q$
(b) $\quad P \mid R \approx Q \mid R$
(c) $\quad ux.P \approx ux.Q$
(c) $\quad *ux.P \approx *ux.Q$
(e) $\quad [u \doteq v]P \approx [u \doteq v]Q.$

Proof sketch: (a) We show by an analysis of possible reductions that

$$\sim \overset{\text{def}}{=} \{\{(\nu a.P, \nu a.Q) \mid a!, a? \notin \text{dom}(H) \land P \approx_{H \cup \delta_{\{a\}}} Q\} \mid H \in \mathcal{H}\}$$

is a bisimulation.

(b) For sets of constants $A = \{a_1, ..., a_n\}$, let $\nu A.P = \nu a_1. ... . \nu a_n.P$. Using (a), we show by an analysis of possible reductions that

$$\sim \overset{\text{def}}{=} \{\{(\nu A.(P \mid R), \nu A.(Q \mid R')) \\ \mid (A! \cup A?) \cap \text{dom}(H) = \emptyset \land P \approx_{H \cup \delta_A} Q \land R \approx_{H \cup \delta_A} R'\} \mid H \in \mathcal{H}\}$$

is a bisimulation.

(c) We distinguish according to the kind of the name u. If u is a constant with output polarity, then $ux.P \approx 0 \approx ux.Q$. Assume now that $u = a?$, for some constant a. Let a *state* S be a parallel composition of zero or more output terms $u_1 v_1 \mid ... \mid u_n v_n$. We show that

$$\sim \overset{\text{def}}{=} \{\{(a?x.P \mid S, a?x.Q \mid S') \\ \mid S \approx_H S' \land \forall u, v.uHv \Rightarrow P[x := u] \approx_H Q[x := u]\} \mid H \in \mathcal{H}\}$$

is a bisimulation, which implies the result. For variables u the result follows from the previous discussion and the definition of bisimilarity on open terms.

(d) The result follows from (b), (c) and the observation that, if $*ux.P \mid Q \twoheadrightarrow R$ then there exist a number $n \geq 0$ and a term R' such that $(ux.P)^n \mid Q \twoheadrightarrow R'$ and $R \equiv *ux.P \mid R'$.

(e) If $u = a!$ and $v = b?$, the result follows directly from the name matching law. If u and v are different constants, or constants with the wrong polarities, we have $[u \doteq v]P \approx 0 \approx [u \doteq v]Q.$. If u or v is a variable, the result follows from the previous discussion and the definition of bisimilarity on open terms. $\square$

Proposition A.4 If $P \approx Q$ then $P[x := u] \approx Q[x := u]$.

Proof: Direct from the definition of $\approx$ for open terms. $\square$

Theorem 4.10 $\approx$ is a congruence.

Proof: From Proposition A.2, Proposition A.3 and Proposition A.4. $\square$

Path Balance Heuristic
for Self-Adjusting Binary Search Trees

R. Balasubramanian and Venkatesh Raman

The Institute of Mathematical Sciences,
C. I. T. Campus, Madras 600 113.
email: {balu, vraman}@imsc.ernet.in

Abstract. In (A. Subramanian, An Explanation of Splaying, *Proceedings of the 14th Foundations of Software Technology and Theoretical Computer Science*, LNCS Springer Verlag 880 354-365), D. Sleator suggested the following heuristic for self adjusting binary search trees: Every time an access is made, restructure the entire path from the root of the search tree to the accessed node into a balanced binary search tree on those nodes, and place all other subtrees rooted at the children of the nodes in the access path in their proper positions. We show that the method has an $O(\log n \log \log n / \log \log \log n)$ amortized complexity.

1 Introduction

A binary search tree is the standard way to implement dictionary operations (insert, search and delete). Traditional approaches to achieve logarithmic worst case time for the operations centered on various ways of balancing binary search trees[5, 3]. On the other hand, instead of maintaining explicit balance conditions, there are heuristics that adjust the tree every time an access or an update is made, using a method that depends only on the structure of the access path. Allen and Munro[1] and Bitner[2] studied such restructuring heuristics, but their methods have linear amortized bounds. Sleator and Tarjan[7] were the first to introduce a method, called the splaying heuristic, which has logarithmic amortized time bound for all dictionary operations. Subramanian[8] has generalized the heuristic to allow any set of *depth reducing*, constant depth rules in the rule set (against the zig-zag, zig-zig and the zig rules of Sleator and Tarjan) as long as the restructuring algorithm satisfies what is called the *progress criterion*. Sleator in [8] suggests the following natural heuristic: After an access, restructure the entire path from the accessed node to the root of the search tree into a balanced binary search tree. He asked whether the heuristic works well. We call the heuristic *path balance heuristic*, and show in this paper that this heuristic has an amortized complexity of $O(\log n \log \log n / \log \log \log n)$. We also show that this bound is tight for the potential function we use, indicating the limitations of the potential function.

2   The Model

We first make the heuristic and the model precise before we state our main result. The basic operation that is performed to convert the access path into a balanced tree is the *rotation* operation, as in the splaying heuristics. Suppose the given binary search tree T has a node x with Y and Z as the left and right subtrees respectively (one or both of them could be empty). Let y and z be the roots of the subtrees Y and Z respectively. Let A and B respectively be the left and right subtrees of y, and C and D respectively be the left and right subtrees of z. Then the operation *rotateright(x, T)* results in the tree in which the subtree rooted at x is converted as follows: the subtree rooted at z remains the same; the subtree rooted at x has now B and the subtree rooted at z as the left and right subtrees respectively. And y becomes the root of the subtree with A and the new subtree rooted at x as the left and right subtrees respectively. The operation *rotateleft(x, T)* can be similarly defined. In the path balance heuristic, after an element is accessed, the entire path from the root of the tree to the accessed node is converted into a balanced binary tree by performing rotations. The subtrees rooted at nodes which are children of nodes on the access path, and themselves not on the access path are placed in their unique positions to maintain the binary search tree property.

When the number of nodes in the access path is not one less than a power of 2, the portion of the resulting tree consisting of those nodes will not be perfectly balanced, and hence the leaves will appear in the last two levels. We adopt the convention that the leaves in that portion of the balanced tree appear consecutively in both levels. We adopt this convention just for the ease of our exposition, but our analysis will work on any heuristic that balances the access path so that the leaves occur in the last two levels. Sleator, Tarjan and Thurston[6] have shown that any binary search tree on n nodes can be converted into any other on those nodes using at most $2n - 6$ rotations. Hence the conversion of the access path into a balanced binary tree requires at most $2h - 6$ rotations if h is the length of (i.e. the number of nodes in) the access path. This number together with h to access the node at depth h, account for the actual time of $3h - 6$ to access a node at depth h of the binary search tree using the path balance heuristic.

Finally, we explain what we mean by the amortized time complexity as stated. Suppose we start with a binary search tree having n nodes and perform m accesses on the tree, balancing the access path after every access. Then the total time for the m accesses is $O(m \log n \log \log n / \log \log \log n + g(n))$ where $g(n)$ is the potential of the initial tree (which, for the potential function we use, turns out to be at most $n \log n / \log \log \log n$). So, when m is at least $n / \log \log n$, the total time for the m accesses is $O(m \log n \log \log n / \log \log \log n)$.

3 The Main Result

Theorem 1. *The amortized complexity of an access in a binary search tree is $O(\log n \log \log n / \log \log \log n)$ if the path balance heuristic is used after every access.*

Proof. We prove the amortized complexity using the potential method[9]. The potential we use is a factor of the potential used by Sleator and Tarjan[7] in their splaying analysis. The size of a node x, size(x) is the number of nodes in the subtree rooted at x, including x. The rank of a node x, rank(x) is the logarithm of size(x). (All logarithms in this paper are to the base 2 unless otherwise specified.) The potential Φ of a search tree having n nodes is $f(n)$ times the sum of the ranks of all nodes in the search tree, where f is a function of n which will be determined later.

Let T_- be the tree (having n nodes) before an access and let T^+ be the tree after the access of a node at depth h of T_-. Then the amortized cost of an access (see [9]) is at most $3h - 6 + \Delta\Phi$ where $\Delta\Phi = \Phi(T^+) - \Phi(T_-)$ is the change in the potential due to the access. In the next two subsections we prove that

$$\Delta\Phi \leq f(n)(3 \log n \log(h+1) + 3 \log n + 1 - \frac{(h-6)}{2} \log((h-6)/2 \log n))$$

$$\text{if } h \geq 2 \log n + 6 \text{ and} \tag{1}$$

$$\leq f(n)(3 \log n \log(h+1) + 3 \log n + 1) \text{ otherwise .} \tag{2}$$

Now, from the above two inequalities, we show that the maximum value (over all h) of the amortized cost $3h - 6 + \Delta\Phi$ is minimized when $f(n)$ is chosen to be $\Theta(1/ \log \log \log n)$ and the optimal value is $\Theta(\log n \log \log n / \log \log \log n)$.

Lemma 2. *Let n be any non-negative integer, and f be any arbitrary function that takes non-negative integers to reals, and for $1 \leq h \leq n$ let*

$$C(h, f, n) = 3h + f(n)(3 \log n \log(h+1) + 3 \log n + 1 - \frac{h-6}{2}(\log((h-6)/2 \log n)))$$

$$\text{for } h \geq 2 \log n + 6 \tag{3}$$

$$= 3h + f(n)(3 \log n \log(h+1) + 3 \log n + 1) \text{ otherwise .} \tag{4}$$

For a given function f, let $\bar{C}(f, n) = max_{1 \leq h \leq n} C(h, f, n)$, and let $C_0(n) = min_f \bar{C}(f, n)$. Then $C_0(n)$ is $\Theta(\log n \log \log n / \log \log \log n)$.

Proof. First we show that by choosing $f(n) = 6 / \log \log \log n$, $\bar{C}(f, n)$ is $O(\log n \log \log n / \log \log \log n)$.

If $h < 2 \log n + 6$, then equation 4 applies and $C(h, f, n)$ is within bounds. Otherwise, for any integer n, let $h = 2k \log n + 6$ for some integer $k > 1$ and $X(k, f) = C(h, f, n)$. Then

$$X(k, f) = 6k \log n + 18 + f(n)(3 \log n \log k + 3 \log n \log \log n + O(\log n) - k \log n \log k).$$

We seek to determine the value of k that maximizes X, given $f(n)$.

$$\frac{dX}{dk} = 6\log n + \frac{f(n)\log n}{\ln 2}(3/k - \ln k - 1)$$
$$\frac{d^2X}{dk^2} = \frac{f(n)\log n}{\ln 2}(-3/k^2 - 1/k) < 0 \text{ when } k > 0$$

So, at the critical value of k, we have

$$\log k = \frac{6}{f(n)} + \frac{3}{k\ln 2} - \frac{1}{\ln 2}. \tag{5}$$

Substituting this expression for $\log k$ in $X(k, f)$ and simplifying, the first and the last terms almost cancel out, and we get the maximum value of X to be

$$Xmax = f(n)\log n(3\log\log n + O(1) + k/\ln 2) + 18(\log n + 1). \tag{6}$$

Since $f(n) = 6/\log\log\log n$, it follows that $k \leq \log\log n$ and $Xmax$ is $O(\log n\log\log n/\log\log\log n)$.

To prove the lower bound, we use the connection between the critical value of k and $f(n)$ in equation 5 ($\log k$ is $6/f(n) + O(1)$) and use it in equation (6) to minimize for $f(n)$. That is, we want to minimize for $f(n)$ in

$$Xmax = f(n)\log n(3\log\log n + O(1) + c2^{6/f(n)}) + 18(\log n + 1)$$

where c is some constant. Clearly the minimum occurs when $3\log\log n = d2^{6/f(n)}$ for some constant d from which it follows that the optimum value of $f(n)$ is $\Theta(1/\log\log\log n)$.

$\square$ (Lemma 2)

Assuming inequalities (1) and (2), this completes the proof of the main theorem.

$\square$ (Theorem 1)

Inequalities (1) and (2) are proved in the next two subsections. We first prove the following lemma which is used in the next subsections to establish the required inequalities.

Lemma 3. *Let* $y_1, y_2, \ldots, y_l$ *be a sequence of integers greater than or equal to 2 such that* $\sum_{i=1}^{l} y_i = n$, *and let*

$$D = \sum_{i=1}^{l}(\log y_i - \log(y_i + y_{i+1} + \ldots + y_l - 1)).$$

Then $D \leq 1$ *for all* l, *and*

$$D \leq 1 - (l-1)\log(\frac{l-1}{\log n})$$

whenever $l - 1 \geq \log n$.

Proof. Since $\log y_l - \log(y_l - 1) \le 1$, and every other term in the summation for D is less than 0, it follows that $D \le 1$.

For $i = 1, 2, \ldots, l$, set $s_i = y_i + y_{i+1} + \ldots + y_l - 1$. Thus, $s_1 = n - 1$ and $s_l = y_l - 1 \ge 1$. As before, we bound the last term in D by 1. For the remaining, we have

$$\sum_{i=1}^{l-1} \log \frac{y_i}{s_i} = \sum_{i=1}^{l-1} \log(1 - \frac{s_{i+1}}{s_i}) \le (l-1)\log(1 - \frac{1}{l-1}\sum_{i=1}^{l-1}\frac{s_{i+1}}{s_i}), \qquad (7)$$

where the inequality holds because the geometric mean is at most the arithmetic mean. On the other hand,

$$\prod_{i=1}^{l-1} \frac{s_{i+1}}{s_i} = \frac{s_l}{s_1} \ge \frac{1}{n-1} \ge \frac{1}{n}.$$

Since Arithmetic Mean $\ge$ Geometric Mean, we have

$$\frac{1}{l-1}\sum_{i=1}^{l-1}\frac{s_{i+1}}{s_i} \ge \left(\prod_{i=1}^{l-1}\frac{s_{i+1}}{s_i}\right)^{1/(l-1)} \ge n^{-1/(l-1)}.$$

Substituting for $\frac{1}{l-1}\sum_{i=1}^{l-1}\frac{s_{i+1}}{s_i}$ in (7), we get

$$\sum_{i=1}^{l-1} \log \frac{y_i}{s_i} \le (l-1)\log(1 - n^{-1/(l-1)}) = (l-1)\log(1 - \exp(-\frac{\ln n}{l-1})).$$

For $x \le 1$, $\exp(-x) \ge 1 - x$. Thus, if $l - 1 \ge \ln n$, then $\exp(-\frac{\ln n}{l-1}) \ge 1 - \frac{\ln n}{l-1}$, and

$$\sum_{i=1}^{l-1} \log \frac{y_i}{s_i} \le (l-1)\log(\frac{\ln n}{l-1}) \le (l-1)\log(\frac{\log n}{l-1}) = -(l-1)\log(\frac{l-1}{\log n}).$$

$\square$

3.1 When the access path is a chain

First, let us assume that the access path is simply a right leaning chain. Similar argument holds when it is a left leaning chain. Let $1, 2, \ldots, h$ be the access path with 1 as the root. Let $g_1, g_2, \ldots, g_h$ be the sizes of the left subtrees of $1, 2, \ldots, h$ respectively, and let g_{h+1} be the size of the right subtree of h. Let $\Phi(T)$ be the potential of the tree T. Then the change in potential due to the access, $\Delta\Phi$ is given by

$$\Delta\Phi \leq f(n)(\log(g_1 + g_2 + 1) + \log(g_3 + g_4 + 1) + \ldots$$
$$+ \log(g_1 + g_2 + g_3 + g_4 + 3) + \log(g_5 + g_6 + g_7 + g_8 + 3) + \ldots$$
$$+ \ldots$$
$$+ \log(g_1 + g_2 + \ldots + g_h + g_{h+1} + h)$$
$$-(\log(g_1 + g_2 + \ldots + g_h + g_{h+1} + h) + \log(g_2 + \ldots + g_h + g_{h+1} + h - 1)$$
$$+ \ldots + \log(g_h + g_{h+1} + 1)))$$

Hence,

$$\Delta\Phi/f(n) \leq \log(g_1 + g_2 + 2) + \log(g_3 + g_4 + 2) + \ldots$$
$$+ \log(g_1 + g_2 + g_3 + g_4 + 4) + \log(g_5 + g_6 + g_7 + g_8 + 4) + \ldots$$
$$+ \ldots$$
$$+ \log n$$
$$-(\log(g_1 + g_2 + \ldots + g_{h+1} + h) + \log(g_2 + \ldots + g_{h+1} + h - 1)$$
$$+ \ldots \log(g_h + g_{h+1} + 2)$$
$$= \log(x_1 + x_2) + \log(x_3 + x_4) + \ldots$$
$$+ \log(x_1 + x_2 + x_3 + x_4) + \log(x_5 + x_6 + x_7 + x_8) + \ldots$$
$$+ \ldots$$
$$+ \log n$$
$$-(\log(x_1 + x_2 + \ldots + x_{h+1} - 1) + \log(x_2 + \ldots + x_{h+1} - 1)$$
$$+ \ldots \log(x_h + x_{h+1} - 1))$$

where $x_i = g_i + 1$.

We break the positive terms of the above inequality into $\lceil \log(h + 1) \rceil$ levels, each level representing the ranks of the nodes in that level of the restructured tree. We upper bound the first term of each level, which are of the form $\log(x_1 + \ldots)$ by $\log n$ to get the term $\lceil \log(h + 1) \rceil \log n$. Then our strategy is to show for every other positive term, the existence of a unique negative term which is larger than the positive term. This will prove inequality (2). Then we appeal to Lemma 3 with some of these terms to show inequality (1).

Note that, for every x_i, there is a negative term of the form $\log(x_i + \ldots + x_{h+1} - 1)$. We first order all the positive terms which are of the form $\log(x_j + \ldots)$ after omitting the first term of each level, in the increasing order of j. If there are several terms of the form $\log(x_j + \ldots)$ for a fixed j, then they are arranged in any order among themselves. Now we inspect each positive term in this order, and identify it with the first negative term of the form $\log(x_k + \ldots + x_{h+1} - 1)$ with $k < j$, which has not been identified earlier with some other positive term. We argue below that such a term exists.

Consider a positive term of the form $\log(x_j + \ldots)$ for a fixed $j > 1$ and suppose each of the positive terms of the form $\log(x_k + \ldots)$ for $k < j, k \neq 1$ has been identified with a negative term larger than itself. The number of negative terms of the form $\log(x_k + \ldots + x_{h+1} - 1), (k < j)$ is exactly $j - 1$. Some of these may have already been identified with positive terms of the form $\log(x_k + \ldots)$ for some $k \leq j, k \neq 1$. (Other positive terms have not been inspected yet.) We show that the number of positive terms of the form $\log(x_k + \ldots), k \leq j, k \neq 1$ is at most $j - 1$. This will prove that we can associate a negative term of the required form to the current positive term.

It is easy to see the argument when h is one less than a power of 2. We handle this case first. In the first (bottom) level, there are $(h + 1)/2$ terms of the form $\log(a + b)$ out of which $\lceil j/2 \rceil$ terms are of the form $\log(x_k + x_{k+1})$ where $k \leq j$. Similarly in the next level where every term is of the form $\log(a+b+c+d)$, $\lceil j/4 \rceil$ terms are of the form $\log(x_k + x_{k+1} + x_{k+2} + x_{k+3})$ where $k \leq j$. Continuing this argument, after omitting the first term of each level, there are totally at most $j - 1$ terms of the form $\log(x_k + \ldots)$ where $k \leq j$ which is what we wanted to show.

When h is not one less than a power of 2, the leaves are at two different levels. In this case also, in the first (bottom) level every term is of the form $\log(a+b)$ and let there be r such terms of the form $\log(x_k + x_{k+1})$, $k \leq j$. Then in the next level, at most $\lceil r/2 \rceil$ terms are of the form $\log(x_a + x_{a+1} + x_{a+2} + x_{a+3})$ (one of them can be of the form $\log(x_u + x_{u+1} + x_{u+2})$) and some s other terms are of the form $\log(x_y + x_{y+1})$ for some $a, u, y \leq j$. Let $t = \lceil r/2 \rceil + s$. From the next level onwards, the number of terms of the form $\log(x_k + \ldots)$ decreases by half (starting from $\lceil t/2 \rceil$). Since the first term of each level is omitted, the total number of terms of the form $\log(x_k + \ldots), k \leq j, k \neq 1$ is at most $(r-1) + (2t-2) \leq 2r + 2s - 1 \leq j - 1$ since $2r + 2s$ is the maximum number of leaves whose subtree indices are at most j.

Thus, for each positive term other than the first terms of each level, we have identified a unique negative term larger than itself. This proves inequality (2). To prove inequality (1), consider the positive terms of the form $\log(x_i + x_{i+1})$ and $\log(x_j + x_{j+1} + x_{j+2})$ in the first (bottom) two levels $(i, j \neq 1)$. There are $\lfloor h/2 \rfloor - 1$ of them with at most one term of the form $\log(a + b + c)$ (if for some i, there is a term $\log(x_i + x_{i+1})$ as well as $\log(x_i + x_{i+1} + x_{i+2})$, then consider only the latter term). The corresponding negative term for $\log(x_i + x_{i+1})$ is $\log(x_k + \ldots + x_{h+1} - 1)$ for some $k < i$. The positive term minus the negative term is clearly at most $\log(x_i + x_{i+1}) - \log(x_i + x_{i+1} + \ldots + x_{h+1})$ (since each x_i is at least 2). Now considering each such positive term as some $\log y_i$ for some y_i, the sequence of y_i's satisfies the hypothesis of Lemma 3 with $l = \lfloor h/2 \rfloor - 1$, and the sum of the positive minus the negative terms is bounded above by D (Since the $\log(x_1 + x_2)$ term is omitted, the $y_i's$ add up to a number less than n. Note in the proof of Lemma 3 that this is sufficient for the Lemma to hold.) Hence by Lemma 3, and by considering the $\lceil \log(h + 1) \rceil \log n$ value accounted by the first term at each level of the positive terms, we have

$$\Delta\Phi/f(n) \leq \log n \lceil \log(h+1)\rceil + 1 - (\lfloor h/2\rfloor - 2)\log((\lfloor h/2\rfloor - 2)/\log n)$$

from which the inequality (1) follows.

Lemma 4. *If $\Delta\Phi$ is the change in potential when the access path that happens to be a chain, is converted into a balanced binary search tree after every access, then $\Delta\Phi \leq f(n)(\log n \log(h+1) + \log n + 1 - \frac{(h-6)}{2}\log((h-6)/2\log n))$.*

3.2 For the general access path

Suppose the access path is not a chain. Let $1, 2, \ldots, h$ be the labels of the nodes in the access path in the ascending order of their values, and let $g_1, g_2, \ldots, g_{h+1}$ be the sizes of the subtrees rooted at the children of these nodes, which are themselves not in the access path (in the inorder traversal). When there is no confusion, we use the same symbols g_i's to denote the trees themselves. Let t be the index such that $g_1, g_2, \ldots, g_t$ are the left subtrees and $g_{t+1}, g_{t+2}, \ldots, g_{h+1}$ are the right subtrees of the nodes in the access path. Now the positive terms in $\Delta\Phi$ will remain the same, and in the negative term $\Phi(T_-)$, the first term is $\log(x_1 + x_2 + \ldots + x_{h+1} - 1)$ as in the previous case (recall that $x_i = g_i + 1$ for all i). However the kth term (for $k \geq 2$), can be defined inductively as follows: Let the $(k-1)st$ term of $\Phi(T_-)$ be $\log(x_l + x_{l+1} + \ldots + x_r - 1)$. Then the k-th term is $\log(x_{l+1} + x_{l+2} + \ldots + x_r - 1)$ if the k-th node of the access path from the root is the right subtree of the $(k-1)st$. Otherwise, the k-th term is $\log(x_l + x_{l+1} + \ldots + x_{r-1} - 1)$.

Now we again break the positive terms into $\lceil \log(h+1)\rceil$ levels and omit the first and the last terms at each level. Also at each level, there will be at most one term consisting of the values g_i and g_j where $1 \leq i \leq t$ and $t + 1 \leq j \leq h + 1$. Omit such a term also, if exists, in each level. Each term omitted can be upper bounded by $\log n$ accounting for $3\lceil \log(h+1)\rceil \log n$ in total. For the rest of the terms, we follow a similar strategy as in the previous subsection.

For every other positive term $\log(x_j + \ldots)$ for $1 \leq j \leq t$, considered in the increasing order of j's, we identify it with the first negative term of the form $\log(x_k + \ldots)$ where $k < j$ which has not been identified earlier with some other positive term. Similarly for a positive term $\log(\ldots + x_r)$ for $t + 1 \leq r \leq h + 1$, considered in the decreasing order of r's, we identify it with a negative term of the form $\log(\ldots + x_s - 1)$ where $s > r$ which has not been identified earlier. It follows from an argument similar to the one in the last subsection that such terms exist and are unique. This proves inequality (2) in this case.

By a similar argument, by considering the positive terms of the form $\log(a+b)$ and $\log(a + b + c)$ and their corresponding negative terms, and by using Lemma 3, we obtain that

$$\Delta\Phi/f(n) \leq 3\log n \lceil \log(h+1)\rceil + 1 - (\lfloor h/2\rfloor - 2)\log((\lfloor h/2\rfloor - 2)/\log n)$$

from which the inequality (1) follows.

Lemma 5. *If $\Delta\Phi$ is the change in potential when the access path is converted into a balanced binary search tree after every access, then*
$$\Delta\Phi \leq f(n)(3\log n \log(h+1) + 3\log n + 1 - \tfrac{(h-6)}{2}\log((h-6)/2\log n)).$$

4 Concluding Remarks and Open Problems

We have analysed the variant of splaying in which after every access, the access path is converted into a balanced binary tree. We have shown that such a method has an $O(\log n \log\log n/\log\log\log n)$ amortized bound. In the classical splaying heuristic, the accessed element is brought to the root immediately after an access which is not the case in this heuristic. However the accessed element is brought to the root (or one level below the root) if it is accessed successively for $\log^* n$ times. In fact this can be achieved in $O(\log n \log\log n/\log\log\log n)$ total time. For, the total time for the $\log^* n$ accesses is dominated by the time for the first one or two accesses. From then on, the depth of the node becomes at most $\log\log n$, and hence the change in potential is $O(\log n)$ which follows from inequality (2). Hence, as in the classical splaying heuristic, similar amortized bounds can be proved for the other dictionary operations: insert, delete, join and split using this heuristic. When all these operations are performed, n is the total number of inserts and joins in the sequence of operations. Also, the potential of a forest of binary search trees is the sum of the potential of individual trees. The insert and delete operations are implemented as in the classical splaying heuristic by converting the appropriate access path into a balanced tree. The bounds change by at most a constant factor. To perform a *join* of two trees T_1 and T_2 (where every value of T_1 is less than or equal to any value of T_2) using this heuristic, it suffices to access the node containing the largest value of T_1 at most twice. This brings that node to a depth of at most $\log\log n$. Now attach the tree T_2 as the right subtree of this node. The increase in potential is at most $O(\log\log n \log n/\log\log\log n)$ since possible potential increase occurs only for nodes along the rightmost path of T_1 (of length at most $\log\log n$). To perform a split along a value i, we can bring the node i to the root or one level below the root as discussed earlier, and perform a join if necessary.

For the potential function we have used, our bound for the operations is tight. To see this, note that the bound in the statement of Lemma 2 is tight. We show the following to prove that the bound in the statement of Lemma 3 is also asymptotically tight.

Lemma 6. *There exists a sequence of reals $y_1, y_2, \ldots y_l$ each greater than or equal to 2, with $\sum_{i=1}^{l} y_i = n$ and $l > 2\log n$ such that*

$$D = \sum_{i=1}^{l-1}(\log y_i - \log(y_i + y_{i+1} + \ldots y_l)) \geq -2l\log(3l/\log n).$$

Proof. Let $y_i = 2(1 + \epsilon)^{l-i}$ where $\epsilon > 1/\sqrt{n}$ is a small positive value less than 1, and $2(1 + \epsilon)^l - 2 = n\epsilon$. In this case,

$$\log y_i - \log(y_i + y_{i+1} + ...y_l) \geq -\log(1 + \epsilon) - \log(1/\epsilon)$$

and hence $D \geq -l \log(1 + \epsilon) - l \log(1/\epsilon)$.

To show that this value is at least $-2l \log(3l/\log n)$, it suffices to show that $\log(1 + \epsilon) + \log(1/\epsilon) \leq 2 \log(3l/\log n)$.

Since $l > 2 \log n, 3l/\log n > 6 > (1 + \epsilon)$ and hence $\log(1 + \epsilon) < \log(3l/\log n)$.

To complete the proof, it suffices to show that $1/\epsilon \leq 3l/\log n$.

Since $2(1 + \epsilon)^l - 2 = n\epsilon$, $\log n\epsilon < l \log(1 + \epsilon) + 1 < 1 + l\epsilon/\ln 2$ (where $\ln x$ is the natural logarithm of x). As $\epsilon > 1/\sqrt{n}$, $\log n\epsilon > (\log n)/2$, and so $\log n < 2 + 2l\epsilon/\ln 2 < 3l\epsilon$ which is what we wanted to show.

$\square$

Hence asymptotic improvement to our bound, if possible, can be obtained only using a different potential function than what we have used. Improving our bound using a different potential function or proving a matching lower bound is an immediate interesting open problem. One useful observation in this context about the path balance heuristic is that after the restructuring following an access, the depth of any node in the access path becomes at most $\log n$, though the depth of every other node could increase by $\log n$. In contrast, for the classical splaying heuristic, the depths of almost all nodes in the access path as well as those of the other subtrees reduce approximately by a constant factor. The depth increase, if exists for any node, is only by a constant.

We note that the same amortized bound using the same potential function (up to constant factors) was proved for the deletemin operation in the multipass variant of Pairing Heaps[4]. The authors had observed that in one of the other variants, the deletemin operation has essentially the same effect as splaying (using the classical splaying heuristic) at the node that contains the largest value in the corresponding "half-ordered binary tree". The deletemin operation in the multipass variant has a similar effect as accessing the same node of the half-ordered binary tree using the path balance heuristic. So, it is not surprising that we obtain similar bounds. In fact, the authors of [4] obtain a term similar to the value D in our Lemma 3, using which they bound the potential difference by an expression similar to that of equation (1). So our observation that our bound is tight for the potential function we use implies the same for their bound.

Sleator and Tarjan have shown that splay trees have several other properties by assigning weights to the nodes of the search tree in many different ways. It would be interesting to see whether any of those results generalizes to the path balance heuristic. One particularly interesting question is to see how close the total access time for a sequence of accesses using this heuristic is, to the total access time in a statically optimal tree built from the frequencies of accesses of the elements.

Finally, as Subramanian[8] asks, what is special about the potential function which is based on the sizes and ranks that seems to give reasonable bounds? It would be interesting to obtain the same results using a potential function that works directly with depths.

5 Acknowledgements

We thank the referees for their comments that simplified some proofs and improved the readability of the paper. We also thank Bob Tarjan for pointing out the connection between this heuristic and the multi pass variant of Pairing Heaps[4].

References

1. B. Allen and I. Munro, Self-organizing Search Trees, *Journal of the ACM* 25 (1978) 526-535.
2. J. R. Bitner, Heuristics that dynamically organize data structures, *SIAM Journal of Computing* 8 (1979) 82-110.
3. T. H. Cormen, C. E. Leiserson, and R. L. Rivest, *Introduction to Algorithms*, The MIT Press, Cambridge, Massachusetts (1990).
4. M. L. Fredman, R. Sedgewick, D. D. Sleator and R. E. Tarjan, The Pairing Heap: A New Form of Self-Adjusting Heap, *Algorithmica* 1 (1986) 111-129.
5. D. E. Knuth, *The Art of Computer Programming, Vol 3: Sorting and Searching*, Addison-Wesley, Reading, Massachusettes (1973).
6. D. D. Sleator, R. E. Tarjan, and Thurston, Rotation distance, triangulations and hyperbolic geometry, *Proceedings of the 18th ACM Symposium on Theory of Computing* (1986) 122-135.
7. D. D. Sleator and R. E. Tarjan, Self-adjusting Binary Search Trees, *Journal of the ACM* 32 (1985) 652-686.
8. A. Subramanian, An Explanation of Splaying, *Proceedings of the 14th Foundations of Software Technology and Theoretical Computer Science*, LNCS Springer Verlag 880 354-365.
9. R. E. Tarjan, Amortized Computational Complexity, *SIAM Journal of Algebraic and Discrete Mathematics*, 6 (2) (1985) 306-318.

Pattern Matching in Compressed Texts [1]

(Preliminary Version)

S. Rao Kosaraju
Department of Computer Science
The Johns Hopkins University
Baltimore, Maryland 21218-2694

Abstract

We consider the problem of pattern matching when the text is in compressed form. As in Amir, Benson and Farach, we assume that the text is compressed by the Lempel-Ziv-Welch scheme. If the compressed text is of length n and the pattern is of length of m, our basic compression algorithm runs in $O(n + m\sqrt{m}\log m)$ steps, as against Amir, et al's bound of $O(n + m^2)$ steps. We extend the basic algorithm into another that achieves, for any $k \geq 1$, $O(nk + m^{1+\frac{1}{k}}\log m)$ steps.

1 Introduction

The need for storing and processing large amounts of data is well recognized. The recent initiatives on digital libraries are a direct result of the need to handle explosive amounts of data. Thus techniques for compressing texts and processing compressed texts are of paramount importance. In this paper we consider the problem of searching for a pattern in a compressed text. As in [2], we assume that the text is compressed by Lempel-Ziv-Welch scheme and this compressed text is of length n. Note that the real text can be significantly longer than n. We first develop, in section 2, algorithms for several problems that appear to be of independent interest. Finally the pattern matching algorithms are given in section 3. We assume that for any set A, $|A|$ denotes the number of elements in A; for any tree A, $|A|$ denotes the number of nodes in A; for any string z, $|z|$ denotes the length of z.

---

[1]Supported by NSF Grants CCR9107293 and CCR9508545

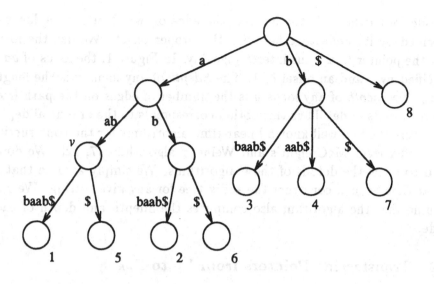

Figure 1: Σ_X for $X = aabbaab$.

2 Preliminary Algorithms

We first review suffix trees for strings, and then develop algorithms for several problems.

2.1 Suffix Trees

For any input $X = x_1 x_2 ... x_n$ and $1 \leq i \leq n$, let $\textit{suffix}_i$, or the ith suffix, be $x_i \cdots x_n \$$ where $\$$ is a special symbol not in the alphabet of X, and $\textit{suffix}_{n+1}$ be $\$$. Let Σ_X be the compact trie for suffixes $1, 2, \cdots, n + 1$. For $X = aabbaab$, Σ_X is shown in Figure 1.

Note that each edge is labeled by a substring of X. In the actual implementation, each label is represented by two pointers into X. The parent and the grandparent of any node u are denoted by $\textit{parent}(u)$ and $\textit{gparent}(u)$, respectively. The $\textit{head}$ of any edge (u, v) is the node v. In addition to the nodes of the tree, it is convenient to be able to refer to each position ($\textit{locus}$) within a label. The $\textit{string for a locus}$, u, is the concatenation of the labels on the path from the root to that locus, and is denoted by σ_u. If u is a node then we say that σ_u $\textit{occurs explicitly}$; otherwise it $\textit{occurs implicitly}$. The $\textit{locus}$ for σ_u is u. Thus in Figure 1, aab occurs explicitly and its locus is v; aa occurs implicitly and its locus is in the edge $(\textit{parent}(v), v)$. The node of any locus u is u itself if u is a node, otherwise it is the head of the edge on which u lies. A locus is specified by its node together with the proper offset. We use the locus and the pointer to a locus interchangeably. In Figure 1, the locus of aa is specified by v and an offset of 1. The *Sdepth* of any locus u is the length of σ_u. The *depth* of any locus u is the number of edges on the path from the root to its node. Thus this notion corresponds to the graphical depth.

There are two well-known linear-time algorithms for the construction of a suffix tree: McCreight's and Weiner's algorithms [7, 12]. We don't need to know the details of these algorithms. We simply assume that in linear-time we can construct the suffix tree for any given string. We also assume that the algorithm also computes the Sdepth and depth of each node.

2.2 Transferring Pointers from Σ_P to Σ_{PR}

Given a pattern $P = p_1 p_2 \cdots p_m$ of length m, suffix trees Σ_P, and Σ_{PR}, and r pointers into Σ_P specifying loci $\tau_1, \tau_2, ..., \tau_r$ we want to find the corresponding pointers into Σ_{PR} whose loci have paths which are reversals of paths in Σ_P (i.e. paths $(\sigma_{\tau_1})^R, ..., (\sigma_{\tau_r})^R$ in Σ_{PR}).

We show that this computation can be performed in $O(m + r)$ steps. The algorithm is based on a nontrivial property of suffix trees. In the following discussion, we assume that in Σ_{PR} the indexing of leaves is not changed; e.g. path of $leaf_i$ in Σ_{PR} is $p_i \cdots p_1 \$$.

For any locus u, let $\hat{u}$ be the maximum k such that $leaf_k$ is in the subtree of u in Σ_P.

Lemma 2.1: *Let u and v be any two loci in Σ_P with Sdepths d_u and d_v, respectively. Let $i = \hat{u} + d_u - 1$ and $i' = \hat{v} + d_v - 1$. Then in Σ_{PR} let the ancestor loci of leaves i and i' (respectively) at Sdepths d_u and d_v (respectively) be w and w' (respectively). Then in Σ_{PR} the paths from w to $leaf_i$ and from w' to $leaf_{i'}$ are disjoint.*

Proof: If neither one of w and w' is an ancestor of the other in Σ_{PR} then the lemma holds trivially. Without loss of generality, let w be an ancestor of w'. If w to $leaf_i$ path doesn't include w' then the lemma holds; otherwise the value of $\hat{v}$ is incorrect. ∎

As a consequence of this lemma, the following algorithm transfers pointers from Σ_P to Σ_{PR} in $O(m + r)$ steps.
Algorithm:

   a) For every pointer u find $\hat{u}$ and Sdepth d_u, and create the pair $(\hat{u} + d_u - 1, d_u)$.

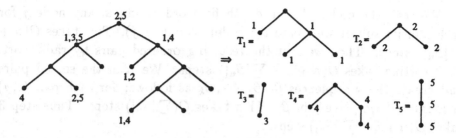

Figure 2: Separating the labels into k trees.

b) Sort the pairs by radix sort.

c) For each i, let $(i, \delta_1)(i, \delta_2) \cdots (i, \delta_k)$ be the sorted substring with i as the first component. In Σ_{PR} climb up from $leaf_i$ and mark loci at Sdepths $\delta_k, ..., \delta_1$ terminating the climb at Sdepth δ_1.

Clearly each of steps a) and b) can be performed in $O(m + r)$ steps. By the previous lemma, all the paths climbed in step c) are disjoint. Hence this step takes $O(m + r)$ steps. Thus the overall algorithm runs in $O(m + r)$ steps.

2.3 Label Separation

Given an n node tree with each node u labeled by a subset, S_u, of $\{1, 2, \cdots, k\}$, we want to construct k trees, $A_1, A_2, \cdots, A_k$, such that each A_i is a compressed tree for the subset of nodes that contain i in its label. An example of this compression is shown in Figure 2.

The following algorithm solves this problem in $O(n + k + \sum |S_u|)$ steps.
Algorithm

1. Label the nodes of A in depth first order.

2. Preprocess A such that any lca (least common ancestor) query can be answered in $O(1)$ steps [8].

3. For each label i, compute the sorted sequence of depth first numbers of nodes that contain i in their subsets. Let this sequence be β_i.

4. Construct A_i making use of β_i and lca queries.

Step 1 can be easily performed in $O(n)$ steps [1]. It is known that step 2 can be performed in $O(n)$ steps [8]. Now we describe the details of step 3.

We scan the nodes of A in depth first order, and at any node j for each i in its subset we create the ordered pair (i, j). This takes $O(n + \sum |S_u|)$ steps. Then we sort the resulting ordered pairs by radix sort. This sorting takes $O(n + k + \sum |S_u|)$ steps. We scan the sorted pairs and create the k sequences $\beta_1, \beta_2, \cdots, \beta_k$ as follows: for each pair (i, j) encountered we place j in β_i. This takes $O(\sum |S_u|)$ steps. Thus step 3 takes $O(n + k + \sum |S_u|)$ steps.

The details of step 4 are as follows. Let $\beta_i = j_1, \cdots, j_p$. We construct the compressed tree for $j_1, \cdots, j_{\ell+1}$ from that for $j_1, \cdots, j_\ell$.

The compressed tree for j_1 is a single node with label j_1 if j_1 is 1 (i.e. the root); otherwise it is a 2 node tree with the non-root labeled by j_1.

Let the compressed tree for $j_1, \cdots, j_\ell$ be denoted by CT_ℓ. Note that for every node in this compressed tree there is a corresponding node in A. For simplicity, we use the same name for a node and its corresponding node. Let the right spine of CT_ℓ be $\alpha_1, \cdots, \alpha_q$, where $\alpha_q = j_\ell$ as shown in Figure 3a. Find $\text{lca}(j_\ell, j_{\ell+1})$ in A. Let it be α. Find the largest x such that α_x is an ancestor of α in A. This can be accomplished by going up CT_ℓ, starting at α_q, and each time testing whether the (node corresponding to) node visited is an ancestor of α in A. Such an x must exist since the node corresponding to the root of CT_ℓ is the root of A. If $\alpha_x = \alpha$, then we make $j_{\ell+1}$ the rightmost child of α_x as shown in Figure 3b; otherwise we create a new node, by splitting the (α_x, α_{x+1}) edge, and then make $j_{\ell+1}$ the right child of the new node as shown in Figure 3c.

We analyze the speed of this algorithm by defining the length of the right spine as the potential of the compressed tree. When we find α_x by climbing up the right spine of the compressed tree starting at α_q, the number of steps spent can be realized as the reduction in potential (within an additive constant). Hence the tree construction for the sequence β_i has a running time of $O(1 + |\beta_i|)$. In addition, it is easy to verify the correctness of this algorithm. Thus step 4 can be performed in $O(\sum |S_u|)$ steps.

2.4 Partial Path Compression

In any n node tree A let $u_1, u_2, ..., u_{k+1}$ be any path. (Each u_i is the parent of u_{i+1}.) Deletion of edges on this path and making each u_i, $2 \leq i \leq k + 1$, a child of u_1 is a *length k path compression* (or simply *k-compression*). We denote this compression as a compression *into node* u_1. Let the number of steps needed to perform this path compression be $k + 1$.

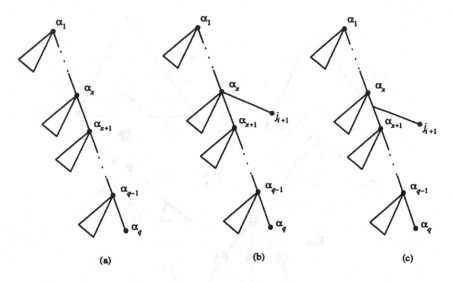

Figure 3: Insertion of j_{l+1}.

Lemma 2.2: *In any n node tree A, the total number of steps needed to perform any $r=\Omega(n\log n)$ path compressions is $O(r)$.*

This lemma is an immediate consequence of the following lemma, which is proved below.

Lemma 2.3: *In any n node tree A if we perform any intermixed sequence of 2-compressions and 3-compressions, the total number of 3-compressions in this sequence is $O(n\log n)$.*

Proof: Let $f(n)$ be the maximum possible sum of 2-compressions into the root and 3-compressions (in an intermixed sequence) we can apply to a tree with n nodes. Thus $f(n)$ counts every 3-compression but only the 2-compressions into the root. It suffices to show that $f(n) = O(n\log n)$.

Observe that if node u is not an ancestor of node v, then u cannot become an ancestor of v due to a path compression.

Let A be an n node tree that realizes $f(n)$. Now in A choose the node w whose subtree contains at least $n/2$ nodes and every one of its children has a subtree with fewer than $n/2$ nodes. It is very easy to handle the case when w is the root of A. In the following we assume that w is not the root of A. Break the tree A up into 2 trees, A_1 and A_2, by detaching the subtree of w as shown in Figure 4. Note that we have created a new root, v, for A_2 and made w its child. We consider v as an extra node that is not in A.

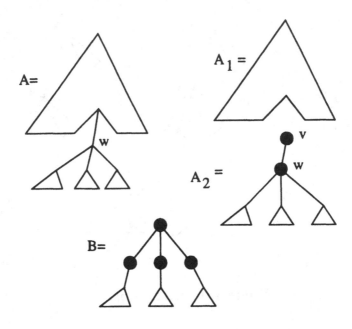

Figure 4: Reassigning 2- and 3-compressions.

As we apply the compressions in A, the original subtree of w breaks up into many individual trees and get attached to a (single) node from A_1. We translate each compression in A into at most one compression in each of A_1 and A_2 as follows.

Case 1: Let the compression be a 2-compression of path $u_1 u_2 u_3$. If both u_2 and u_3 belong to either A_1 or A_2, we simply apply a 2-compression to the corresponding tree. Otherwise, u_2 and u_3 must belong to A_1 and A_2, respectively. (Recall that no new ancestor, descendant pairs can be created.) In this case, no compression is required. However, if u_1 is the root of A_1, then we assign a unit charge to *EXTRA*. (Note that 2-compressions into the root need to be counted.) Observe that when EXTRA gets charged, the root of A acquires an additional child from the nodes of A_2. Hence EXTRA cannot get charged more than $n - 2$ times by this case.

Case 2: Let the compression be a 3-compression of path $u_1 u_2 u_3 u_4$. As in the previous case, if both u_2 and u_4 belong to either A_1 or A_2 we can apply a 3-compression to the corresponding tree. Now let u_2 and u_4 belong to A_1 and A_2, respectively. If u_3 is from A_2, then we 2-compress the path $v u_3 u_4$ in A_2. (This 2-compression into the root of A_2 can be counted toward the original 3-compression.) If u_3 is from A_1, then we 2-

compress the path $u_1 u_2 u_3$ in A_1. This 2-compression doesn't contribute a unit charge if u_1 is not the root. We assign a unit charge to EXTRA in such a case. Observe that, in this case, the path length from the root of A to the node in A_1 to which the trees from A_2 get attached decreases by 1. In addition, the original path length of the attachment node is no more than $|A_1| - 1$. Hence EXTRA cannot be charged more than $\frac{n}{2} - 1$ times by this case.

Now we transform tree A_2 into tree B as shown in Figure 4. (The node w of A_2 is split into separate nodes for the subtrees.) We can easily establish that the total charges to A_2 cannot be more than the total charges to B. In addition, each subtree of the root of B can be considered independently. In B, the subtree of each child of the root is of size no more than $n/2$. Hence if w has k children, then

$f(n) \leq f(n_0) + f(n_1) + \cdots + f(n_k) + 1.5n - 3$ in which $n_0 \leq n/2$ (corresponding to A_1), $1 \leq k \leq n-2$, for each $1 \leq i \leq k$ $1 \leq n_i \leq n/2+1$, and $n_1 + n_2 + \cdots + n_k \leq n/2 + k$. (Note that EXTRA cannot be charged more than $1.5n - 3$ times.)

Now we can easily establish that $f(n) = O(n \log n)$. ∎

This proof is significantly different from that of [9] developed in connection with the Turn Conjecture. It is also different from the proof of Theorem 4 of [10]. Our result appears to be entirely new.

2.5 Pointer Location

Let P be any pattern of length m, and let each $leaf_i$ of Σ_P contain a set, S_i, of positive integers. For every j in S_i we want to compute a pointer to the ancestor of $leaf_i$ that is at Sdepth j.

Lemma 2.4: If $\sum |S_i| = \Omega(m \log m)$, then all the ancestors can be computed in $O(\sum |S_i|)$ steps.

Proof: For each $j \epsilon S_i$ we create value j pointing to $leaf_i$, and then sort the values (by radix sort) in decreasing order. These sorted values are processed as follows. Pick the largest remaining value, say j, and go to the corresponding locus. Climb up from that locus until a locus of Sdepth j is reached. Then mark that locus and associate with that locus all the sets encountered during the climb. We repeat this process until the sorted sequence is exhausted. ∎

2.6 Maximal Path Tracing

In this problem we are given two trees A_1 and A_2 of sizes n_1 and n_2, respectively. Each edge of the trees is labeled by a substring of the pattern, P. Each label is specified by 2 pointers: the starting position and the ending position of the substring, into P. As in suffix trees, we define any node or any point separating consecutive symbols of any edge label to be a *locus*. As before we also define the string for any locus u, denoted σ_u, as the concatenation of the labels on the path from the root to u.

In this problem we need to find for the string of each leaf in A_1 the maximum prefix that is the string of a locus in A_2 and pointers to the corresponding loci in A_1 and A_2. Thus for the trees shown in Figure 5, the pointers in A_2 are: u_2 with offset 2, u_2 with offset 2, u_3 with offset 1, u_5 with offset 3, u_5 with offset 3, and u_6.

Algorithm:

We first construct the suffix tree for P, and then preprocess it so that we can perform each lca query in $O(1)$ steps. We then scan A_1 in depth first order and at any instant we keep track of the corresponding loci in A_1 and A_2. Suppose, at any instant, the corresponding loci in A_1 and A_2 are u_1 and u_2, respectively. (It will be the case that at least one of these is a node unless we are about to backtrack in the depth first search.) We traverse the 2 trees simultaneously such that we exhaust one edge of A_1 or A_2 after spending $O(1)$ steps. There are many cases to consider. A typical case is when u_1 is not a node and u_2 is a node. We then choose the outgoing edge of u_2 whose label's first symbol matches with the first symbol following u_1 on its edge. Let ℓ_1 be the length of the remaining string of the edge of u_1, and let ℓ_2 be the length of the label of the chosen edge going out of u_2. Let ℓ be the minimum of ℓ_1 and ℓ_2. Then by performing an lca query we can find out whether the remaining label on the edge of u_1 and the label of the chosen edge out of u_2 match in the first ℓ symbols. If so, we advance both u_1 and u_2 by that length. The rest of the details are easy to produce.

It is easy to prove that excluding the construction of Σ_P and its preprocessing, the rest of the effort is $O(|A_1| + |A_2|)$ steps.

2.7 Pattern Search Problem

Given a pattern P of length m, its suffix tree Σ_P, and q pointers, $\tau_1, \tau_2, ...,$ and τ_q, each pointing to a locus in Σ_P, we want to develop an algorithm for the following. Does $\sigma_{\tau_1} \sigma_{\tau_2} ... \sigma_{\tau_q}$ contain P as a substring? (Even though we pose this as a decision problem, we will convert it into the

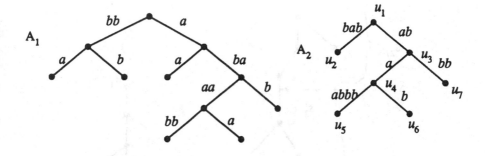

Figure 5: Maximal paths.

problem of computing the leftmost occurrence of P in the final version.)

2.7.1 Basic Algorithm

We first consider the case when $q = 2m^{1.5}$. For each pair τ_{2j-1} and τ_{2j} of pointers we compute the following information: (For simplicity, for every i let $s_i = \sigma_{\tau_i}$. We simply refer to the string of a locus as a path.) Let $\delta_j = s_{2j-1} s_{2j}$.

a) Does δ_j contain P as a substring?

   Otherwise, compute the following information.

b) If δ_j is a path in Σ_P compute a pointer to its locus.

   Otherwise, compute the following information.

c1) Compute the maximum length prefix, and a pointer to its locus, of δ_j that is a path in Σ_P, and

c2) Compute the maximum length suffix, and a pointer to its locus, of δ_j that is a path in Σ_P.

   This information is significantly harder to compute than the information computed in [2]. Note that a suffix tree for a string of length m can have m^2 loci and hence m^4 locus pairs. No simple table calculation technique will be sufficiently fast. An advantage of our approach is that it can be easily implemented as a parallel algorithm. It appears to us that the approach in [2] is inherently sequential.

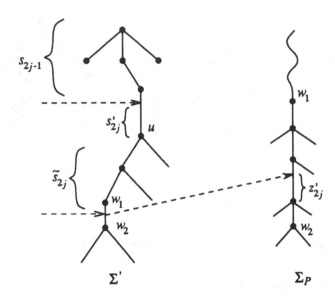

Figure 6: Processing $s_{2j-1}s_{2j}$.

Consider Σ_P as an ordered tree in any arbitrary way. (That is, arbitrarily order the children at each node.) We mark every $\sqrt{m}^{th}$ leaf (based on pre-order) and compress Σ_P into a $\sqrt{m}$ leaf tree, denoted Σ', by compressing out the unmarked leaves and the chains that result [5]. By the standard rake - compress arguments, each edge of Σ' corresponds to a subtree (within which a subtree of a node is removed) of Σ_P with no more than $2\sqrt{m}$ nodes. Let us denote nodes of Σ_P that correspond to nodes of Σ' as *big nodes*. Note that the number of big nodes in Σ_P is $O(\sqrt{m})$.

We first show how to compute a), b), and c1). We preprocess Σ_P such that each lca query can be computed in $O(1)$ steps. The processing of $s_{2j-1}s_{2j}$ is shown schematically in Figure 6.

Using lca queries, for each j, in $O(1)$ steps, we compute the shortest prefix of s_{2j}, denoted s'_{2j}, such that $s_{2j-1}s'_{2j}$ is the locus of a big node or the node in one of the compressed out subtrees from which there is no path to a big node. (It is easy to handle the case when no such prefix exists.) We consider the interesting case in which the locus is a big node. Let $s_{2j} = s'_{2j}s''_{2j}$. For each big node, u, we collect all the j values such that the locus of $s_{2j-1}s'_{2j}$ is u. Each remaining s''_{2j} is specified by 2 loci pointers into Σ_P. From the Sdepths of the 2 loci, we can easily specify s''_{2j} by a depth value at a leaf such that the depth value is $|s''_{2j}|$ and the $|s''_{2j}|$

length prefix of the suffix string specified by the leaf matches s''_{2j}. Then we apply the construction of section 2.5 and mark the loci corresponding to all such s''_{2j}. Then by applying the construction of section 2.6 ($\Sigma_P = A_1$ and $\Sigma' = A_2$), we find for each such j the maximum prefix of s''_{2j}, denoted $\tilde{s}_{2j}$, such that $\sigma_u \tilde{s}_{2j}$ is a path in Σ'. We also find the corresponding locus. Let $s''_{2j} = \tilde{s}_{2j} z_{2j}$. If the locus is not a big node, by an lca query, we find the shortest prefix of z_{2j}, denoted z'_{2j}, such that extending the locus by z'_{2j}, in Σ_P, results in a node of a compressed out subtree. Let $z_{2j} = z'_{2j} z''_{2j}$. As was done earlier for s''_{2j}, we mark the loci corresponding to all z''_{2j}.

At this stage we have transformed each $s_{2j-1} s_{2j}$ into a path to a node in a compressed out tree followed by a string which is represented by a pointer into Σ_P. Thus the task is reduced to the application of the remaining path to a smaller tree (of size $\leq 2\sqrt{m}$). There are in total $O(m)$ nodes in the compressed out trees. (Each tree contains no more than $2\sqrt{m}$ nodes.) For each node in each compressed out tree, we find all the pairs that lead to that node and the pointers specifying the remaining paths. Now by applying the construction of section 2.3 we create a tree with $O(n_u)$ nodes for each node u with n_u associated pointers. Now we apply this tree to the compressed out tree starting at node u as in section 2.6. (The $O(n_u)$ node tree is A_1 and the subtree of u within the corresponding compressed out tree is A_2.) This easily facilitates the computation of a), b), and c1). The details will be provided in the final version.

To compute c2), we transfer all the s_i pointers to Σ_{PR}, by applying the construction in section 2.2. Then we apply the above procedure to Σ_{PR} and compute c2) in Σ_{PR}. Finally we transfer the pointers back to Σ_P, as in section 2.2.

We will show that the complete algorithm runs in $O(m^{1.5})$ steps.

2.7.2 Extension

We extend the previous algorithm for any number of pointers. We process the pointers in batches of $m^{1.5}$. After considering the previous batch, all the previous pointers have been reduced to no more than $m^{1.5}$ pointers. We combine these with the next batch of $m^{1.5}$ pointers. All these pointers are reduced back to no more than $m^{1.5}$ pointers, as in the previous procedure, by applying $O(m^{1.5})$ steps. Thus we are able to handle a new batch of $m^{1.5}$ pointers at the expense of $O(m^{1.5})$ steps. Finally we apply the above procedure $O(\log m)$ times each time halving the number of pointers remaining.

Thus the overall algorithm runs in $O(q + m^{1.5} \log m)$ steps.

2.7.3 Improved Algorithm

We can achieve further improvement in speed by performing a k stage compression of Σ_P. At the beginning of each ith stage, the tree is of size $n^{1-\frac{i-1}{k}}$. In the ith stage we choose $n^{1-\frac{i}{k}}$ equally spaced leaves and compress out the other leaves. Note that each edge of the compressed tree represents a compressed out subtree of size no more than $2n^{\frac{1}{k}}$.

The details of handling pointers in this k stage compression will be provided in the final version.

3 Pattern Matching in Compressed Texts

As a consequence of the basic algorithm of subsection 2.7.2, by following the analysis of [2] we can show that the pattern matching, for a pattern of length m, in a compressed text of length n can be performed in $O(n + m^{1.5} \log m)$ steps. As a consequence of the refined algorithm of subsection 2.7.3, a time bound of $O(nk + m^{1+\frac{1}{k}})$ can be achieved, for any constant $k \geq 1$. The details will be provided in the final version.

Acknowledgements

We sincerely thank Professor Art Delcher for his many constructive comments.

References

[1] A.V. Aho, J.E. Hopcroft, and J.D. Ullman. The design and analysis of computer algorithms. *Addison-Wesley Publishing Co.*, Reading, Mass., 1974.

[2] A. Amir, G. Benson, and M. Farach. Let sleeping files lie: Pattern matching in Z-compressed files. *Proc. of 5th Annual ACM-SIAM Symp. on Discrete Algorithms*, pages 705-714, 1994.

[3] T. Eilam-Tsoreff and U. Vishkin. Matching patterns in a string subject to multilinear transformations. *Proc. of International Workshop on Sequences, Combinatorics, Compression, Security and Transmission, Salerno, Italy*, June 1988.

[4] M. Farach and M. Thorup. Pattern matching in Lempel-Ziv compressed strings. *Proc. of 27th Annual ACM Symp. on Theory of Computing*, pages 703-712, 1995.

[5] J. JaJa. An introduction to parallel algorithms. *Addison Wesley Publishing Co.*, Reading, Mass., 1992.

[6] D. E. Knuth, J. H. Morris, and V. R. Pratt. Fast pattern matching in strings. *SIAM J. on Computing*, pages 323-350, 1977.

[7] E. M. McCreight. A space-economical suffix tree construction algorithm. *J. of the ACM*, pages 262-272, 1976.

[8] B. Schieber and U. Vishkin. On finding lowest common ancestors: Simplification and parallelization. *SIAM J. on Computing*, pages 1253-1262, 1988.

[9] R. Sundar. Twists, turns, cascades, deque conjecture, and scanning theorem. *Proc. of 30th Annual IEEE Symp. on Foundations of Computer Science*, pages 555-559, 1989.

[10] R. Tarjan. Efficiency of a Good But Not Linear Set Union Algorithm. *J. of ACM*, pages 215-225, 1975.

[11] R. Tarjan. Sequential access in splay trees takes linear time. *Combinatorica*, pages 367-378, 1985.

[12] P. Weiner. Linear pattern matching algorithm. *Proc. of 14th Annual IEEE Symp. on Switching and Automata Theory*, pages 1-11, 1973.

[13] T. A. Welch. A technique for high-performance data compression. *IEEE Computer*, pages 8-19, 1984.

[14] J. Ziv and A. Lempel. A universal algorithm for sequential data compression. *IEEE Trans. on Information Theory*, pages 337-343, 1977.

All-Pairs Min-Cut in Sparse Networks*

Srinivasa R. Arikati   Shiva Chaudhuri   Christos D. Zaroliagis

Max-Planck-Institut für Informatik
Im Stadtwald, D-66123 Saarbrücken, Germany
E-mail: {arikati, shiva, zaro}@mpi-sb.mpg.de

Abstract. Algorithms for the all-pairs min-cut problem in bounded tree-width and sparse networks are presented. The approach used is to preprocess the input network so that, afterwards, the value of a min-cut between any two vertices can be efficiently computed. A tradeoff between the preprocessing time and the time taken to compute min-cuts subsequently is shown. In particular, after $O(n \log n)$ preprocessing of a bounded tree-width network, it is possible to find the value of a min-cut between any two vertices in constant time. This implies that for such networks the all-pairs min-cut problem can be solved in time $O(n^2)$. This algorithm is used in conjunction with a graph decomposition technique of Frederickson to obtain algorithms for sparse networks. The running times depend upon a topological property γ of the input network. The parameter γ varies between 1 and $\Theta(n)$; the algorithms perform well when $\gamma = o(n)$. The value of a min-cut can be found in time $O(n + \gamma^2 \log \gamma)$ and all-pairs min-cut can be solved in time $O(n^2 + \gamma^4 \log \gamma)$.

1 Introduction

Network flows are of fundamental importance in computer science, engineering and operations research, to name a few areas. The textbook [1] is an exhaustive reference on the subject. A central problem in network flows is that of computing an s-t min-cut. We are given a directed graph (called a network) with nonnegative capacities edges, and two distinguished vertices s and t. An s-t cut in this graph is a partition of the vertices into two parts, one containing s and the other containing t. The capacity of the cut is the sum of the capacities of the edges going from the part containing s to the part containing t. An s-t min-cut is a cut of minimum capacity among all s-t cuts.

An s-t flow in a network is an assignment of a value less than the capacity to each edge such that the net flow out of each node except s and t is zero, where the net flow out of a node is the sum of flows on edges leaving the node minus the sum of flows on edges entering the node. It follows that the net flows out of s and t sum to zero. An s-t max-flow is a flow that maximizes the net flow out of s, which is called the value of s-t max-flow. The max-flow min-cut theorem

---

* This work was partially supported by the EU ESPRIT Basic Research Action No. 7141 (ALCOM II).

[11] states that the capacity of an s-t min-cut in a network is equal to the value of an s-t max-flow.

In this paper, we are concerned with the all-pairs min-cut problem (APMC problem, for brevity). The problem is to compute the value of an s-t min-cut for each pair of vertices s, t in the network. Since the value of an s-t min-cut can be computed by solving an s-t max-flow problem, the naive solution to the APMC problem solves $n(n-1)$ max-flow problems on n-vertex networks. It was shown by Gomory and Hu [15] that in undirected networks, the APMC problem can be solved by solving $n-1$ well-chosen max-flow problems. Thus the APMC problem on undirected network takes $O((n-1)F(n, m))$ time ($F(n, m)$ is the time required to solve a max-flow problem on an n-vertex, m-edge network). For directed networks, the method of Gomory and Hu does not apply and nothing better than the naive solution (taking $O(n^2 F(n, m))$ time) is known.

The time taken to compute a max-flow when nothing is known about the structure of the network is $O(\min\{n^3/\log n, nm \log n\})$ [9, 16]. However, one can do better when the structure of the input network is known. Recently, it was shown that the max-flow problem in directed or undirected *bounded tree-width* networks can be solved in $O(n)$ time [14]. The tree-width is a parameter that, intuitively, indicates how close the structure of the network is to a tree (see Section 2.2 for a formal definition). The class of bounded tree-width networks includes (among others) outerplanar networks, series-parallel networks, networks with bounded bandwidth or cutwidth [3, 6]. Thus giving better algorithms for this class of networks is an important step in the development of better algorithms for sparse networks, i.e. networks with $O(n)$ edges. For sparse networks, in general, the best max-flow algorithm runs in time $O(n^2 \log n)$. For the APMC problem in the undirected case, substituting the values of $F(n, m)$ yields running times of $O(n^3 \log n)$ for sparse networks and $O(n^2)$ for bounded tree-width networks. For directed networks, the corresponding running times are $O(n^4 \log n)$ and $O(n^3)$ respectively. From now on, we consider only directed networks.

The starting point of this paper is a new algorithm for the APMC problem in bounded tree-width networks that runs in $O(n^2)$ time, improving upon the previous algorithm for directed networks by a factor of n. The approach used is completely different from previous approaches. Instead of computing a number of separate max-flows from scratch, our approach is to preprocess the network so that, subsequently, the value of an s-t max-flow can be efficiently computed for any pair of vertices s and t. We show a tradeoff between the amount of preprocessing required and the time required to compute the value of an s-t max-flow subsequently. The tradeoff is: after $O(nI_k(n))$ preprocessing, the value of an s-t max-flow can be computed in $O(k)$ time, for each integer $k \geq 1$. The function $I_k(n)$, defined formally in Section 2.3, decreases rapidly as k increases; for example, $I_1(n) = \lceil \log n \rceil$ and $I_2(n) = \log^* n$. If the preprocessing is restricted to $O(n)$, then the value of an s-t max-flow can be computed in $O(\alpha(n))$ time (where $\alpha(n)$ is the inverse-Ackermann function; see Section 2.3).

We use the algorithm for bounded tree-width networks to develop an algorithm for sparse networks, based on a decomposition of a sparse network into networks of bounded tree-width. Frederickson [12] showed how to decompose a sparse graph into a number of edge-disjoint outerplanar subgraphs, called hammocks, each of which is connected with the rest of the graph via at most 4 vertices. (An outerplanar graph has tree-width 2.) The number of hammocks obtained, γ, depends on the topological properties of the graph and varies between 1 and $\Theta(n)$. We give an algorithm that computes an s-t max flow in a sparse network in time $O(n + \gamma^2 \log \gamma)$. Thus this algorithm is always competitive with the $O(n^2 \log n)$-time algorithm in [16] and does better if $\gamma = o(n)$. We also show how to solve the APMC problem in time $O(n^2 + \gamma^4 \log \gamma)$ on a sparse network.

The above algorithms output the value of of a max-flow or min-cut. In case the actual min-cut is desired, we show how to output the edges crossing a min-cut in time linear in the size of the output. Specifically, for bounded tree-width networks, we show that, for each $k \geq 1$, after $O(nI_k(n))$ preprocessing, the edges crossing an s-t min-cut can be output in time $O(k + L)$, where L is the number of edges crossing the cut. After $O(n)$ preprocessing this can be done in time $O(\alpha(n) + L)$.

Necessary and sufficient conditions (called *external flow inequalities*) for realizable flows in multi-terminal networks are derived in [14]. An important lemma in that paper shows how to combine the flow inequalities of a number of subnetworks to obtain a single set of flow inequalities for the combined network. Their proof uses linear programming. We give a simple and direct proof of the same result and our proof avoids linear programming.

Our algorithms use the construction of a small network that "mimics" the flow behaviour of a large network. This idea was developed in [14]. The structure of the algorithms for bounded tree-width networks is derived from an algorithm used to solve shortest path queries [7]. The hammock decomposition technique has been used in shortest path problems (see e.g. [10, 12, 13]). To our knowledge, this paper is the first application of this technique to a different problem.

2 Preliminaries

2.1 Mimicking networks of multi-terminal networks

A *network* is a directed graph $G = (V, E)$ with a nonnegative real capacity c_e associated with each edge $e \in E$. The *terminals* of G are a distinguished subset, Q, of its vertices. A *flow* in G is an assignment of a nonnegative real value f_e not greater than c_e to each edge e such that the net flow out of each nonterminal vertex is zero, where the net flow out of a vertex is the sum of flows on edges leaving the vertex minus the sum of flows on edges entering the vertex. An *external flow* $x = (x_1, \ldots, x_{|Q|})$ is an assignment of a real value x_p to each terminal p. A *realizable external flow* is an external flow such that there exists a flow in which the net flow out of each terminal p is x_p. A *cut* is defined by a subset S of V, called its *defining subset*. The *capacity* of a cut is the sum of capacities of edges going from vertices in S to vertices in $V \backslash S$. For a subset R of Q, an *R-separating cut* is a cut with defining subset S where $Q \cap S = R$.

In the special case of a network with two terminals s and t, an s-t max-flow is a flow that maximizes the value of the net flow out of s, which is called the value of the max-flow. An $\{s\}$-separating cut is called an s-t cut. The max-flow min-cut theorem states that the value of an s-t max-flow is equal to the capacity of an s-t min-cut. A direct consequence of this theorem is the following.

Corollary 1. *If f is an s-t flow and C is an $\{s\}$-separating cut such that the value of f equals the capacity of C, then f is an s-t max-flow and C is an $s - t$ min-cut.*

Let G be a network with terminal set Q. Network $M(G)$ with terminal set Q' is a *mimicking* network for G if there exists a bijection between Q and Q' such that every realizable external flow in G is also realizable in $M(G)$ and vice versa. In [14], it is shown that for any G, there exists a mimicking network with 2^{2^q} vertices, where q is the number of terminals of G. Henceforth, when we speak of mimicking networks, we will require that they have no more than this many vertices.

The mimicking network of [14] is constructed by finding 2^q min-cuts in G, namely, a minimum R-separating cut, for each $R \subseteq Q$. Those vertices of G that are on the same side of all these cuts form equivalence classes. Induction on q shows that there can be at most 2^{2^q} equivalence classes. $M(G)$ is constructed by replacing each equivalence class with a single vertex. The edge between two vertices of $M(G)$ in a given direction has capacity equal to the sum of the capacities of the edges in G between the corresponding equivalence classes, taking direction into account. For a given $R \subseteq Q$, a minimum R-separating cut is computed by the standard method of introducing a new source, connected to each vertex in R with infinite capacity edges, and a new sink to which each vertex in $Q \backslash R$ is similarly connected, and computing a max-flow from the source to the sink. Thus we have the following result.

Proposition 2. *A mimicking network of a network G with q terminals can be computed in $O(2^q F(G))$ time, where $F(G)$ is the time required to compute a max-flow in G.*

Suppose we are given the mimicking networks of a number of networks. A number of pairs are specified, each pair consisting of two terminals belonging to different networks. We are asked to combine the different networks by identifying the specified pairs of terminals. Finally, we are given a subset of all the terminals, and asked to find the mimicking network of the combined network at this new set of terminals. Note that in the combined network, the set of terminals of each subnetwork is an *attachment set* for that subnetwork, where an attachment set for a subnetwork is a set of vertices whose deletion disconnects the subnetwork from the rest of the network. Using Proposition 2 and computing max-flows with an $O(n^3)$ algorithm (see e.g. [1]), we can show the the following (which is a reformulation of the result in [14]).

Lemma 3. *Let $G = G_1 \cup \ldots \cup G_m$, where the G_i's are edge disjoint, and let G_i have attachment set C_i. Given the mimicking networks $M(G_i)$ for each G_i at terminals Q_i satisfying $C_i \subseteq Q_i$, and a set $Q \subseteq \cup_{i=1}^m Q_i$, we can compute the mimicking network $M(G)$ for G at terminals Q in time $O(2^q \cdot (\sum_{i=1}^m 2^{2^{q_i}})^3)$, where $q_i = |Q_i|$ and $q = |Q|$.*

2.2 Tree-width

A *tree decomposition* of a (directed or undirected) graph $G = (V(G), E(G))$ is a pair (X, T), where $T = (V(T), E(T))$ is a tree and X is a family $\{X_i : i \in V(T)\}$ of subsets of $V(G)$ that cover $V(G)$, and the following conditions hold:

- *(edge mapping)* $\forall (v, w) \in E(G)$, there exists an $i \in V(T)$ with $v \in X_i$ and $w \in X_i$.
- *(continuity)* $\forall i, j, k \in V(T)$, if j lies on the path from i to k in T, then $X_i \cap X_k \subseteq X_j$, or equivalently: $\forall v \in V(G)$, the nodes $\{i \in V(T) : v \in X_i\}$ induce a connected subtree of T.

The *width* of the tree decomposition is $\max_{i \in V(T)} |X_i| - 1$. The *tree-width* of G is the minimum width over all possible tree decompositions of G.

Bodlaender [5] gave a linear-time algorithm to compute a constant width tree decomposition of a graph with constant tree-width. In [4] a linear-time algorithm is given to convert a tree decomposition of (constant) width t into another one of tree-width $3t + 2$, in which the tree is binary. We call such a tree decomposition a *binary tree decomposition*.

Let G be an n-vertex graph of constant tree-width and let (X, T) be its tree decomposition of constant width. The edge mapping condition ensures that the endpoints of each edge in G appear together in some set $X_i \in X$, belonging to vertex i of T. Thus, in a sense, each edge is represented in at least one vertex of T. For our applications, we need to explicitly associate each edge of G with exactly one vertex of T. We will, therefore, compute an *augmenting function* $h : E(G) \longrightarrow V(T)$, satisfying the property that both endpoints of an edge are present in the set belonging to the vertex that the edge is mapped to by h. More precisely, $\forall (v, w) \in E(G), \{v, w\} \subseteq X_{h((v,w))}$. Any augmenting function will suffice for our applications. It is easy to compute one such function, by doing a traversal of T and assigning $h((v, w)) = i$ for each $i \in V(T)$, if $\{v, w\} \subseteq X_i$, $(v, w) \in E(G)$ and $h((v, w))$ has not yet been assigned a value. This takes time proportional to $\sum_{i \in V(T)} |X_i|^2$, which is $O(n)$, since the tree decomposition is of constant width. The resulting tree decomposition with the values $h((v, w)), \forall (v, w) \in E(G)$, is called an *augmented tree decomposition*. The discussion above is summarized:

Proposition 4. *Given an n-vertex graph G of constant tree-width t, in $O(n)$ time we can compute an augmented binary tree decomposition of G of width $O(t)$.*

2.3 Tree products

For a function g let $g^{(1)}(n) = g(n)$; $g^{(i)}(n) = g(g^{(i-1)}(n))$, $i > 1$. Define $I_0(n) = \lceil \frac{n}{2} \rceil$ and $I_k(n) = \min\{j \mid I_{k-1}^{(j)}(n) \le 1\}$, $k \ge 1$. The functions $I_k(n)$ decrease rapidly as k increases; in particular, $I_1(n) = \lceil \log n \rceil$ and $I_2(n) = \log^* n$. Define $\alpha(n) = \min\{j \mid I_j(n) \le j\}$.

The following theorem was proved in [2, 8].

Theorem 5. *Let $\bullet$ be an associative operator defined on a set S, such that for $q, r \in S$, $q \bullet r$ can be computed in constant time. Let T be a tree with n nodes such that each edge is labeled with an element from S. Then: (i) for each integer $k \ge 1$, after $O(nI_k(n))$ preprocessing, the composition of labels along any path in the tree can be computed in $O(k)$ time; and (ii) after $O(n)$ preprocessing, the composition of labels along any path in the tree can be computed in $O(\alpha(n))$ time.*

3 Bounded tree-width networks

Let G be a network of bounded tree-width and (X, T) its augmented binary tree decomposition. For a subtree T' of T, we define the subgraph G' *spanned by T'*, as follows. The vertices of G' are the vertices in the sets associated with the vertices of T', i.e. $V(G') = \cup_{i \in V(T')} X_i$. The edges of G' are those edges that the augmenting function maps to vertices in T', i.e. $E(G') = \{e \in E(G) : h(e) \in V(T')\}$. It is easy to check that vertex-disjoint subtrees span edge-disjoint subgraphs. (In fact it is only to ensure this property that we introduce the augmenting function.)

For $i, j \in V(T)$ let $path(i, j)$ denote the unique path from i to j in T. Deleting the first and last edges on this path breaks up T into three components T_i, T_j, the ones containing i and j respectively, and the remaining component T_{ij}. If $path(i, j)$ is an edge, then the first and last edges on the path are the same; consequently, the component T_{ij} is empty. The vertices in T_{ij} that are adjacent to i and j are denoted n_i and n_j respectively.

Define a set $U = \{P_{ij} = (M_i, M_j, M_{ij}) : \forall i, j \in V(T), i \ne j\}$, where M_i and M_j are the mimicking networks for the subgraphs spanned by T_i and T_j at terminals X_i and X_j respectively, and M_{ij} is the mimicking network for the subgraph spanned by T_{ij} at terminals $X_{n_i} \cup X_{n_j}$. If T_{ij} is empty, then M_{ij} is empty.

Define the following operator $\bullet$ on U. For $i, j, k \in V(T)$, $P_{ij} \bullet P_{jk} = P_{ik}$, if the tree path from i to k includes the vertex j, and $P_{ij} \bullet P_{jk} = \emptyset$ otherwise. It follows easily from the definition that $\bullet$ is associative: If i, j, k, l are vertices (appearing in that order) on a simple path in T, then $(P_{ij} \bullet P_{jk}) \bullet P_{kl} = P_{ik} \bullet P_{kl} = P_{il}$ and $P_{ij} \bullet (P_{jk} \bullet P_{kl}) = P_{ij} \bullet P_{jl} = P_{il}$. If i, j, k, l are not on a simple path in T, then $(P_{ij} \bullet P_{jk}) \bullet P_{kl} = P_{ij} \bullet (P_{jk} \bullet P_{kl}) = \emptyset$. In general, the product $P_{i_1 i_2} \bullet P_{i_2 i_3} \cdots \bullet P_{i_{m-1} i_m} = P_{i_1 i_m}$ iff $i_1, \ldots, i_m$ is a path in T.

Suppose we have computed P_{ab} for every a and b such that (a, b) is an edge in T. Then the operator $\bullet$ can be implemented by combining networks as follows.

Suppose the path from i to k passes through vertex j and we wish to compute $P_{ik} = P_{ij} \bullet P_{jk}$, given P_{ij} and P_{jk}. Then, since T is binary, j has at most one neighbour x apart from its neighbours on the path from i to k. Let T_x be the component of T containing x, obtained by deleting edge (j, x). Let n_i and n_k be the neighbours of i and k on the path from i to k. (See Figure 1.)

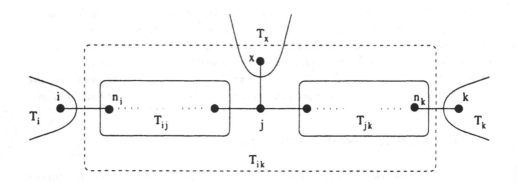

Fig. 1. Computation of P_{ij}.

The value P_{ik} consists of the three mimicking networks M_i, M_k and M_{ik}, for the subgraphs spanned by T_i, T_k and T_{ik} respectively. The former two are already available as part of the values P_{ij} and P_{jk}. Hence we need to compute only M_{ik}. The component T_{ik} is the union of components T_{ij}, T_{jk}, T_x, and vertex j, which are pairwise vertex-disjoint. By supposition, we have the mimicking network for the subgraph spanned by T_x, as part of the value P_{jx}. The mimicking networks for the subgraphs spanned by T_{ij} and T_{jk} are available in the values P_{ij} and P_{jk}. The mimicking network for the subgraph spanned by j can be computed using Proposition 2. From the continuity property of tree decompositions, it follows that the set of terminals for each of the subgraphs is an attachment set for the subgraph and that the final set of terminals desired, namely $X_{n_i} \cup X_{n_k}$, is a subset of all the terminals. Combining the above mimicking networks using Lemma 3 yields M_{ik}. Since the total number of terminals is constant, we have the following result.

Lemma 6. *Let G be a network and let (X, T) be its augmented binary tree decomposition of constant width. Given $P_{ab}, \forall (a, b) \in E(T)$, and P_{ij}, P_{jk} for some $i, j, k \in V(T)$, $P_{ij} \bullet P_{jk}$ can be computed in constant time.*

We now show how to compute P_{ij} for each edge (i, j) in T. Root T at any vertex. For a vertex i, let S_i be the subtree rooted at i. Consider an edge (i, j) such that i is a child of j. Then P_{ij} consists of two values M_i and M_j, where M_i is the mimicking network for the subgraph spanned by S_i, with terminals

X_i, and M_j is the mimicking network for the subgraph spanned by $T \backslash S_i$, with terminals X_j. We compute P_{ij} in two phases. In the first phase we compute M_i for each edge (i, j) with i a child of j. In the second phase, we compute M_j for each such edge.

During the first phase, suppose we are at an edge (i, j), with i a child of j. Suppose also that we have computed the mimicking network M_l and M_r for the (at most) two edges connecting i to its children. Then, to obtain M_i, use Lemma 3 to combine the mimicking networks M_l, M_r and the mimicking network for the subgraph spanned by i, retaining the terminals X_i. A postorder traversal of T with this operation performed at each edge completes the first phase.

During the second phase, suppose we are at edge (i, j), with i a child of j. Let p and c be the parent of j and the sibling of i respectively (if they exist). Suppose we have already computed M_p, the mimicking network for the subgraph spanned by $T \backslash S_j$. In the first phase, we have computed M_c, the mimicking network for the subgraph spanned by the subtree rooted at c. Then, use Lemma 3 to combine M_p, M_c and the mimicking network for the subgraph spanned by j, retaining terminals X_j. This yields M_j, the mimicking network for the subgraph spanned by $T \backslash S_i$. A preorder traversal of T with this operation performed at each edge completes the second phase.

Each time Lemma 3 is invoked, it combines a constant number of networks, each with a constant number of terminals, hence taking constant time. Since the lemma is invoked twice for each edge, we have proved the following result.

Lemma 7. *Let G be an n-vertex network and let (X, T) be its augmented binary tree decomposition of constant width. Then, in time $O(n)$ we can compute P_{ab} for all edges $(a, b) \in E(T)$.*

We can now prove the main result of this section.

Lemma 8. *Let G be an n-vertex network and let (X, T) be its augmented binary tree decomposition of constant width. For each integer $k \geq 1$, after $O(nI_k(n))$ preprocessing, we can find the mimicking network for G at terminals $X_i \cup X_j$ in time $O(k)$, for any $i, j \in V(T)$. Further, after $O(n)$ preprocessing, we can find these mimicking networks in time $O(\alpha(n))$.*

Proof. For each edge (a, b) of T, compute P_{ab} using Lemma 7. Use Theorem 5 to preprocess T, with the P_{ab} values associated with its edges, so that queries asking for the product of P values along paths in T can be answered. A query for the product on the path from i to j returns the value $P_{ij} = (M_i, M_j, M_{ij})$. Combine these three mimicking networks using Lemma 3, with the desired set of terminals being $X_i \cup X_j$. This yields the mimicking network for G with these terminals.

Theorem 9. *Let G be an n-vertex network of constant tree-width. For each integer $k \geq 1$, after $O(nI_k(n))$ preprocessing, we can find the value of an s-t max-flow in time $O(k)$, for each $s, t \in V(G)$. Further, after $O(n)$ preprocessing, we can find the value of an s-t max-flow in time $O(\alpha(n))$.*

Proof. First, compute a constant-width augmented binary tree decomposition (X, T) of G using Proposition 4. Preprocess G and (X, T) using Lemma 8.

Let $s \in X_i$ and $t \in X_j$, for some $i, j \in V(T)$. A single query returns the mimicking network for G at terminals $X_i \cup X_j$. Now simply compute the value of an s-t max-flow in this mimicking network. Since the size of the mimicking network is constant, the entire computation after the query takes constant time, implying the time bounds in the theorem.

In order to solve the APMC problem in a bounded tree-width network, simply apply Theorem 9 with $k = 2$, i.e. perform $O(n \log n)$ preprocessing so that an s-t max-flow can be computed in constant time. Thus the APMC problem can be solved by querying for s-t max-flows, for each pair s, t in the network. This proves the following result.

Corollary 10. *The all-pairs min-cut problem can be solved for bounded tree-width networks in time $O(n^2)$.*

4 Sparse networks

Frederickson [12] shows how to decompose a sparse graph G into γ outerplanar subgraphs (called hammocks), each of which is connected to the rest of the graph via at most 4 vertices, called attachment vertices. The parameter γ is $O(g + p)$ where g is the genus of G and p is the minimum number of faces that cover all vertices of G, over all possible cellular embeddings into an orientable surface of genus g. Note that $g + p$ is the minimum possible number of hammocks in such a decomposition. It is known that γ can vary between 1 and $\Theta(n)$. The algorithm in [12] runs in linear time and does not require an embedding to be provided with the input. In this section, we give algorithms whose running times depend on γ, and which perform well when $\gamma = o(n)$.

Let G be a sparse graph which is decomposed into hammocks $H_1, \ldots, H_\gamma$. Let A_i be the set of (at most 4) attachment vertices of H_i. We now show how to preprocess G so that max-flows can be efficiently found. Preprocess each hammock H_i as follows. First, find an augmented binary tree decomposition (X', T) of H_i, of constant width (outerplanar graphs have tree-width 2). Replace each set of $X'_j \in X'$ by $X_j = X'_j \cup A_i$, i.e. add the attachment vertices to each set. Let X be the collection of sets so obtained. Then (X, T) is also an augmented binary tree decomposition of H_i of constant width. We will work with this new tree decomposition. Use Lemma 7 to preprocess H_i in $O(|H_i|)$ time, so that for each edge $(a, b) \in T$, the mimicking network for H_i at terminals $X_a \cup X_b$ can be found using a single query.

Now, (i) the mimicking network for H_i at terminals A_i can be found in constant time, and (ii) for any $s, t \in V(H_i)$ the mimicking network for H_i at terminals $\{s, t\} \cup A_i$ can be found in time $O(\alpha(n))$. The first claim follows from the fact that the values P_{ab}, for each edge $(a, b) \in T$, are computed during preprocessing. $P_{ab} = (M_a, M_b, \emptyset)$, where M_a and M_b are the mimicking networks for the subgraphs of H_i spanned by the two components of T obtained by deleting edge (a, b). Recall that $A_i \subseteq X_a$ and $A_i \subseteq X_b$. Combining M_a and M_b and retaining terminals A_i yields the desired mimicking networks. The second claim follows by selecting $c, d \in V(T)$ such that $s \in X_c, t \in X_d$, applying Lemma 8 and retaining the desired terminals.

We can now find the value of an s-t max-flow as follows. Let $s \in V(H_i)$ and $t \in V(H_j)$. Define G_{ij} to be the network obtained by replacing each hammock $H_k, k \notin \{i, j\}$, by its (constant size) mimicking network at terminals A_k and deleting H_i and H_j except their attachment vertices. The terminals of G_{ij} are $A_i \cup A_j$. Note that G_{ij} has $O(\gamma)$ vertices and edges and can be constructed in time $O(\gamma)$, since each mimicking network can be found in constant time. Construct G_{ij} and find the mimicking network for G_{ij} at terminals $A_i \cup A_j$ using Proposition 2. Find the mimicking network for H_i at terminals $\{s\} \cup A_i$ and H_j at terminals $\{t\} \cup A_j$ in time $O(\alpha(n))$, as described above. (If $i = j$, then find the mimicking network for H_i at terminals $\{s, t\} \cup A_i$.) Combining these networks yields the mimicking network for G at terminals $\{s, t\} \cup A_i \cup A_j$. Now the value of s-t max-flow can be found using the method described in the proof of Theorem 9. To estimate the time complexity, once the hammocks have been preprocessed and the mimicking networks for G_{ij} found, the remaining computation takes constant time. Preprocessing the hammocks takes $O(n)$ time and finding the mimicking network for G_{ij} takes $O(\gamma^2 \log \gamma)$ time, when we apply Proposition 2 with a max-flow algorithm (see e.g. [1]) for which $F(G) = O(nm \log n)$ on an n-vertex, m-edge network G. We summarize the above discussion:

Theorem 11. *The value of an s-t max-flow in an n-vertex sparse network G can be computed in time $O(n + \gamma^2 \log \gamma)$, where γ is the number of hammocks of G.*

To solve the APMC problem, preprocess the H_i's using $O(|H_i| \cdot \log |H_i|)$ time so that the mimicking network for H_i at the appropriate terminals (as in (ii) above) can be found in constant time. For each $i, j \in \{1, 2, \ldots, \gamma\}$, construct G_{ij} and find its mimicking network. Now for each $s, t \in V(G)$, such that $s \in V(H_i)$ and $t \in V(H_j)$, find the mimicking network for H_i at terminals $\{s\} \cup A_i$ and for H_j at terminals $\{t\} \cup A_j$. (If $i = j$, then find the mimicking network for H_i at terminals $\{s, t\} \cup A_i$.) Combine these mimicking networks with the mimicking network for G_{ij} and find the value of s-t max-flow, as before. Once the H_i's have been preprocessed and the mimicking networks for the G_{ij}'s found, computing an s-t max flow takes constant time for each pair s, t. Hence, the following result is proved.

Theorem 12. *The all-pairs min-cut problem for an n-vertex sparse network G can be solved in $O(n^2 + \gamma^4 \log \gamma)$ time, where γ is the number of hammocks of G..*

5   Computing s-t min-cut in bounded tree-width networks

In this section we outline an extension of the methods in Sections 2.1 and 3 that allows us to output the edges crossing an s-t min-cut in time linear in the number of edges in the cut.

The essential feature is the computation of supplementary information when a mimicking network is computed. Let G be a network and let $M(G)$ be its mimicking network, as computed in Section 2.1. In this construction, each vertex of $M(G)$ represents a subset of the vertices of G and each edge (u, v) of $M(G)$ represents a subset of the edges of G, namely, the edges between the subsets of vertices of G represented by u and v. During the construction of $M(G)$, for each edge e of $M(G)$ we compute a value $trace(e)$, which is a list of the edges of G that e represents. It is easily verified that distinct edges of $M(G)$ represent disjoint subsets of edges of G.

For every mimicking network computed in Section 3 we will also compute the trace information associated with their edges. For edges of the input graph, the trace value of an edge is simply the edge itself. For reasons of efficiency, which will become clear later, we have one special condition: if an edge e of $M(G)$ represents a single edge e' of G, then $trace(e)$ is defined to be the same as $trace(e')$. In other words, instead of a singleton list containing e, $trace(e)$ is the same list as $trace(e')$. This condition ensures that except for edges of the original input graph, the trace value of each edge is a list with at least two elements. Regarding the elements in the trace value of an edge as the children of the edge, we have that each edge e is the root of a tree defined by the trace values, whose leaves are edges of the input graph. We call this tree the *trace subtree* of e. It is not hard to see that the leaves of the trace subtree are exactly those edges of of the input graph that e represents. Further, the condition above ensures that every non-leaf vertex in the trace subtree has at least two children.

We now perform preprocessing as described in Lemma 8, computing trace information for each mimicking network constructed in the process. Suppose we now wish to compute an s-t min-cut for some s, t. Then, as in the proof of Theorem 9, we compute a mimicking network $M(G)$ of constant size, whose terminals include s and t, for the input graph G. We compute an s-t min-cut in $M(G)$, which corresponds to an s-t min-cut in G in the natural way. Each edge crossing the cut in $M(G)$ represents a subset of edges crossing the cut in G, i.e. the leaves of the trace subtree of the edge. Any standard tree traversal algorithm will output the leaves of the trace subtree in time linear in the size of the tree, which is linear in the number of leaves, since each non-leaf vertex has at least two children. Doing this for each edge crossing the cut in $M(G)$ outputs in linear time all the edges crossing the cut in G. This yields the following result.

Theorem 13. *Let G be an n-vertex network of constant tree-width. For each integer $k \geq 1$, after $O(nI_k(n))$ preprocessing, we can output the edges crossing an s-t min-cut in time $O(k + L)$, where L is the number of edges crossing the cut. Further, after $O(n)$ preprocessing, we can output the edges crossing an s-t min-cut in time $O(\alpha(n) + L)$.*

6 Characterization of flows in multi-terminal networks

In [14] necessary and sufficient conditions are derived for an external flow to be realizable:

Lemma 14. *An external flow* $(x_1, \ldots, x_{|Q|})$ *is realizable in a network* G *with terminals* Q *iff (i)* $\sum_{p \in Q} x_p = 0$ *and (ii)* $\sum_{r \in R} x_r \leq b_R$, $\forall R \subseteq Q$, *where* b_R *is the minimum capacity of an* R-separating cut.

Thus the realizable external flows of a network with q terminals can be characterized by the above system of 2^q linear inequalities, where each inequality is represented by the pair (R, b_R). A system of inequalities for a network G, of the form as in Lemma 14, is called the *external flow inequalities* of G at terminals Q.

Given a network $G = (V, E)$ with terminals Q, it is possible to compute its external flow inequalities by following the method described in Section 2.1 for computing the capacities of minimum R-separating cuts in G, for every $R \subseteq Q$, in time $O(2^q F(G))$, where $q = |Q|$.

Using linear programming methods, it is shown in [14] how to combine the external flow inequalities of a number of networks. Here we give a simpler proof of this result; our proof is based on the max-flow min-cut theorem (see Corollary 1) and avoids linear programming. We note that the proof in [14] results in an algorithm with running time exponential in the square of the total number of terminals, whereas our proof results in a time that is exponential in twice the total number of terminals.

Let G be the union of two edge-disjoint networks G_1 and G_2. Observe that the attachment sets for both networks are the same. Assume that the attachment set is a subset of the terminals of both networks. Given the external flow inequalities of these two networks, we wish to find the external flow inequalities of G at a subset of all the terminals. We solve this problem in two steps. In the first step, we find the external flow inequalities at all the terminals, and in the second step we drop some terminals.

We now find the capacity of a minimum R-separating cut, where R is a subset of all the terminals. Let S and T be intersections of R with the terminals of G_1 and G_2, respectively. Further, let X and Y be the defining subsets of a minimum S-separating cut in G_1 and a minimum T-separating cut in G_2. Since the attachment set is a subset of the terminals of both networks, observe that the intersections of the attachment set with X and Y are equal. Hence $X \cup Y$ is a defining subset of an R-separating cut in G. Moreover, the capacity of this cut is the sum of the capacities of the cuts defined by X and Y, because G_1 and G_2 are edge-disjoint. On the other hand, by adding the corresponding max-flows in G_1 and G_2, we obtain a flow from R to the rest of the terminals in G and the value of this flow is equal to the sum of the values of the two former flows. We have thus proved (by Corollary 1) that $X \cup Y$ is a defining subset of a minimum R-separating cut and that the capacity of this cut is equal to the sum of the above two capacities. The latter two capacities are already available in the given external flow inequalities. Since this computation is done for each subset of all the terminals, the time taken in this step is exponential in the total number of terminals.

In the second step, we wish to drop some terminals and find the capacity of a minimum P-separating cut, where P is a subset of the remaining terminals. A subset R of all the terminals is said to be consistent with P if it contains every terminal in P and doesn't contain any other remaining terminal. For any such R, observe that every R-separating cut is also a P-separating cut and hence the capacity of a minimum P-separating cut is at most the capacity of a minimum R-separating cut. On the other hand, the intersection of all the terminals with the defining subset of a minimum P-separating cut is consistent with P. We have thus proved that the capacity of a minimum P-separating cut is equal to the minimum capacity of all R-separating cuts, where R is consistent with P. In the first step, we have computed the latter capacities. Thus we can compute the former capacity in time exponential in the total number of terminals, by considering all possible consistent subsets. Since this computation is done for all subsets of the remaining terminals, the time taken in this step is at most an exponential in twice the total number of terminals.

The proof of the following lemma now comes by a simple induction on m.

Lemma 15. Let $G = G_1 \cup \ldots \cup G_m$, where the G_i's are edge-disjoint, and let C_i be the attachment set of G_i. Assume that C_i is a subset of the terminals Q_i in G_i, for all i. Given the external flow inequalities for each G_i at terminals Q_i, and a set $Q' \subseteq \cup_{i=1}^{m} Q_i$ of terminals, we can compute the external flow inequalities for G at terminals Q' in time $O(2^{2(q_1 + \cdots + q_m)})$, where $q_i = |Q_i|$.

7 Remarks

We presented efficient algorithms for the all-pairs min-cut problem on bounded tree-width networks and sparse networks. The constants in the running time of the algorithms are not small, however. For example, in the algorithm for networks of tree-width t, the constant is $2^{2^{O(t)}}$. An open problem is to design more practical algorithms for these problems.

References

1. R. Ahuja, T. Magnanti and J. Orlin, "Network Flows", Prentice-Hall, 1993.
2. N. Alon and B. Schieber, "Optimal Preprocessing for Answering On-line Product Queries", Tech. Rep. No. 71/87, Tel-Aviv University, 1987.
3. S. Arnborg, "Efficient Algorithms for Combinatorial Problems on Graphs with Bounded Decomposability - A Survey", *BIT*, 25, pp.2-23, 1985.
4. H. Bodlaender, "NC-algorithms for Graphs with Small Treewidth", *Proc. 14th WG'88*, LNCS 344, Springer-Verlag, pp.1-10, 1989.
5. H. Bodlaender, "A Linear Time Algorithm for Finding Tree-decompositions of Small Treewidth", *Proc. 25th ACM STOC*, pp.226-234, 1993.

6. H. Bodlaender, "A Tourist Guide through Treewidth", *Acta Cybernetica*, Vol.11, No.1-2, pp.1-21, 1993.

7. S. Chaudhuri and C. Zaroliagis, "Shortest Path Queries in Digraphs of Small Treewidth", *Proc. 22nd Int. Col. on Automata, Languages and Prog. (ICALP'95)*, LNCS 944, Springer-Verlag, pp. 244–255.

8. B. Chazelle, "Computing on a Free Tree via Complexity-Preserving Mappings", *Algorithmica*, 2, pp.337-361, 1987.

9. J. Cheriyan, T. Hagerup and K. Mehlhorn, "Can a maximum flow be computed on $o(nm)$ time?", *Proc. 17th Intl. Col. on Automata, Languages and Prog. (ICALP'90)*, LNCS 443, Sringer-Verlag, pp.235-248, 1990.

10. H. Djidjev, G. Pantziou and C. Zaroliagis, "On-line and Dynamic Algorithms for Shortest Path Problems", *Proc. 12th Symp. on Theor. Aspects of Comp. Sc. (STACS'95)*, LNCS 900, Springer-Verlag, pp.193-204, 1995.

11. L.R. Ford and D.R. Fulkerson, "Maximal flow through a network", *Canadian Journal of Mathematics*, 8, pp.399-404, 1956.

12. G.N. Frederickson, "Using Cellular Graph Embeddings in Solving All Pairs Shortest Path Problems", *Proc. 30th Annual IEEE Symp. on FOCS*, 1989, pp.448-453.

13. G.N. Frederickson, "Searching among Intervals and Compact Routing Tables", *Proc. 20th Int. Col. on Automata, Languages and Prog. (ICALP'93)*, LNCS 700, Springer-Verlag, pp.28-39, 1993.

14. T. Hagerup, J. Katajainen, N. Nishimura and P. Ragde, "Characterizations of k-Terminal Flow Networks and Computing Network Flows in Partial k-Trees", *Proc. 6th ACM-SIAM Symp. on Discr. Alg. (SODA'95)*, pp.641-649, 1995.

15. R.E. Gomory and T.C. Hu, "Multi-terminal network flows", *Journal of SIAM*, 9, pp.551-570, 1961.

16. D.D. Sleator and R.E. Tarjan, "A Data Structure for Dynamic Trees", *Journal of Computer and System Sciences*, 26, pp.362-391, 1983.

17. K. Weihe, "Maximum (s,t)-Flows in Planar Networks in $O(|V|\log|V|)$ Time", *Proc. 35th Annual IEEE Symp. on FOCS*, 1994, pp.178-189.

Minimizing Space Usage in Evaluation of Expression Trees

Sandip K. Biswas* and Sampath Kannan

Department of CIS
University of Pennsylvania
Philadelphia, PA 19104
{sbiswas@saul,kannan@central}.cis.upenn.edu

Abstract. An important issue in the evaluation of expression trees (and more generally, directed acyclic graphs (dags)) is the order in which the nodes of the graph are evaluated. In the *register sufficiency problem*, the evaluation begins with none of the nodes of the dag being placed in registers. On the other hand, if an expression dag, containing variables, is being evaluated in an environment then the nodes of the graph, with out-degree zero, are variables, whose values are contained in the environment. Hence space occupied before the evaluation begins equals the size of the environment. Thus, the strategy to minimize space usage during the evaluation of the dag has to take into account the initial space occupied by the nodes of out-degree zero.

By a simple reduction from the *register sufficiency problem* it is shown that, for an arbitrary dag, the problem of selecting an optimal evaluation strategy which minimizes space is NP-Complete. While the *register sufficiency problem* for a tree has a trivial linear time algorithm, the problem of minimizing space usage for a tree is non-trivial: it involves solving a *sequencing problem*. We present an $O(n \log n)$ algorithm for solving the problem.

1 Introduction

In many programming languages, particularly functional languages, the entire program is an expression to be evaluated. Since code size of a compiled expression is fixed, the initial size of an expression is considered to be the size of the environment containing the values of the free variables. If the size of the value returned from the computation of each subexpression is known, the problem we seek to address is that of finding an order of evaluation of the subexpressions such that the space consumed at any intermediate level does not cause an overflow of the heap.

A lot of research effort has gone into looking for solutions of the *register sufficiency problem* on expressions structured as dags. A more general and abstract presentation of the problem, in the form of pebble games, has been studied

* This research was supported by NSF Grant CCR94-15443 and ONR Contract N00014-89-J-3155

by Sethi in [7]. The problem we seek to address does not fit into this general framework because of two reasons:

- The framework assumes all values in the computation to be of the same constant size, namely 1. We would like to address computations in which different values have different sizes, represented by different positive integers.
- The framework starts computation from an initial state in which no registers have been no allocated. On the contrary, evaluation of an expression, containing variables, occupies some initial space, which equals the size of the environment containing the values of the free variables in the expression.

1.1 Motivation

To give the reader a feel of the importance of choosing the right strategy for evaluating expressions, we present a small example from [1] showing how a change in the order of evaluation of an expression can change the asymptotic space complexity of the expression. The following program is written in the functional language Standard ML (SML), [5].

```
(* The function sl takes an argument n and         *)
(* returns a list [1,...,n].                        *)
(* The function length computes the length of a list. *)

fun f n =  if (n=0)
           then 1
           else (f (n-1)) + (length (sl n))
```

Consider a modified version of the function f, where the order of evaluation is changed,

```
fun f1 n = if (n=0)
           then 1
           else  let val l = (sl n)
                 in (f1 (n-1)) + (length l)
                 end
```

For any implementation, a call to f with argument n uses $O(n)$ space. This is because there will be $O(n)$ activation records in the dump, corresponding to the n recursive calls made, each pointing to an environment of constant size, containing the current value of the integer n.

In f1, as l is a list of integers whose size depends on n, a call to f1 with argument n uses $O(n^2)$ space. This is because there will be $O(n)$ activation records in the dump, each pointing to an environment of size $O(n)$, corresponding to the list l constructed.

Hence, the order of evaluation of the sub-expressions of an expression is extremely important, as far as the space complexity is concerned.

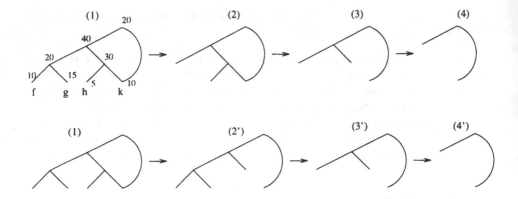

Fig. 1. Expression Evaluation

1.2 A Simple Presentation of the Problem

Consider the following expression,

((f g)(h k)) k

The body of the expression, **((f g)(h k)) k** , may be represented by the dag (1) in Fig 1. Every node of the dag is annotated by the size of the object computed at that node, e.g. **(f g)**, when computed, is an object of size 20. A dag is computed in an environment which binds objects to the nodes of out-degree 0. (We assume throughout that the edges of the dag are oriented from the node representing an operator to the nodes representing the operands. All the internal nodes of this dag represent the application operator.) Thus the size of the environment required for the computation of a dag is at least the sum of the sizes of the objects at nodes, of out-degree 0. Hence dag (1), is evaluated in an environment of size 40. Once the computation **(f g)** is over, the size of the environment required for computation of dag (2) shrinks to 35.

For the computation sequence, (1), (2), (3), (4), the maximum intermediate space required is 60, in stage (3). For the computation sequence, (1), (2'), (3'), (4'), the maximum intermediate space required is 65, in stage (2'). Hence the strategy for evaluation of an expression determines the space required for the computation.

The problem we seek to address is as follows: if the sizes of all intermediate values in a computation, structured as a dag, are known, can we decide on a strategy which minimizes the peak size of the environment required for the computation?

Another important application of this problem is the following: Given a collection of databases, it may be necessary to restructure them into a different collection of databases. In this case, we need a strategy which ensures that the disk space used in the intermediate computation is minimum.

2 An Abstract Formulation

In an abstract setting, an expression can be visualized as a directed acyclic graph. Since our primary concern here is to analyze space, let us assume that every node of the graph is annotated with an integer *size*, giving the size of the value computed at that node. When the size of all objects is 1 we get the following *register sufficiency problem* which was formalized in [3].

2.1 Abstract Register Sufficiency Problem

INSTANCE: Directed Acyclic Graph $G = (V, E)$, positive integer K.
QUESTION: Is there a computation for G that uses K or fewer registers, i.e. an ordering v_1 , v_2 , $\ldots$, v_n of the vertices in V, where $n = |V|$, and a sequence S_0 , S_1 , $\ldots S_n$ of subsets of V, each satisfying $|S_i| \leq K$, such that S_0 is empty, S_n contains all the vertices with in-degree 0 in G, and , for $1 \leq i \leq n$, $v_i \in S_i$, $S_i - \{v_i\} \subseteq S_{i-1}$, and S_{i-1} contains all vertices u for which $\langle v_i, u \rangle \in E$?

As proved in [6], the problem stated above is NP-Complete.

2.2 Abstract Space Minimization Problem

This is an abstract formulation of the *space minimization problem* [2]:

INSTANCE: Directed Acyclic Graph $G = (V, E)$, <u>a function *size*: $V \to Z$</u> and a positive integer K.
QUESTION: Is there a computation for G that uses K or less units of space, i.e an ordering v_1 , v_2 , $\ldots$, v_n of the vertices in V, where $n = |V|$ and <u>all vertices with out-degree 0 are placed before vertices with non-zero out-degree</u>, and a sequence S_0 , S_1 , $\ldots S_n$ of subsets of V, each satisfying <u>$size(S_i) \leq K$</u>, such that S_0 is empty, S_n contains all the vertices with in-degree 0 in G, and, for $1 \leq i \leq n$, $v_i \in S_i$, $S_i - \{v_i\} \subseteq S_{i-1}$, and S_{i-1} contains all vertices u for which $\langle v_i, u \rangle \in E$?

It is trivial to show that this problem is NP-complete.

Theorem 2.1 *The Abstract Space Minimization Problem is NP-complete.*

Proof. Consider an instance R of the Register Sufficiency Problem:
$G = (V, E)$, a positive integer K. We build an instance M of the Abstract Space Minimization Problem:

- $G' = (V \cup \{u\}$, $E \cup \{\langle v, u \rangle \, | \, v \in V$ is a vertex of out-degree 0 $\})$
- $size(v) = \begin{cases} 0 & \text{if } v = u \\ 1 & \text{otherwise} \end{cases}$

---

[2] The underlined clauses indicate the differences from the Abstract Register Sufficiency Problem.

– positive integer K.

The problem instance M has a solution iff R does.

If a solution to R imposes the order $(v_1, v_2, \ldots, v_n)$ and involves the sets $(\emptyset, S_1, S_2, \ldots S_n)$. Then $(u, v_1, v_2, \ldots, v_n)$ and $(\emptyset, \{u\}, S_1, S_2, \ldots S_n)$ is a solution to M.

Similarly, any solution to the Abstract Space Minimization Problem M must be of the form $(u, v_1, v_2, \ldots, v_n)$ and $(\emptyset, \{u\}, S_1, S_2, \ldots S_n)$. This provides $(v_1, v_2, \ldots, v_n)$ and $(\emptyset, S_1 - \{u\}, S_2 - \{u\}, \ldots S_n - \{u\})$ as a solution to R.

3 The Domain of Tree Expressions

Given that the *abstract space minimization* for dags is NP-Complete, we now investigate whether there are important sub-classes of dags for which the solution is polynomially computable. The class of tree expressions is a very important sub-class of dags.

The *register sufficiency problem* for tree expressions has a trivial linear time algorithm. In contrast, the *space minimization problem* for tree expressions requires non-trivial analysis.

3.1 Register Sufficiency For Tree Expressions

Consider the tree T with sub-trees T_1 and T_2. By induction, we can compute the minimum number of registers required for the computation of T_1 and T_2. Let these minimum numbers be n_1, n_2. Without loss of generality, we can assume that $n_1 \geq n_2$. If the computation is performed by first completing the allocation for T_1, followed by T_2 then the number of registers required for the entire computation is $Maximum(n_1, n_2 + 1)$. This is the minimum number of registers required for the computation of T.

A priori we cannot make the assumption that the strategy of completing one sub-tree and then proceeding to complete the next, is a strategy that uses the the minimum number of registers. Consider an arbitrary strategy for register allocation. Let T_{i_1}, T_{i_2} be the order in which the sub-trees reach configurations in which they use maximum number of registers. Let the maximum number of registers used for each sub-tree, in this arbitrary strategy, be m_{i_1}, m_{i_2}. From this we can infer that the number of registers required for the computation is at least $Maximum(m_{i_1}, m_{i_2} + 1)$. But we know that $n_{i_1} \leq m_{i_1}, n_{i_2} \leq m_{i_2}$ and $n_1 \geq n_2$. Hence, the number of registers used is at least $Maximum(n_1, n_2 + 1)$.

3.2 Space Minimization for Tree Expressions

Given a tree $T = (V, E)$ and a function $size : V \rightarrow Z^+$, computation on a tree expression can be viewed as re-writing the tree T into another tree T', (denoted by $T \Longrightarrow T'$).

Definition 1. For any tree $T = (V, E)$, $\mathbf{T} \Longrightarrow \mathbf{T'}$ iff

- $\exists\, u \in V$ such that all its children are leaves of T.
- $T' = (V - S,\ E|_{V-S})$, where $S = \{v \mid v \in V \text{ and } \langle u, v \rangle \in E\}$

Definition 2. For any tree $T = (V, E)$ and a function $size : V \to Z^+$,

$$\mathbf{space}(T) = \sum \{size(u) \mid u \text{ is a leaf of } T\}$$

The *space minimization problem* for tree expressions can then be recast as the following problem.

PROBLEM 1: Given a tree $T = (V, E)$, with root r, and a function $size :$ $V \to Z^+$, can a rewriting sequence, $\delta_0 \equiv [T \Longrightarrow T_1 \Longrightarrow \ldots \Longrightarrow T_n \equiv (\{r\}, \emptyset)]$, be defined such that, $Max\_Space(\delta_0) = Min\{Max\_Space(\delta) \mid \delta \in \mathcal{R}\}$,

where, $Max\_Space(T \Longrightarrow T_1 \Longrightarrow \ldots \Longrightarrow T_m) = Max\{space(T_i) \mid 1 \le i \le m\}$,

$\mathcal{R} = \{\delta \mid \delta \text{ is any rewriting sequence from } T \text{ to } (\{r\}, \emptyset)\}$

Definition 3. For any tree $T = (V, E)$ and a function $size : V \to Z^+$,

$$s(\mathbf{v}) = \begin{cases} 0 & \text{if } v \text{ is a leaf in } T \\ size(v) - P & \text{where } P = \sum\{size(u) \mid \langle v, u \rangle \in E\} \end{cases}$$

Note:

- If $T \Longrightarrow T'$, then $size(T') = size(T) + s(u)$, where u is the node whose children have been deleted in the transition.
- Any sequence of rewrites, $T_1 \Longrightarrow \ldots \Longrightarrow T_n$, is completely specified by the sequence of nodes whose children are to be deleted at each step.

Definition 4. Given a tree $T = (V, E)$, a function $s : V \to Z$ and a sequence $\Pi \equiv v_1, v_2, \ldots, v_n$ elaborating the vertices in V then

$$\mathbf{Peak}(\Pi) \equiv Max\left\{\sum_{j=1}^{i} s(v_j) \mid 1 \le i \le n\right\}$$

PROBLEM 2: Given a tree $T = (V, E)$ and a function $s : V \to Z$, can a topological sort, Π_0, of the tree T be computed such that,

$Peak(\Pi_0) = Minimum\{Peak(\Pi) \mid \Pi \text{ is a topological sort of } T\}$

Note:

- It is easy to see that any instance $(T, size)$ of Problem 1, is an instance (T, s) of Problem 2, where s is computed from the function $size$, based on Definition 3.
- If $s : V \to Z^+$, i.e. the size of the result of every computation exceeds the sum of the size of its arguments, then any topological sort is the optimal answer because for every topological sort Π, $Peak(\Pi) = \sum\{s(v) \mid v \in V\}$.

– If we are given a set of numbers V with no partial order imposed on them, then the sorting the numbers in increasing order produces the minimum *Peak* value over all possible permutations. Actually we need not even sort the numbers, merely placing the negatives before the positives is good enough. Let us assume $v_1, \ldots, v_{i-1}, v_i, \ldots, v_n$ is an elaboration of the numbers such that $v_{i-1} > v_i$. Then the elaboration $v_1, \ldots, v_i, v_{i-1}, \ldots, v_n$ has a *Peak* value less than or equal to the original.

From the observations made in the note above, a possible strategy to a solution of Problem 2 would be to obtain topological sorts in which vertices with negative s values are pushed as close as possible to the start.

4 The Solution

Problems of this nature, involving precedence constraints, have been studied by Monma and Sidney in [4], in the more general framework of series-parallel graphs.

A *sequencing problem* on a set of jobs J with a cost function f, is to find a permutation of J, contained in a set of feasible permutations F, which minimizes f, i.e

$$Minimize\ f(\Pi)$$
$$\Pi \in F$$

If the set of feasible permutations is the set of topological sorts on a series-parallel graph and f is inductively defined, then Monma and Sidney provide an algorithm for the sequencing problem, if we can define a total order, called preference order on J, which satisfies a set of equations.

Defining such a preference order for the space minimization problem for trees is non-trivial. Showing that this order satisfies the requisite set of equations requires a description of their framework and a definition of their terminology. Hence, we directly present our algorithm, for the minimization problem, and a succinct proof of its correctness.

In their research on testing correctness of concurrent data structures, Bruno et al [2] have also developed very general algorithms for linearizations of events, satisfying a partial order.

4.1 A more general problem

For a graph $G = (V, E)$, a given function s and a given permutation $\Pi \equiv v_1, v_2, \ldots, v_n$, of the vertices, let us look at the *Peak* function on a sequence defined inductively in terms of subsequences.

Let Π_i^j denote the sequence $v_i, \ldots, v_j$.

Let $T(\Pi_i^j)$ denote the sum of the s values of the sequence $v_i, \ldots, v_j$, $\sum_{k=i}^{j} s(v_k)$.

The following identities are easily seen to hold because of the associativity of *Max*.

- $\forall i,\ Peak(\Pi_1^{i+1}) = Max(Peak(\Pi_1^i),\ T(\Pi_1^{i+1}))$
- $\forall i,\ Peak(\Pi_1^n) = Max(Peak(\Pi_1^i),\ T(\Pi_1^i) + Peak(\Pi_{i+1}^n))$
- $\forall i, j$ st $1 < i < j < n,\ Peak(\Pi_1^n) =$
  $$Max(Peak(\Pi_1^i),\ T(\Pi_1^i) + Peak(\Pi_{i+1}^j),\ T(\Pi_1^j) + Peak(\Pi_{j+1}^n))$$
- If $Max(P_{i+1}^n,\ T(\Pi_{i+1}^n) + P_1^i) < Max(P_1^i,\ T(\Pi_1^i) + P_{i+1}^n)$,

  where $P_1^i = Peak(\Pi_1^i)$ and $P_{i+1}^n = Peak(\Pi_{i+1}^n)$,

  then the sequence $\Pi' \equiv [v_{i+1},\ \ldots,\ v_n,\ v_1,\ \ldots,\ v_i]$ has a smaller *Peak* value.

We will refer to any ordered sequence of vertices as a *block*. The above identities involving blocks, each with a given total size t and a *Peak* value p, suggests that we should actually address a more general problem of arranging blocks with their t values provided by a function T and p values provided by a function P. Intuitively, a block is a blackbox representing an expression tree. The t value of a block stands for the size of the object returned at the end of the computation of the tree. The p value of a block stands for the peak space consumed during the computation of the tree. Let us first address the simple problem in which there is no partial order between the blocks.

Definition 5. Given a set B of blocks, functions $T,\ P : B \longrightarrow Z$ and a sequence $\Pi \equiv b_1,\ b_2,\ \ldots,\ b_n$ elaborating the set B then

$$\mathbf{Peak}(\Pi) \equiv Max\{Total(\Pi_1^{i-1}) + P(b_i) \mid 1 \le i \le n\}$$

$$\mathbf{Total}(\Pi_1^k) \equiv \sum_{j=1}^{k} T(b_j)$$

PROBLEM 3: Given a set of blocks B and functions $T,\ P : B \longrightarrow Z$, can we compute a permutation, Π_0, of B, such that

$$Peak(\Pi_0) = Minimum\{Peak(\Pi) \mid \Pi \text{ is a permutation of } B\}$$

Henceforth, a block b will be interchangeably referred to by the tuple $(t \equiv T(b),\ p \equiv P(b))$. An optimal order for a set of blocks, is one which has the minimum value of *Peak*, over all possible permutations.

Definition 6. $(\mathbf{t}, \mathbf{p}) \preceq (\mathbf{t'}, \mathbf{p'})$ iff $Max(p,\ t + p') \le Max(p',\ t' + p)$ i.e. $(t, p), (t', p')$ is the optimal order between the two blocks.

Lemma 4.1 *For any set of blocks B and functions $T,\ P : B \longrightarrow Z$, if Π is a permutation of the blocks such that there exists two adjacent blocks, $b_i,\ b_{i+1}$, with $(t_{i+1}, p_{i+1}) \preceq (t_i, p_i)$, then for Π', the permutation obtained by flipping b_i and b_{i+1}, $Peak(\Pi') \le Peak(\Pi)$.*

Proof. Given a set of blocks B and a permutation $\Pi \equiv [b_1, b_2, \ldots, b_n]$, let b_i, b_{i+1} be two adjacent blocks in Π such that $(t_{i+1}, p_{i+1}) \preceq (t_i, p_i)$. From the definition of *Peak* and *Total*, it is seen that,

$$Peak(\Pi) = Max(\ Peak(\Pi_1^{i-1}), \boxed{Total(\Pi_1^{i-1}) + p_i,\ Total(\Pi_1^i) + p_{i+1},}$$
$$Total(\Pi_1^{i+1}) + Peak(\Pi_{i+2}^n))$$
$$= Max(\ Peak(\Pi_1^{i-1}), \boxed{Total(\Pi_1^{i-1}) + Max(p_i, t_i + p_{i+1}),}$$
$$Total(\Pi_1^{i+1}) + Peak(\Pi_{i+2}^n))$$

But by assumption, $Max(p_{i+1}, t_{i+1} + p_i) \leq Max(p_i, t_i + p_{i+1})$. Hence the permutation Π', in which the order of b_i and b_{i+1} is flipped, must have a smaller *Peak* value. This is because,

$$Peak(\Pi') = Max(\ Peak(\Pi_1^{i-1}),\ Total(\Pi_1^{i-1}) + Max(p_{i+1}, t_{i+1} + p_i),$$
$$Total(\Pi_1^{i+1}) + Peak(\Pi_{i+2}^n))$$

The above lemma shows that the $\preceq$ relation, is the relation which defines whether two adjacent blocks are in the right order. Unfortunately, this relation cannot be used to sort blocks to obtain an optimal permutation. This is because it is not transitive: adjacent elements of a sequence may be sorted but this does imply that if $a \preceq b$ then a occurs before b. For example, $(4, 9) \preceq (0, 3) \preceq (-2, 6)$ but $(4, 9) \npreceq (-2, 6)$. Hence, we define a total order $\ll$ which is transitive and is contained in the $\preceq$ relation, i.e. if $(t_1, p_1) \ll (t_2, p_2)$ then $(t_1, p_1) \preceq (t_2, p_2)$.

Definition 7. $(t_1, p_1) \ll (t_2, p_2) = \begin{cases} (p_1 - t_1) \geq (p_2 - t_2) & \text{if } t_1, t_2 \geq 0 \\ p_1 \leq p_2 & \text{if } t_1, t_2 < 0 \\ (t_1 \leq 0)\&\&(t_2 \geq 0) & \text{otherwise} \end{cases}$

Lemma 4.2 *The order $\ll$ is transitive, i.e. if $(t_1, p_1) \ll (t_2, p_2) \ll (t_3, p_3)$ then $(t_1, p_1) \ll (t_3, p3)$.*

Proof. If all the t values are of the same sign then $(t_1, p_1) \ll (t_3, p_3)$. This is because, we will be using $\geq$ on $(p - t)$, or $\leq$ on p, both of which are transitive.

If all the t values do not have the same sign then t_1 is negative and t_3 is positive. Then, by the definition of $\ll$, $(t_1, p_1) \ll (t_3, p3)$.

Lemma 4.3 *For any two blocks (t_1, p_1), (t_2, p_2), If $(t_1, p_1) \ll (t_2, p_2)$ then $(t_1, p_1) \preceq (t_2, p_2)$, i.e.*

1. If $t_1, t_2 \geq 0$ then,
*   if $(p_1 - t_1) \geq (p_2 - t_2)$ then $(t_1, p_1) \preceq (t_2, p_2)$.*
2. If $t_1, t_2 < 0$ then,
*   if $p_1 \leq p_2$ then $(t_1, p_1) \preceq (t_2, p_2)$.*
3. if $t_1 < 0, t_2 \geq 0$ then $(t_1, p_1) \preceq (t_2, p_2)$.

Proof. 1. By definition $(t_1, p_1) \preceq (t_2, p_2)$ iff $Max(p_1, t_1+p_2) \leq Max(p_2, t_2+p_1)$.

Since $t_2 \geq 0$, $p_1 \leq (t_2 + p_1)$.

If $(p_1 - t_1) \geq (p_2 - t_2)$ then $(p_1 + t_2) \geq (p_2 + t_1)$.

Hence, $Max(p_2, t_2 + p_1) \geq Max(p_2, p_1, t_2 + p_1) \geq Max(p_1, p_2 + t_1)$.

2. As $t_1 < 0$, $Max(p_1, t_1 + p_2) \leq Max(p_1, p_2)$.

Since $p_1 + t_2 < p_1 \leq p_2$, $Max(p_1, p_2) = Max(p_2, t_2 + p_1) = p_2$.

3. $Max(p_1, t_1 + p_2) \leq Max(p_1, p_2) \leq Max(t_2 + p_1, p_2)$.

In the terminology of Monma and Sidney [4], the function *Peak* and the order $\ll$ have the *adjacent pairwise interchange (API) property*.

Theorem 4.1 *Given a set of blocks B and functions $T, P : B \longrightarrow Z$, any permutation Π, in which the blocks are sorted by the $\ll$ order, is optimal.*

Proof. Consider any arbitrary *optimal* permutation Π. If there are two adjacent blocks b_i, b_{i+1} such that $b_{i+1} \ll b_i$, then by Lemma 4.3, $b_{i+1} \preceq b_i$. By Lemma 4.1, the permutation Π', in which b_i, b_{i+1} are swapped, still has the optimal value of *Peak*. This process may be repeated to obtain an optimal permutation Π_0 which is sorted by $\ll$.

It is important to note that $b_i \ll b_j, b_j \ll b_i \nRightarrow i = j$. Hence, there may be many permutations in which the blocks are sorted by $\ll$. Using Lemma 4.1, it can be shown that all these permutations have the same value of *Peak*.

4.2 A Solution to PROBLEM 2

For convenience, Problem 2 is re-stated below.

PROBLEM 2: Given a tree $T = (V, E)$ and a function $s : V \rightarrow Z$, can a topological sort, Π_0, of the tree T, be computed such that

$$Peak(\Pi_0) = Min\{Peak(\Pi) \mid \Pi \text{ is a topological sort of } T\}$$

where $Peak(\Pi) = Max\{\sum_{j=1}^{i} s(v_j) \mid 1 \leq i \leq n\}$

Without loss of generality, we may assume the tree to be a binary tree. Each vertex v of the tree T may be considered to be a block, consisting of a single vertex, with $t = p = s(v)$. For a given tree, the algorithm computes a sequence of blocks, each consisting of a sequence of vertices, such that the flattened list is a topological sort and has the minimum value of *Peak*. The (t, p) value of each block in the sequence, returned by the algorithm, can be computed from the definition. Such blocks are ordered by the $\ll$ relation. The importance of such a solution is that the order imposed by the tree is not relevant any more: the $\ll$ order on blocks respects the partial order imposed by the tree.

The way the algorithm works is as follows: For a tree $T \equiv (v, T_l, T_r)$, it makes recursive calls to obtain the solution for sub-trees T_l and T_r. It then merges the sequence of blocks, returned from the solution of T_l and T_r, based on the order relation $\ll$. At the tail of the merged sequence, Π, the block $[v]$

is placed to obtain a new sequence Π'. If the last two blocks in Π' are sorted by $\ll$ then Π' is sorted and hence, a block-structured solution for T. If the last two blocks in Π' are out of order then they are coalesced into one block, while keeping the rest of the sequence the same, to obtain a new sequence Π''. Again the last two elements of Π'' are compared and the process repeated.

The algorithm is described in Figure 2. For a tree $T = (V, E)$, with a root r, and function $s : V \to Z$, a solution to Problem 2 can be obtained by making a call to SORT(r, s). The algorithm has the same complexity as merge sort, $O(n \log(n))$. It should be noted that for programming languages, with built-in list primitives, **append** and **pop** are not constant time functions. If such languages are used to implement the algorithm, it is best to do a reverse sort, i.e. by the order $\gg$. Then **append** and **pop** are not required: constant time functions, **car** and **cdr** suffice.

4.3   Proof of Correctness

The algorithm partitions a sequence of vertices into blocks and then computes *Peak* in terms of the (t, p) value of the blocks. It should be noted that for any permutation Π, of vertices, $Peak(\Pi)$ equals $Peak([\Pi_1^{k_1}, \Pi_{k_1+1}^{k_2}, \ldots, \Pi_{k_{m-1}+1}^n])$, for any partition of Π into m blocks of vertices, where the ith block contains k_i vertices.

Lemma 4.4 *Given three blocks arranged in the order* Π : b_l, b, b_r, *where* $b_r \ll b_l$, *a permutation with* b_l, b_r *adjacent can be found such that its* Peak *value is less than or equal to that of* Π.

Proof. If $b_l \ll b$ then $b_r \ll b$. Hence, by Theorem 4.1, the permutation b_l, b_r, b is at least as good.

If $b \ll b_l$ then, by Theorem 4.1, the permutation b, b_l, b_r is at least as good.

Theorem 4.2 (Correctness of the Algorithm)
(a) *Given a tree* $T = (V, E)$ *and a function* s, *the algorithm computes an optimal order of vertices which minimizes* Peak *over all possible topological sorts. The solution also partitions the optimal permutation into blocks, which are ordered by* $\ll$ *on the* (t, p) *value of the block, i.e the optimal permutation* $\Pi \equiv [v_1, v_2, \ldots, v_n]$ *is partitioned into blocks* $B_1, B_2, \ldots, B_m$.

$$\boxed{v_1 \ldots v_{k_1}} \ll \boxed{v_{k_1+1} \ldots v_{k_2}} \ll \ldots \ll \boxed{v_{k_{m-1}+1} \ldots v_n}$$
$$\quad B_1 \qquad\qquad B_2 \qquad\qquad\qquad\qquad B_m$$

(b) *Let* T *be any tree containing a forest,* $T_1, T_2, \ldots, T_m$, *of disjoint sub-trees. Let* $\text{Block}(T_i) \equiv B_1^i, \ldots, B_{m_i}^i$ *be the sequence of blocks in the solution of* T_i *generated by the algorithm. There is an optimal permutation for* T, *in which the block structure of the solution of each sub-tree* T_i *is retained, i.e*

- *For every tree* T_i, *vertices belonging to a block* B_j^i *are adjacent.*
- *For every tree* T_i, *vertices belonging to* B_{j+1}^i *occur after those of* B_j^i.

```
function SORT(v : V , s : V -> int) :  (V list * int * int) list        */
/*  The SORT function returns a list of triples: The first component    */
/*  gives the list of vertices in the block. The second is the t value  */
/*  of the block  and the third is the p value of the block.            */
   begin
    if leaf(v)
    then [([v],s(v),s(v))]
    else { Tl = left_sub-tree(v) ;
           Tr = right_sub-tree(v) ;
           Ll = SORT(Tl , s);
           Lr = SORT(Tr , s);
           L  = MERGE(Ll,Lr);
           LL = L @ [(v,s(v),s(v))] ;
           LLL = BLOCK_STRUCTURE(LL);
           return (LLL)
         }
   end
function LESSER((v1,t1,p1),(v2,t2,p2) : V * int *int): bool
/* This function defines the order in Definition 7 */
   begin
    if (t1 < 0)
    then return ((t2 <= 0) && (p1 <= p2))
    else return ((t2 >= 0) && (p1-t1 >= p2-t2))
   end
function MERGE(Ll,Lr:(V list*int*int)list):(V list*int*int) list
   begin
    if (Ll == nil)
    then Lr
    else if (Lr == nil)
         then Ll
         else { h1 = car Ll;
                h2 = car Lr;
                if LESSER(h1,h2)
                then return h1::(MERGE(cdr Ll , Lr))
                else return h2::(MERGE(Ll , cdr Lr))
              }
   end
function BLOCK_STRUCTURE(L:(V list*int*int)list):(V list*int*int) list
/* The function pop takes in a list and returns a tuple */
/* consisting  of the last element of the list and       */
/* the list itself, with its last element deleted.       */
   begin
    if (length(L) < 2)
    then L
    else { (h1 as (vl1,t1,p1) , L1) = pop(L);
           (h2 as (vl2,t2,p2) , L2) = pop(L1);
           if LESSER(h2,h1)
           then return(L)
           else BLOCK_STRUCTURE(L2 @ (vl2@vl1,t2+t1,Max(p2,t2+p1)))
         }
   end
```

Fig. 2. The Algorithm

Proof. The proof of (a) is given by induction on the height of the tree. The proof of (b) is given by induction on the maximum height of the disjoint sub-trees considered.

The base cases are trivial, hence omitted.

(a) Consider a tree T, of height n, with sub-trees T_1 and T_2. As T_1, T_2 have heights strictly less than n, we can apply induction hypothesis (a). Let the optimal ordering for T_1 consist of blocks A_1, A_2, ..., A_{m_1}, where $A_1 \ll A_2 \ll \ldots \ll A_{m_1}$. Similarly, let the optimal ordering for T_2 consist of blocks B_1, $B_2 \ldots$, B_{m_2}, where $B_1 \ll B_2 \ll \ldots \ll B_{m_2}$.

Since T_1 and T_2 are disjoint sub-trees, of height less than n, of the tree T, we can apply induction hypothesis (b). By this hypothesis, there is an optimal permutation Π for vertices in T, in which all vertices belonging to any A_i[similarly B_j] are adjacent and vertices belonging to A_{i+1}[similarly B_{j+1}] occur after those of A_i[similarly B_j].

In this optimal permutation Π, if there is a sequence of vertices forming a block A_a right next to a sequence of vertices forming the block B_b, such that $B_b \ll A_a$, then the entire sequence forming A_a may be flipped with the sequence forming B_b. By Lemma 4.1, this new permutation Π' also has the optimal value of peak. Hence by this bubble sort approach, there is an optimal ordering for T where the blocks $\{A_1, \ldots A_{m_1}, B_1, \ldots, B_{m_2}\}$ are ordered by $\ll$, followed by v: the root of T. This is a valid topological sort. Let this be denoted by $C_1, C_2, \ldots, C_{m_1+m_2}, [v]$. The blocks C_i are in increasing order. If $C_{m_1+m_2} \ll [v]$ then the partition of vertices in the optimal solution for T is $C_1, C_2, \ldots, C_{m_1+m_2}, [v]$. If $[v] \ll C_{m_1+m_2}$ then the blocks $C_{m_1+m_2}$ and $[v]$ are made into one block C. C is now compared with $C_{m_1+m_2-1}$ and the process is repeated till we obtain a block partition which is sorted in increasing order.

(b) Let $T_1, T_2, \ldots, T_m$ be disjoint sub-trees, of height less than or equal to n, of a tree T. Let T_{i_l}, T_{i_r} be the left and right sub-trees of the tree T_i. We can now apply the induction hypothesis (b) on the tree T and the forest $\Upsilon = \{T_{1_l}, T_{1_r}, \ldots, T_{m_l}, T_{m_r}\}$, of disjoint sub-trees of height strictly less than n. By this hypothesis there is an optimal solution, Π for T, in which the vertices belonging to any block of a sub-tree in Υ are clustered together. Also, in Π, blocks belonging to any sub-tree in Υ are arranged in sorted order.

Let us look at the permutation Π and locate the closest pair of blocks B_1 and B_2, from T_{i_l} and T_{i_r}, such that they are not order-correct. Between B_1 and B_2 there are no blocks from T_{i_l} and T_{i_r}. This is because they were selected to be the closest such blocks and Π has blocks from T_{i_l}, T_{i_r} individually in sorted order. By Lemma 4.4, we can shift what lies between B_1 and B_2 to either beyond B_2 or in front of B_1, without affecting the value of the *Peak*. This is still a topological sort of T because the tree order imposed by T does not require any element to be present between T_{i_l} and T_{i_r}. This shift does not affect any relevant existing orders between other blocks. Now B_1 and B_2 may be flipped to get the order correct. This new permutation, Π', respects all other previous orders. By

Lemma 4.1, Π' has the same value of *Peak* since Π was optimal. This bubble sort approach ensures that blocks in Block(T_{i_l}) ∪ Block(T_{i_r}) are in sorted order.

Let v_i be the root of the sub-tree T_i. Let B_{m_i} be the last block in the sequence obtained after this bubble sort. If $[v_i] \ll B_{m_i}$ then by Lemma 4.4, what lies between v_i and B_{m_i} may be pushed out and B_{m_i}, v_i may be grouped into a block B, while still retaining the topological order, the value of *Peak* and the relevant orders between other blocks. Now B_{m_i-1} is compared with this block B and this process is repeated.

5 Conclusion

The *space minimization problem* for dags and the *register sufficiency problem* for dags are reducible to each other. Solving the *space minimization problem* for trees is significantly different from the *register sufficiency problem* for trees. The former requires an $O(n \log n)$ algorithm while the latter has an $O(n)$ algorithm. The authors conjecture that the algorithm for trees is optimal, in terms of complexity.

There are other important sub-classes of dags which have simple internal structure, e.g. interval graphs, series-parallel graphs. It may be worthwhile investigating the complexity of the *space minimization problem* on such graphs.

Acknowledgements

The authors would like to thank Phil Gibbons for bringing to their notice the paper by Monma and Sidney.

References

1. A. Appel. *Compiling with Continuations.* Cambridge University Press, 1992.
2. J.L. Bruno, P.B. Gibbons, and S. Phillips. Testing concurrent data structures. Technical Report BL011211-941201-29TM, AT&T Bell Laboratories, 1994.
3. M. R. Garey and D.S. Johnson. *Computers and Intractibility.* W.H. Freeman And Company, 1979.
4. Clyde L. Monma and Jeffrey B. Sidney. Sequencing with series-parallel precedence constraints. *Mathematics of Operations Research*, 4:215–224, 1979.
5. L. C. Paulson. *ML for the Working Programmer.* Cambridge University Press, 1991.
6. Ravi Sethi. Complete register allocation problems. *SIAM Journal of Computing*, 4:226–248, 1975.
7. Ravi Sethi. Pebble games for studying storage sharing. *Theoretical Computer Science*, 19:69–84, 1982.

Smooth Surfaces for Multi-Scale Shape Representation*

Herbert Edelsbrunner†

Abstract

The *skin* of a set of weighted points in $\mathbf{R}^d$ is defined as a differentiable and orientable $(d-1)$-manifold surrounding the points. The skin varies continuously with the points and weights and at all times maintains the homotopy equivalence between the dual shape of the points and the body enclosed by the skin. The variation allows for arbitrary changes in topological connectivity, each accompanied by a momentary violation of the manifold property. The skin has applications to molecular modeling and docking and to geometric metamorphosis.

1 Introduction

The main purpose of this paper is the introduction of a new concept combining the intuitive ideas of geometric shape and continuous deformation. It is based on the combinatorial notions of Voronoi and Delaunay complexes [3, 19] developed in the area of discrete and computational geometry.

Shape and shape change. The intuitive idea of a geometric shape seems clear or at least sufficiently clear so any further effort towards a precise definition is frequently discarded as a waste of precious time. Indeed, such contemplative effort too often results in the awkward realization that there is little hope for an exact and generally satisfying definition of shape. Too complex and too varied are the geometric sets and phenomena one would hope to encompass and classify.

*Research reported in this paper is partially supported by the Office of Naval Research, grant N00014-95-1-0691, and by the National Science Foundation through the Alan T. Waterman award, grant CCR-9118874.

†Department of Computer Science, University of Illinois at Urbana-Champaign, Urbana, Illinois 61801, USA.

The resulting common practice is the treatment of the word 'shape' as the descriptor of a vague and fuzzy concept. In contradiction to this tendency, the development of alpha shapes [8, 9] is an attempt to rigorously and unambiguously specify the shape of a finite point set. To be more accurate, a one-parametric family of polyhedra is suggested to represent the shape of the set. The parameter, α, specifies the degree of detail or the desired compromise between crude and fine description. The purpose of this development has always been partially computational. Efficient algorithms have been described in the above references and publically available software exists for 2 and 3 dimensions [1].

The alpha shape is a polyhedron embedded in the same Euclidean space as the point data, say in $\mathbf{R}^3$. It changes as discrete values of α, and a typical change adds or removes a vertex, an edge, a triangle, or a tetrahedron. The piecewise linear and discretely changing nature of the concept admits a purely combinatorial treatment with all associated computational advantages. The remaining quest for a differentiable and continuously changing surface that represents shape is partly motivated by aesthetic consideration. The purpose of this paper is the introduction of such a surface that remains faithful to the idea of alpha shapes.

Illustration and summary of results. The main new concept in this paper is the alpha skin of a finite set of points with weights. Think of a weighted point as a spherical ball with the point as center and the weight as radius. Denote the resulting set of balls by B. The parameter, α, is used to simultaneously increase or decrease all radii and to change B to B_α. A ball $b \in B_\alpha$ is *redundant* if the Voronoi complex defined by B_α is the same as that of $B_\alpha - \{b\}$. Let $\mathrm{ucl}\, B_\alpha$ be maximal so the subsets of non-redundant balls in $\mathrm{ucl}\, B_\alpha$ and in B_α are the same. $\mathrm{ucl}\, B_\alpha$ of course contains infinitely many balls. Now shrink every ball $b \in \mathrm{ucl}\, B_\alpha$ by a factor $1/\sqrt{2}$ towards its center. The union of the infinitely many reduced balls is the α-*body*, R_α, and

$$S_\alpha = \mathrm{bd}\, R_\alpha$$

is the α-*skin* of B.

As an example consider figure 1. The alpha shape is the subset of $\mathbf{R}^2$ covered by a subcomplex of the Delaunay complex. The alpha skin is a collection of simple closed curves surrounding the alpha shape. There is a closed curve for each component and for each hole of the alpha shape. The collection of closed curves decomposes $\mathbf{R}^2$ into regions, exactly one of which is unbounded. The α-body, R_α, consists of all regions separated from the unbounded region by an odd number of closed curves. Each component of R_α contains a connected piece of the alpha shape, and each component of the complement contains a connected piece of a complementary shape covered by a subcomplex of the Voronoi complex. A short list of noteworthy properties enjoyed by the alpha skin and body follows.

(1) In the non-degenerate case, S_α is a differentiable $(d-1)$-manifold consisting of finitely many patches of low algebraic degree.

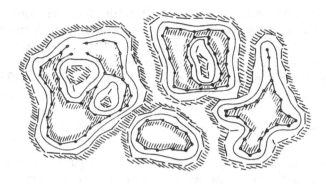

Figure 1: In the non-degenerate 2-dimensional case, the alpha skin consists of pairwise disjoint simple closed curves. They surround the alpha shape and separate it from a complementary shape representing the background. The schematic drawing distinguishes the alpha shape from the complementary shape by indicating vertices as tiny circles.

(2) There are finitely many values of α at which the non-degeneracy assumption does not apply. For each such α, S_α has finitely many points violating the manifold requirement. This is where and when the topology of S_α changes.

(3) The alpha body, R_α, is homotopy equivalent to the alpha shape.

(4) R_α changes continuously with α, that is, if the difference between two values of α is small then the Hausdorff distance between the two α-bodies is small.

(5) Locally, S_α looks the same from both sides. That is, there is a set of (weighted) points on the other side whose skin is the same as S_α.

Style and outline. An effort is made to describe all ideas in a concise and reasonably self-contained manner. The reader uncomfortable with the resulting style might find help in textbooks discussing similar material in a more leisurely fashion [11, 12]. Some of the concepts in this paper are well known and others are new. For example, the Voronoi and Delaunay complexes discussed in section 2 are among the most comprehensively studied data structures in computational geometry [5, 17]. On the other hand, the mixed complex of section 4 whose cells are sums of corresponding Voronoi and Delaunay cells is new. As a general policy, formal proofs are omitted and claims requiring lengthy arguments are left unjustified. The emphasis in this paper is to illustrate that Voronoi and Delaunay complexes form the foundation of a rich topological and combinatorial theory suitable to computing, studying, and describing general geometric shapes.

Section 2 introduces Voronoi and Delaunay complexes. Section 3 introduces alpha shapes, orthogonal balls, and complementary shapes. Section 4 constructs the alpha skin as the half-way point of a deformation retracting a union of balls to the complement of the union of orthogonal balls. Section 5 studies the evolution of the alpha skin over the range of alpha values. Section 6 briefly mentions possible applications to molecular modeling and docking and to geometric metamorphosis.

2 Proximity Complexes

This section reviews Voronoi and Delaunay complexes forming the combinatorial basis of all our constructions. For greater generality, the complexes are defined for sets of spherical balls (or points with weights) rather than for unweighted points.

Voronoi cells. The *(Euclidean) norm* of a point $x \in \mathbf{R}^d$ is the root of the sum of its coordinates squared, $\|x\| = \langle x, x \rangle^{\frac{1}{2}}$. The *(Euclidean) distance* between two points $x, y \in \mathbf{R}^d$ is $|xy| = \|x - y\|$. The (d-dimensional spherical) *ball* with *center* $z \in \mathbf{R}^d$ and *radius* ζ is

$$b(z, \zeta) = \{x \in \mathbf{R}^d \mid |xz|^2 \leq \zeta^2\}.$$

Negative values of ζ^2 are important in this paper. We thus choose ζ from the set of positive square roots of real numbers: $\zeta \in \mathbf{R}^{\frac{1}{2}}$ and therefore $\zeta^2 \in \mathbf{R}$. This convention has only formal consequences and all formulas are written in terms of ζ^2 rather than ζ. By definition, $b(z, \zeta)$ is empty if $\zeta^2 < 0$. The *weighted distance* of a point $x \in \mathbf{R}^d$ from $b = b(z, \zeta)$ is

$$\begin{aligned}\pi_b(x) &= |xz|^2 - \zeta^2 \\ &= \|x\|^2 - 2\langle x, z \rangle + \|z\|^2 - \zeta^2.\end{aligned}$$

Observe $\pi_b(x) > 0$ if $x \notin b$, $\pi_b(x) = 0$ if $x \in \mathrm{bd}\, b$, and $\pi_b(x) < 0$ if $x \in \mathrm{int}\, b$. The points with equal weighted distance from two balls lie on a hyperplane. Let B be a finite set of balls in $\mathbf{R}^d$. The *Voronoi region* of $b \in B$ is

$$\nu_b = \{x \in \mathbf{R}^d \mid \pi_b(x) \leq \pi_a(x) \text{ for all } a \in B\}.$$

Voronoi regions are convex polyhedra which overlap at most along their boundaries, and if they overlap they intersect in convex polyhedra of dimension $d - 1$ or less. A *Voronoi k-cell* is a k-dimensional convex polyhedron of the form

$$\nu_X = \bigcap_{b \in X} \nu_b.$$

We assume $X \subseteq B$ is maximal generating the same common intersection, and we adopt the convention that $\nu_X = \emptyset$ for non-maximal sets X. V_B^k is the set of Voronoi k-cells, and

$$\mathsf{V}_B = \bigcup_{k=0}^{d} \mathsf{V}_B^k.$$

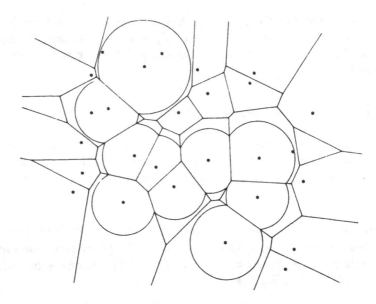

Figure 2: The Voronoi cells of 27 disks (balls in $\mathbf{R}^2$). Only the 10 disks with real radii are shown; the other 17 disks have imaginary radii and only their centers are drawn. There is one disk (with imaginary radius) whose Voronoi region lies outside the window captured by the figure. The disks are in general position, so every non-empty intersection of 2 cells is an edge, every non-empty intersection of 3 cells is a vertex, and no 4 cells share a point.

is the *Voronoi complex* of B, see figure 2. Recall $b \in B$ is redundant if $V_B = V_{B-\{b\}}$. It is not difficult to see b is redundant iff its Voronoi region, ν_b, has dimension $d-1$ or less. V_B satisfies the properties of a complex: all cells are convex, the intersection of any two cells is either empty or again a cell in V_B, and the boundary of every cell is the union of other cells in V_B.

Delaunay cells. A maximal set of Voronoi regions defines a Voronoi cell by intersection and another cell by taking the convex hull of ball centers. The *Delaunay cell* defined by a set $X \subseteq B$ with $\nu_X \in V_B$ is the convex hull of the centers of balls in X:

$$\delta_X \;=\; \mathrm{conv}\,\{z \mid b(z,\zeta) \in X \text{ for some } \zeta\}.$$

The dimension of δ_X is $\dim \delta_X = d - \dim \nu_X$. D_B^k is the set of Delaunay k-cells, and

$$D_B \;=\; \bigcup_{k=0}^{d} D_B^k$$

is the *Delaunay complex* of B, see figure 3. Observe the Delaunay vertices are the centers of all non-redundant balls in B. We will see shortly that D_B is indeed a complex.

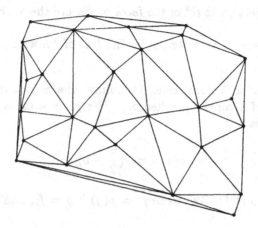

Figure 3: The Delaunay complex of the 27 disks in figure 2. All 2-cells are triangles because the disks are in general position.

Piecewise linear convex functions. Voronoi cells can be interpreted as projections of faces of a convex polyhedron in $\mathbf{R}^{d+1}$. Identify $\mathbf{R}^d$ with the subspace spanned by the first d coordinate axes. The *vertical* direction is defined by the $(d+1)$-st axis. For a point $p \in \mathbf{R}^{d+1}$ we write $p = (x, \xi)$ if x is the projection to $\mathbf{R}^d$ and ξ is the $(d+1)$-st coordinate. We also write $\mathrm{prj}\, p = x$. The projection of a set $A \subseteq \mathbf{R}^{d+1}$ is $\mathrm{prj}\, A = \{\mathrm{prj}\, p \mid p \in A\}$ and its *upward closure* is

$$\mathrm{ucl}\, A = \{(x, \eta) \mid \eta \geq \xi \text{ for some } (x, \xi) \in A\}.$$

Clearly, $\mathrm{prj}\, \mathrm{ucl}\, A = \mathrm{prj}\, A$.

A *linear function* is a map $f : \mathbf{R}^d \to \mathbf{R}$ defined by $f(x) = \langle u, x \rangle + v$, for some $u \in \mathbf{R}^d$ and $v \in \mathbf{R}$. A point $p = (x, \xi) \in \mathbf{R}^{d+1}$ lies *on* f if $\xi = f(x)$. Similarly, p lies *above* f if $\xi > f(x)$ and p lies *below* f if $\xi < f(x)$. The *graph* of f is the set of points on f; it is a non-vertical hyperplane in $\mathbf{R}^{d+1}$. We generally ignore the difference between a function and its graph. For a ball $b = b(z, \zeta)$ define

$$f_b(x) = 2\langle z, x \rangle - \|z\|^2 + \zeta^2,$$

and observe that $f_b(x) = \|x\|^2 - \pi_b(x)$. It follows that x belongs to the Voronoi region of $b \in B$ iff $f_b(x) \geq f_a(x)$ for every $a \in B$. This suggests we define

$$F_B(x) = \max\{f_b(x) \mid b \in B\}.$$

F_B is a piecewise linear convex map. Its graph is the boundary of a convex polyhedron: $F_B = \mathrm{bd}\, \mathrm{ucl}\, F_B$. A hyperplane *supports* F_B if its intersection with

F_B is non-empty and no point of the hyperplane lies above F_B. The *faces* of F_B are the intersections with supporting hyperplanes. The above discussion implies the following easy and useful result.

FACT 1. The projection to $\mathbf{R}^d$ of the faces of F_B are the Voronoi cells of B.

To express the bijection we write φ_X for the face of F_B with $\mathrm{prj}\,\varphi_X = \nu_X \in V_B$.

Polarity. Similar to Voronoi cells, Delaunay cells can be interpreted as projections of faces of a convex polyhedron. The *polar point* of a linear function $g(x) = \langle u, x \rangle + v$ is

$$p(g) \;=\; (\frac{u}{2}, -v),$$

which is a point in $\mathbf{R}^{d+1}$. We write $p_b = p(g)$ if $g = f_b$, and for $b = b(z, \zeta)$ we have

$$
\begin{aligned}
p_b \;&=\; (z, \|z\|^2 - \zeta^2) \\
\;&=\; (z, f_b(z) - 2\zeta^2).
\end{aligned}
$$

The following is easy to prove.

FACT 2. Let g_1, g_2 be linear functions and $p_1 = p(g_1)$, $p_2 = p(g_2)$ the corresponding polar points.

   (i) $p_1 \in g_2$ iff $p_2 \in g_1$.

   (ii) p_1 lies above g_2 iff p_2 lies above g_1.

It follows every point above F_B corresponds to a linear function below all points p_b, $b \in B$. Let Z be the set of centers of balls in B and construct $G_B : \mathrm{conv}\, Z \to \mathbf{R}$ defined by

$$G_B(x) \;=\; \max\{g(x) \mid p(g) \in \mathrm{ucl}\, F_B\}.$$

Intuitively, G_B is the lower boundary of the convex hull of all points p_b, $b \in B$. The *faces* of G_B are the intersections with supporting hyperplanes. By fact 2, f supports F_B iff $p(f) \in G_B$, and g supports G_B iff $p(g) \in F_B$. This implies the following useful result.

FACT 3. The projection to $\mathbf{R}^d$ of the faces of G_B are the Delaunay cells of B.

Since G_B is a function, D_B is indeed a complex. To express the bijection we write γ_X for the face of G_B with $\mathrm{prj}\,\gamma_X = \delta_X \in D_B$.

Balls with centers at infinity. An inessential difference between Voronoi and Delaunay complexes is that $\bigcup V_B = \mathbf{R}^d$ while $\bigcup D_B = \operatorname{conv} Z$ is a compact subset of $\mathbf{R}^d$. The difference fades when we grow some balls and simultaneously move their centers to infinity. Such a growing ball can be specified e.g. by fixing a point on its boundary and the tangent hyperplane through this point. The ball approaches the half-space bounded by the said hyperplane.

As the balls grow, their centers drag Delaunay cells with them. In the limit, the centers form Delaunay vertices *at* infinity and we have two types of Delaunay cells other than the usual ones with finite vertices only. The first type of cell *extends to* infinity and has at least one but not all vertices at infinity. It resembles an unbounded Voronoi cell. The second type of cell is *at* infinity and has all its vertices at infinity. It has no counterpart in the ordinary Voronoi complex.

How do the growing balls influence the Voronoi complex? For each ball we have a hyperplane that contains a d-dimensional face of F_B. As the balls grow, the hyperplanes get steeper and eventually become vertical. The vertical hyperplanes bound the Voronoi regions of the balls with finite radii. If the limitation is to within a compact subset of $\mathbf{R}^d$ then the Voronoi complex resembles an ordinary Delaunay complex, albeit for a different set of balls.

It will be convenient to call a Voronoi complex, V_B, and a Delaunay complex, D_C, *equal* if after removing all cells at infinity V_B and D_C are equal as sets of cells. In other words, all cells in the symmetric difference of V_B and D_C are at infinity.

3 Shapes and Channels

General shapes can be defined and generated by taking subcomplexes of either a Delaunay or a Voronoi complex. This section discusses the notion of orthogonality between balls and describes a general method for selecting cells, see also [6].

Orthogonal balls. Let $b = b(z, \zeta)$ and $c = b(y, \eta)$ be two balls in $\mathbf{R}^d$ and define

$$\pi_{b,c} = |yz|^2 - \eta^2 - \zeta^2.$$

b and c are *orthogonal* if $\pi_{b,c} = 0$, and they are *further than orthogonal* if $\pi_{b,c} > 0$. Indeed, $\pi_{b,c} = 0$ iff $\pi_b(y) = \eta^2$ iff $\pi_c(z) = \zeta^2$. The spheres bounding two orthogonal balls thus meet at a right angle, see figure 4. Observe that

$$\pi_{b,c} = \|y\|^2 - 2\langle y, z \rangle + \|z\|^2 - \eta^2 - \zeta^2$$
$$= \|y\|^2 - \eta^2 - f_b(y).$$

In words, $\pi_{b,c}$ is the difference between the $(d+1)$-st coordinate of p_c and $f_b(y)$. It follows b and c are orthogonal iff $p_c \in f_b$ iff $p_b \in f_c$.

Additional properties of linear functions and polar points are derived from their interaction with the graph of $\varpi : \mathbf{R}^d \to \mathbf{R}$ defined by $\varpi(x) = \|x\|^2$. This

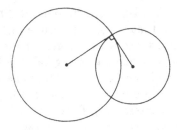

Figure 4: Two disks with positive radii are orthogonal iff their circles meet at a right angle.

graph is a paraboloid of revolution and denoted ϖ, same as the map. The projection to $\mathbf{R}^d$ of the intersection with f_b is $\mathrm{prj}\,(\varpi \cap f_b) = \mathrm{bd}\,b$:

$$\|x\|^2 - 2\langle z, x\rangle + \|z\|^2 \;=\; \zeta^2.$$

Consider the linear function f_x defined for a point $b(x,0) = x \in \mathrm{bd}\,b$. b and x are orthogonal and therefore $p_b \in f_x$. f_x is tangent to ϖ and touches ϖ in point $(x, \|x\|^2)$. It follows the hyperplanes f_x, $x \in \mathrm{bd}\,b$, are the tangents of a cone from p_b to ϖ. To formally state this observation call $y \in \varpi$ *visible* from a point p below ϖ if the line segment $py = \mathrm{conv}\,\{p, y\}$ meets ϖ in y only. By definition, no point is visible from a point on or above ϖ. Let $\mathrm{vis}\,p$ denote the set of points visible from p.

FACT 4. A point $x \in \mathbf{R}^d$ is contained in a ball b iff $(x, \|x\|^2) \in \varpi$ lies on or below f_b iff $(x, \|x\|^2) \in \mathrm{vis}\,p_b$.

Dual complex. Recall D_B is the set of cells δ_X with $\nu_X = \bigcap_{b \in X} \nu_b \neq \emptyset$. We define a subcomplex by selecting δ_X only if ν_X contains points of the ball union. The *dual complex* of B is

$$\mathsf{K}_B \;=\; \{\delta_X \mid \nu_X \cap \textstyle\bigcup B \neq \emptyset\},$$

and the *dual shape* is $\bigcup \mathsf{K}_B$, see figure 5. Clearly, $\mathsf{K}_B \subseteq \mathsf{D}_B$ is a complex itself. Observe $\nu_b \cap \bigcup B = \nu_b \cap b$ for every $b \in B$, and therefore

$$\nu_X \cap \textstyle\bigcup B \;=\; \bigcap_{b \in X} (\nu_b \cap b).$$

To develop a $(d + 1)$-dimensional interpretation of the dual complex recall the bijections relating $\delta_X \in \mathsf{D}_B$ with $\nu_X \in \mathsf{V}_B$, $\varphi_X \subseteq F_B$, and $\gamma_X \subseteq G_B$. By fact 4, $\nu_X \cap \bigcup B \neq \emptyset$ iff $\varphi_X \cap \mathrm{ucl}\,\varpi \neq \emptyset$. Let g be a linear function with polar point $p(g) \in \varphi_X \cap \mathrm{ucl}\,\varpi$. By construction, g supports G_B, g contains the face γ_X, and

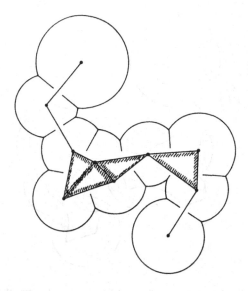

Figure 5: The Voronoi cells decompose the union of 10 disks into convex regions. The dual complex indicates which subsets of the set of 10 regions have non-empty common intersection.

all points of g lie on or below ϖ. The existence of g implies that $\delta_X \in K_B$ only if all points of γ_X lie on or below ϖ. The reverse is not necessarily true. We state the resulting characterizations of the dual complex for later reference.

FACT 5. $\delta_X \in K_B$ iff $\varphi_X \cap \mathrm{ucl}\, \varpi \neq \emptyset$ iff there exists a supporting hyperplane g of points on or below ϖ with $g \cap G_B = \gamma_X$.

Because of the symmetry between the Delaunay and Voronoi complex, we can expect the idea of the dual complex also applies to Voronoi complexes. To see this is indeed the case we construct a new set of balls, C, so V_B is D_C.

Orthogonal complement. We are interested in the set of balls orthogonal to or further than orthogonal from all balls $b \in B$:

$$B_\perp = \{c \in \mathbf{R}^d \times \mathbf{R} \mid \pi_{b,c} \geq 0 \text{ for all } b \in B\}.$$

This is the set of balls c with $p_c \in \mathrm{ucl}\, F_B$. The set $B_\perp$ can thus be represented by a finite set of balls generating the vertices of F_B; all other balls are implied by the upward closure operation. To reproduce the unbounded faces of F_B we also choose vertices at the infinite ends of the unbounded edges. For each point $y \in \mathbf{R}^d$, let η_y^2 be the maximum η^2 with $(y, \eta) \in B_\perp$. The *orthogonal complement* of B is the set C of balls (y, η_y) over all points y that are either vertices in V_B or lie at the infinite ends of unbounded edges in V_B. The balls of the latter

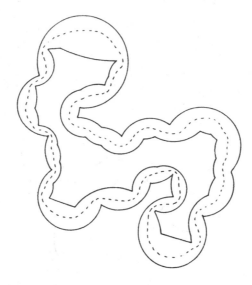

Figure 6: The disks in the orthogonal complement of the 27 disks in figure 2 cover the entire outside: $\overline{\bigcup B} \subseteq \bigcup C$. Symmetrically, $\overline{\bigcup C} \subseteq \bigcup B$. Only the boundary of $\bigcup C$ is shown, which is a simple closed curve surrounded by the boundary of $\bigcup B$. The two curves bound the inner channel from the outside and inside. The $\frac{1}{2}$-skin or simply skin of B is drawn schematically in the middle of the inner channel.

type have infinite radius and are really half-spaces, as discussed at the end of section 2. Recall we agreed to call two complexes equal if they are the same after removing cells at infinity. With this understanding we have two sets of balls with interchanged Voronoi and Delaunay complexes:

FACT 6. $D_C = V_B$, $V_C = D_B$, $G_C = F_B$, and $F_C = G_B$.

If the balls in B are in general position then the Voronoi cells are simple poly-hedra and the Delaunay cells are simplices. It follows the balls in C are not in general position. Indeed, the degeneracy in C suffices to generate simplicial Voronoi cells and simple Delaunay cells. Since F_C and G_C remain unchanged even if balls $c \in B_\perp$ are added to C, fact 6 implies

$$\bigcup C = \bigcup B_\perp. \tag{1}$$

As a consequence, the balls in B and C together cover the entire d-dimensional space, see figure 6:

FACT 7. $\bigcup B \cup \bigcup C = \mathbf{R}^d$.

Indeed, if $y \in \mathbf{R}^d$ lies outside this union we have $\eta_y^2 > 0$. Adding (y, η_y) to C would thus increase $\bigcup C$, contradicting (1).

Channels. The cells of the dual complex are contained in the union of the balls: $\bigcup K_B \subseteq \bigcup B$. To see this note $G_{\mathbf{R}^d} = \varpi$, where $\mathbf{R}^d$ denotes the set of balls $b(x, 0)$ with zero radius. Similarly, $G_{B \cup \mathbf{R}^d}$ is the boundary of the convex hull of $G_B \cup \varpi$. With this preparation

$$
\begin{aligned}
\bigcup K_B &\subseteq \operatorname{prj}(G_B - \operatorname{int} \operatorname{ucl} \varpi) \\
&\subseteq \operatorname{prj}(G_{B \cup \mathbf{R}^d} - \operatorname{int} \operatorname{ucl} \varpi) \\
&= \operatorname{prj} \bigcup_{b \in B} \operatorname{vis} p_b \\
&= \bigcup B.
\end{aligned}
$$

Observe the projection to $\mathbf{R}^d$ of the points on G_B that lie below ϖ is the complement of $\bigcup C$. What we just proved is thus slightly stronger than the claim, namely that the dual shape of B is contained in the closure of the complement of $\bigcup C$, which in turn is contained in $\bigcup B$. By symmetry, $\bigcup B$ is contained in the closure of the complement of the dual shape of C:

FACT 8. $\bigcup K_B \subseteq \operatorname{cl} \overline{\bigcup C} \subseteq \bigcup B \subseteq \operatorname{cl} \overline{\bigcup K_C}$.

In spite of fact 8, it is possible the two dual shapes share common points, see figure 7. The common points are indicative of degeneracies where the spheres

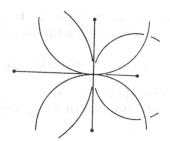

Figure 7: If two disks in B touch at a point on the boundary of $\bigcup B$ then the edge connecting their centers belongs to K_B. C contains two orthogonal disks that touch at the same point and the edge connecting their centers belongs to K_C. The two edges intersect at the point common to the circles bounding the four disks.

bounding 2 or more balls intersect in a single point.

The skin of B will be constructed between the middle two sets in fact 8. Define the *outer channel* as the set of points that neither belong to $\bigcup K_B$ nor to $\bigcup K_C$. The *inner channel* is

$$
I_B = \operatorname{int} \bigcup B \cap \operatorname{int} \bigcup C,
$$

see figure 6. Both channels are open sets and the outer channel contains the inner channel. By symmetry, the inner channel of B is the same as the inner channel of C, and we write $I = I_B = I_C$.

There is an important difference between the outer and the inner channels. Consider the balls in B changing continuously. The outer channel changes abruptly as cells appear or disappear. In contrast, the inner channel changes gradually. We will exploit this property and construct the skin of B using the closures of the inner channel as the possible location for its points.

4   Skin Through Mixing

This section defines the skin of a set of balls by intersecting ϖ with the graph of a convex map and projecting the intersection. If B is a non-degenerate set of balls in $\mathbf{R}^d$, then the skin is a differentiable $(d-1)$-dimensional manifold separating the dual complexes of B and C.

Mixed volume construction. The projection to $\mathbf{R}^d$ of the points on G_B that lie vertically below ϖ is the complement of $\bigcup C$. We seek a continuous deformation of G_B so eventually the points below ϖ project to $\operatorname{int} \bigcup B$. To describe the deformation define

$$
\begin{aligned}
P_0 &= \operatorname{ucl} G_B \text{ and} \\
P_1 &= \operatorname{ucl} G_{B \cup \mathbf{R}^d}.
\end{aligned}
$$

P_0 and P_1 are convex bodies and the points of P_1 below ϖ indeed project to $\operatorname{int} \bigcup B$. Observe the set of points common to the boundaries of P_0 and P_1 is

$$
\operatorname{bd} P_0 \cap \operatorname{bd} P_1 = G_B \cap G_{B \cup \mathbf{R}^d},
$$

and its projection to $\mathbf{R}^d$ is the dual shape of B.

The deformation is most naturally defined using weighted Minkowski sums. For each $t \in [0, 1]$ define

$$
P_t = (1-t) \cdot P_0 + t \cdot P_1.
$$

Since $P_0 \subseteq P_1$ we have $P_t \subseteq P_u$ whenever $0 \le t \le u \le 1$. Also note the points in common the boundaries of P_t and P_u, $t \ne u$, are the same as the points common to the boundaries of P_0 and P_1. For $t \in [0,1)$, the t-body of B is

$$
R_{B,t} = \operatorname{prj}(P_t \cap \varpi),
$$

and for $t = 1$ we supplement $R_{B,1} = \bigcup B$. By inheritance from the P_t we have $R_{B,t} \subseteq R_{B,u}$ whenever $0 \le t \le u \le 1$. The t-skin of B is the boundary of the t-body:

$$
S_{B,t} = \operatorname{bd} R_{B,t}.
$$

In section 5 we will settle for a single representative each of the family of t-bodies and the family of t-skins, both chosen for $t = \frac{1}{2}$: $R_B = R_{B,\frac{1}{2}}$ is the *body* and $S_B = S_{B,\frac{1}{2}}$ is the *skin* of B.

The union of all t-skins, over $t \in [0, 1]$, is the closure of the inner channel. In the non-degenerate case, the t-skin is a manifold, and even in degenerate cases all but a finite number of points $x \in S_{B,t}$ have a neighborhood homeomorphic to $\mathbf{R}^{d-1}$.

Non-degenerate skin. The non-degenerate case can be characterized by the absence of any tangent hyperplane f of ϖ that supports P_t and intersects ϖ and $\mathrm{bd}\, P_t$ in a common point:

$$f \cap \varpi \cap \mathrm{bd}\, P_t \neq \emptyset.$$

Such a hyperplane would indicate the existence of a subset of balls in B whose bounding spheres meet in a single point. In the non-degenerate case the t-skins satisfy a few nice properties violated in the degenerate case. For example, the points common to the boundaries of $P_t \neq P_u$ lie strictly below ϖ. It follows $S_{B,t} \cap S_{B,u} = \emptyset$ whenever $t \neq u$.

Since the t-skins are differentiable for all $t \in (0, 1)$, we can use the normal direction to construct a fibration of $\mathrm{cl}\, I$. Each fiber is a simple curve with one endpoint on $S_{B,0} = \mathrm{bd} \bigcup C$ and the other on $S_{B,1} = \mathrm{bd} \bigcup B$. In between its endpoints the fiber meets all t-skins orthogonally. The fibers are pairwise disjoint. This is clear in the inner channel where the t-skins are differentiable. An argument for the disjointness of fibers is needed at the points where $S_{B,0}$ and $S_{B,1}$ are not differentiable. Such an argument is omitted.

FACT 9. In the non-degenerate case, $\bigcup_{t \in [0,1]} S_{B,t} = \mathrm{cl}\, I$ forms an isotopy between $\mathrm{bd} \bigcup C$ and $\mathrm{bd} \bigcup B$.

The isotopy in fact 9 implies the t-skins are topologically all the same. It certainly follows the bodies $R_{B,t}$ are all homotopy equivalent. The homotopy equivalence between $R_{B,1} = \bigcup B$ and $\bigcup K_B$ has been established in [6], and by transitivity every $R_{B,t}$ is homotopy equivalent to $\bigcup K_B$, as stated in section 1, (4).

Distributing Minkowski sums. It is possible to find a simpler convex body than P_t whose boundary intersects ϖ in the same set of points. First, we rewrite P_t by observing that the Minkowski sum distributes over the upward closure of the convex hull of the union. To state this more clearly define

$$Y \sqcup Z = \mathrm{ucl\, conv}\, (Y \cup Z)$$

for any two sets $Y, Z \subseteq \mathbf{R}^{d+1}$. Let X be another set in $\mathbf{R}^{d+1}$.

FACT 10. $X + (Y \sqcup Z) = (X + Y) \sqcup (X + Z)$.

The proof is elementary. Set $s = 1 - t$ to simplify the application of fact 10 to the definition of P_t:

$$
\begin{aligned}
P_t &= s \cdot P_0 + t \cdot P_1 \\
&= s \cdot P_0 + t \cdot (P_0 \sqcup \varpi) \\
&= (s \cdot P_0 + t \cdot P_0) \sqcup (s \cdot P_0 + t \cdot \varpi) \\
&= P_0 \sqcup \bigsqcup\nolimits_{b \in B} (s \cdot p_b + t \cdot \varpi).
\end{aligned}
$$

In words, P_t is the upward closure of the convex hull of P_0 and a number of translates of $t \cdot \varpi$, one per ball in B. Dropping P_0 from the expression we get

$$
Q_t = \bigsqcup\nolimits_{b \in B} (s \cdot p_b + t \cdot \varpi). \tag{2}
$$

Union of balls. We need a few structural properties of Q_t, P_t, and how they intersect ϖ. Let $q = (z, \zeta)$ be a point in $\mathbf{R}^{d+1}$ and $\varpi_q = s \cdot q + t \cdot \varpi$ a homothetic copy of ϖ, see figure 8. By fact 4, the projection to $\mathbf{R}^d$ of the set of points on

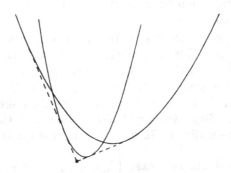

Figure 8: The cone with apex q tangent to ϖ is tangent to every homothetic copy $s \cdot q + t \cdot \varpi$ of ϖ.

ϖ visible from q is the ball $b(z, \zeta)$. Similarly, the projection of the set of visible points on ϖ_q is the ball $b(z, t \cdot \zeta)$. Now consider the intersection of ϖ and ϖ_q. Its projection to $\mathbf{R}^d$ is a sphere bounding a somewhat larger cocentric ball, namely

$$
b_{q,t} = b(z, t^{\frac{1}{2}} \cdot \zeta).
$$

Since $b_{q,t}$ contains the first ball, for all points $q \in P_0$ and all $t \in [0,1]$, the difference $P_t - Q_t$ is disjoint from ϖ. This implies

$$
\begin{aligned}
R_{B,t} &= \mathrm{prj}\,(P_t \cap \varpi) \\
&= \mathrm{prj}\,(Q_t \cap \varpi)
\end{aligned}
$$

for all $t \in [0,1]$. We can thus specify the t-body of B as a union of balls, as anticipated in section 1.

FACT 11. $R_{B,t} = \bigcup_{q \in P_0} b_{q,t}$.

Mixed complex. The boundary of Q_t can be decomposed into faces whose projection to $\mathbf{R}^d$ define a natural and useful complex. Recall the points p_b are the vertices of G_B. We can therefore replace the p_b in (2) by the faces of G_B they span:

$$Q_t = \bigsqcup_{X \subseteq B}(s \cdot \gamma_X + t \cdot \varpi).$$

For each face γ_X of G_B, let Γ_X be a parabolic cylinder defined by the points in the affine hull of γ_X:

$$\Gamma_X = \mathrm{bd}\,(s \cdot \mathrm{aff}\,\gamma_X + t \cdot \varpi).$$

The d-dimensional face of Q_t defined by X is

$$\psi_X = \mathrm{bd}\,Q_t \cap \Gamma_X.$$

The d-faces cover $\mathrm{bd}\,Q_t$ and overlap at most along their boundaries. Let $\mu_X = \mathrm{prj}\,\psi_X$ be the projection to $\mathbf{R}^d$.

FACT 12. For all $0 \leq t < 1$, $\mu_X = s \cdot \delta_X + t \cdot \nu_X$.

The μ_X are the d-dimensional cells of a complex in $\mathbf{R}^d$. To include lower-dimensional cells we need to add weighted sums of δ_X with faces of ν_X and vice versa. The *t-mixed complex* of B is

$$\mathsf{M}_{B,t} = \{s \cdot \delta_X + t \cdot \nu_Y \mid X \subseteq Y\},$$

see figure 9. For $t = 0$, we have $\mathsf{M}_{B,0} = \mathsf{D}_B$. Although $\mathsf{M}_{B,t}$ can be obtained from Q_t only for $t < 1$, the above definition is fine also for $t = 1$, in which case $\mathsf{M}_{B,1} = \mathsf{V}_B$. Observe the symmetry in the definition which implies $\mathsf{M}_{B,t} = \mathsf{M}_{C,s}$. For $t = \frac{1}{2}$ the complexes are the same for B and C, and we call $\mathsf{M}_B = \mathsf{M}_C = \mathsf{M}_{B,1/2} = \mathsf{M}_{C,1/2}$ the *mixed complex* of B (and C).

We remark $\mathsf{M}_{B,t}$ is in general different from the complex obtained by projecting the faces of $s \cdot \mathrm{ucl}\,F_B + t \cdot \mathrm{ucl}\,G_B$. Although the latter complex generally shares quite a few cells with $\mathsf{M}_{B,t}$, there are differences related to Delaunay cells, δ_X, disjoint from their corresponding Voronoi cells, ν_X.

Patches and symmetry. The t-mixed complex decomposes the t-skin into patches of low algebraic degree. Consider a d-face ψ_X of Q_t and the parabolic cylinder Γ_X that contains it. The type of the cylinder is determined by $k = \dim\gamma_X$, which ranges from 0 through d. Γ_X is the sum of a k-flat and a paraboloid in the orthogonal $(d-k)$-dimensional linear subspace of $\mathbf{R}^{d+1}$.

As stated in fact 11, the t-body is the union of infinitely many balls. The same is true for the projection to $\mathbf{R}^d$ of the set of points on Γ_X on or below ϖ. Let this projection be $R_{X,t} = \mathrm{prj}\,(\mathrm{ucl}\,\Gamma_X \cap \varpi)$, and recall the definition of $b_{q,t}$. Then

$$R_{X,t} = \bigcup_{q \in \mathrm{aff}\,\gamma_X} b_{q,t}.$$

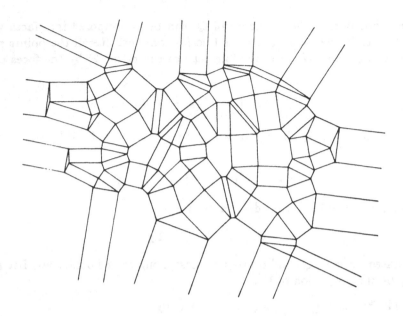

Figure 9: The mixed complex defined by the set of (real and imaginary) disks in figure 2. There are three types of 2-cells. The first type are reduced copies of Voronoi 2-cells, see figure 2. The second type are reduced copies of Delaunay 2-cells, see figure 3. The third type are reduced rectangles, each the Minkowski sum of a Voronoi and a Delaunay edge.

By construction, the patch where the boundary of $R_{X,t}$ coincides with the t-skin, S_t, is the part that projects to the corresponding cell in the t-mixed complex:

FACT 13. $S_t \cap \mu_X = \operatorname{bd} R_{X,t} \cap \mu_X$.

We can do the same exercise the other way round, starting with a Voronoi cell, ν_X, and the corresponding face φ_X of F_B. In this approach the cylinder is

$$\Phi_X = \operatorname{bd}(t \cdot \varphi_X + s \cdot \varpi),$$

and the projection to $\mathbf{R}^d$ of $\operatorname{ucl} \Phi_X \cap \varpi$ is

$$R'_{X,t} = \bigcup_{p \in \operatorname{aff} \varphi_X} b_{p,s}.$$

By construction, all balls $b_{q,1}$ with $q \in \operatorname{aff} \gamma_X$ are orthogonal to all balls $b_{p,1}$ with $p \in \operatorname{aff} \varphi_X$. After shrinking two such orthogonal balls by factors $t^{\frac{1}{2}}$ and

$s^{\frac{1}{2}}$ towards their respective centers, they are either disjoint or touch in a single point. The latter occurs iff the ratio between the original radii is $t^{\frac{1}{2}}$ to $s^{\frac{1}{2}}$. One of the two affine hulls, say aff γ_X, contains only points on or below w. For each point $p \in \varphi_X$ there is a point $q \in \gamma_X$ so the ratio of radii of $b_{q,1}$ and $b_{p,1}$ is $t^{\frac{1}{2}}$ to $s^{\frac{1}{2}}$. It follows the balls $b_{q,t}$ and $b_{p,s}$ cover $\mathbf{R}^d$ without leaving any gap. Since the union of the $b_{q,t}$ shares no interior points with the union of the $b_{p,s}$, they must share the same boundary:

FACT 14. $R'_{X,t} = \text{cl}\,\overline{R_{X,t}}$ and $S'_{X,t} = R_{X,t} \cap R'_{X,t}$.

Observe fact 14 implies the skins of B and C, which are defined for $s = t = \frac{1}{2}$, are the same: $S_B = S_C$. The bodies are complementary: $R_B = \text{cl}\,\overline{R_C}$.

5 Size Variation with Alpha

The preceding sections considered a static set of balls, B, and defined the orthogonal complement, C, the dual complexes, K_B and K_C, and the body and skin, R_B and S_B. This section studies the dynamic situation where balls are allowed to grow.

Weighted growth model. Of all possible growth models we consider the one that keeps the Voronoi complex invariant. The same model has been used in the extension of alpha shapes from unweighted to weighted points [1, 6]. For a ball $b = (z, \zeta)$ and a parameter $\alpha^2 \in \mathbf{R}$ define $b_\alpha = (z, (\zeta^2 + \alpha^2)^{\frac{1}{2}})$. For a set of balls B define $B_\alpha = \{b_\alpha \mid b \in B\}$. The difference between the weighted distances of a point x from b and b_α is

$$\pi_b(x) - \pi_{b_\alpha}(x) = |xz|^2 - \zeta^2 - |xz|^2 + \zeta^2 + \alpha^2$$
$$= \alpha^2.$$

Since the change in distance is the same for every point and ball pair, we have $\mathsf{V}_{B_\alpha} = \mathsf{V}_B$ and $\mathsf{D}_{B_\alpha} = \mathsf{D}_B$ for every $\alpha^2 \in \mathbf{R}$.

The situation is somewhat different for the dual complexes, K_B and K_C. For positive α^2 the union of balls grows and so does the dual complex of B. In spite of the additional cells, K_{B_α} remains a subcomplex of D_B. We call $\mathsf{K}_\alpha = \mathsf{K}_{B_\alpha}$ the α-complex of B and $\bigcup \mathsf{K}_\alpha$ the α-shape. Similarly, we call $R_\alpha = R_{B_\alpha}$ the α-body of B and $S_\alpha = S_{B_\alpha}$ the α-skin.

How does the growth of the balls in B affect the balls in C? The centers remain the same because the Voronoi vertices do not change. To maintain orthogonality, the balls in C must shrink following the same model, and more specifically the orthogonal complement of B_α is $C_{\alpha\sqrt{-1}}$. As mentioned earlier, the skins of a set of balls and of its orthogonal complement are the same:

FACT 15. The α-skin of B is the $(\alpha\sqrt{-1})$-skin of C.

Skin evolution. Although the Voronoi and Delaunay complexes do not change with α, the associated convex maps F and G do. In particular,

$$
\begin{aligned}
F_{B_\alpha}(x) &= F_B(x) + \alpha^2, \text{ and} \\
G_{B_\alpha}(x) &= G_B(x) - \alpha^2,
\end{aligned}
$$

see section 2. It is convenient to change notation so the index indicates the amount of growth rather than the deformation value, which from now on is fixed to $t = \frac{1}{2}$. We thus define

$$
Q_\alpha = \frac{1}{2} \cdot P_0 + \frac{1}{2} \cdot \varpi,
$$

where $P_0 = \text{ucl}\, G_{B_\alpha}$ is defined with respect to B_α. The boundary of Q_α is the graph of a convex map $\text{bd}\, Q_\alpha : \mathbf{R}^d \to \mathbf{R}$. The vertical motion experience by Q_α is half the motion for P_0, and therefore

$$
\text{bd}\, Q_\alpha(x) = \text{bd}\, Q_0(x) - \frac{\alpha^2}{2}.
$$

We can understand the evolution of the α-skin through a vertical motion of $\text{bd}\, Q_0$. The intersection with ϖ sweeps out the paraboloid, and its projection sweeps out $\mathbf{R}^d$ with a continuous succession of skins. The same effect can be obtained by keeping $\text{bd}\, Q_0$ fixed and moving ϖ, also vertically but in opposite direction.

Shape history. A possibly more intuitive picture can be painted by introducing another map, $H = H_B : \mathbf{R}^d \to \mathbf{R}$, called the *history map* of B and defined by

$$
H(x) = \text{bd}\, Q_0(x) - \varpi(x).
$$

By construction, the skin of B is the zero-set: $S_B = \{x \in \mathbf{R}^d \mid H(x) = 0\}$. Similarly, the body of B is the projection of the past history: $R_B = \{x \in \mathbf{R}^d \mid H(x) \leq 0\}$. The α-skin and α-body can be obtained analogously at height $\frac{\alpha^2}{2}$:

FACT 16. $S_\alpha = \{x \in \mathbf{R}^d \mid H(x) = \frac{\alpha^2}{2}\}$ and $R_\alpha = \{x \in \mathbf{R}^d \mid H(x) \leq \frac{\alpha^2}{2}\}$.

We have arrived at a fairly satisfying picture for the succession of α-skins and α-bodies representing the shape of a set of balls at all levels of detail: the history map is an everywhere differentiable map, and the succession of S_α and R_α can be recovered by sweeping H with a horizontal hyperplane from bottom to top.

   Similar to the skin, H consists of patches of low algebraic degree. There is one patch per d-cell of the mixed complex, M_B. While exchanging B and C leaves the skin invariant, it causes a reflection of the history map:

$$
H_C(x) = -H_B(x)
$$

for all $x \in \mathbf{R}^d$. This is consistent with the earlier claim that the skin of B and C is common to the complementary bodies, R_B and R_C, see fact 14.

As already mentioned in section 1, the α-skin has the ability to change its topology with changing α. This happens when the hyperplane passes through a critical value $\xi = H(x) \in \mathbf{R}$ defined by a point x where all partial derivatives vanish. Right at the moment the sweep reaches ξ, the skin develops one or more singularities violating the manifold requirement, and these singular points disappear immediately after passing ξ. Each such singularity forms a local transition of connectivity. The detailed study of this phenomenon is the topic of Morse theory [16]. The algorithmic issues in maintaining the betti numbers of the the α-body are the same as in the incremental construction of complexes [4].

6 Applications

Applications of the concepts in this paper are most interesting for the 3-dimensional real space. In $\mathbf{R}^3$ the skin is a closed surface. The idea of defining a surface by placing points and choosing weights is reminiscent to the method of splines, see e.g. [10]. It seems worthwhile to study the use of skin in the generic construction of surfaces in computer aided geometric design. The rest of this section considers two specific problems involving surfaces.

Molecular modeling and docking. The union of 3-dimensional spherical balls is a standard representation for molecules with known relative atom positions, see e.g. [15]. The application of this representation to the problem of molecular docking [2] suggests the construction of a similar representation for the outside or complement. This idea has been pursued by Kuntz and coworkers, see e.g. [14]. They identify regions in the complement that are of particular interest and fill these regions with spherical balls. After removing the balls of the molecule, they are left with an approximate negative imprint, which is then used to search for matching molecules through shape similarity. The spherical balls filling the complementary region have the same purpose as the balls in the orthogonal complement, see section 3. The (exact) complementarity of the bodies of B and of C is relevant in this context as it asserts there is a perfect geometric match between B, e.g. representing a biomolecule, and C, e.g. representing a ligand.

In order to limit the investigation to regions of interest, we can either use a priori knowledge to restrict the orthogonal complement to an active site, or we can compute pockets in the complement where interesting molecular interaction is likely to happen. [7] offers an unambiguous definition of a 3-dimensional pocket of a union of balls, $\bigcup B$. Intuitively, this is a region or subset of the complement with limited accessibility from the outside. By construction, part of the orthogonal complement fills up the pocket, and we may be tempted to consider that part the matching protrusion of $\bigcup C$. By symmetry, we just obtained a definition of a *protrusion* of $\bigcup B$.

We note the similarity between the skin of B and the molecular surface of $\bigcup B$ as defined in [18]. The molecular surface suffers from occasional self-intersections, whereas the skin is the boundary of a proper body in $\mathbf{R}^3$. Another difference between the two surfaces is the relation between inside and outside. The skin is locally symmetric in the sense it can be defined by a set of balls inside and alternatively by a set of balls outside. The molecular surface is asymmetric and an exact match with the molecular surface of a docking ligand cannot exist.

Geometric metamorphosis. Suppose we are given two geometric objects or shapes, X and Y. The problem of geometric metamorphosis or morphing asks for the construction of some kind of continuous deformation that gradually changes X to Y. The 3-dimensional instance of this problem has been studied in the computer graphics literature, and various methods for its solution have been proposed, see e.g. [13]. The currently published methods suffer from a variety of limitations. The most difficult obstacle to overcome seems to be the meaningful planning of changing topological connectivity. Such change is necessary unless X and Y have the same homotopy type.

The author of this paper is collaborating on ideas to use alpha complexes and alpha skins in the construction of a deformation that changes X to Y. Methodologically, this means geometry is given priority over topology, which follows by the uniqueness of the alpha complex of a given input set. After constructing the succession of complexes specifying the deformation, the skin is used to obtain a gradually changing geometric representation that follows through the same sequence of topological connectivity.

Acknowledgements

The author thanks Ho-Lun Cheng, Siu-Wing Cheng, and Ping Fu for discussions on the topic of this paper.

References

[1] N. AKKIRAJU, H. EDELSBRUNNER, M. FACELLO, P. FU, E. P. MÜCKE AND C. VARELA. Alpha shapes: definition and software. *In* "Proc. Internat. Comput. Geom. Software Workshop", ed. N. Amenta, Geometry Center Res. Rept. GCG-80, 1995.

[2] J. M. BLANEY AND J. S. DIXON. A good ligand is hard to find: automated docking methods. *Perspective in Drug Discovery and Design* 1 (1993), 301–319.

[3] B. DELAUNAY. Sur la sphère vide. *Izv. Akad. Nauk SSSR, Otdelenie Matematicheskii i Estestvennyka Nauk* 7 (1934), 793–800.

[4] C. J. A. DELFINADO AND H. EDELSBRUNNER. An incremental algorithm for betti numbers of simplicial complexes. *In* "Proc. 9th Ann. Sympos. Comput. Geom. 1993", 232–239.

[5] H. EDELSBRUNNER. *Algorithms in Combinatorial Geometry.* Springer-Verlag, Heidelberg, Germany, 1987.

[6] H. EDELSBRUNNER. The union of balls and its dual shape. *László Fejes Tóth Festschrift*, eds. I. Bárány and J. Pach, *Discrete Comput. Geom.* **13** (1995), 415–440.

[7] H. EDELSBRUNNER, M. FACELLO AND J. LIANG. On the definition and the construction of pockets in macromolecules. Manuscript, 1995.

[8] H. EDELSBRUNNER, D. G. KIRKPATRICK AND R. SEIDEL. On the shape of a set of points in the plane. *IEEE Trans. Inform. Theory* **IT-29** (1983), 551–559.

[9] H. EDELSBRUNNER AND E. P. MÜCKE. Three-dimensional alpha shapes. *ACM Trans. Graphics* **13** (1994), 43–72.

[10] G. FARIN. *Curves and Surfaces for Computer Aided Geometric Design.* Academic Press, Boston, 1988.

[11] P. J. GIBLIN. *Graphs, Surfaces and Homology.* 2nd edition, Chapman and Hall, London, 1981.

[12] M. HENLE. *A Combinatorial Introduction to Topology.* Freeman, San Francisco, 1979.

[13] J. R. KENT, W. E. CARLSON AND R. E. PARENT. Shape transformation for polyhedral objects. *Computer Graphics* **26** (1992), 47–54.

[14] I. W. KUNTZ. Structure-based strategies for drug design and discovery. *Science* **257** (1992), 1078–1082.

[15] B. LEE AND F. M. RICHARDS. The interpretation of protein structures: estimation of static accessibility. *J. Mol. Biol.* **55** (1971), 379–400.

[16] J. MILNOR. *Morse Theory.* Princeton Univ. Press, New Jersey, 1969.

[17] F. P. PREPARATA AND M. I. SHAMOS. *Computational Geometry – an Introduction.* Springer-Verlag, New York, 1985.

[18] F. M. RICHARDS. Areas, volumes, packing, and protein structure. *Ann. Rev. Biophys. Bioeng.* **6** (1977), 151–176.

[19] G. VORONOI. Nouvelles applications des paramètres continus à la théorie des formes quadratiques. *J. Reine Angew. Math.* **133** (1907), 97–178.

On Parallel Complexity of Planar Triangulations

Christos Levcopoulos[1], Andrzej Lingas[1] and Cao Wang[2]

[1] Department of Computer Science, Lund University,
Box 118, S-221 00 Lund, Sweden.
[2] Department of Computer Science, Memorial University of Newfoundland,
St. John's, Canada A1C 5S7.

Abstract. The greedy triangulation of a finite planar point set is obtained by repeatedly inserting a shortest diagonal that doesn't intersect those already in the plane. We show that the problem of constructing the greedy triangulation of a finite set of points with integer coordinates in the plane is P-complete. This is the first known geometric P-complete problem where the input is given as a set of points. On the other hand, we provide general NC-methods for testing whether a given triangulation of a set of points and/or line segments can be built by inserting the diagonals in a given partial order, and for constructing such triangulations for simple polygons. As corollaries, we obtain NC-algorithms for testing whether a triangulation is respectively the greedy triangulation or the so called sweep-line triangulation, and for constructing respectively the greedy triangulation or the sweep-line triangulation of a simple polygon. The latter result solves the open problem posed by Atallah et al.

1 Introduction

Triangulations belong to the most useful structures in computational geometry. Several different kinds of triangulations in the plane and higher dimensions have been studied. For many of them efficient polynomial time algorithms have been derived [2, 14]. Since triangulations, and, in particular, planar triangulations are so useful it is natural to ask whether fast and processor feasible parallel algorithms for their construction are available.

Often, by a fast and processor feasible algorithm one means an NC-algorithm, i.e., an algorithm running in polylog time and using a polynomial number of processors. The class of all problems solvable by such algorithms is denoted by NC [6, 8]. We can rephrase our question by asking whether planar triangulations computable in (sequential) polynomial-time admit NC-algorithms.

In 1990, Atallah et al. [1, 6], and independently, Hershberger [7, 6], were first to show that certain geometric problems solvable in polynomial-time are so called P-complete. A problem is P-complete if it admits a polynomial-time algorithm and each problem solvable in polynomial time admits an NC-reduction to this problem [6, 8]. Showing that a P-complete problem admits an NC-algorithm is equivalent to proving the class P of problems solvable in polynomial time to be equal to NC.

Atallah *et al.* showed that it is P-complete to construct what they called sweep-line triangulation of a set of line segments [1]. Their sweep-line triangulation * is sequentially constructed by sweeping the plane from top to bottom with a horizontal line L, and whenever L encounters a vertex v, drawing from v all possible diagonals that do not cross previously drawn ones . They also posed the problem of whether in case the line segments form a simple polygon (i.e., a polygon without holes) the sweep-line triangulation could be constructed by an NC-algorithm.

In this paper, we consider in particular one of the classical triangulations, called the greedy triangulation. It is obtained by repeatedly inserting a shortest diagonal of the input set of points and/or line segments that doesn't intersect those already in the plane. Originally, the greedy triangulation has been proposed as a polynomial-time heuristic for the minimum length triangulation [2]. Since then, several interesting properties of the greedy triangulation have been proved [11, 12]. Also, $O(n \log n)$-time algorithms for constructing the greedy triangulation have been claimed [10, 16].

We show that the problem of constructing the greedy triangulation of a finite set of points with integer coordinates in the plane is P-complete. This is the first known geometric P-complete problem where the input is given as a set of points. The method of Atallah *et al.* [1] suitable for proving P-completeness for line-segment problems is merely a starting point in our involved P-completeness proof.

On the other hand, we provide the first known NC-methods for testing whether a given triangulation of a set of points and/or line segments in the plane can be built by inserting the diagonals in a given partial order. Combining these methods with the parallel dynamic programming method from [13], we obtain an NC-method for constructing such triangulations in the simple polygon case. Consequently, we obtain NC-algorithms for testing whether a triangulation is the greedy one, and for constructing the greedy triangulation of a simple polygon. Similarly, we obtain NC-algorithms for testing whether a triangulation is the sweep-line triangulation, and an NC-algorithm for constructing the sweep-line triangulation of a simple polygon. In this way, we affirmatively answer the open problem posed in [1, 6].

The structure of our paper is as follows. In Section 2, we formalize the notion of planar triangulation. In Section 3, we show the problem of constructing the greedy triangulation of a planar point set to be P-complete. In Section 4, we provide NC-methods for testing whether a given triangulation can be built in a given partial order. In Section 5, we combine the test methods of Section 4 with known parallel dynamic methods to derive NC-algorithms for such triangulations (in particular, the greedy triangulation and the sweep-line triangulation) of a simple polygon.

---

* It shouldn't be mixed with the known sweep-line triangulation method which consists in decomposing the input configuration into monotone polygons which are triangulated separately [14].

2 Definitions

Generally, we shall consider planar figures which are called *planar straight-line graphs* (PSLG for short) [14].

A PSLG G is a pair (V, E) such that V is a set of points in the plane and E is a set of nonintersecting, open straight-line segments whose endpoints are in V. The points in V are called vertices of G, whereas the segments in E are called edges of G. If G is a simple cycle, it is a (simple) polygon. If G has no edges, it is a planar point set. A *diagonal* of G is an open straight-line segment with endpoints in V neither intersecting any edge of G nor including any vertex of G. A triangulation of G is a maximal set of nonintersecting diagonals of G. The greedy triangulation of G is obtained by repeatedly drawing a shortest diagonal of G that doesn't intersect those already in the plane. The sweep-line triangulation of G is obtained by sweeping the plane from top to bottom with a horizontal line L, and whenever L encounters a vertex v, drawing from v all possible diagonals of G that do not cross previously drawn ones.

3 Greedy Triangulation is P-Complete

In this section, we prove that the problem of constructing the greedy triangulation of a finite set of points with integer coordinates in the plane is P-complete. As this problem is sequentially solvable in polynomial time by its definition it remains to prove its P-hardness. Our proof is by an NC-reduction of the so called planar circuit value problem (PCVP) to the greedy triangulation problem. In fact, we shall use a variant of PCVP where an instance is given as a planar circuit composed of or-gates and inverters already embedded in a grid. Atallah *et al.* showed this variant to be P-complete in [1].

3.1 Gadgets

An instance of the "embedded" variant of PCVP consists of an alternating sequence of horizontal routing layers and logic layers. Routing layers are built of standard size "left-shift", "right-shift", "fan-out-gate" and "vertical wire" boxes. Logic layers consist of standard size "or-gate" and "inverter" boxes. Input boxes containing input logical values are placed on the top. For details see Section 2 in [1].

First of all, we need to construct "gadgets" from which all the above components of an embedded planar circuit could be easily assembled. We assume all points forming our gadgets to be on an integer grid (i.e., a grid with integer coordinates).

Our basic gadget is a directed wire. A wire is a sligthly convex, almost vertical or horizontal chain of points with monotonously increasing inter-distances in the wire direction. See Fig. 1. The points on a wire chain are alternately colored with black and white. A greedy triangulation of our point configuration containing such a wire necessarily includes the diagonals connecting the neighboring

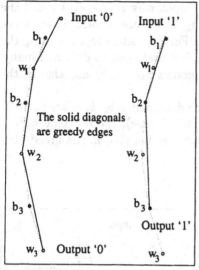

Fig. 1. An illustration for wires.

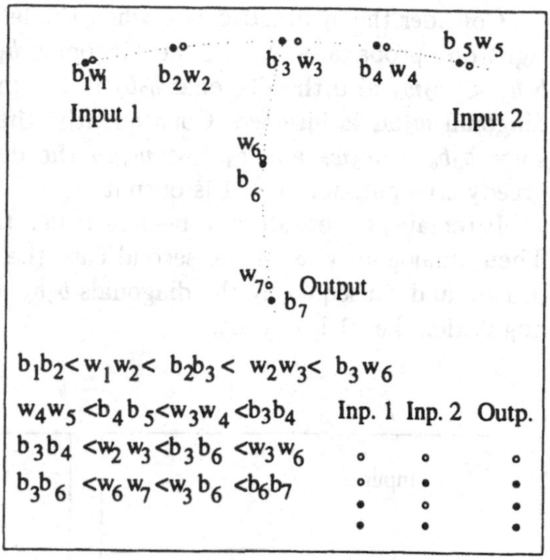

Fig. 2. An example of or-gate.

white and black points on the wire chain and either the diagonals connecting the consecutive white points or the diagonals connecting the consecutive black points. The mutually excluding presence of the white diagonals and the black diagonals respectively represents the 0 signal, and the 1 signal.

Our most complicated gadget is an or-gate. See Fig. 2. It consists of two almost horizontal input wires whose union is sligthly convex from below, and an almost vertical output wire, slightly convex from either left or right. The last points of the two input wires induce a short horizontal segment called the gap of the gate.

Lemma 3.1 *If the two input wires of an or-gate propagate 0, i.e., are triangulated by white diagonals, then the output wire propagates also 0, i.e., it is triangulated by white diagonals. Otherwise, the output wire propagates 1, i.e., it is triangulated by black diagonals.*

Proof. We shall adhere to Fig. 2 where the length of a straight-line segment with endpoints a and b is simply denoted by ab.

Suppose that both input wires propagate 0, i.e., the diagonals w_1w_2 and w_4w_5 are drawn. Then, since $w_3w_4 < b_3b_4$ the diagonal w_3w_4 is drawn. Further, since $w_2w_3 < b_3w_6$ the diagonal w_2w_3 is inserted. Consequently, the diagonal w_3w_6 is drawn. Finally, since $w_6w_7 < w_3b_6$ and $w_6w_7 < b_6b_7$ the diagonal w_6w_7 is also in the greedy triangulation, i.e., 0 is output.

Suppose in turn that the left input wire propagates 0 and the right wire propagates 1, i.e., the diagonals w_1w_2, b_4b_5 are drawn. Then, since $b_3b_4 < w_2w_3$ and $b_3b_4 < w_3w_6$ the diagonal b_3b_4 is inserted. Finally, since $b_3b_6 < w_6w_7$ the diagonals b_3b_6, b_6b_7 are drawn, i.e., 1 is output.

Consider the symmetric case when the left input wire propagates 1 and the right one propagates 0, i.e., the diagonals b_1b_2, w_4w_5 are drawn. Then, since $b_2b_3 < w_2w_3$ also the diagonal b_2b_3 is drawn. Further, since $w_3w_4 < b_3b_4$ the diagonal w_3w_4 is inserted. Consequently, the diagonal b_3w_6 is drawn. Finally, since $b_3b_6 < w_3w_6$ and $b_3b_6 < w_6w_7$ the diagonals b_3b_6, b_6b_7 are also in the greedy triangulation, i.e., 1 is output.

It remains to consider the $1-1$ case, i.e., the diagonals b_1b_2, b_4b_5 are present. Then, analogously as in the second case (i.e., $0-1$ case), the diagonal b_3b_4 is drawn, and consequently the diagonals b_3b_6, b_6b_7 are inserted in the greedy triangulation, i.e., 1 is output. $\square$

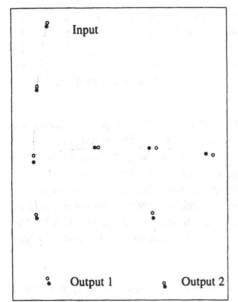

Fig. 3. Right fan-out.

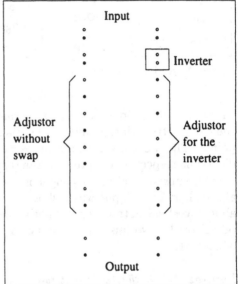

Fig. 4. The idea for inverter and adjustors (the slight bendings are not shown)

By taking the mirror image of our or-gate and exchanging the colors we also obtain an and-gate. The proof of the correctness of its operation is analogous to that for our or-gate given in Lemma 3.1.

Note that the wires comprising our or-gates and and-gates are asymmetric with respect to the distances between the consecutive white-black and black-white pairs in a wire direction. In fact, they are composed of the three following types of wires:

(1) "vertical" wires convex from the right side with "short" white-black distances and "long" black-white distances in the top-down direction;

(2) "vertical" wires convex from the left side with "short" black-white distances and "long" white-black distances in the top-down direction;

(3) "horizontal" wires convex from the bottom with "short" black-white distances and "long" white-black distances in the left-to-right direction.

We can also use our or-gate (or, and-gate) gadget to implement a left or right wire turn. Simply, it is sufficient to start respectively the right input wire or the left one at a white point so it will bring 0 (respectively, at a black point so it will bring 1). Fig. 3 contains some turn examples.

By a composition of appropriate turns, we can also obtain a left or right shift such that the vertical output wire of the shift is of type i if its input vertical wire is of type $i \in \{1, 2\}$. For instance, if the input wire is of type 1 we can implement the right shift as follows. First, we turn to the left (in the direction of the wire) by an and-gate, then to the right by and-gate, next to the right sufficiently long by an or-gate, and finally to the left by an or-gate.

Also, by using the idea of our or-gate or and-gate we can implement a fan-out gate. Simply, we let the single input wire go over and continue on the other "input" side, and we turn the output wire of the gate respectively. See Fig. 3. By choosing the or-gate or and-gate as a basis, the output wires can be ensured to be of type 1 or 2. We can also use the structure of our fan-out gate to change from a wire of type (1) to a wire of type (2), or *vice versa*. For example, if we use an "and-gate" instead of the final "or-gate" in Fig. 3, then Output 2 comes out as a wire of type (2). By arguing similarly as in the $0 - 0$ and $1 - 1$ cases in the proof of Lemma 3.1, we show the correctness of our fan-out gate.

Lemma 3.2 *If the input wire of a fan-out gate propagates z, $z \in \{0, 1\}$, then its two output wires also propagate z.*

To obtain an inverter we simply color two consecutive points on a vertical wire with the same color, see Fig 4. This however changes the type of wire so it doesn't belong to type 1-2 any more. For this reason, below each inverter, we shall place an adjustor, see Fig. 4. It expands the "short" distance and keeps the "long" one stable so they finally can be swapped. In this way, a "wrong" type wire is put back to a proper type. The interpoint distances of wires of "good" type passing down in parallel also have to be proportionally increased by applying another kind of adjustor without the aforementioned swap. In such an adjustment zone, after increasing the vertical interpoint distances the horizontal distances between wires also have to be expanded proportionally through shifts in order to make layout possible (see next section).

We set the input values to the circuit by starting the corresponding input wires with a white or black point respectively. The output value of the circuit is collected from the output wire.

3.2 Layout

It is not sufficient to insert our components in the corresponding boxes of an instance of the "embedded" variant of PCVP because of three following requirements: (1) the wires of wrong type have to be corrected by insertion of adjustment zones which cause horizontal and vertical expansion of the components below, (2) the distances between alternating points in wires and other gadgets have to monotonously increase along any path from an input box to an output box, (3) our wires are not straight but slightly convex.

To start with, we assume temporarily that our wires in both routing and logical components are straight-line vertical or horizontal segments. Consequently, for each component we produce its simplified version with straight wire parts, satisfying the required inequalities, and insert its copies into the corresponding boxes of the embedded instance of PCVP. See Fig. 5. Since each of the components contains a constant number of points and the number of different (types of) components is constant we can easily standardize them so the distances between two consecutive wire points with the same color at their tops, or their bottoms, are respectively equal. By δ we shall denote the constant difference between the aforementioned distances respectively at the top and the bottom of a component. Note that we can both fit the copies of our components into the corresponding boxes and make δ very (even exponentially) small by appropriately constructing the components on a much (respectively, exponentially) denser grid.

To satisfy (1) we also produce simplified versions of the two types of adjustors with straight-line vertical segments so the distance between two consecutive points at their tops is equal to that at the tops of the standard versions of our other components. Next, below each logical layer containing inverters, we insert appropriately scaled-up parallel copies of the two types of adjustors. They respectively extend the "bad" output wires of no-gates and other "good" wires coming from or through the logical layer (see previous section). We call the parallel scaled-up copies an *adjustment slice* further. It increases the distances between the alternating points for all vertical wires going down through it by a constant multiplicative factor, say s. Below the adjustment slice we place an additional routing layer consisting of a linear number of non-overlapping right shifts. It moves each vertical wire (passing through it) to the right multiplying the X-coordinate of the wire by s. The adjustment slice together with the additional routing layer form an adjustment zone.

It follows that the interpoint distances have to be scaled up by s^i between the i-th and the next (if it exists) adjustment zone. Thus, in particular the $i+1$-th adjustment zone has to be rescaled up by the s^i factor. Now, knowing the thickness of consecutive rescaled zones and the exponential increase factor for the consecutive areas between consecutive zones we can easily compute the new Y-coordinates for wire points between and in the zones by parallel prefix sums [8]. The corresponding X-coordinates are simply obtained by multiplying by s^i the old ones, where i is the number of adjustment zones above.

Further, we cut each routing layer into slices by drawing horizontal lines through the top and bottom edges of fan-out and shift boxes. Here we may assume w.l.o.g. that none of the horizontal lines properly intersects another fan-out or shift box of the embedded PCVP transformed by the adjustment zones [1]. In this way, each routing slice as well as each logical layer (called logical slice further), and each adjustment slice is simply a collection of parallel disconnected copies of components and vertical wire pieces with $O(1)$ points. The latter are also made standard with respect to the distance of consecutive points with the same color respectively at their tops and bottoms. Summarizing, the distances between the consecutive points with the same color at the top, and respectively at the bottom of any routing or logical slice, are integers $s^i c_t$, $s^i c_b$, where i is the number of adjustment slices above and $c_t - c_b = \delta$. In case of the $i + 1$-st adjustment slice the corresponding integers are respectively $s^{i+1} c_b$ and $s^i c_b$.

Also, the requirement (2) is satisfied within each slice and the total number of points in all slices is easily seen to be $O(n^2)$ where n is the size of the input embedded PCVP.

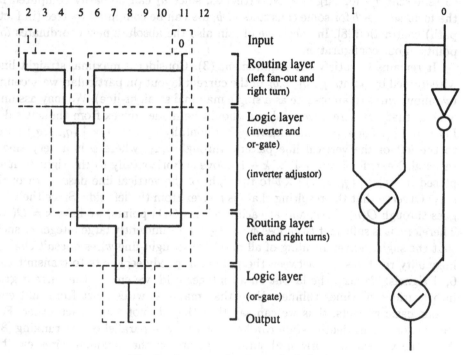

Fig. 5. An example of a preliminary layout.

To satisfy (2) globally we proceed as follows. Let d_{l-1} stand for the distance between two consecutive points with the same color at the bottom of slice $l - 1$. Then, it is sufficient to increase the distance between any two consecutive wire points in l by $g_l = (d_{l-1} - s^i c_b)/2$. To estimate the vertical movements caused by these increases, let $t(l)$ denote the maximum number of vertical pairs of consecutive points on a wire path from the the top to the bottom of l. Then, l is moved by $\sum_{j=1}^{l-1} g_j t(j)$ down, and a horizontal buffer strip of height $g_l t(l)$ has to be added. Since the slice l is a collection of disconnected $O(1)$-point components, we can trivially compute the new coordinates for all points in l in terms of d_{l-1}. In case any point originally at the bottom of l doesn't reach the bottom of the buffer strip we move it down there. After that, we can easily compute d_l in terms of d_{l-1}. Note that $d_l \leq s^i c_b + g_l t(l)$. By induction on i, l, we obtain $d_l = s^i c_b + s^{i+O(l)} \delta$ in case l is a routing or logical slice. By choosing $\delta \leq \frac{c_b}{s^{dn}}$ for a sufficiently large integer constant d, we can preserve the correct operation of the components in spite of the vertical and horizontal movements. In the case of an or-gate or an and-gate, we additionally rely on the symmetry between the input wires and the fact that the relatively large gap between them serves as a buffer for their slight horizontal movements in opposite directions. In case l is an adjustment slice, the operation of the vertical (wire) adjustors is trivially preserved.

Note that one may need nearly a linear number of $O(\log n)$-bit machine words to represent $d_l's$ for large $l's$. Nevertheless, since d_l can be easily computed in the form $a d_{l-1} + b$ for some constants a, b, $d_l's$ can be computed by tree (in fact, path) contraction [8]. In effect, we obtain also the absolute new coordinates for points in our configuration.

It remains to satisfy the requirement (3). Consider a maximal straight-line wire formed by points q_0, q_1, ...q_k in the current layout (in particular, we account two input wires to an or-gate as a single maximal straight-line). We may assume w.l.o.g. that our wire is vertical and should be made convex from the left side. Let $r = \lfloor \frac{k}{2} \rfloor$. For $i = 1, ..., r$, move q_i horizontally by $(r - i) \times | (q_{i-1}, q_i) | \times \phi$ to the left of the vertical line passing through q_{i-1}, where ϕ is a very small rational. Next, for $i = r + 1, ..., k - 1$, move q_i horizontally to the right so it is placed $(i - r) \times | (q_{i-1}, q_i) | \times \phi$ to the right of the vertical line passing through q_{i-1}. Observe that the resulting chain is convex from the left side. Since the wire goes through $O(n)$ slices, and at each level has $O(1)$ points, we have $k = O(n)$. Therefore, it is sufficient to choose $\phi = \frac{1}{c_1 n^2}$ for a sufficiently large integer c_1 such that the slight convex bending of all maximal straight-line wires doesn't change inequality relationships between the potential greedy segments to transmit the 0, 1 signals. To keep the layout on an integer grid, we embed the current grid into a grid $c_1 n^2$-times thinner. Note that maximal wires start from and end at our basic gadgets, thus we can say that they do not touch each other. For this reason, their identification can be done easily in parallel by list ranking [8]. Also, the vertical or horizontal shifts of points on the maximal wires can be easily computed by parallel sums.

In spite of increasing the density of the original grid $2^{O(n)}$ times totally, we have been able to keep a polynomial number of points in our configuration simulating the input PCVP instance, and to show that this configuration can be constructed by an NC-algorithm. Hence, we obtain our main result:

Theorem 3.3 *The problem of constructing the greedy triangulation of a finite set of points with integer coordinates in the plane is P-complete.*

4 Parallel Triangulation Test

In this section, we consider the problem of efficiently testing in parallel whether a triangulation of a PSLG belongs to a specific class of triangulations. We are concerned with classes of triangulations implied by a partial order on diagonals easily extensible to a linear order. In particular, we show that the greedy triangulation and the sweep-line triangulation admit NC-test (even in the general case of PSLG) in spite of their P-completeness.

4.1 Parallel triangulation test for PSLG

Let R be a partial order among straight-line segments in the plane. An R-triangulation of a PSLG is obtained by repeatedly inserting a minimal (in the R-order) diagonal of the graph that doesn't intersect those already in the plane.

The following lemma is useful in testing whether a triangulation is an R triangulation.

Lemma 4.1 *A triangulation T of a PSLG G is its R-triangulation if and only if for any diagonal d of G not in T there is a diagonal in T which properly intersects d and doesn't follow d in the R-order.*

Proof. The necessity of the above condition immediately follows from the definition of an R-triangulation. To show its sufficiency, sort topologically the diagonals in T in the R-order. For $k = 1, 2, ..,$ let T_k be the partial triangulation consisting of the first k diagonals in T in the sorted order. By induction on k, we show that T_k can be produced in k steps of the definitional algorithm for R-triangulation. Consider the $k + 1$-st diagonal e_{k+1} in T in the sorted order. Denote the PSLG $G \cup T_k$ by G_k. Let D be the set of all diagonals of G_k that precede e_{k+1} in the R-order. None of them can occur on the sorted list after e_{k+1}. Hence, the diagonal sets D and T are disjoint. Consequently, for each diagonal d in D there is a diagonal d' in T that doesn't follow d in the R-order and intersects d. By the transitivity of R, such a diagonal d' has to occur before e_{k+1} on the sorted list. Hence, any diagonal d in D intersects a diagonal in T_k. By the definition of D, we conclude that $D = \emptyset$. Thus, T_k can be augmented to T_{k+1} by the definitional algorithm. $\qquad\qquad\square$

By Lemma 4.1, we obtain the following theorem on parallel R-triangulation test.

Theorem 4.2 *Suppose that one can determine whether a line segment precedes another line segment in the R order in constant time using a single processor. One can test whether a triangulation of a PSLG G on n vertices is an R-triangulation in constant time using a CRCW PRAM with $O(n^3)$ processors, or in $O(\log n)$-time using a CREW PRAM with $O(\frac{n^3}{\log n})$ processors.*

Proof. First, in parallel, we test each potential diagonal d of G for intersection with the edges of G. Whenever such an intersection exists e is not a diagonal of G. It takes constant time on a CRCW PRAM with $O(n)$ processors. Since there are $O(n^2)$ potential diagonals it takes $O(n^3)$ processors totally.

Next, in parallel, we test each of the diagonals d of G for intersection with the $O(n)$ diagonals in the input triangulation T. Whenever there is such an intersection with a diagonal in T that doesn't follow d in the R-order, d is deactivated. By our assumptions, it again takes a constant time on a CRCW PRAM with $O(n^3)$ processors totally.

If there is any left active diagonal of G the input triangulation is reported not to be an R-triangulation by Lemma 4.1. Otherwise, the answer is positive.

By dividing the edges of G and diagonals in T into groups of logarithmic size, we can also easily implement this algorithm in logarithmic time on a CREW PRAM with $O(\frac{n^3}{\log n})$ processors. $\qquad\square$

To apply Theorem 4.2 to test whether a triangulation of a PSLG is its greedy triangulation it is sufficient to assume that R is the partial order imposed by line segment length.

Corollary 4.3 *One can test whether a triangulation of a PSLG is the greedy triangulation in constant time using a CRCW PRAM with $O(n^3)$ processors.*

To apply Theorem 4.2 to test whether a triangulation of a PSLG is the sweep-line triangulation it is sufficient to assume that R is the partial order imposed by the Y-coordinates of the top diagonal endpoints.

Corollary 4.4 *One can test whether a triangulation of a PSLG is the sweep-line triangulation in constant time using a CRCW PRAM with $O(n^3)$ processors.*

4.2 Parallel triangulation test for simple polygons

Here we assume the partial order R among straight-line segments to be induced by an equivalence relation among straight-line segments with linearly ordered equivalence classes. The length equality and the equality of the Y coordinates of top endpoints are examples of such equivalence relations respectively inducing the partial order of the greedy and the sweep-line triangulations.

We shall build a data structure which allows for any polygon diagonal to find quickly the minimum (in the sense of the linear order of equivalence classes) polygon diagonal that intersects it. In effect, we can substantially decrease the number of processors used in the R-triangulation test based on Lemma 4.1 in the simple polygon case.

Theorem 4.5 *Let R be a partial order among straight-line segments induced by an equivalence relation among straight-line segments with linearly ordered equivalence classes. Suppose that one can determine whether a line segment precedes another line segment in the R order in constant time using a single processor. One can test whether a triangulation of a simple polygon is its R-triangulation in $O(\log n)$-time using a CREW PRAM with $O(\frac{n^2}{\log n})$ processors.*

Proof. Let P, T respectively denote the input polygon and the input triangulation.

Our data structure is based on the hierarchical decomposition of P resulting from the recursive application of the polygon cutting theorem [3]. The decomposition has the form of a binary tree with the root labeled with P, each non-leaf node labeled with a subpolygon that is the union of the subpolygons labeling its children, and each leaf labeled with a subpolygon having $O(1)$ vertices.

To obtain such a decomposition we construct the so called centroid decomposition of the tree dual to T. It can be easily done within the resource bounds given in the theorem (for example, see [4]).

We preprocess each of the subpolygons Q labeling the decomposition tree for answering the following query in constant time:

for a diagonal d of P coming through an edge a of Q in T (or, starting from a vertex a of Q) and going out through an edge b of Q in T (or, ending at a vertex b of Q) report the minimum (in the R-order) diagonal of Q in T or edge of Q in T intersecting d.

Our preprocessing has a form of an $O(size(Q)) \times O(size(Q))$ table A_Q with rows and columns corresponding to the vertices of Q and the diagonals in T on the perimeter of Q. Since in the query d is assumed to be a diagonal of P, we do not need to check for any entry $T_Q(a, b)$ whether there exists a diagonal of P passing through a and b. Therefore, for each of the $O(n)$ leaf-subpolygons Q, the table A_Q can be easily built in constant time by a single CREW PRAM processor.

Further, in a bottom up fashion, for each non-leaf node m of the decomposition tree, we construct such a table for the corresponding subpolygon from the tables for the subpolygons corresponding to the children of m. The construction consists of a quadratic number of copying operations and binary minimum (in the R order) operations. By our assumptions, it can be done in constant time on a CREW PRAM with a quadratic number of processors. Consequently, we can compute the tables for all subpolygons with $O(\frac{n}{\log n})$ vertices labeling the tree in logarithmic time on a CREW PRAM with $O(\log n(\frac{n}{\log n})^2)$ processors. To keep the total number of processors $O(\frac{n^2}{\log n})$, we slow down evaluating the tables for subpolygons with $\Omega(\frac{n}{\log n})$ vertices from the tables of the child subpolygons by the Brent principle [8]. For a subpolygon with $O(\frac{2^i n}{\log n})$ vertices, $i = 1, 2, ..., \lceil \log\log n \rceil$, we compute it in $O(2^i)$-time using $O(\frac{2^i n^2}{\log^2 n})$ processors. We conclude that our data structure in the form of the table for P can be computed in logarithmic time on a CREW PRAM with $O(\frac{n^2}{\log n})$ processors.

Also, we can list all diagonals of P within logarithmic time on a CREW PRAM with $O(\frac{n^2}{\log n})$ processors by [5]. Now, by Lemma 4.1, it is sufficient to query the table for P with the diagonals of P outside T. If for any of these diagonals, say d, the minimum diagonal in T that intersects d follows d in the R-order, T is reported not to be an R-triangulation. Otherwise, the answer is positive. Clearly, the querying and reporting can be done in logarithmic time on a CREW PRAM with $O(\frac{n^2}{\log n})$ processors (by assigning $O(\log n)$ diagonals to a processor). $\qquad\square$

By combining Theorems 5.1, 4.5, we obtain the following corollaries.

Corollary 4.6 *One can test whether a triangulation of a simple polygon is its greedy triangulation in $O(\log n)$-time using a CREW PRAM with $O(\frac{n^2}{\log n})$ processors.*

Corollary 4.7 *One can test whether a triangulation of a simple polygon is its sweep-line triangulation in $O(\log n)$-time using a CREW PRAM with $O(\frac{n^2}{\log n})$ processors.*

5 Simple Polygon Case

It is known that several variants of diagonal partitions of a simple polygon (e.g., minimum weight triangulation [9]) can be obtained by dynamic programming. The idea is to consider a family of subpolygons of the input polygon obtained by a finite number of cuts along polygon diagonals. The construction of the partition for a subpolygon in the family is reduced to the construction of the partitions of subpolygons of the subpolygon belonging to the family. Unfortunately, the recursion depth in this reduction approach may be linear. To achieve a logarithmic recursion depth it is sufficient to find such a reduction where the subpolygons of the subpolygon have size smaller by a constant fraction.

For example, if the partition is a triangulation we may consider the $O(n^4)$ subpolygons of the input polygon (on n vertices) with at most two polygon diagonals on their perimeter. Any triangulation of such a subpolygon contains a diagonal that splits the subpolygon into two subpolygons, each having at least one third of vertices. The problem is that the balanced diagonal not necessarily separates the at most two diagonals on the perimeter of the subpolygon. In result, a subpolygon with three diagonals on its perimeter may be created. To avoid this note that there is always a triangle in the triangulation that separates the three diagonals. The edges of the triangle and the balanced diagonal split the subpolygon into four subpolygons, each with at most two diagonals on its perimeter and the size smaller by a constant fraction! A straight-forward fast parallel implementation of this approach requires $O(n^4 \times n^2 \times n^3)$ processors (the same asymptotic processor bound for the corresponding algebraic and grammar problems has been given in [15]).

In [13], a more refined, bottom-up method of parallelization of such dynamic programming algorithms has been presented in the dual setting of trees. For example, it has been concluded that a minimum weight triangulation of a simple polygon can be constructed in $\log^2 n$-time on a CREW PRAM with $O(\frac{n^6}{\log n})$ processors. Analogous conclusions could be derived for many other known optimal triangulations, e.g., minimizing maximum diagonal length [2] *etc.*

The method from [13] can be also easily adapted to the problem of constructing other triangulations of simple polygons, say belonging to a class C, provided that

1. the test of whether a triangulation of a simple polygon is its C-triangulation (i.e., belongs to the class C) admits an NC algorithm,
2. cutting the polygon along any diagonal in its C-triangulation and producing a C-triangulation of the two resulting subpolygons always yields a C-triangulation of the original polygon.

Then, instead of, e.g., summing the weights of the component triangulations, we simply test whether the union of the component triangulations and the separating diagonals is a C-triangulation of the current subpolygon. To keep the "minimum" framework of [13], we may report 0 in the positive case and 1 otherwise. Usually, the test requires examining the component triangulations and therefore it takes more time and processors than the simple summing. Hence, the recursion depth bound and the processor bound from [13] are multiplied by the time and processor bounds for the test in the following theorem.

Theorem 5.1 *Let C be a class of triangulations of simple polygons, satisfying the condition 2. Suppose that there exists an $t(n)$-time and $p(n)$-processor CREW PRAM algorithm for the C-triangulation test. A C-triangulation of a simple polygon can be constructed in $O(\log n(\log n + t(n)))$-time on a CREW PRAM with $O(\frac{n^6 p(n)}{\log n})$ processors.*

It is easy to see that for any partial order R, the R-triangulation (see Section 4) has the property required in the above theorem. It also admits efficient parallel tests by Section 4. By combining Theorem 5.1 with Theorems 4.2, 4.5, we obtain the general result of this section.

Theorem 5.2 *Let R be a partial order among straight-line segments. Suppose that one can determine whether a line segment precedes another line segment in the R order in constant time using a single processor. One can construct the R-triangulation of a simple polygon in $O(\log^2 n)$-time using a CREW PRAM with $O(\frac{n^9}{\log^2 n})$ processors.*
Moreover, if R is induced by an equivalence relation among line segments with linearly ordered equivalence classes then the R-triangulation of a simple polygon can be constructed in $O(\log^2 n)$-time using a CREW PRAM with $O(\frac{n^8}{\log^2 n})$ processors.

Corollary 5.3 *The greedy triangulation of a simple polygon can be constructed in $O(\log^2 n)$-time using a CREW PRAM with $O(\frac{n^8}{\log^2 n})$ processors.*

The next corollary solves the open problem of the parallel complexity status of the sweep-line triangulation of a simple polygon posed in [1].

Corollary 5.4 *The sweep-line triangulation of a simple polygon can be constructed in $O(\log^2 n)$-time using a CREW PRAM with $O(\frac{n^8}{\log^2 n})$ processors.*

6 Final Remarks

The problem of constructing the greedy triangulation of planar point sets is arguably a more "natural" geometric problem than the three problems shown to be P-complete in [1, 7] (in particular the sweep-line triangulation problem). Hence, our result should be of interest as a very simple example of a P-complete geometric problem.

Interestingly, we can decrease the high upper bound on the work needed to construct the R-triangulation in Theorem 5.2 dramatically if we increase the time bound substantially, e.g., to $O(\sqrt{n})$.

References

1. M.J. Atallah, P. Callahan, M.T. Goodrich. P-Complete Geometric Problems. In International Journal of Computational Geometry and Applications. Vol. 3, No. 4 (1993) 443-462. (Also in Proc. SPAA'90, 317-326.)
2. M. Bern, D. Eppstein. Mesh Generation and Optimal Triangulation. CSL-92-1, Palo Alto Research Center.
3. B. Chazelle. A theorem on polygon cutting with applications. Proc. 23rd Annual Symposium on Foundations of Computer Science, 1982, 339-349.
4. H. ElGindy, M.T. Goodrich. Parallel Algorithms for Shortest Path Problems in Polygons. The Visual Computer 3(6), 1988, 371-378.
5. M.T. Goodrich, S.B. Shauck, S. Guha. Parallel Methods for Visibility and Shortest Path Problems in Simple Polygons. Proc. ACM Symp. on Comp. Geom., 1990.
6. R. Greenlaw, H.J. Hoover, W.L. Ruzzo A Compedium of Problems Complete for P. Manuscript. 1991.
7. J. Hershberger. Upperenvelope onion peeling. In Proc. 2nd Scandinavian Workshop on Algorithm Theory, LNCS Vol. 447, Springer Verlag, 1992, 368-379.
8. R. M. Karp and V. Ramachandran. Parallel algorithms for shared memory machines. In: Handbook of Theor. Computer Sc., Vol. A (Elsevier, 1990) 869-941.
9. G.T. Klincek. Minimal triangulations of polygonal domains. Ann. Disc. Math. 9, 1980, 121-123.
10. C. Levcopoulos and D. Krznaric. The greedy triangulation can be computed from the Delaunay triangulation in linear time. Technical report LU-CS-TR:94-136, Lund University.
11. C. Levcopoulos and A. Lingas. On approximation behavior of the greedy triangulation for convex polygons. Algorithmica 2, 1987, pp. 175-193.
12. C. Levcopoulos and A. Lingas. C-sensitive Triangulations Approximate the Min-Max Length Triangulation. Proc. of FST-TCS, New Delhi, December, LNCS 652, 104-115, Springer Verlag, 1992.
13. W. Rytter. On Efficient Parallel Computations for Some Dynamic Programming Problems. In Theoretical Computer Science 59 (1988) 297-307.
14. F.P. Preparata and M.I. Shamos. Computational Geometry, An Introduction. Texts and Monographs in Computer Science, Springer Verlag, New York, 1985.
15. L. Valiant, S. Skyum, S. Berkowitz and C. Rackoff. Fast parallel computation of polynomials using few processors. SIAM J. Comput. 12(4) (1983) 641-644.
16. C.A. Wang. An optimal algorithm for greedy triangulation of a set of points. Proc. of 6th CCCG, Saskatchewan, August, 1994, pp. 332-338.

Computing a Largest Empty Anchored Cylinder, and Related Problems

Frank Follert[1], Elmar Schömer[1], Jürgen Sellen[1],
Michiel Smid[2], Christian Thiel[2]

[1] Universität des Saarlandes, Fachbereich 14, Informatik, Lehrstuhl Prof. Hotz, Im Stadtwald, D-66041 Saarbrücken, Germany. E-mail: {follert,schoemer,sellen}@cs.uni-sb.de. ***
[2] Max-Planck-Institut für Informatik, Im Stadtwald, D-66123 Saarbrücken, Germany. E-mail: {michiel,thiel}@mpi-sb.mpg.de. †

Abstract. Let S be a set of n points in $\mathbb{R}^d$, and let each point p of S have a positive weight $w(p)$. We consider the problem of computing a ray R emanating from the origin (resp. a line l through the origin) such that $\min_{p \in S} w(p) \cdot d(p, R)$ (resp. $\min_{p \in S} w(p) \cdot d(p, l)$) is maximal. If all weights are one, this corresponds to computing a silo emanating from the origin (resp. a cylinder whose axis contains the origin) that does not contain any point of S and whose radius is maximal. For $d = 2$, we show how to solve these problems in $O(n \log n)$ time, which is optimal in the algebraic computation tree model. For $d = 3$, we give algorithms that are based on the parametric search technique and run in $O(n \log^5 n)$ time. The previous best known algorithms for these three-dimensional problems had almost quadratic running time. In the final part of the paper, we consider some related problems.

1 Introduction

Geometric optimization problems in low-dimensional spaces have received great attention. See e.g. [1, 3, 7, 9]. Such problems often occur in practical situations. Consider the following example from the field of neurosurgery: A surgeon wants to remove tissue samples from the brain of a patient for diagnosis purposes. This is done by inserting a probe through a small hole in the skullcap of the patient. In order to minimize the exposure to danger, the point of entry has to be chosen in such a way that the trajectory of the probe stays away from certain brain areas. If we model this trajectory as a ray, and the brain areas we want to avoid by weighted points in three-dimensional space, then we want to find a ray R emanating from the position at which we want to remove the tissue sample such that the minimal weighted distance from any of the points to R is maximal.

We denote the Euclidean distance between a point p and the origin by $\|p\|$. Also, the Euclidean distance between two points p and q is denoted by $d(p,q)$. If p

*** Frank Follert was supported by a Graduiertenkolleg Fellowship from DFG, Germany.
† These authors were supported by the ESPRIT Basic Research Actions Program, under contract No. 7141 (project ALCOM II).

is a point in $\mathbb{R}^d$, and R is a closed subset of $\mathbb{R}^d$, then the distance between p and R is defined as $d(p, R) := \min\{d(p, q) : q \in R\}$. Finally, we define an *anchored ray* as a ray that emanates from the origin. The above mentioned optimization problem is the three-dimensional version of the following problem.

Problem 1. Let S be a set of n points in $\mathbb{R}^d$, and let each point p of S have a weight $w(p)$, which is a positive real number. Compute an anchored ray R for which $\min_{p \in S} w(p) \cdot d(p, R)$ is maximal.

We get an obvious generalization if we ask for a line through the origin instead of an anchored ray:

Problem 2. Let S be a set of n points in $\mathbb{R}^d$, and let each point p of S have a weight $w(p)$, which is a positive real number. Compute a line l through the origin for which $\min_{p \in S} w(p) \cdot d(p, l)$ is maximal.

Let R be any ray, and let $\delta \geq 0$. The set of all points in $\mathbb{R}^d$ that are at distance at most δ from R is called a *silo* with *axis* R and *radius* δ.

If each point of S has weight one, then Problem 1 asks for the silo whose axis starts in the origin, that does not contain any point of S in its interior, and that has maximal radius. Also, in this case, Problem 2 asks for the cylinder of maximal radius whose axis contains the origin and that does not contain any point of S in its interior.

Problem 1 and the application described above were posed by Prof. Hotz, and appeared for the first time in Follert's Master Thesis [3]. He shows how to solve this problem in $O(n\alpha(n) \log n)$ time when $d = 2$, and in $O(n^{2+\epsilon})$ expected time when $d = 3$. Here, $\alpha(n)$ denotes the inverse of Ackermann's function, and ϵ is an arbitrarily small positive constant.

Follert also considers Problem 2. For $d = 2$, he shows how to solve this problem in $O(n \log n)$ time. Moreover, he reduces problem Max-Gap-on-a-Circle to Problem 2. (See also Lee and Wu [9].) Hence, Problem 2 has time complexity $\Omega(n \log n)$ in the algebraic computation tree model. For $d = 3$, Follert gives an algorithm that solves Problem 2 in $O(n\lambda_6(n) \log n)$ time, where $\lambda_6(n)$ is the maximal length of any Davenport-Schinzel sequence of order six over an alphabet of size n. It is known that $\lambda_6(n)$ is slightly superlinear. (See Agarwal *et al.* [2].) Hence, Follert's algorithm has almost quadratic running time.

1.1 Our contribution

In Section 2, we prove some preliminary results. First, we show that we can assume w.l.o.g. that all points have weight one, i.e., it suffices to consider the unit-weight versions of Problems 1 and 2. (This observation appears already in [3, 9].) Then we show that the time complexity of Problem 2 is bounded above by that of Problem 1.

In Section 3, we consider the two-dimensional version of Problem 1. We give an extremely simple algorithm that solves this problem in $O(n \log n)$ time. This algorithm uses the lower envelope of some appropriately chosen curves. A careful analysis shows that this lower envelope has linear combinatorial complexity.

The results of Section 2 imply that the two-dimensional version of Problem 2 can also be solved in $O(n \log n)$ time. Since Follert [3] proved an $\Omega(n \log n)$ lower bound for this problem, it follows that our algorithms for solving the planar versions of Problems 1 and 2 are optimal in the algebraic computation tree model.

In Section 4, we consider the three-dimensional version of Problem 1. The appropriate technique to apply seems to be Megiddo's parametric search [10]. We show that this is indeed true. In particular, we show that it suffices to design sequential and parallel algorithms for the following problem: Given a set of n disks on the unit sphere, decide whether these disks cover the sphere. Then, Megiddo's technique immediately solves Problem 1. Our algorithms for solving the sphere cover problem are based on the following topological fact: The boundary of the union of the n disks has combinatorial complexity $O(n)$ and can be computed by a divide-and-conquer algorithm. (See Kedem et al. [8].) The overall algorithm for solving Problem 1 has running time $O(n \log^5 n)$. Another solution based on the parametric search technique, which requires only running time $O(n \log^4 n)$, is proposed in [4]. In contrast to this approach, which uses additional geometric properties, our solution uses only topological information and can be easier modified for other obstacles.

By the results of Section 2, the three-dimensional version of Problem 2 can be solved within the same time bound. Compared with the previous almost quadratic time bounds of [3], these are drastic improvements.

In Section 5, we consider some related problems. In particular, the dual of the three-dimensional version of Problem 1, which asks for an anchored ray R for which $\max_{p \in S} w(p) \cdot d(p, R)$ is minimal, can be solved in $O(n \log^5 n)$ time using basically the same approach as in Section 4. We also discuss the dual of the three-dimensional version of Problem 2, which seems to be much more difficult. Finally, for $d = 3$, we show how to compute a plane H through the origin such that $\max_{p \in S} w(p) \cdot d(p, H)$ is minimal, in $O(n \log n)$ time. It was proved in [9] that the planar version of the latter problem has an $\Omega(n \log n)$ lower bound in the algebraic computation tree model. Hence, our algorithm is optimal in this model.

2 Some preliminary results

Let S be a set of points in $\mathbb{R}^d$. If S contains the origin, then any anchored ray R (resp. any line l through the origin) is a solution to Problem 1 (resp. 2). Therefore, from now on, we assume that set S does not contain the origin.

Lemma 3. Let $p = (p_1, p_2, \ldots, p_d)$ be a point in $\mathbb{R}^d$, let w be a positive real number, and let R be an anchored ray in $\mathbb{R}^d$. Let $p' := (wp_1, wp_2, \ldots, wp_d)$. Then $w \cdot d(p, R) = d(p', R)$.

Corollary 4. Let $T(n)$ denote the complexity of the unit-weight version of Problem 1. Then the weighted version of Problem 1 has complexity $O(T(n))$.

Lemma 5. *Let $T(n)$ be the complexity of Problem 1. Then the complexity of Problem 2 is bounded by $O(T(2n))$.*

Proof: Let S be a set of n points in $\mathbb{R}^d$, and let each point p of S have a positive weight $w(p)$. We want to compute a line l through the origin for which $\min_{p \in S} w(p) \cdot d(p, l)$ is maximal. Let $S' := S \cup -S$, where $-S := \{(-p_1, -p_2, \ldots, -p_d) : (p_1, p_2, \ldots, p_d) \in S\}$. We give each point in $-S$ the weight of the corresponding point of S. Let R^* be the anchored ray such that $\min_{p \in S'} w(p) \cdot d(p, R^*)$ is maximal. Let l^* be the line that supports R^*. Then, l^* is a solution to Problem 2 for the set S. □

3 Problem 1: the two-dimensional case

Let S be a set of n points in the plane, and let each point p of S have a positive weight $w(p)$. We want to compute an anchored ray R such that $\min_{p \in S} w(p) \cdot d(p, R)$ is maximal. By Corollary 4, we can assume w.l.o.g. that $w(p) = 1$ for all points p. Define

$$\delta^* := \max\{\min_{p \in S} d(p, R) : R \text{ is an anchored ray}\}.$$

Let δ_l^* (resp. δ_r^*) denote the analogous quantity where we only consider anchored rays that lie on or to the left (resp. right) of the y-axis. It is clear that $\delta^* = \max(\delta_l^*, \delta_r^*)$. We show how to compute δ_r^*. The value δ_l^* can be computed in a symmetric way.

Let $\delta_{min} := \min\{\|p\| : p \in S\}$. For each $\delta \geq 0$ and each point p of S, let D_p^δ denote the disk with center p and radius δ. For $0 \leq \delta \leq \delta_{min}$ and $p \in S$, let C_p^δ denote the cone consisting of all anchored rays that intersect or touch the disk D_p^δ. (Since $\delta \leq \delta_{min}$, D_p^δ does not contain the origin. Therefore, C_p^δ really is a cone.) Note that C_p^δ has the origin as its apex.

Observation 1 *Using these notations, we have*

1. *δ_r^* is the maximal value of δ, $0 \leq \delta \leq \delta_{min}$, such that there is an anchored ray in the halfplane $x \geq 0$ that does not intersect the interior of any disk D_p^δ, $p \in S$.*
2. *$0 \leq \delta_r^* \leq \delta_{min}$.*
3. *δ_r^* is the minimum of δ_{min} and the minimal value of δ, $0 \leq \delta \leq \delta_{min}$, such that the cones C_p^δ, $p \in S$, cover the halfplane $x \geq 0$.*

Let $0 \leq \delta \leq \delta_{min}$ and let $p \in S$. Consider the intersection of the cone C_p^δ with the halfplane $x \geq 0$. Let $I_p(\delta)$ be the interval of slopes spanned by all anchored rays that lie in this intersection. We represent each slope by the angle between the ray and the positive x-axis. Hence, $I_p(\delta) \subseteq [-\pi/2, \pi/2]$. We can easily write down this interval explicitly:

Let p have coordinates (p_1, p_2), and let φ_p, $-\pi < \varphi_p \leq \pi$, be the angle between the vector $\mathbf{p}$ and the positive x-axis. Then, $\sin \varphi_p = p_2/\|p\|$. Also, for

$0 \leq \delta \leq \delta_{min}$, let α_p^δ be the angle between p and an anchored ray that is tangent to the disk D_p^δ. (There are two such tangents, but both define the same angle.) Then, $0 \leq \alpha_p^\delta \leq \pi/2$ and $\sin \alpha_p^\delta = \delta/\|p\|$. (See Figure 1.) If $p_1 \geq 0$, then

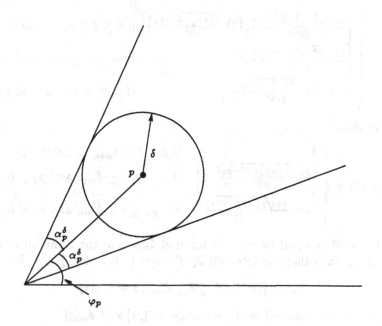

Fig. 1. Illustration of the angles φ_p and α_p^δ.

$$I_p(\delta) = \begin{cases} [\varphi_p - \alpha_p^\delta, \varphi_p + \alpha_p^\delta] & \text{if } 0 \leq \delta \leq \delta_{min} \text{ and } \delta \leq p_1, \\ [\varphi_p - \alpha_p^\delta, \pi/2] & \text{if } p_1 \leq \delta \leq \delta_{min} \text{ and } p_2 \geq 0, \\ [-\pi/2, \varphi_p + \alpha_p^\delta] & \text{if } p_1 \leq \delta \leq \delta_{min} \text{ and } p_2 \leq 0. \end{cases}$$

If $p_1 \leq 0$, then

$$I_p(\delta) = \begin{cases} \emptyset & \text{if } 0 \leq \delta \leq \delta_{min} \text{ and } \delta \leq -p_1, \\ [\varphi_p - \alpha_p^\delta, \pi/2] & \text{if } -p_1 \leq \delta \leq \delta_{min} \text{ and } p_2 \geq 0, \\ [-\pi/2, \varphi_p + \alpha_p^\delta] & \text{if } -p_1 \leq \delta \leq \delta_{min} \text{ and } p_2 \leq 0. \end{cases}$$

It is clear that the cones C_p^δ, $p \in S$, cover the halfplane $x \geq 0$ if and only if the intervals $I_p(\delta)$, $p \in S$, cover $[-\pi/2, \pi/2]$. Hence, δ_r^* is the minimum of (1) δ_{min}, and (2) the minimal value of δ, $0 \leq \delta \leq \delta_{min}$, such that the intervals $I_p(\delta)$ cover $[-\pi/2, \pi/2]$.

Using the intervals $I_p(\delta)$ has the disadvantage that we need non-algebraic functions. In order to stay within the algebraic computation tree model, our algorithm works with the intervals

$$J_p(\delta) := \sin(I_p(\delta)) = \{\sin \gamma : \gamma \in I_p(\delta)\}.$$

Note that $I_p(\delta) \subseteq [-\pi/2, \pi/2]$ and that the function $\sin(\cdot)$ is increasing on $[-\pi/2, \pi/2]$. Using the relations $\sin \varphi_p = p_2/\|p\|$, $\cos \varphi_p = p_1/\|p\|$, $\sin \alpha_p^\delta = \delta/\|p\|$, $\cos \alpha_p^\delta = \sqrt{p_1^2 + p_2^2 - \delta^2}/\|p\|$, and $\sin(x + y) = \sin x \cos y + \cos x \sin y$, we get the following expressions for $J_p(\delta)$. If $p_1 \geq 0$, then

$$
J_p(\delta) = \begin{cases} \left[\dfrac{p_2\sqrt{p_1^2+p_2^2-\delta^2}-p_1\delta}{\|p\|^2}, \dfrac{p_2\sqrt{p_1^2+p_2^2-\delta^2}+p_1\delta}{\|p\|^2} \right] & \text{if } 0 \leq \delta \leq \delta_{min} \text{ and } \delta \leq p_1, \\[3mm] \left[\dfrac{p_2\sqrt{p_1^2+p_2^2-\delta^2}-p_1\delta}{\|p\|^2}, 1 \right] & \text{if } p_1 \leq \delta \leq \delta_{min} \text{ and } p_2 \geq 0, \\[3mm] \left[-1, \dfrac{p_2\sqrt{p_1^2+p_2^2-\delta^2}+p_1\delta}{\|p\|^2} \right] & \text{if } p_1 \leq \delta \leq \delta_{min} \text{ and } p_2 \leq 0. \end{cases}
$$

If $p_1 \leq 0$, then

$$
J_p(\delta) = \begin{cases} \emptyset & \text{if } 0 \leq \delta \leq \delta_{min} \text{ and } \delta \leq -p_1, \\[3mm] \left[\dfrac{p_2\sqrt{p_1^2+p_2^2-\delta^2}-p_1\delta}{\|p\|^2}, 1 \right] & \text{if } -p_1 \leq \delta \leq \delta_{min} \text{ and } p_2 \geq 0, \\[3mm] \left[-1, \dfrac{p_2\sqrt{p_1^2+p_2^2-\delta^2}+p_1\delta}{\|p\|^2} \right] & \text{if } -p_1 \leq \delta \leq \delta_{min} \text{ and } p_2 \leq 0. \end{cases}
$$

The value of δ_r^* is equal to the minimum of δ_{min} and the minimal value of δ, $0 \leq \delta \leq \delta_{min}$, such that the intervals $J_p(\delta)$ cover $[-1, 1]$. For $p \in S$, let

$$
R_p := \{(x, \delta) : 0 \leq \delta \leq \delta_{min}, x \in J_p(\delta)\}.
$$

The region R_p is contained in the rectangle $[-1, 1] \times [0, \delta_{min}]$.

Observation 2 δ_r^* *is the minimum of (1)* δ_{min}, *and (2) the minimal value of* δ, $0 \leq \delta \leq \delta_{min}$, *such that the horizontal segment with endpoints* $(-1, \delta)$ *and* $(1, \delta)$ *is completely contained in* $\bigcup_{p \in S} R_p$.

Let l_p be the lower envelope of R_p. Then, l_p is the graph of a continuous function on a subinterval of $[-1, 1]$. Finally, let L be the lower envelope of the graphs l_p, $p \in S$, and the line segment with endpoints $(-1, \delta_{min})$ and $(1, \delta_{min})$.

Observation 3 δ_r^* *is the y-coordinate of a highest vertex of* L.

We now analyze the lower envelope L. Let B_l, B_r, B_t and B_b be the left, right, top and bottom side of the rectangle $[-1, 1] \times [0, \delta_{min}]$, respectively. Let $p = (p_1, p_2)$ be a point of S, and consider the graph l_p. If $p_1 \geq 0$, then l_p consists of a decreasing part l_p^- that has $(p_2/\|p\|, 0)$ as its lowest and rightmost endpoint, and an increasing part l_p^+ that has $(p_2/\|p\|, 0)$ as its lowest and leftmost endpoint. Moreover, l_p^- (resp. l_p^+) has its leftmost (resp. rightmost) endpoint on B_l or B_t (resp. B_r or B_t). If $p_1 \leq 0$ and $p_2 \geq 0$, then l_p is decreasing from some point on B_t to some point on B_r. Finally, if $p_1 \leq 0$ and $p_2 \leq 0$, then l_p is increasing from some point on B_l to some point on B_t.

Lemma 6. *Let* $p = (p_1, p_2)$ *and* $q = (q_1, q_2)$ *be two distinct points of* S. *The graphs* l_p *and* l_q *intersect at most twice.*

Proof: We give a geometric explanation for this claim. In the full paper, we give a rigorous proof. Assume that p_1 and q_1 are both positive and that $\varphi_q > \varphi_p$. For $0 \le \delta \le \delta_{min}$, let $U_p(\delta)$ (resp. $L_p(\delta)$) be the anchored ray that is upper (resp. lower) tangent to the disk D_p^δ. Define $U_q(\delta)$ and $L_q(\delta)$ analogously.

Intersections of l_p and l_q are in one-to-one correspondence with values of δ such that $\{U_p(\delta), L_p(\delta)\} \cap \{U_q(\delta), L_q(\delta)\} \ne \emptyset$.

Consider what happens when we grow δ from 0 to δ_{min}. Initially, $U_p(\delta) = L_p(\delta)$ and $U_q(\delta) = L_q(\delta)$. If δ increases, then the tangents $U_p(\delta)$ and $L_p(\delta)$ move in opposite directions. Similarly, the tangents $U_q(\delta)$ and $L_q(\delta)$ move in opposite directions. (See Figure 2.) Clearly, there is exactly one δ_0 such that $L_q(\delta_0) = U_p(\delta_0)$. This corresponds to an intersection of l_q^- and l_p^+. Also, for $\delta < \delta_0$, there are no intersections between l_p and l_q. Now we grow δ further, from δ_0 to the next "time" δ_1 at which $\{U_p(\delta_1), L_p(\delta_1)\} \cap \{U_q(\delta_1), L_q(\delta_1)\} \ne \emptyset$. (If there is no such time, then the graphs l_p and l_q intersect exactly once, and we are done.) Then, $U_p(\delta_1) = U_q(\delta_1)$ or $L_p(\delta_1) = L_q(\delta_1)$. Assume w.l.o.g. that at time δ_1, $U_p(\delta_1) = U_q(\delta_1)$. This corresponds to the second intersection between l_p and l_q; more precisely, an intersection between l_p^+ and l_q^+. Note that then $U_p(\delta)$ must move faster than $U_q(\delta)$. Hence, for $\delta > \delta_1$, these two tangents never coincide any more. That is, l_p^+ and l_q^+ intersect only once. Now look at $L_p(\delta)$ and $L_q(\delta)$: Since $L_p(\delta)$ and $U_p(\delta)$ (resp. $L_q(\delta)$ and $U_q(\delta)$) move at the same, but opposite, velocities, $L_q(\delta)$ will never overtake $L_p(\delta)$. That is, l_p^- and l_q^- do not intersect. $\qquad\square$

Lemma 7. *The lower envelope L consists of $O(n)$ vertices.*

Proof: We will show that the names of the points that correspond to the edges of L, when we traverse L from left to right, form a Davenport-Schinzel sequence of order two. This will prove the claim. (See e.g. [6].) Hence, we must show that for any pair p and q of distinct points of S, this sequence of names does not contain a subsequence of the form $p \ldots q \ldots p \ldots q$. But this follows from the fact that l_p and l_q intersect at most twice, and from the restrictions on the endpoint of these graphs. $\qquad\square$

Now we are ready to give the algorithm for computing δ_r^* and a corresponding ray R^*.

1. Compute the graphs l_p, $p \in S$.
2. Compute the lower envelope L of the graphs l_p, $p \in S$, and the horizontal segment with endpoints $(-1, \delta_{min})$ and $(1, \delta_{min})$.
3. Walk along L and find a highest vertex on it. Let this vertex have coordinates (a, δ).
4. Output δ and the anchored ray $R := \{(x, ax/\sqrt{1-a^2}) : x \ge 0\}$.

To prove the correctness of this algorithm, consider the vertex (a, δ) that is found in Step 3. Observation 3 implies that $\delta = \delta_r^*$. Let φ be the angle such that $-\pi/2 \le \varphi \le \pi/2$ and $\sin \varphi = a$. Let R^* be the anchored ray that makes an angle of φ with the positive x-axis. Then $\delta = \min_{p \in S} d(p, R^*)$. It is easy to see that $R = R^*$.

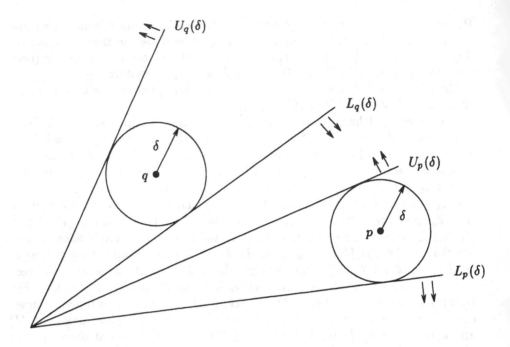

Fig. 2. Growing δ from 0 to δ_{min}.

Next we analyze the running time of our algorithm. Step 1 takes $O(n)$ time. The lower envelope L can be computed by a divide-and-conquer algorithm. (See e.g. [6].) Since L has linear size, this algorithm, and hence Step 2, takes $O(n \log n)$ time. Step 3 takes $O(n)$ time, and Step 4 takes $O(1)$ time. We have proved the following result.

Theorem 8. *Let S be a set of n points in the plane, and let each point p of S have a positive weight $w(p)$. In $O(n \log n)$ time, we can compute an anchored ray R^* for which $\min_{p \in S} w(p) \cdot d(p, R^*)$ is maximal.*

Corollary 9. *Let S be a set of n points in the plane, and let each point p of S have a positive weight $w(p)$. In $O(n \log n)$ time, we can compute a line l^* through the origin for which $\min_{p \in S} w(p) \cdot d(p, l^*)$ is maximal.*

The results of Theorem 8 and Corollary 9 are optimal in the algebraic computation tree model. Follert [3] proves an $\Omega(n \log n)$ lower bound for Problem 2. It follows from Lemma 5 that this lower bound holds for Problem 1 as well.

4   Problem 1: the three-dimensional case

We briefly recall Megiddo's parametric search technique [10]. Suppose we are given a fixed set of n data items, such as points in $\mathbb{R}^3$. Let $\mathcal{P}(t)$ be a decision problem whose value depends on the n data items and a real parameter t. Assume that $\mathcal{P}$ is monotone, meaning that if $\mathcal{P}(t_0)$ is true for some t_0, then $\mathcal{P}(t)$ is also true for all $t < t_0$. Our aim is to find the maximal value of t for which $\mathcal{P}(t)$ is true. We denote this value by t^*.

Assume we have a sequential algorithm A_s that, given the n data items and t, decides if $\mathcal{P}(t)$ is true or not. The control flow of this algorithm is governed by comparisons, each of which involves testing the sign of some low-degree polynomial in t. Let T_s and C_s denote the running time and the number of comparisons made by algorithm A_s, respectively. Note that by running A_s on input t, we can decide if $t \leq t^*$ or $t > t^*$: we have $t \leq t^*$ iff $\mathcal{P}(t)$ is true.

The parametric search technique simulates A_s on the unknown value t^*. Whenever A_s reaches a branching point that depends on a comparison operation, the comparison can be reduced to testing the sign of a suitable low-degree polynomial $f(t)$ at $t = t^*$. The algorithm computes the roots of this polynomial and checks each root a—by running A_s on input a—to see if it is less than or equal to t^*. In this way, the algorithm identifies two successive roots between which t^* must lie and thus determines the sign of $f(t^*)$. Hence, we get an interval I that contains t^*. Also the comparison now being resolved, the execution can proceed. As we proceed through the execution, each comparison that we resolve results in constraining I further and we get a sequence of progressively smaller intervals each known to contain t^*. The simulation will run to completion and we are left with an interval I that contains t^*. It can be shown that for any real number $r \in I$, $\mathcal{P}(r)$ is true. Therefore, t^* must be the right endpoint of I.

Since A_s makes at most C_s comparisons during its execution, the entire simulation and, hence, the computation of t^* take $O(C_s T_s)$ time. To speed up this algorithm, Megiddo replaces A_s by a parallel algorithm A_p that uses P processors and runs in T_p parallel time. At each parallel step, let A_p make a maximum of W_p independent comparisons. Then our algorithm simulates A_p sequentially, again at the unknown value t^*. At each parallel step, we get at most W_p low-degree polynomials in t. We compute the roots of all of them and do a binary search among them using repeated median finding to make the probes for t^*. For each probe, we run the sequential algorithm A_s. In this way, we get the correct sign of each polynomial in t^*, and our algorithm can simulate the next parallel step of A_p.

For the simulation of each parallel step, we spend $O(W_p)$ time for median finding. Hence, the entire simulation of this step takes time $O(W_p + T_s \log W_p)$. As a result, the entire algorithm computes t^* in time $O(W_p T_p + T_s T_p \log W_p)$. Since $W_p \leq P$, the running time is bounded by $O(P T_p + T_s T_p \log P)$.

4.1 Applying the parametric search technique

Let S be a set of n points in $\mathbb{R}^3$. Each point p of S has a positive weight $w(p)$. Define

$$\delta^* := \max\{\min_{p \in S} w(p) \cdot d(p, R) : R \text{ is an anchored ray}\}.$$

Our goal is to compute δ^* together with the corresponding ray R^*. We saw already that we may assume w.l.o.g. that $w(p) = 1$ for all points p.

In order to apply the parametric search technique, we have to solve the following decision problem $\mathcal{P}(\delta)$: Given the set S and the real number $\delta \geq 0$, is there an anchored ray R such that $\min_{p \in S} d(p, R) \geq \delta$. It is clear that $\mathcal{P}$ is monotone, and δ^* is the maximal δ for which $\mathcal{P}(\delta)$ is true.

We reformulate the decision problem $\mathcal{P}(\delta)$ in the following way. Let $\delta \geq 0$. For each point p of S, let B_p^δ denote the ball with center p and radius δ. Then $\mathcal{P}(\delta)$ is true if and only if there is an anchored ray R that does not intersect the interior of any of these balls.

Let $\delta_{min} := \min\{\|p\| : p \in S\}$. Then $\mathcal{P}(\delta)$ is clearly false for $\delta > \delta_{min}$.

For $0 \leq \delta \leq \delta_{min}$, let C_p^δ denote the circular cone consisting of all anchored rays that intersect or touch the ball B_p^δ. This cone intersects the unit sphere—i.e., the surface of the ball of radius one centered at the origin—in a disk. We denote this disk by D_p^δ.

Let $0 \leq \delta \leq \delta_{min}$. It is clear that $\mathcal{P}(\delta)$ is true if and only if there is a point x on the unit sphere that is not contained in the interior of any of these n disks. If there is such a point x, then the ray R that starts in the origin and contains x satisfies $\min_{p \in S} d(p, R) \geq \delta$. In other words, $\mathcal{P}(\delta)$ is true if and only if the interiors of these n disks do not cover the unit sphere.

In the next two subsections, we give sequential and parallel algorithms that decide the latter condition.

A sequential algorithm that decides the covering problem Let $D_1, D_2, \ldots, D_n$ be a set of n disks on the unit sphere. We want to decide if the interiors of these disks cover the unit sphere. Clearly, we can use the arrangement of the disks for deciding this. This arrangement, however, may have size $\Omega(n^2)$.

Let I_i (resp. γ_i) denote the interior (resp. boundary) of D_i, $1 \leq i \leq n$, and let $I := \bigcup_{i=1}^n I_i$. (Note that there may be $i \neq j$ such that $\gamma_i = \gamma_j$.) We denote the closure of I by $cl(I)$. The boundary B of I is equal to

$$B = cl(I) \setminus I = \left(\bigcup_{i=1}^n D_i\right) \setminus \left(\bigcup_{i=1}^n I_i\right).$$

The interiors of the disks $D_1, D_2, \ldots, D_n$ cover the unit sphere if and only if B is empty. Hence, our problem can be solved by computing the boundary B rather than the entire arrangement of the n disks.

The boundary B is a planar graph on the unit sphere. Each edge of this graph is part of a circle γ_i for some i, and each vertex is an intersection point of at least two distinct circles. We choose an arbitrary point p_i on each circle γ_i, $1 \leq i \leq n$, with the restriction that $p_i = p_j$ if $\gamma_i = \gamma_j$. Then, if γ_i does not intersect any other circle, it forms an edge of B with both endpoints equal to p_i. Note that B can have isolated vertices: If three circles intersect in one point x, and there is an arbitrarily small disk α (not equal to any of $D_1, D_2, \ldots, D_n$) centered at x such that $\alpha \setminus \{x\}$ is contained in the union of the interiors of these three circles, then x is a vertex of B, and x is not incident to any edge of B.

Lemma 10 [8]. *The boundary B is a planar graph on the unit sphere, and, if $n \geq 3$, it contains at most $6n - 12$ vertices.*

In [8], an algorithm is given that computes the boundary of the union of n regions in the plane, each of which is bounded by a simple closed Jordan curve. This algorithm follows the divide-and-conquer paradigm, and the merge step is implemented by using a plane sweep algorithm of Ottmann, Widmayer and Wood [11] for computing the boundary of the union of superimposed polygonal planar regions. This plane sweep algorithm also works if the edges of the planar regions are curved. We can easily modify this algorithm such that it computes the boundary B:

Consider the disks $D_1, D_2, \ldots, D_n$. Recursively compute the boundary B_1 (resp. B_2) of the union of the interiors of $D_1, D_2, \ldots, D_{n/2}$ (resp. $D_{n/2+1}, D_{n/2+2}, \ldots, D_n$). Note that B_1 and B_2 are planar graphs on the unit sphere. Let l and r be the points on the unit sphere with minimal and maximal y-coordinate, respectively. Using the algorithm of [11], we compute the boundary B from B_1 and B_2 by sweeping a circular arc with endpoints l and r around the unit sphere. Let b_1 and b_2 denote the number of edges of B_1 and B_2, respectively, and let t denote the number of intersections between B_1 and B_2. Then this sweep algorithm runs in time $O((b_1 + b_2 + t) \log(b_1 + b_2))$. It follows from Lemma 10 that $b_1 + b_2 = O(n)$. Since each intersection point between B_1 and B_2 is a vertex of B, Lemma 10 also implies that $t = O(n)$. Hence, the entire sweep algorithm runs in time $O(n \log n)$. This shows that the entire divide-and-conquer algorithm for computing the boundary B takes $O(n \log^2 n)$ time. The interiors of the input disks $D_1, D_2, \ldots, D_n$ cover the unit sphere if and only if the graph B is empty. If B is not empty, then any vertex of B is a point on the unit sphere that is not contained in the interior of any disk. We have proved the following result.

Lemma 11. *Let $D_1, \ldots, D_n$ be a set of disks on the unit sphere. In $O(n \log^2 n)$ time, we can decide if the union of the interiors of these disks cover the unit sphere. If this is not the case, then the algorithm finds a point on the unit sphere that is not contained in the interior of any disk.*

A parallel algorithm that decides the covering problem Now we give a parallel algorithm for computing the boundary B. Consider again the disks $D_1, D_2, \ldots, D_n$. The algorithm uses n processors. The first (resp. last) $n/2$ processors compute the boundary B_1 (resp. B_2) of the union of the interiors of $D_1, D_2, \ldots, D_{n/2}$ (resp. $D_{n/2+1}, D_{n/2+2}, \ldots, D_n$). It remains to describe the merge step. That is, given B_1 and B_2, how to compute the boundary B of the union of the interiors of the n input disks.

Rüb [12] gives a parallel algorithm based on a segment tree, that computes the intersections among red and blue curved segments in the plane. The interiors of the red (resp. blue) segments are assumed to be pairwise disjoint. Also, each segment is assumed to be x-monotone, meaning that any vertical line intersects a segment at most once. Finally, it is assumed that each red-blue pair of segments intersect at most a constant number of times. If n denotes the total number of red and blue segments, and t denotes the total number of intersection points among the red-blue pairs of segments, then Rüb's algorithm runs on a CREW-PRAM in time $O(\log n + t/n)$ using n processors.

This algorithm can be used to compute the boundary B from B_1 and B_2: In our case, the slabs that define the segment tree are bounded by circular arcs on the unit sphere with two fixed diametral endpoints. In order to guarantee that each curved edge of B_1 and B_2 is monotone, we cut each of them into at most two parts. Note that, by Lemma 10, $t = O(n)$. Hence, using Rüb's algorithm, we compute all intersections of B_1 and B_2 in $O(\log n)$ time using n processors.

Then, for each edge e of B_1, we sort the intersection points on this edge. This gives the arrangement A of the union B_1 and B_2. Given this arrangement, we compute the boundary B by removing the appropriate vertices and edges from A. All this can be done in $O(\log n)$ time using n processors.

Hence, the entire merge step of our parallel divide-and-conquer algorithm takes $O(\log n)$ time and uses n processors. This proves:

Lemma 12. *Let $D_1, D_2, \ldots, D_n$ be a set of disks on the unit sphere. There is a CREW-PRAM algorithm that decides if the union of the interiors of these disks cover the unit sphere. If this is not the case, then the algorithm finds a point on the unit sphere that is not contained in the interior of any disk. The algorithm takes $O(\log^2 n)$ time and uses n processors.*

Lemmas 11 and 12, and the parametric search technique immediately provide a solution for Problem 1:

Theorem 13. *Let S be a set of n points in $\mathbb{R}^3$, and let each point p of S have a positive weight $w(p)$. In $O(n \log^5 n)$ time, we can compute an anchored ray R^* for which $\min_{p \in S} w(p) \cdot d(p, R^*)$ is maximal.*

Corollary 14. *Let S be a set of n points in $\mathbb{R}^3$, and let each point p of S have a positive weight $w(p)$. In $O(n \log^5 n)$ time, we can compute a line l^* through the origin for which $\min_{p \in S} w(p) \cdot d(p, l^*)$ is maximal.*

5 Some related problems

Problem 15. Let S be a set of n points in $\mathbb{R}^d$, and let each point p of S have a weight $w(p)$, which is a positive real number. Compute an anchored ray R for which $\max_{p \in S} w(p) \cdot d(p, R)$ is minimal.

As before, we can assume w.l.o.g. that all points have weight one. In [9], Lee and Wu show how to solve this problem in $O(n \log n)$ time when $d = 2$. We show how to solve it for $d = 3$. Let B_p^δ denote the ball with center p and radius δ. Then we want to compute the minimal real number $\delta \geq 0$ such that there is an anchored ray that intersects all balls B_p^δ, $p \in S$. We find this minimal δ using the parametric search technique.

Let $\delta \geq 0$. We need sequential and parallel algorithms for deciding if there is an anchored ray that intersects all balls B_p^δ, $p \in S$. Clearly, we do not have to consider those balls that contain the origin. Using the same approach as in Section 4, we arrive at the following problem: Given a set of at most n disks on the unit sphere, decide if their intersection is empty. This intersection has combinatorial complexity $O(n)$. Moreover, it can be computed by basically the same approaches as in Sections 4.1 and 4.1.

Theorem 16. *Let S be a set of n points in $\mathbb{R}^3$, and let each point p of S have a positive weight $w(p)$. In $O(n \log^5 n)$ time, we can compute an anchored ray R for which $\max_{p \in S} w(p) \cdot d(p, R)$ is minimal.*

Problem 17. Let S be a set of n points in $\mathbb{R}^d$, and let each point p of S have a weight $w(p)$, which is a positive real number. Compute a line l through the origin for which $\max_{p \in S} w(p) \cdot d(p, l)$ is minimal.

For $d = 2$, this problem can be solved in $O(n \log n)$ time, which is optimal in the algebraic computation tree model. See [9]. The three-dimensional version seems to be much harder. Follert [3] solves this problem in $O(n \lambda_6(n) \log n)$ time.

A *symmetric slab* is defined as the region between two parallel planes in $\mathbb{R}^3$ that are at the same distance from the origin. If we intersect a symmetric slab with the unit sphere, then we get a *symmetric slab on the unit sphere*. A natural approach to solve the three-dimensional version of Problem 17 is to use the parametric search technique. Then we have to design sequential and parallel algorithms for the following decision problem: Given a set of n symmetric slabs on the unit sphere, do they cover the unit sphere.

This decision problem resembles the following problem: Given a circle C and a set of n slabs, both in the plane, decide whether these slabs cover C. Gajentaan and Overmars [5] proved that this problem is n^2-hard, which indicates that it is probably very hard to find a subquadratic algorithm for it.

Open problem 1 *Decide if the problem "Given a set of n symmetric slabs on the unit sphere, do they cover the unit sphere", is n^2-hard, or if it can be solved in subquadratic time. Note that if this problem is n^2-hard, that then also the three-dimensional version of Problem 17 is n^2-hard.*

Problem 18. Let S be a set of n points in $\mathbb{R}^d$, and let each point p of S have a weight $w(p)$, which is a positive real number. Compute a hyperplane H through the origin for which $\max_{p \in S} w(p) \cdot d(p, H)$ is minimal.

Lee and Wu [9] proved an $\Omega(n \log n)$ lower bound for the planar version of this problem. We show how to solve the three-dimensional version of Problem 18 in $O(n \log n)$ time. We can assume w.l.o.g. that all points of S have weight one.

Our problem is equivalent to that of computing the symmetric slab of minimal width that contains all points of S. Let $S' := S \cup -S$, where $-S := \{(-p_1, -p_2, -p_3) : (p_1, p_2, p_3) \in S\}$. For any plane H through the origin, we have $H = -H$. Therefore, $d(p, H) = d(-p, -H) = d(-p, H)$. As a result, it suffices to solve our problem for the set S'. Since this set is symmetric w.r.t. the origin, the width of the minimal symmetric slab containing S' is equal to the width of S', which is defined as the minimal width of *any* slab containing this set.

The best known algorithm for computing the width of an arbitrary set of n points in $\mathbb{R}^3$ has running time $O(n^{17/11+\epsilon})$, where ϵ is an arbitrarily small positive constant. (See Agarwal *et al.* [1].) In our case, however, the set of points has a special form.

Houle and Toussaint [7] observed that the width of a set of points in $\mathbb{R}^3$ is the minimum distance between parallel planes of support passing through either an antipodal vertex-face pair or an antipodal edge-edge pair of the convex hull of the set.

It is not difficult to see that in order to compute the width of our set S', we only have to consider parallel planes of support passing through an antipodal vertex-face pair of the convex hull of S', and take the minimum distance between any such pair of planes. This minimum distance can be computed in $O(n \log n)$ time. (See [7].)

Theorem 19. *Let S be a set of n points in $\mathbb{R}^3$, and let each point p of S have a positive weight $w(p)$. In $O(n \log n)$ time, we can compute a plane H through the origin for which $\max_{p \in S} w(p) \cdot d(p, H)$ is minimal. This is optimal in the algebraic computation tree model.*

Acknowledgement

The authors thank Prof. Hotz for posing the problems, and the application to neurosurgery, that were considered in this paper. They also thank Christine Rüb and Stefan Schirra for several helpful discussions.

References

1. P.K. Agarwal, B. Aronov, and M. Sharir. *Computing enveloped in four dimensions with applications.* Proc. 10th Annual ACM Conf. on Comp. Geom., 1994, pp. 348–358.

2. P.K. Agarwal, M. Sharir, and P. Shor. *Sharp upper and lower bounds for the length of general Davenport-Schinzel sequences.* J. Combin. Theory, Ser. A **52** (1989), pp. 228–274.

3. F. Follert. *Lageoptimierung nach dem Maximin-Kriterium.* Master's Thesis. Department of Computer Science, Universität des Saarlandes, Saarbrücken, 1994.

4. F. Follert. *Maxmin location of an anchored ray in 3-space and related problems,* submitted to CCCG95

5. A. Gajentaan and M.H. Overmars. n^2-*Hard problems in computational geometry.* Tech. Rep. RUU-CS-93-15, Department of Computer Science, University of Utrecht, 1993.

6. L. Guibas and M. Sharir. *Combinatorics and algorithms of arrangements.* In: New Trends in Discrete and Computational Geometry, Ed. J. Pach. Springer-Verlag, Berlin, 1993, pp. 9–36.

7. M.E. Houle and G.T. Toussaint. *Computing the width of a set.* IEEE Trans. Pattern Anal. Mach. Intell., PAMI-10 (1988), pp. 761–765.
8. K. Kedem, R. Livne, J. Pach, and M. Sharir. *On the union of Jordan regions and collision-free translational motion amidst polygonal obstacles.* Discrete Comput. Geom. **1** (1986), pp. 59–71.
9. D.T. Lee and Y.F. Wu. *Geometric complexity of some location problems.* Algorithmica **1** (1986), pp. 193–211.
10. N. Megiddo. *Applying parallel computation algorithms in the design of serial algorithms.* J. ACM **30** (1983), pp. 852–865.
11. T. Ottmann, P. Widmayer, and D. Wood. *A fast algorithm for Boolean mask operations.* Comput. Vision Graph. Image Process. **30** (1985), pp. 249–286.
12. C. Rüb. *Computing intersections and arrangements for red-blue curve segments in parallel.* Proc. 4th Canadian Conf. on Comp. Geom., 1992, pp. 115–120.

Computing Hierarchies of Clusters from the Euclidean Minimum Spanning Tree in Linear Time*

Drago Krznaric and Christos Levcopoulos

Department of Computer Science
Lund University, Box 118, S-221 00 Lund, Sweden.

Abstract. A new hierarchical clustering method for point sets is presented, called diameter clustering, whose clusters belong to most other natural clusterings. For each cluster it holds that its diameter is small compared to the distance to a nearest point outside the cluster. Given a Euclidean minimum spanning tree of the input point set, it is shown that the diameter clustering can be computed in linear time. In addition we derive a nice property of this hierarchy which makes it particularly useful as a building block. It is shown in this paper that it can be employed to obtain a good approximation for the known single linkage clustering in roughly linear time. Other examples of its usefulness include computing the greedy triangulation, the complete linkage hierarchy, and a data structure for faster range queries.

1 Introduction

Hierarchical clustering algorithms are important for structuring and interpreting data in domains such as biology, medicine, geographical information systems, and image processing [2, 4, 5]. In addition, they can be used for speeding up and organizing algorithms in computational geometry, by performing local computation for each cluster in the hierarchy. There are several algorithms for hierarchical clustering, of which the so-called single linkage and complete linkage clustering are two well-known examples (see [1, 5] and Section 4).

Let S be any set of n points (called vertices) in the plane. We say that D, $D \subset S$, is a diameter cluster iff the distance between vertices in D and $S - D$ is greater than the diameter of the smallest isothetic rectangle that encloses D. This property causes the clusters to become well separated and form a unique hierarchy which can be described by a rooted tree. This hierarchy is defined more precisely in Section 2, and in Section 3.2 we develop an algorithm that, given a Euclidean minimum spanning tree of S, computes it in $O(n)$ time.

The diameter clustering has also a natural extension to more than two dimensions. In k dimensions, the diameter clustering can be computed in $O(n)$ time if k is regarded as a constant, by just slightly modifying the 2-dimensional algorithm. We discuss the k-dimensional case in Section 3.3.

---

* This paper was partially supported by TFR. E-mail: {drago,christos}@dna.lth.se.

In Section 3.4 we show the following property: for any (diameter) cluster D with $m \geq 2$ children, the ratio between the longest and the shortest distance between the children of D is less than 3^{m-1}. This property is useful for obtaining the following new results: (1) the greedy triangulation can be computed from the Voronoi diagram in $O(n)$ time [9], and (2) the complete linkage clustering can be approximated in $O(n)$ time given the Voronoi diagram [8]. In addition, the diameter clustering is helpful as a building block for reducing the time to compute the complete linkage clustering from quadratic to $O(n \log^2 n)$ [8]. It plays also an important role in order to develop a hierarchical data structure which has the ability to answer certain types of non-trivial range queries in constant time [7] (that data structure is also used as a subroutine for the above three mentioned algorithms).

To give a detailed example of its usage, we show in Section 4 how it can be employed to obtain an approximation of the single linkage clustering. Given a Euclidean minimum spanning tree, we compute an approximation in time $O(n \, \alpha(n, n))$, where $\alpha(n, n)$ is the inverse of Ackermann's function. For an arbitrary constant c, we obtain an approximation of error less than $\frac{1}{n^c}$. In addition, we show that the single linkage clustering generally requires $\Omega(n \log n)$ time even if a Euclidean minimum spanning tree is given.

2   Definitions and Preliminaries

Let S be the input vertex set consisting of n points in the plane. We define the *rectangular diameter* of a vertex set D, abbreviated rdiam(D), to be the diameter of the smallest rectangle with sides parallel to the coordinate axes which contains all vertices in D. Note that rdiam(D) can be computed in constant time if we know the (at most) four extreme vertices of D in x- and y direction. These vertices shall be called the *xy-extremes* of D. (The diameter of a point set is usually defined as the distance between the two points that are farthest apart, but we gain some computational time by using the rectangular diameter, and the only consequence is that the clusters become slightly more separated.)

Definition 1. A subset D of S is a *diameter cluster* (d-cluster) if and only if the distance between vertices of D and vertices of $S - D$ is greater than rdiam(D) or D equals S.

Lemma 2. *Any two different d-clusters D and D' are either disjoint or one of them is a proper subset of the other.*

Proof. Assume the contrary, so there are vertices u, v, w contained in, respectively, $D - D', D \cap D', D' - D$. By definition, the distance from v to w is greater than rdiam(D), since $v \in D$ and $w \notin D$. Consequently, since D' contains both v and w, rdiam(D') > rdiam(D). In an analogous manner we have that rdiam(D) $\geq |v, u|$ > rdiam(D'), which is a contradiction.　　□

The above observation tells us that we can in a natural way decompose S into a unique hierarchy of d-clusters, which can be described by the following rooted tree.

Definition 3. The *diameter clustering tree* (DC-tree) of S is a rooted tree whose nodes correspond to distinct d-clusters, where the root corresponds to S and the leaves to single vertices in S. Let a be any internal node and let A be its corresponding d-cluster. Then the children of a correspond to all d-clusters C such that $C \subset A$ and there is no d-cluster B such that $C \subset B \subset A$.

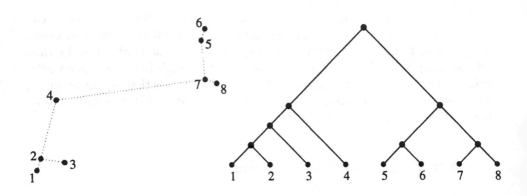

Fig. 1. A vertex set and its DC-tree (dotted lines illustrate the EMST).

Let EMST(S) stand for any Euclidean minimum spanning tree of S. We shall next mention some properties of EMST(S) that are helpful in order to compute the DC-tree. By $V(G)$ we denote the vertex set of a graph G.

Definition 4. Let G be a subgraph of EMST(S). Then the set $I(G)$ consists of those edges of EMST(S) which are incident to exactly one vertex of G.

The following is straightforward.

Observation 5. *Let D be a proper subset of S and let G be the subgraph of EMST(S) induced by D. Then D is a d-cluster if and only if all edges in $I(G)$ have length greater than rdiam($V(G)$). If D is a d-cluster then G is a (connected) subtree of EMST(S).*

It is easily seen that any vertex of S may be incident to at most 6 edges of EMST(S). The following is a similar observation.

Lemma 6. *For any subtree G of EMST(S) there is at most a constant number of edges in $I(G)$ that have length greater than rdiam($V(G)$).*

Proof. Let $d = \text{rdiam}(V(G))$, let P be the set of edges in $I(G)$ of length $> d$, and let C be the smallest circle enclosing all vertices of G. So C has diameter $\leq d$ and every edge in P intersects C in exactly one point. Let $e_0, e_1, \ldots, e_k$ be all edges in P according to the clockwise order given by their intersections with C. Let $v_0, \ldots, v_k$ be the endpoints of $e_0, \ldots, e_k$ that are not in $V(G)$. In the following we use i modulo $(k+1)$. First we note that $|v_i, v_{i+1}| \geq |e_i|$ and $|v_i, v_{i+1}| \geq |e_{i+1}|$ (for example, if $|v_i, v_{i+1}| < |e_i|$, we could remove e_i to break EMST(S) into two components and then reconnect them by adding (v_i, v_{i+1}), which would yield a new spanning tree of less total length than EMST(S)).

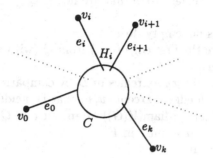

Fig. 2. For the proof of Lemma 6.

Denote by H_i the part of C that starts at e_i and continues clockwise to e_{i+1}. Clearly only a constant number of H_i's can have length $\Omega(d)$ since the circumference of C is at most πd long. Therefore it is enough to show that only a constant number of H_i's can have length less than, say, $d/10$. Suppose that H_i is shorter than $d/10$ for some i (see Figure 2). It is then straightforward to realize that the angle between e_i and e_{i+1} is greater than some constant (they cannot be almost parallel since $|v_i, v_{i+1}| \geq |e_i| > d$ and $|v_i, v_{i+1}| \geq |e_{i+1}| > d$). Thus after a constant number of such H_i's we have turned 2π. $\qquad\square$

3 Computing the Diameter Clustering Tree

3.1 Finding a Parent Node

Let w be any vertex of S. To give some intuition for the computation of the DC-tree, we shall describe how the parent of w in the DC-tree can be computed (recall that w is a d-cluster itself and that the parent of w is the minimal d-cluster of at least two vertices that contains w). To compute the parent of w we augment a subtree G of EMST(S) that is always a subtree of this parent. (In the continuation of this section we will by a d-cluster refer to its corresponding subtree of EMST(S) rather than to its vertex set only.) Initially G consists of the vertex w only. The edges of $I(G)$ are partitioned into two parts: one part is kept in a queue Q and corresponds to those edges that can be added safely to

G (they have length $\leq$ rdiam($V(G)$)); the rest of $I(G)$ is kept in a pool P. The algorithm works by repeatedly moving an edge from Q to G; and whenever Q becomes empty, all safe edges in P are moved to Q. The xy-extremes of $V(G)$ are kept in a set D and they are updated accordingly whenever a new vertex is added to G. Less informally we have

Algorithm: PDC(EMST(S), w) (*** parent d-cluster of w ***)
1. $V(G) \leftarrow \{w\}$
2. $D \leftarrow \{w\}$
3. $Q \leftarrow$ the shortest edge incident upon w
4. $P \leftarrow$ all edges incident to w that are not in Q
5.   **loop**
6.   **while** Q is nonempty **do**
7.   remove the first edge (u, v) from Q (with $u \in V(G)$)
8.   add (u, v) to G
9.   update the xy-extremes in D by comparing them with v
10.   **for** each edge e except (u, v) that is incident to v **do**
11.   **if** $|e| \leq$ rdiam($V(G)$) **then** put e in Q
12.   **else** put e in P
13.   **endif**
14.   **endfor**
15.   **endwhile**
16.   move all edges in P of length $\leq$ rdiam($V(G)$) to Q
17.   **if** Q is empty **then**
18.   **return** the parent d-cluster G
19.   **endif**
20.   **endloop**
end PDC

It is immediate from Observation 5 that this algorithm correctly finds the parent d-cluster of w. To analyze its running time, we first note that each execution of lines 1-4 and lines 7-14 takes constant time, because a vertex is incident to at most 6 edges of EMST(S). Further, by Lemma 6, all edges but a constant number are removed from P at line 16, and every one of these edges will be added to G during the while-loop. Hence, since exactly one vertex is added to G at each iteration of the while-loop, the total running time for PDC is linear in the number of vertices of the parent d-cluster of w.

3.2 Finding All Nodes

To avoid reconsiderations of edges within an already computed d-cluster, we augment a partial d-cluster by considering edges of a tree T. Initially T is equal to EMST(S). But whenever a d-cluster G has been computed it is shrunk to a single vertex of T. This means that a new vertex v enters T and all vertices of G that correspond to endpoints of edges in $I(G)$ are replaced by v. The new vertex

v receives a pointer to the data structure of G, which consists of the tree G and the xy-extremes of $V(G)$. For each new edge we remember the corresponding edge in (the original) EMST(S), and the length of a new edge is defined to be the length of the corresponding edge in EMST(S). (It would be strictly more correct to call it "weight", but since it refers to the length of a geometric edge, we prefer to call it "length".)

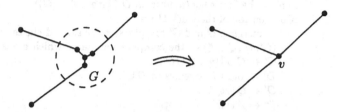

Fig. 3. A d-cluster G is shrunk to a single vertex v of T.

To obtain the hierarchy of d-clusters we organize the computation so that all children of a d-cluster G are computed before G. To accomplish this we try to compute a new d-cluster whenever we encounter a vertex that we suspect belongs to a child of the d-cluster presently in computation. More precisely, if we can add an edge (u,v) to a partial d-cluster G, with u in G, then the vertex v may belong to a child of G only if there is an (unadded) edge (v,w) such that $|u,v| > |v,w|$. Therefore, if $|u,v| > |v,w|$, then the computation of G is declared idle, and we try to compute a new d-cluster containing v. We declare G idle by attaching to u a pointer to a record, called *computation status* for G, and thereafter pushing u onto a stack. The computation status for G contains all the relevant computations made on G so far. This can be described by the quadruple (G, D, Q, P) corresponding to, respectively, the partial d-cluster G, the xy-extremes of $V(G)$, the queue Q, and the pool P. (The stack could be removed by having a recursive variant of the algorithm.)

Suppose that $|u,v| > |v,w|$, so G is declared idle by pushing u onto the stack together with a pointer to (G, D, Q, P), and we try to compute a new d-cluster containing v. In due time we will during the computation of some d-cluster, say G', try to add u to G', and this is discovered since u is then on top of the stack (this is so because we traverse a tree). At this moment we know that G and G' are parts of the same d-cluster, and so all the computations made on G and G' are merged, and we proceed with computing this d-cluster. Note that if we have succeeded in computing another d-cluster that contains v, then the edge (u,v) has already been replaced by an edge (u,v') where v' is a shrunk vertex corresponding to a child of G.

The main refinements on Algorithm PDC of the previous section needed to make it compute the whole hierarchy of d-clusters have now been described, and we summarize them below by giving the complete algorithm for hierarchical diameter clustering (the input to this algorithm is EMST(S)).

Algorithm: HDC(T) (* hierarchical diameter clustering *)

1.   let w be a vertex of T

2.   attach to w a pointer to Initiate(w) and push w onto the stack

3.   **loop 1**

4.   pop w from the stack

5. $(G, D, Q, P) \leftarrow$ the computation status which w has a pointer to

6.   **loop 2**

7.   **while** Q is nonempty **do**

8.   remove the first edge (u, v) from Q (with $u \in V(G)$)

9.   **if** v is on top of the stack **then**

                    (* merging: G and G' are parts of the same d-cluster *)

10. $(G', D', Q', P') \leftarrow$ the computation status which v has a pointer to

11. $G' \leftarrow G \cup \{(u, v)\} \cup G'$

12. $D' \leftarrow$ the xy-extremes of $D \cup D'$

13. $Q' \leftarrow Q \cup Q'$

14. $P' \leftarrow P \cup P'$

15.   **exit loop 2**

16.   **else if** there is a shorter edge than (u, v) incident to v **then**

                    (* v may belong to a child of G *)

17.   attach to u a pointer to (G, D, Q, P) and push u onto the stack

18.   attach to v a pointer to Initiate(v) and push v onto the stack

19.   **exit loop 2**

20.   **else** (* v belongs to G *)

21.   add (u, v) to G

22.   update the xy-extremes in D by comparing them with v

23.   **for** each edge e except (u, v) that is incident to v **do**

24.   **if** $|e| \leq \text{rdiam}(V(G))$ **then** put e in Q

25.   **else** put e in P

26.   **endif**

27.   **endfor**

28.   **endif**

29.   **endwhile**

30.   move all edges in P of length $\leq \text{rdiam}(V(G))$ to Q

31.   **if** Q is empty **then** (* G is now a complete d-cluster *)

32.   shrink G into a single vertex v of T

33.   **if** no edge is incident upon v **then**

34.   **exit HDC**

35.   **else**

36.   attach to v a pointer to Initiate(v) and push v onto the stack

37.   **exit loop 2**

38.   **endif**

39.   **endif**

40.   **endloop 2**

41.   **endloop 1**

end HDC

To avoid repeating the same lines of code several times, we use in Algorithm HDC above a function *Initiate*, which takes a vertex w as an input and returns the computation status for the new partial d-cluster consisting only of w (it is the edge-less tree which is to be augmented until it becomes the parent d-cluster of w, so it corresponds to the computation status of PDC(T, w) after executing its first four lines).

Algorithm: Initiate(w)
    $V(G) \leftarrow \{w\}$
    $D \leftarrow$ the xy-extremes of w (* w may be a shrunken vertex *)
    $Q \leftarrow$ the shortest edge incident to w
    $P \leftarrow$ all edges incident to w that are not in Q
    return (G, D, Q, P)
end Initiate

Referring to Algorithm HDC above, the while-loop differs from the while-loop of Algorithm PDC by performing two checks before an edge (u, v) may be added to a partial d-cluster G. First it checks whether G and the partial d-cluster on top of the stack are parts of the same d-cluster (line 9), in which case a merging is performed (lines 10-14). If the first check fails, then it checks whether the vertex to be added to G may belong to a child of G (line 16). If so, then it declares the computation of G idle (line 17), and starts a new computation from this vertex (lines 18-19). If both checks fail then it adds (u, v) to G in the same manner as in Algorithm PDC (lines 21-27). Further, when a d-cluster G has been computed, it is shrunk to a single vertex v of T (line 32). Thereafter the algorithm starts to compute the parent of G in the DC-tree (this is done by pushing v onto the stack together with a pointer to Initiate(v)). If v has no edges incident then the algorithm terminates (line 34) because v corresponds now to the root of the DC-tree.

The DC-tree is constructed implicitly by Algorithm HDC. To see this consider the final d-cluster G produced by HDC, which corresponds to the root of the DC-tree. The children of the root are the vertices of G, and since each shrunken vertex has a pointer to its corresponding d-cluster, it follows that any d-cluster can be reached from G.

Theorem 7. *Let S be any set of n vertices in the plane. Given a Euclidean minimum spanning tree of S, we can compute the diameter clustering tree of S in $O(n)$ time using $O(n)$ space.*

Proof. When a d-cluster G has been shrunk to a single vertex v of T, only a constant number of edges may be incident to v by Lemma 6, because all edges incident to v have length greater than rdiam($V(G)$). This means that each execution of lines 23-27 takes constant time, and so it is straightforward to realize that each execution of lines 8-28 takes constant time.

Recall that for any configuration of T, each edge of T corresponds to an edge of (the original) EMST(S). Therefore, for the sake of analysis, we consider an edge of T to be the same as the corresponding edge of EMST(S), although its endpoints may no longer be single vertices of S but shrunken vertices corresponding to d-clusters. So, in this sense, let e be any edge of EMST(S). Consider the first time e is removed from some queue Q at line 8. Suppose that e is not added to some partial d-cluster during this iteration of the while-loop, so one of its endpoints is pushed onto the stack (i.e. lines 17-19 are performed). Consider now the second time e is removed from some queue Q at line 8. Clearly the endpoint of e that was previously pushed onto the stack is now on top of the stack. This means that e will be added to some partial d-cluster during this iteration of the while-loop (line 11). Hence, since each edge of EMST(S) is added to exactly one d-cluster (the minimal d-cluster containing the edge), the lines 8-28 are executed at most twice per edge. Thus lines 8-28 takes totally $O(n)$ time.

Next we observe that each execution of lines 32-38 takes constant time since a shrunken vertex has a constant number of edges incident. Further, a d-cluster is produced each time these lines are executed, so they are executed as many times as there are internal nodes in the DC-tree. The latter sums up to at most $n - 1$, and so lines 32-38 takes total time $O(n)$.

Let us finally consider line 30. Clearly we cannot remove more edges from some pool P than the number of edges we put into P. Further, an edge is put into some pool P only at line 25 and by the function Initiate, which we already have shown may happen at most $O(n)$ times. Hence, line 30 takes total time $O(n)$. □

3.3   Extensions of the Diameter Clustering

A natural extension of the diameter clustering is to compute the hierarchy of d-clusters of a set S of n points in the k-dimensional Euclidean space. We only need to redefine the rectangular diameter of a vertex set D to become the diameter of the smallest hyperrectangle in E^k containing all vertices in D and having sides parallel to the coordinate axes. Consequently, we should keep track of an extreme vertex in each of the $2k$ coordinate directions. Having modified Algorithm HDC in the previous subsection accordingly, we can compute the hierarchy of d-clusters in the same manner as in the 2-dimensional case. But for a Euclidean minimum spanning tree in k dimensions, a vertex can be incident to $2^{\Theta(k)}$ edges (it equals the so-called Hadwiger number for k-hyperspheres [11], which is not larger than $2^{0.401k(1+o(1))}$). Similarly, for a d-cluster G in k dimensions, we can prove that the number of edges in $I(G)$ having length greater than the diameter of G is $2^{\Theta(k)}$ in the worst case. However, if k is regarded as a constant, the statement in Lemma 6 still holds, and so we obtain an $O(n)$ time algorithm in this case.

3.4 A Useful Property of Diameter Clusters

Below we show a property that has been proved useful in those contexts where the diameter clustering has been employed (see Section 1 and the next section).

Lemma 8. *Let D be an arbitrary d-cluster with $m \geq 2$ children. Among all diagonals with endpoints in distinct children of D, let l be the length of a shortest and let l' be the length of a longest such diagonal. Then the ratio between l' and l is less than 3^{m-1}.*

Proof. Let A and B be the children of D that are closest to each other, so l is the distance between A and B. By definition of d-clusters, both $\mathrm{rdiam}(A)$ and $\mathrm{rdiam}(B)$ are less than l. Hence, if D has only two children then $\mathrm{rdiam}(D) = \mathrm{rdiam}(A \cup B) < 3l$. Suppose now that D has three children and let C be the third child. Since $\mathrm{rdiam}(A \cup B) < 3l$, C has to be within distance $< 3l$ from $A \cup B$, because $A \cup B$ would otherwise constitute a d-cluster by itself. Further, since C is a d-cluster, $\mathrm{rdiam}(C)$ is less than $3l$. Thus, $\mathrm{rdiam}(D) = \mathrm{rdiam}(A \cup B \cup C) < 3^2 l$. By repeating the above scenario, we infer that $\mathrm{rdiam}(D) < 3^{m-1}l$ if D has m children. Hence, since l' is not greater than $\mathrm{rdiam}(D)$, the ratio between l' and l is less than 3^{m-1}. $\qquad\square$

4 Approximating the Single Linkage Clustering

The single linkage (s-link) clustering [5] produces a hierarchy of clusters as follows: Initially each vertex in S constitutes a cluster. As long as there is more than one cluster, the two closest clusters are merged, where the distance between two clusters is defined as the distance between the two closest vertices, one from each cluster. We define the accuracy of an approximation as follows.

Definition 9. We say that a hierarchy of clusters is an ϵ-*approximation* of a single linkage hierarchy if its clusters can be produced by a sequence of mergings in the following way. After an arbitrary number of mergings, let d be the distance between the two closest clusters. If d' is the distance between the two clusters merged in the next step, then it holds that $\frac{d'}{d} \leq 1 + \epsilon$.

It is easy to show that each d-cluster is also a cluster in any s-link hierarchy. Thus a s-link hierarchy can be created by a depth-first search from the root of the DC-tree, and, for each d-cluster D, clustering the children of D when backtracking from D. In this way it is enough to consider the problem of computing a s-link hierarchy of a d-cluster D, under the assumption that we have computed it within each child of D.

In [3] the following simple strategy for constructing a s-link hierarchy was proposed. Let L be a list of all edges of $\mathrm{EMST}(S)$ sorted by increasing length. As long as L is nonempty, remove the first (a shortest) edge and merge the two clusters which are connected by the edge. Next we show a lower bound for this approach.

Theorem 10. *Computing the single linkage clustering from a Euclidean minimum spanning tree requires $\Omega(n \log n)$ time in a comparison based model of computation.*

Proof. We show that sorting is linear-time transformable to this problem. Consider a sequence $r_1, r_2, \ldots, r_n$ of n distinct reals, and assume w.l.o.g. that each r_i is in $(0, 0.1)$. For each r_i we have points p_i and q_i: both have x-coordinate i, but p_i has y-coordinate 0 whereas q_i has y-coordinate $(-1)^i(1 + r_i)$. (This is illustrated in Figure 4.)

Fig. 4. The linear-time transformation.

It is straightforward to construct the Euclidean minimum spanning tree of this point set in $O(n)$ time. According to the definition of the s-link clustering, all p_i's are first merged into a common cluster. Thereafter, all q_i's are merged one after one with the main cluster. More precisely, the s-link clustering will merge q_i with the main cluster before q_j if and only if $r_i < r_j$. Thus, in the tree that describes the s-link hierarchy, q_i is a leaf at depth k if and only if r_i is the k'th largest among the reals. $\square$

The above theorem motivates an algorithm that, given EMST(S), computes an ϵ-approximation of a s-link hierarchy in time $o(n \log n)$ for sufficiently small ϵ.

Let D be an arbitrary d-cluster, and let $L(D)$ be a list of all edges of EMST(S) that connect the children of D. The edges in $L(D)$ are kept approximately sorted according to their lengths, in such a way that two edges may be in the wrong order only if the ratio between the longest and the shortest of them is less than $1 + \epsilon$. Having computed an ϵ-approximation within each child of D, we can obtain an ϵ-approximation of D by repeatedly removing the first edge from $L(D)$ and merging the two clusters that the edge connects. To find the clusters in which an edge of $L(D)$ has its endpoints, we can use the standard union-find algorithm from [12]. In this way, we obtain an ϵ-approximation of the single linkage hierarchy of S in time $O(n \alpha(n, n))$ plus the time needed to create the list $L(D)$ of every d-cluster D, where $\alpha(n, n)$ is the inverse of Ackermann's function.

Let c be an arbitrary constant greater than 1. We will show that the list $L(D)$ of every d-cluster D can be created in total time $O(n)$ if $\epsilon = \frac{1}{n^c}$. Let D

be any d-cluster with $m \geq 2$ children. By Lemma 8, the ratio between a longest and a shortest edge in $L(D)$ is less than 3^{m-1}. Thus we can in $O(m)$ time create a sequence $b_1, b_2, \ldots, b_{m-1}$ of buckets so that b_i contains all edges in $L(D)$ of length in $[3^{i-1}l, 3^i l)$, where l is the length of a shortest edge in $L(D)$. Next, to each edge e in $L(D)$ we assign the integer $\lceil \frac{|e|n^c}{3^{i-1}l} \rceil$ if e is contained in bucket b_i. This integer shall be called the *order number* for e. Having assigned an order number to each edge of EMST(S), we can use traditional radix sort to sort them according to their order numbers in $O(n)$ time, because the greatest order number is less than $3n^c$. After this, the sorted list is scanned, inserting each edge in its bucket in ascending order according to its order number. In this way we ensure that the edges in each bucket are sorted with respect to their order numbers.

It remains to show that two edges e_1 and e_2 contained in a bucket b_i may be in the wrong order only if the ratio between the longest and the shortest of them is less than $1 + \frac{1}{n^c}$. Suppose therefore that $|e_1| > |e_2|$ but that e_1 has equal or smaller order number than that of e_2. This means that $\frac{|e_1|n^c}{3^{i-1}l} - \frac{|e_2|n^c}{3^{i-1}l} < 1$, and so $|e_1| < |e_2| + \frac{3^{i-1}l}{n^c}$. Consequently, since $|e_2| \geq 3^{i-1}l$, the ratio $|e_1|/|e_2|$ is less than

$$\frac{|e_2| + 3^{i-1}l/n^c}{|e_2|} = 1 + \frac{3^{i-1}l}{n^c|e_2|} \leq 1 + \frac{1}{n^c}$$

Thus we obtain

Theorem 11. *Let S be any set of n vertices in the plane, and let c be an arbitrary constant greater than 1. Given a Euclidean minimum spanning tree of S, we can obtain an $\frac{1}{n^c}$-approximation of the single linkage hierarchy of S in time $O(n\,\alpha(n,n))$, where $\alpha(n,n)$ is the inverse of Ackermann's function.*

Remark. The above theorem can be extended to the case when c is not a constant as follows. Recall that the order numbers that we sort are n integers in the range $[n^c..3n^c)$. If n^c fits in a constant number of machine words, we can use the technique in [6] to sort the order numbers in time $O(n \log c)$, thus yielding a total running time in $O(n\,(\alpha(n,n) + \log c))$. Otherwise, we need at most $O(c)$ machine words to store an order number (usually it is assumed that a machine word has at least $\Omega(\log n)$ bits). Hence, in this case we can use radix sort to sort the order numbers in time $O(nc)$, and so the total running time of the algorithm would become $O(n\,(\alpha(n,n) + c))$.

References

1. F. Aurenhammer. Voronoi diagrams—a survey of a fundamental geometric data structure. *ACM Computing Surveys*, 23(3):345–405, 1991.
2. A. D. Gordon. *Classification*. Chapman and Hall, 1981.
3. J. C. Gover and G. J. S. Ross. Minimum spanning trees and single linkage-linkage cluster analysis. *Applied Statistics*, 18:54–64, 1969.
4. J. A. Hartigan. *Clustering algorithms*. Wiley, New York, 1975.

5. A. K. Jain and R. C. Dubes. *Algorithms for clustering data.* Prentice-Hall, New Jersey, 1988.
6. D. Kirkpatrick and S. Reisch. Upper bounds for sorting integers on random access machines. *Theoretical Computer Science*, 28:263–276, 1984.
7. D. Krznaric and C. Levcopoulos. Computing a threaded quadtree (with links between neighbors) from the delaunay triangulation in linear time. In *7th Canadian Conference on Computational Geometry*, pages 187–192, 1995.
8. D. Krznaric and C. Levcopoulos. The first subquadratic algorithm for complete linkage clustering. In *6th International Symposium on Algorithms and Computation*, Lecture Notes in Computer Science. Springer-Verlag, 1995. To appear.
9. C. Levcopoulos and D. Krznaric. The greedy triangulation can be computed from the delaunay in linear time. Technical Report LU-CS-TR:94-136, Department of Computer Science, Lund University, Lund, Sweden, 1994.
10. F. P. Preparata and M. I. Shamos. *Computational geometry: an introduction.* Springer-Verlag, New York, 1985.
11. G. Robins and J. S. Salowe. Low-degree minimum spanning trees. *Discrete & Computational Geometry*, 14(2):151–165, 1995.
12. R. E. Tarjan. Efficiency of a good but not linear disjoint set union algorithm. *Journal of the ACM*, 22(2):215–225, 1975.

Determinizing Büchi Asynchronous Automata

Nils Klarlund[1], Madhavan Mukund[2], Milind Sohoni[3]

[1] BRICS[†], Aarhus University, Ny Munkegade,
DK 8000 Aarhus C, Denmark. E-mail: klarlund@daimi.aau.dk
[2] School of Mathematics, SPIC Science Foundation, 92 G N Chetty Rd
Madras 600 017, India. E-mail: madhavan@ssf.ernet.in
[3] Dept of Computer Sc and Engg, Indian Institute of Technology
Bombay 400 076, India. E-mail: sohoni@cse.iitb.ernet.in

Abstract. Büchi asynchronous automata are a natural distributed machine model for recognizing ω-regular *trace languages*. Like their sequential counterparts, these automata need to be non-deterministic in order to capture all ω-regular languages. Thus complementation of these automata is non-trivial. Complementation is an important operation because it is fundamental for treating the logical connective "not" in decision procedures for monadic second-order logics.

In this paper, we present a direct determinization procedure for Büchi asynchronous automata, which generalizes Safra's construction for sequential Büchi automata. As in the sequential case, the blow-up in the state space is essentially that of the underlying subset construction.

Introduction

Finite-state automata are, by definition, sequential. To describe finite-state *concurrent* computations, Zielonka introduced *asynchronous automata* [Zie1]. An asynchronous automaton consists of a set of independent processes which cooperate to read their input. Each letter a in the alphabet is associated with a subset $\theta(a)$ of processes which jointly decide on a move when a is read.[1] The distribution function θ introduces an *independence relation* I between letters: $(a, b) \in I$ iff a and b are read by disjoint sets of processes.

Earlier, Mazurkiewicz had proposed a framework for studying concurrent systems where the alphabet Σ is equipped with a pre-specified independence relation I, describing the concurrency in the system [Maz]. Two words w and w' describe the same computation iff w' can be obtained from w by a finite sequence of permutations of adjacent independent letters. This gives rise to an equivalence relation on words over Σ. The equivalence class $[w]$ containing w is called a *trace*. A set of words L is a *trace language* if it obeys the equivalence relation generated by I— for each word w in L, all of $[w]$ is contained in L.

[†] Basic Research in Computer Science, Centre of the Danish National Research Foundation.

[1] Calling these automata *asynchronous* is, in a sense, misleading. The processes communicate synchronously. The asynchrony refers to the fact that different components of the network can proceed independently while reading the input.

Zielonka proved that any regular trace language over a *concurrent* alphabet (Σ, I) can be recognized by a deterministic asynchronous automaton over a *distributed* alphabet (Σ, θ), such that the independence relation generated by θ is exactly I. Gastin and Petit have extended the connection between asynchronous automata and trace languages to the setting of infinite inputs. In [GP], they introduce the class of Büchi asynchronous automata which accept precisely the class of ω-regular trace languages.

Like automata over infinite strings, Büchi asynchronous automata have close connections to logic [EM, Thi]. In order to exploit these connections—for instance, to automate verification of formulae defined using these logics—we need to develop techniques for manipulating these automata. Basic operations include complementation and determinization.

As in the sequential case, complementing Büchi asynchronous automata is non-trivial, since they are necessarily non-deterministic: deterministic Büchi asynchronous automata cannot recognize all ω-regular trace languages [GP]. With a Muller acceptance condition, deterministic automata suffice [DM], but a direct determinization procedure has so far been elusive.

Contributions of this paper

We extend the subset construction for asynchronous automata [KMS1] to a direct determinization construction for Büchi asynchronous automata, based on Safra's technique for determinizing Büchi automata on infinite strings [Saf]. The determinized automaton we construct has an acceptance condition described in terms of "Rabin pairs". As in the sequential case, we can efficiently simulate the complement of the determinized automaton using a non-deterministic Büchi asynchronous automaton. So, we also have a direct complementation construction for Büchi asynchronous automata. In both the determinized Rabin automaton and the complementary Büchi automaton, the number of local states of each process is exponential in the number of global states of the original automaton. As in Safra's original construction, this blow-up is essentially that of the underlying subset construction for these automata.

In related work, Muscholl [Mus] has described a complementation construction for Büchi asynchronous *cellular* automata, which are an alternative distributed model for recognizing trace languages [Zie2]. Her construction does not involve determinization—she makes use of progress measures [Kla] and directly constructs a non-deterministic complement automaton.

An asynchronous cellular automaton allocates a separate process for each letter in the input alphabet—even when the underlying system is completely sequential, a cellular automaton will have a number of components. Processes communicate using a non-standard variant of a shared memory. As a result, though both the approaches are formally equivalent, asynchronous automata seem to be more natural models for describing distributed systems.

Converting between asynchronous automata and asynchronous cellular automata involves a blow-up in the state space of each process which is exponential in $|\Sigma|$, the size of the input alphabet. However, since $|\Sigma|$ could itself be exponential in the size of the global state space of the automaton, there is effectively a double exponential blow-up in this translation. So, complementing asynchronous automata directly using our construction can be significantly more efficient than complementing them indirectly via the construction described in [Mus].

In general, it appears to be advantageous to work directly with asynchronous automata for automating decision procedures in logic, instead of using asynchronous cellular automata. Incorporating the alphabet into the state space of the automaton is known to be expensive in such applications—for example, the decision procedure for monadic second-order logic on strings generates alphabets that are exponential in the number of free variables in the input formula; see [HJJ] for techniques which allow automata with exponentially sized alphabets to be represented and manipulated within polynomial bounds.

Working directly with asynchronous automata is also relevant to *model checking* —a technique for mechanically verifying if a program satisfies a property specified in a logical language. If the same kind of automata are used both for describing the program and for checking satisfiability, the model checking problem reduces to a simple intersection problem involving the automata [VW]. Since asynchronous automata are a natural model for distributed programs, automata-theoretic model checking can be applied to the logics considered in [EM, Thi].

The paper is organized as follows. We begin with some definitions regarding asynchronous automata. In Section 2 we introduce Büchi and Rabin asynchronous automata and formulate the problem. The next three sections recapitulate some basic techniques developed in [KMS1, MS] for manipulating asynchronous automata. In Section 6 we show how to apply these techniques to determinize Büchi asynchronous automata. Due to space limitations we have been forced to omit many examples and proofs. For a more detailed exposition, the reader is referred to [KMS2].

1 Preliminaries

The following definitions are essentially those of [KMS1] adapted to the setting of infinite inputs.

Distributed alphabets Let $\mathcal{P}$ be a finite set of processes, where the size of $\mathcal{P}$ is N. A *distributed alphabet* is a pair (Σ, θ) where Σ is a finite set of *actions* and $\theta : \Sigma \to 2^{\mathcal{P}}$ assigns a non-empty set of processes to each $a \in \Sigma$.

State spaces With each process p, we associate a finite set of states denoted V_p. Each state in V_p is called a *local state*. For $P \subseteq \mathcal{P}$, V_P denotes the product $\prod_{p \in P} V_p$. An element $\vec{v}$ of V_P is a tuple or *joint state* that determines a local state for each p in P. We refer to a joint state from V_P as a *P-state*. A $\mathcal{P}$-state is also called a *global state*.

Given $\vec{v} \in V_P$, and $P' \subseteq P$, $\vec{v}_{P'}$ denotes the projection of $\vec{v}$ onto $V_{P'}$. Also, $\vec{v}_{\overline{P'}}$ abbreviates $\vec{v}_{P-P'}$. For a singleton $p \in P$, we write $\vec{v}_p$ for $\vec{v}_{\{p\}}$. For $a \in \Sigma$, we write V_a to mean $V_{\theta(a)}$ and $V_{\overline{a}}$ to mean $V_{\overline{\theta(a)}}$. Similarly, if $\vec{v} \in V_P$ and $\theta(a) \subseteq P$, we write $\vec{v}_a$ for $\vec{v}_{\theta(a)}$ and $\vec{v}_{\overline{a}}$ for $\vec{v}_{\overline{\theta(a)}}$.

Asynchronous automata An *asynchronous automaton* $\mathfrak{A}$ over (Σ, θ) is of the form $(\{V_p\}_{p \in \mathcal{P}}, \{\rightarrow_a\}_{a \in \Sigma}, \mathcal{V}_0)$, where $\rightarrow_a \subseteq V_a \times V_a$ is the *local transition relation* for a, and $\mathcal{V}_0 \subseteq V_{\mathcal{P}}$ is a set of *initial* global states. Each relation $\rightarrow_a$ specifies how the processes $\theta(a)$ that meet on a may decide on a joint move. Other processes do not change their state. Thus we define the *global transition relation* $\Rightarrow \subseteq V_{\mathcal{P}} \times \Sigma \times V_{\mathcal{P}}$ by $\vec{v} \overset{a}{\Longrightarrow} \vec{v}'$ if $\vec{v}_a \rightarrow_a \vec{v}'_a$ and $\vec{v}_{\overline{a}} = \vec{v}'_{\overline{a}}$.

$\mathfrak{A}$ is called *deterministic* if the global transition relation of $\mathfrak{A}$ is a function from $V_{\mathcal{P}} \times \Sigma$ to $V_{\mathcal{P}}$ and if the set of initial states $\mathcal{V}_0$ is a singleton.

Runs Let α be an infinite word over Σ. It is convenient to think of α as a function of time; i.e., $\alpha : \mathbf{N} \rightarrow \Sigma$. (We use $\mathbf{N}$ to denote the set $\{1, 2, \ldots\}$ and $\mathbf{N}_0$ for $\{0, 1, 2, \ldots\}$.) A *global run* of $\mathfrak{A}$ on $\alpha : \mathbf{N} \rightarrow \Sigma$ is a function $\rho : \mathbf{N}_0 \rightarrow V_{\mathcal{P}}$ such that $\rho(0) \in \mathcal{V}_0$ and, for $i \in \mathbf{N}$, $\rho(i-1) \overset{\alpha(i)}{\Longrightarrow} \rho(i)$.

Analogously, we represent a finite word $u \in \Sigma^*$ of length m as a function $u : [1..m] \rightarrow \Sigma$, where $[i..j]$ abbreviates the set $\{i, i+1, \ldots, j\}$. A global run of $\mathfrak{A}$ on $u : [1..m] \rightarrow \Sigma$ is a function $\rho : [0..m] \rightarrow V_{\mathcal{P}}$ such that $\rho(0) \in \mathcal{V}_0$ and, for $i \in [1..m]$, $\rho(i-1) \overset{u(i)}{\Longrightarrow} \rho(i)$.

Let ρ be a global run. For $P \subseteq \mathcal{P}$, ρ_P denotes the projection of ρ onto the P-components. So, ρ_P is a sequence of P-states. As usual, $inf(\rho_P)$ denotes the set of P-states which occur infinitely often in ρ_P; $inf(\rho_P) = \{\vec{v} \in V_P \mid$ for infinitely many i, $\rho_P(i) = \vec{v}\}$.

2 Asynchronous automata on infinite inputs

To define how an asynchronous automaton accepts an infinite word α, we have to examine the communication between processes in the limit, as α is read.

Limit graphs With each infinite word α, we associate an undirected graph $\mathcal{G}_\alpha = (\mathcal{P}, E_\alpha)$ called the *limit graph* of α. The graph has an edge between processes p and q provided they synchronize infinitely often while $\mathfrak{A}$ processes α. In other words, $(p, q) \in E_\alpha$ iff for infinitely many i, $\{p, q\} \subseteq \theta(\alpha(i))$. Let $Conn_\alpha$ denote the maximal connected components of $\mathcal{G}_\alpha$.

Let $Finite_\alpha$ denote the set of processes which move only finitely often while $\mathfrak{A}$ reads α—i.e., p belongs to $Finite_\alpha$ if there are only finitely many i such that $p \in \theta(\alpha(i))$. Clearly, if $p \in Finite_\alpha$ then the singleton $\{p\}$ belongs to $Conn_\alpha$.

Büchi asynchronous automata A *Büchi asynchronous automaton* is a pair $B\mathfrak{A} = (\mathfrak{A}, \mathcal{T}_B)$ where $\mathfrak{A}$ is an asynchronous automaton and $\mathcal{T}_B$ is a *Büchi acceptance table*. The table $\mathcal{T}_B$ is a list $(\tau_1, \tau_2, \ldots, \tau_k)$. Each entry τ_i in $\mathcal{T}_B$ is of the form $(\mathcal{C}, T, \{(p_C, G_C)\}_{C \in \mathcal{C}})$, where $\mathcal{C}$ is a partition of $\mathcal{P}$, T is a subset of $\mathcal{P}$ and, for each subset $C \in \mathcal{C}$, p_C is a designated process from C and G_C is a set of p_C-states. We call the processes $\{p_C\}_{C \in \mathcal{C}}$ the *signalling processes* in τ_i.

A run ρ of the automaton $B\mathfrak{A} = (\mathfrak{A}, \mathcal{T}_B)$ on an input α is said to *satisfy* an entry $\tau = (\mathcal{C}, T, \{(p_C, G_C)\}_{C \in \mathcal{C}})$ in $\mathcal{T}_B$ provided $\mathcal{C} = Conn_\alpha$, $T = Finite_\alpha$ and, for each signalling process p_C, $inf(\rho_{p_C}) \cap G_C \neq \emptyset$. The automaton accepts α if there is a run ρ on α and a table entry τ such that ρ satisfies τ.

Recall that every process p in $Finite_\alpha$ constitutes a separate singleton component in $Conn_\alpha$. For a signalling process $p \in T$, the set G_p denotes the set of possible *terminating states* for p. On the other hand, for a signalling process p which does not belong to T, G_p is a set of *recurring states*, one of which must be visited infinitely often by $\mathfrak{A}$ for ρ to satisfy τ.

Our definition of Büchi asynchronous automata is adapted from [Mus] and differs from the original formulation of Gastin and Petit [GP]. The crucial part of our definition is the extra information we record about $Conn_\alpha$ in each entry of the acceptance table. This allows us to separate the processes in $\mathfrak{A}$ into independent groups. After a finite prefix of α has been read, there will be no further synchronizations between processes in different connected components of $\mathcal{G}_\alpha$. So, in the limit, each subset $C \in Conn_\alpha$ moves as a separate, independent unit.

Since non-deterministic Büchi asynchronous automata are strictly more powerful than their deterministic counterparts [GP], to determinize these automata we have to strengthen the acceptance condition. We shall work with a generalization of the "pairs" condition proposed by Rabin [Rab].

Rabin asynchronous automata A *Rabin asynchronous automaton* is a pair $\mathbf{R}\mathfrak{A} = (\mathfrak{A}, \mathcal{T}_R)$ where $\mathfrak{A}$ is an asynchronous automaton and $\mathcal{T}_R$ is a *Rabin acceptance table*. The table $\mathcal{T}_R$ is a list $(\tau_1, \tau_2, \ldots, \tau_k)$. Each entry τ_i in $\mathcal{T}_R$ is of the form $(\mathcal{C}, T, \{(p_C, pairs_C)\}_{C \in \mathcal{C}})$, where $\mathcal{C}$, T and p_C are as in a Büchi acceptance table and, for each signalling process p_C, $pairs_C$ is a list $\{(G_C^j, R_C^j)\}_{j \in [1..k_C]}$ such that for each pair (G_C^j, R_C^j), both G_C^j and R_C^j are subsets of V_{p_C}.

The automaton $\mathbf{R}\mathfrak{A} = (\mathfrak{A}, \mathcal{T}_R)$ accepts an input α if there is a run ρ of $\mathfrak{A}$ on α such that for some entry $\tau = (\mathcal{C}, T, \{(p_C, pairs_C)\}_{C \in \mathcal{C}})$ in the table $\mathcal{T}_R$, $\mathcal{C} = Conn_\alpha$, $T = Finite_\alpha$ and, for each signalling process p_C, there is an entry (G_C^j, R_C^j) in $pairs_C$ such that $inf(\rho_{p_C}) \cap G_C^j \neq \emptyset$ and $inf(\rho_{p_C}) \cap R_C^j = \emptyset$.

The problem For a given non-deterministic Büchi asynchronous automaton $\mathbf{B}\mathfrak{A} = (\mathfrak{A}, \mathcal{T}_B)$ over (Σ, θ), construct a deterministic Rabin asynchronous automaton $\mathbf{R}\mathfrak{B} = (\mathfrak{B}, \mathcal{T}_R)$ over (Σ, θ), such that $\mathbf{B}\mathfrak{A}$ and $\mathbf{R}\mathfrak{B}$ accept the same set of infinite words over Σ.

Notice that an asynchronous automaton where $\mathcal{P}$ is a singleton $\{p\}$ is just a conventional sequential finite state automaton. Further, if $\mathcal{P} = \{p\}$, our definitions of Büchi and Rabin asynchronous automata reduce to the standard formulations of these automata in the setting of infinite strings [Tho].

For sequential Büchi automata, Safra has described an elegant determinization construction which is an extension of the classical subset construction for finite automata [Saf]. We do not have space to describe this construction here. However, it is possible to understand the salient features of our construction *without* getting into the precise details of how Safra's construction works.

To determinize Büchi asynchronous automata, we shall apply Safra's construction in a distributed setting. Let $\mathbf{B}\mathfrak{A} = (\mathfrak{A}, \mathcal{T}_B)$ be a Büchi asynchronous automaton. Our strategy will be to construct a deterministic Rabin automaton $\mathbf{R}\mathfrak{B}_\tau = (\mathfrak{B}_\tau, \mathcal{T}_{R_\tau})$ corresponding to each entry τ in the table $\mathcal{T}_B$. The automaton $\mathbf{R}\mathfrak{B}_\tau$ accepts an input α provided there is a run ρ of $\mathbf{B}\mathfrak{A}$ which satisfies τ.

We can then combine the individual automata $\{R\mathfrak{B}_\tau\}_{\tau \in \mathcal{T}_B}$ into a deterministic Rabin automaton $R\mathfrak{B}$ which accepts exactly the same infinite strings as $B\mathfrak{A}$.

To construct the automaton $R\mathfrak{B}_\tau$ for the table entry $\tau = (\mathcal{C}, T, \{(p_C, G_C)\}_{C \in \mathcal{C}})$ we have to check that for each signalling process p_C, $inf(\rho_{p_C}) \cap G_C \neq \emptyset$. To do this, we run Safra's construction for each signalling process p_C, using the subset construction for asynchronous automata [KMS1] in place of the classical subset construction for sequential automata.

The catch is that each signalling process p_C may meet its recurring set G_C infinitely often along a different run. So, we have to further ensure that the accepting runs detected by the independent copies of Safra's construction at each signalling process are mutually consistent. This will involve some analysis of the way information is passed between the connected components in $\mathcal{G}_\alpha$ *before* they branch out as independent groups.

3 Local and global views

We represent words over a distributed alphabet as labelled partial orders. The notions we use are essentially those of trace theory [Maz].

Events With $\alpha : \mathbf{N} \to \Sigma$, we associate a set of *events* $\mathcal{E}_\alpha$. Each event $(i, \alpha(i))$ consists of a letter $\alpha(i)$ together with the time i of its occurrence. In addition, we define an *initial event* denoted 0. The initial event marks the beginning when all processes synchronize and agree on an initial global state. Usually, we will write $\mathcal{E}$ for $\mathcal{E}_\alpha$. If $e = (i, a)$ is an event, then we may use e instead of a in abbreviations such as V_e, which stands for V_a, i.e., $V_{\theta(a)}$, or $\to_e$, which is just $\to_a$. For $p \in \mathcal{P}$ and $e = (i, a)$, we write $p \in e$ to denote that $p \in \theta(a)$. When $e = 0$, we define $p \in e$ to hold for all $p \in \mathcal{P}$. If $p \in e$, then we say that e is a *p-event*.

Ordering relations on $\mathcal{E}$ The word α naturally imposes a total order $\leq$ on events: $e \leq f$ if e happens at time i and f happens at time j with $i \leq j$.

Each process p imposes a total order $\leq_p$ on the events in which it participates. Thus $e \leq_p f$ if p participates in both e and f and $e \leq f$. If e is the p-event that immediately precedes the p-event f, then we write $e \lessdot_p f$. Thus $e \lessdot_p f$ if $e \leq_p f$ and no g with $e < g < f$ is a p-event.

The asynchronous nature of the automaton is reflected more accurately by the partial order generated by the relations $\{\lessdot_p\}_{p \in \mathcal{P}}$ than by the temporal order $\leq$. We say that e is an immediate predecessor of f and write $e \lessdot f$ if $e \lessdot_p f$ for some p. Let $\sqsubseteq$ be the reflexive and transitive closure of $\lessdot$. If $e \sqsubseteq f$, then we say e is *below* f. Note that the initial event 0 is below any event. The set of events below e is denoted $e\!\downarrow$. They represent the only synchronizations that may have affected the state of $\mathfrak{A}$ at e.

Ideals An *ideal* I is any set of events closed with respect to $\sqsubseteq$. Ideals represent possible partial computations of the system. We assume that every ideal I we consider is non-empty—i.e., 0 always belongs to I. Let α_m denote the prefix of α of length m. Then the events $\{(i, \alpha(i)) \mid i \leq m\} \cup \{0\}$ form an ideal. Conversely, every ideal gives rise to a subword of α—if I is the finite ideal $\{0, (i_1, a_1), (i_2, a_2), \ldots, (i_m, a_m)\}$, then $\alpha[I] : [1..m] \to \Sigma$ is the word $\alpha(i_1)\alpha(i_2) \cdots \alpha(i_m) = a_1 a_2 \cdots a_m$. Similarly, we can associate infinite ideals in $\mathcal{E}$ with infinite subsequences of α. Even when I is finite, $\alpha[I]$ is not, in general, a prefix of α because of the asynchronous manner in which α is processed. Clearly the entire set $\mathcal{E}$ is an ideal, as is the set $e{\downarrow}$ for any event $e \in \mathcal{E}$.

P-views Let I be an ideal. The *p-view of* I, $\partial_p(I)$, is the set $\{e \in I \mid \exists f \in I. \ p \in f \text{ and } e \sqsubseteq f\}$. So, $\partial_p(I)$ is the set of all events in I which p can "see". If the number of p-events in I is finite—for instance, if I itself is finite—it is easy to see that $\partial_p(I) = max_p(I){\downarrow}$, where $max_p(I)$ is the $\sqsubseteq$-maximum p-event in I.

For $P \subseteq \mathcal{P}$, the *P-view of* I, denoted $\partial_P(I)$, is $\bigcup_{p \in P} \partial_p(I)$. Notice that $\partial_P(I)$ is always an ideal. In particular, we have $\partial_{\mathcal{P}}(I) = I$.

4 Local runs and histories

For the rest of this section, we fix a (non-determinstic) asynchronous automaton $\mathfrak{A} = (\{V_p\}_{p \in \mathcal{P}}, \{\to_a\}_{a \in \Sigma}, V_0)$.

Neighbourhoods The *neighbourhood* of an event e, $nbd(e)$, consists of e together with its immediate predecessors; i.e., $nbd(e) = \{e\} \cup \{f \mid f \lessdot e\}$. Notice that if $e \in \partial_P(I)$ for some $P \subseteq \mathcal{P}$, then $nbd(e) \subseteq \partial_P(I)$ as well.

Local runs A *local run* on an ideal I assigns a joint state to each event in I in such a way that all neighbourhoods are consistently labelled. More precisely, a local run on I is a function r that assigns to each $e \in I$ an e-state—i.e., a state in V_e—such that $r(0) \in V_0$ and for all $e \neq 0$, r is consistent with $\to_e$ in $nbd(e)$ in the following sense: suppose that $\vec{v}$ is the e-state whose p-component, for each $p \in e$, is the same as the p-component of $r(f_p)$, where f_p is the immediate p-predecessor of e. In other words, for each $p \in e$, $\vec{v}_p = r(f_p)_p$, where $f_p \lessdot_p e$. Then r is such that $\vec{v} \to_e r(e)$. Given a local run r, there is a natural "last" global state $\vec{v}$ defined by $\vec{v}_p = r(max_p(I))$ for all p. We say that $\vec{v}$ is a state of $\mathfrak{A}$ *on* I. Similarly, a *P-state* of $\mathfrak{A}$ on I is $\vec{v}_P$, where $\vec{v}$ is a state of $\mathfrak{A}$ *on* I.

Let $\mathcal{R}(I)$ denote the set of all local runs on I. The following is easy to verify.

Proposition 1. *Let* $\alpha : \mathbf{N} \to \Sigma$ *and* $I \subseteq \mathcal{E}_\alpha$ *an ideal. Then, there is a 1-1 correspondence between* $\mathcal{R}(I)$ *and the set of global runs of* $\mathfrak{A}$ *on* $\alpha[I]$.

Histories A history on an ideal I is a partial function h that assigns joint states to some events in I. Thus $dom(h) \subseteq I$ and when $h(e)$ is defined it denotes a tuple in V_e. A history is *reachable* if there is some local run r on I such that $h(e) = r(e)$ for e in $dom(h)$. A set of histories H is *consistent* if each pair of histories h and h' in the set agree on all common events; i.e., for each $h, h' \in H$, for each e in $dom(h) \cap dom(h')$, $h(e) = h(e')$.

History Products Let $\mathcal{I} = \{I_1, I_2, \ldots, I_n\}$ be a set of ideals with $J = \bigcup_{j \in [1..n]} I_j$. Let $\{h_1, h_2, \ldots, h_n\}$ be a consistent set of histories such that h_j is a history over I_j for each $j \in [1..n]$. We define the *product* $h = \bigotimes_{j \in [1..n]} h_j$ as follows:

$dom(h) = \{e \in J \mid \text{for all } j \in [1..n], \text{if } e \in I_j \text{ then } e \in dom(h_j)\}$

$h(e) = h_k(e)$, where k is such that $e \in dom(h_k)$ (the choice of k does not matter since $\{h_j\}_{j \in [1..n]}$ is consistent).

In other words, h is a history over J which inherits its values from the set $\{h_j\}_{j \in [1..n]}$. The value $h(e)$ is defined whenever $h_j(e)$ is defined *for all* j such that $e \in I_j$. This means that if e is in $I_j \cap I_k$ for some pair $\{I_j, I_k\} \subseteq \mathcal{I}$ and $e \in dom(h_j)$ but $e \notin dom(h_k)$, then $e \notin dom(h)$.

We extend the notion of product to sets of histories spanning a set of ideals. Let $\mathcal{I} = \{I_1, I_2, \ldots, I_n\}$ be a set of ideals, with $J = \bigcup_{j \in [1..n]} I_j$. Let $\mathcal{H}_{\mathcal{I}} = \{H_1, H_2, \ldots, H_n\}$ where H_j is a set of histories over I_j for each $j \in [1..n]$. A *choice* from $\mathcal{H}_{\mathcal{I}}$ is a set $\{h_j\}_{j \in [1..n]}$ which picks out a history $h_j \in H_j$ for each $j \in [1..n]$. The choice is consistent if the set $\{h_j\}_{j \in [1..n]}$ is. We then define

$$\bigotimes_{j \in [1..n]} \mathcal{H}_{\mathcal{I}} = \{ \bigotimes h_j \mid \{h_j\}_{j \in [1..n]} \text{ is a consistent choice from } \mathcal{H}_{\mathcal{I}}\}.$$

So, $\bigotimes \mathcal{H}_{\mathcal{I}}$ contains all histories on J that may be pieced together from mutually consistent histories in the collection $\mathcal{H}_{\mathcal{I}}$.

Products of histories play a crucial role in the subset construction for asynchronous automata [KMS1]. Suppose X_p and X_q are the sets of possible states of p and q on ideal I. The set of possible joint $\{p, q\}$-states on I is *not*, in general, the naïve product $X_p \times X_q$. To determine which states from $X_p \times X_q$ are valid $\{p, q\}$-states on I, p and q have to record additional information about the runs leading to each state in the current subsets X_p and X_q. Since the amount of information that a process can store is bounded, it can at best record histories defined over a finite subset of the events it has seen.

In the subset construction of [KMS1], on an ideal I, each process p maintains the set H_p of all reachable histories over a specific bounded subset of $\partial_p(I)$. This subset includes $max_p(I)$, so H_p has, in particular, information about all the possible states that p can be in on I. Suppose a subset $P \subseteq \mathcal{P}$ synchronizes. In terms of the notation above, we have $\mathcal{I} = \{\partial_p(I)\}_{p \in P}$, $J = \partial_P(I)$ and $\mathcal{H}_{\mathcal{I}} = \{H_p\}_{p \in P}$. The goal is to ensure that $\bigotimes \mathcal{H}_{\mathcal{I}}$ generates all possible consistent "joint" histories of P over an appropriate subset of $\partial_P(I)$. This will allow the processes in P to jointly compute all the possible moves they can make on reading the new letter from the input.

The key step is to characterize when the product of a set of reachable histories $\{h_j\}_{j \in [1..n]}$ over $\mathcal{I} = \{I_1, I_2, \ldots, I_n\}$ remains a reachable history over the joint ideal $J = \bigcup_{j \in [1..n]} I_j$. For this, we need the notion of a frontier.

Frontiers Let I and J be ideals and p a process. We say that event e of I is an *p-sentry* for I relative to J if e is also in J and its p-successor is in J but not in I. Thus the process p "leaves" I at e. Let $border(I, J)$ be the set of all such sentries. Note that there is at most one p-sentry for each p, so there are at most N events in $border(I, J)$—recall that $N = |\mathcal{P}|$. In general,

$border(I, J) \neq border(J, I)$. We are normally interested in the two sets together, which we denote $frontier(I, J)$; i.e., $frontier(I, J) = border(I, J) \cup border(J, I)$. It is clear that $frontier(I, J) = frontier(J, I)$ and $frontier(I, I) = \emptyset$. We then have the following crucial result which is proved in [KMS1].

Lemma 2. *Let* $\mathcal{I} = \{I_1, I_2, \ldots, I_n\}$ *be a set of ideals and* $\{h_j\}_{j \in [1..n]}$ *a consistent set of histories such that for each* $j \in [1..n]$:

(i) h_j *is a reachable history over* I_j; *and*
(ii) $dom(h_j)$ *includes* $\bigcup_{k \in [1..n]} frontier(I_j, I_k)$.

Then $h = \bigotimes_{j \in [1..n]} h_j$ *is a reachable history over* $J = \bigcup_{j \in [1..n]} I_j$.

So, whenever the reachable histories $\{h_j\}_{j \in [1..n]}$ span all the frontiers between the ideals in $\mathcal{I} = \{I_1, I_2, \ldots, I_n\}$, their product is also reachable.

Recall that each process p maintains H_p, the set of all reachable histories over a specific subset of $\partial_p(I)$. Suppose that this specific subset of $\partial_p(I)$ includes $frontier_p(I)$, where $frontier_p(I) = \bigcup_{q \in \mathcal{P}} frontier(\partial_p(I), \partial_q(I))$. Then, if $\mathcal{I} = \{\partial_p(I)\}_{p \in P}$ and $\mathcal{H}_{\mathcal{I}} = \{H_p\}_{p \in P}$, the previous lemma guarantees that every history in $\bigotimes \mathcal{H}_{\mathcal{I}}$ is reachable in $\partial_P(I)$.

The problem now is for a process p to compute the bounded set of events $frontier_p(I)$. This can be done using slightly larger, but still bounded, sets of events called primary and secondary information, which between them subsume the frontiers.

5 Primary and secondary information

Primary information Let I be a finite ideal. Recall that $max_p(I)$ denotes the $\sqsubseteq$-maximum p-event in I. The *primary information* of I, $primary(I)$, is the set of events $\{max_p(I)\}_{p \in \mathcal{P}}$. We can define $primary(I)$ analogously for infinite ideals as well, where we include the events $max_p(I)$ for only those processes p such that there are only finitely many p-events in I.

Secondary and tertiary information Let I be a finite ideal. The *secondary information* in I, $secondary(I)$, is the set of events $\bigcup_{p \in \mathcal{P}} primary(\partial_p(I))$. The *tertiary information* in I, $tertiary(I)$, is the set $\bigcup_{p, q \in \mathcal{P}} primary(\partial_p(\partial_q(I)))$.

The primary information of I represents the latest information available in I about each process in the system. Similarly, the secondary information $primary(\partial_p(I))$ is the latest information that process p has in I about the other processes in the system, while the tertiary information $primary(\partial_q(\partial_p(I)))$ is the latest information that p has about the primary information of q in I.

It is clear that every event in $primary(I)$ also belongs to $secondary(I)$, since $max_p(\partial_p(I)) = max_p(I)$ for all $p \in P$. Similarly, every event in $secondary(I)$ belongs to $tertiary(I)$.

Let I and J be ideals. If I and J satisfy a simple condition, the events in $frontier(I, J)$ can be characterized in terms of the primary and secondary information of I and J, as described in the following lemma.

Lemma 3 [KMS1]. *Let I and J be ideals such that $I = \partial_P(K)$ and $J = \partial_Q(K)$, where K is an ideal and $P, Q \subseteq \mathcal{P}$ are sets of processes. Let e be a p-sentry for I with respect to J. Then $e = max_p(I)$ and, for some process q, $e = max_p(\partial_q(J))$. Thus, $e \in primary(I) \cap secondary(J)$.*

Let I be an ideal. From the previous lemma, it is clear that for a process P to maintain reachable histories over $frontier_p(I)$, it is sufficient for p to maintain reachable histories over $secondary(\partial_p(I))$. Processes can unambiguously keep track of their primary and secondary information by using time-stamps.

Time-stamps and the subset construction Let I be a finite ideal. Then, there are at most N^3 distinct events in $tertiary(I)$. We can thus use a finite set $\mathcal{L}$ of labels to *time-stamp* each event in this set—let this assignment of time-stamps be denoted by a function $\lambda : tertiary(I) \to \mathcal{L}$. For $p \in \mathcal{P}$, let λ_p denote the restriction of λ to $\partial_p(I)$. It turns out that the processes in $\mathcal{P}$ can locally maintain and update the functions λ_p so that, overall, the events in $tertiary(I)$ are assigned consistent time-stamps.

Theorem 4 Time-stamping [MS]. *For any distributed alphabet (Σ, θ), we can fix a finite set of labels $\mathcal{L}$ and construct a deterministic asynchronous automaton $\mathfrak{A}_T$ over (Σ, θ) in which, on any finite ideal I, each process p maintains $\lambda_p : secondary(\partial_p(I)) \to \mathcal{L}$, where λ_p is the restriction to $\partial_p(I)$ of a consistent labelling $\lambda : tertiary(I) \to \mathcal{L}$. Process p maintains λ_p as a function from $\mathcal{P} \times \mathcal{P}$ to $\mathcal{L}$. The value $\lambda_p(q, r)$ is the label assigned to the event $max_r(max_q(\partial_p(I)))$.*

The automaton $\mathfrak{A}_T$ allows each process to maintain reachable histories over the set $secondary(\partial_p(I))$—each history h is maintained as a partial function assigning joint states to labels in $\mathcal{L}$ such that whenever $\lambda(e) = \ell$ for some event $e \in secondary(\partial_p(I))$, $h(\ell)$ is defined and yields an e-state. In conjunction with Lemmas 2 and 3, this yields the following result.

Theorem 5 Subset construction [KMS1]. *Let $\mathfrak{A}$ be a non-deterministic asynchronous automaton over (Σ, θ). Then, we can construct a deterministic asynchronous automaton $\mathfrak{A}_S$ over (Σ, θ) such that for any finite ideal I, the unique global state $\vec{v}$ reached by $\mathfrak{A}_S$ on I has the following properties:*

(i) For each process p, $\vec{v}_p$ contains H_p, the set of all reachable histories over $secondary(\partial_p(I))$.

(ii) For any subset P of $\mathcal{P}$, we can compute from the information in the P-state $\vec{v}_P$ the set of all possible P-states of $\mathfrak{A}$ on I. In particular, from $\vec{v}$ we can recover all possible global states of $\mathfrak{A}$ on I.

6 Determinizing Büchi asynchronous automata

We now have enough machinery at hand to apply Safra's construction in a distributed setting. Recall that we are initially given a non-deterministic Büchi asynchronous automaton $\mathbf{B}\mathfrak{A} = (\mathfrak{A}, \mathcal{J}_B)$. Our goal is to construct a deterministic

Rabin asynchronous automaton $\mathbf{R}\mathfrak{B} = (\mathfrak{B}, \mathcal{T}_R)$ which accepts the same set of infinite strings that $\mathbf{B}\mathfrak{A}$ does.

As we remarked earlier, our strategy is to construct a separate deterministic Rabin automaton $\mathbf{R}\mathfrak{B}_\tau = (\mathfrak{B}_\tau, \mathcal{T}_{R_\tau})$ for each entry τ in the Büchi table $\mathcal{T}_B$ such that $\mathbf{R}\mathfrak{B}_\tau$ accepts an input α iff there is a run ρ of $\mathbf{B}\mathfrak{A}$ on α which satisfies τ. We shall then combine these individual automata $\{\mathbf{R}\mathfrak{B}_\tau\}_{\tau \in \mathcal{T}_B}$ into a single automaton $\mathbf{R}\mathfrak{B}$ which accepts the same inputs as $\mathbf{B}\mathfrak{A}$.

Let $\tau = (\mathcal{C}, T, \{(p_C, G_C)\}_{C \in \mathcal{C}})$ be an entry from $\mathcal{T}_B$. We first describe how to construct the corresponding Rabin automaton $\mathbf{R}\mathfrak{B}_\tau$. For simplicity, we assume that $T = \emptyset$—i.e., a run ρ of $\mathbf{B}\mathfrak{A}$ on an input α can satisfy τ only if $Finite_\alpha = \emptyset$. In other words, every process moves infinitely often as $\mathfrak{A}$ reads α. Later, we shall see how to eliminate this "progress" assumption.

The automaton $\mathbf{R}\mathfrak{B}_\tau$ has to check that there is a run ρ of $\mathfrak{A}$ on α such that along ρ, each signalling process p_C visits some recurring state from G_C infinitely often. Each process p_C can detect whether there is some local run ρ_C of $\mathfrak{A}$ on α which meets G_C infinitely often by running Safra's construction locally. However, we have to check that the individual runs $\{\rho_C\}_{C \in \mathcal{C}}$ are mutually consistent.

Let $\mathcal{I}_\alpha = \{I_C\}_{C \in Conn_\alpha}$ be the set of ideals such that $I_C = \partial_C(\mathcal{E})$ for each $C \in Conn_\alpha$. (Recall that each set C is a subset of $\mathcal{P}$, so the C-view of $\mathcal{E}$ is well-defined.) If there is a run of $\mathfrak{A}$ satisfying τ, it must be the case that $\mathcal{C} = Conn_\alpha$, so we can alternatively regard $\mathcal{I}_\alpha$ as the collection $\{\partial_C(\mathcal{E})\}_{C \in \mathcal{C}}$.

Let I_α^{joint} denote the set of events which occur in more than one ideal in the collection $\mathcal{I}_\alpha$—i.e., $I_\alpha^{joint} = \{e \in \mathcal{E} \mid \exists C, C' \in \mathcal{C}. \ C \neq C' \text{ and } e \in I_C \cap I_{C'}\}$. Since $\mathcal{C} = Conn_\alpha$, it must be the case that I_α^{joint} is finite—"above" I_α^{joint}, the ideals in $\mathcal{I}_\alpha$ are pairwise disjoint. Moreover, the union $\cup \mathcal{I}_\alpha$ is the entire set $\mathcal{E}$. So, if we can ensure that the local runs $\{r_C\}_{C \in \mathcal{C}}$ agree on the events in I_α^{joint}, they can be "pasted" together to form a global run ρ of $\mathfrak{A}$ satisfying τ.

Actually, it is not necessary that the local runs $\{r_C\}_{C \in \mathcal{C}}$ agree on the *entire* set I_α^{joint} in order to synthesize a global run ρ satisfying τ. It is sufficient for these local runs to agree along the frontiers of the ideals in $\mathcal{I}_\alpha$.

Lemma 6. *Let $\alpha : \mathbf{N} \to \Sigma$ be an infinite word. For $I_C \in \mathcal{I}_\alpha$, let $frontier(I_C, \mathcal{I}_\alpha)$ denote the set of events spanning the frontiers of I_C with respect to all the ideals in $\mathcal{I}_\alpha$—i.e., $frontier(I_C, \mathcal{I}_\alpha) = \bigcup_{C' \in \mathcal{C}} frontier(I_C, I_{C'})$.*

Let $\mathcal{R} = \{r_C\}_{C \in \mathcal{C}}$ be a set of local runs of $\mathfrak{A}$ on α such that:

(i) For $C \in \mathcal{C}$, r_C is a local run over I_C.

(ii) For each pair $C, C' \in \mathcal{C}$, the local runs r_C and $r_{C'}$ agree on $frontier(I_C, I_{C'})$.

Then, there is a local run r of $\mathfrak{A}$ over $\mathcal{E}$ which agrees with each run $r_C \in \mathcal{R}$ for all events $e \in I_C$ "above" $frontier(I_C, \mathcal{I}_\alpha)$. In other words, for each $C \in \mathcal{C}$, for each $e \in I_C$, if there exists $f \in frontier(I_C, \mathcal{I}_\alpha)$ such that $f \sqsubseteq e$, then $r(e) = r_C(e)$.

Proof For $C \in \mathcal{C}$, let h_C be the history generated by restricting r_C to the set $\{e \in I_C \mid \exists f \in frontier(I_C, \mathcal{I}_\alpha). f \sqsubseteq e\}$. It is easy to check that the histories in $\{h_C\}_{C \in \mathcal{C}}$ satisfy the assumptions of Lemma 2. So $h = \bigotimes_{C \in \mathcal{C}} h_C$ is a reachable history over $\cup \mathcal{I}_\alpha = \mathcal{E}$. Let r be the local run extending h to all of $\mathcal{E}$. $\quad\square$

So, if the local runs $\{r_C\}_{C \in \mathcal{C}}$ detected by the copies of Safra's construction agree along the frontiers in $\mathcal{I}_\alpha$, we can synthesize a local run r over $\mathcal{E}$ which agrees with each local run r_C outside I_α^{joint}. It is clear that the global run ρ of $\mathfrak{A}$ on α which corresponds to r does in fact satisfy τ. Of course, to check the conditions of the previous lemma, we have to verify that, in the limit, the local runs detected by each signalling process agree on the frontier events. In principle, this involves an infinite amount of computation. However, since there is only a finite amount of communication across the ideals in $\mathcal{I}_\alpha$, the frontier events of interest get "frozen" at some finite stage.

Lemma 7. *Let* $\alpha : \mathbf{N} \to \Sigma$ *be an infinite word. Let* J *be an ideal such that* $I_\alpha^{joint} \subseteq J \subseteq \mathcal{E}$. *Then, for each pair* $\{C, C'\}$ *in* $Conn_\alpha$, $frontier(\partial_C(J), \partial_{C'}(J)) = frontier(\partial_C(\mathcal{E}), \partial_{C'}(\mathcal{E}))$.

Proof Observe that for every J such that $I_\alpha^{joint} \subseteq J$, and for every pair $\{C, C'\}$ in $Conn_\alpha$, $\partial_C(J) \cap \partial_{C'}(J) = \partial_C(\mathcal{E}) \cap \partial_{C'}(\mathcal{E})$. The result then follows. $\square$

Let α be an infinite word and C, C' be components in $Conn_\alpha$. From Lemma 3, we know that the events in $frontier(\partial_C(\mathcal{E}), \partial_{C'}(\mathcal{E}))$ are contained in $secondary(\partial_C(\mathcal{E}))$ and $secondary(\partial_{C'}(\mathcal{E}))$.

Let p_C and $p_{C'}$ be processes in C and C' respectively. From the definition of $Conn_\alpha$, it follows that $\partial_C(\mathcal{E}) = \partial_{p_C}(\mathcal{E})$ and $\partial_{C'}(\mathcal{E}) = \partial_{p_{C'}}(\mathcal{E})$. So, $frontier(\partial_C(\mathcal{E}), \partial_{C'}(\mathcal{E}))$ is, in fact, contained in the secondary information of both $\partial_{p_C}(\mathcal{E})$ and $\partial_{p_{C'}}(\mathcal{E})$.

Let $e \in secondary(\partial_{p_C}(\mathcal{E}))$. From the definition of secondary information, $e = max_r(\partial_q(\partial_{p_C}(\mathcal{E})))$ for some pair of processes q and r. In other words, there are only finitely many r-events in $\partial_q(\partial_{p_C}(\mathcal{E}))$. There are two possibilities:

- The ideal $\partial_q(\partial_{p_C}(\mathcal{E}))$ is itself finite, in which case $q \in (\mathcal{P} - C)$.
- The ideal $\partial_q(\partial_{p_C}(\mathcal{E}))$ is infinite, but the number of r-events in $\partial_q(\partial_{p_C}(\mathcal{E}))$ is finite. This means that $q \in C$ but $r \in (\mathcal{P} - C)$.

This observation prompts the following definition:

Stable information Let $\alpha : \mathbf{N} \to \Sigma$ be an infinite word and let $p_C \in C$ for some connected component $C \in Conn_\alpha$. For any ideal I, the *stable information* of p_C in I, $stable\text{-}info_{p_C}(I)$ is the subset of $secondary(\partial_{p_C}(I))$ given by

$$\{max_r(\partial_q(\partial_{p_C}(I))) \mid q \notin C, r \in \mathcal{P}\} \cup \{max_r(\partial_q(\partial_{p_C}(I))) \mid q \in C, r \notin C\}.$$

The events in $stable\text{-}info_p(I)$ are frozen once I grows beyond the finite initial portion I_α^{joint} in $\mathcal{E}$. In other words, for any ideal $J \supseteq I_\alpha^{joint}$, $stable\text{-}info_{p_C}(J) = stable\text{-}info_{p_C}(\mathcal{E})$. By our earlier observations, this means that for any $J \supseteq I_\alpha^{joint}$, $stable\text{-}info_{p_C}(J)$ subsumes the events lying in the sets $\bigcup_{C' \in Conn_\alpha} frontier(\partial_C(\mathcal{E}), \partial_{C'}(\mathcal{E}))$.

Let us get back to our distributed version of Safra's construction corresponding to an entry $\tau = (C, T, \{(p_C, G_C)\}_{C \in \mathcal{C}})$ in $\mathcal{J}_B$. Suppose that each signalling process p_C ensures that it has crossed the finite portion $\mathcal{I}_\alpha^{joint}$ before starting

Safra's construction. Then, along with each successful run r_c on $\partial_C(\mathcal{E})$ that it detects, it can record the value of r_c on $stable\text{-}info_{p_c}(\mathcal{E})$. If the successful runs $\{r_c\}_{C \in \mathcal{C}}$ agree on the stable information across all the signalling processes, we know that the runs satisfy the assumption of Lemma 6, which means that there is some global run of $\mathfrak{A}$ on α which satisfies τ.

The catch is that the signalling processes have no way of knowing when the finite portion I_α^{joint} is over. However, since $\mathfrak{B}_\tau$ includes the subset automaton for $\mathfrak{A}$, $\mathfrak{B}_\tau$ also incorporates the time-stamping automaton $\mathfrak{A}_T$ which maintains consistent labels across $tertiary(I)$ at the end of any ideal I. If the time-stamps assigned by $\mathfrak{A}_T$ to the events in $stable\text{-}info_{p_c}(I)$ change, the process p_c knows that I_α^{joint} is *not* yet over.

So, we adopt the following strategy. Initially, each signalling process p_c starts off Safra's construction. Whenever it detects that $stable\text{-}info_{p_c}(I)$ has changed, it "kills" the old copy of Safra's construction and restarts a new copy. In fact, the process starts a separate copy of Safra's construction for each distinct history over $stable\text{-}info_{p_c}(I)$. So, in the limit, p_c can signal whether or not there is an accepting local run r_c for each history over its stable information.

The structure of $\mathbf{R}\mathfrak{B}_\tau$. Let $\tau = (\mathcal{C}, T, \{(p_c, G_c)\}_{C \in \mathcal{C}})$. The local state of each signalling process p_c in $\mathfrak{B}_\tau$ consists of the following information:

(i) The local state of the subset automaton for $\mathfrak{A}$. This includes the set H_{p_c} of all reachable histories over $secondary(\partial_{p_c}(I))$ at the end of any ideal I. This component incorporates the local state of the time-stamping automaton $\mathfrak{A}_T$, which stores the labels of events in $secondary(\partial_{p_c}(I))$ as a function $\lambda_{p_c} : \mathcal{P} \times \mathcal{P} \to \mathcal{L}$. The time-stamps assigned to the events in $stable\text{-}info_{p_c}(I)$ are the values $\lambda_p(q, r)$ where either $q \notin C$ or $(q \in C$ and $r \notin C)$.

(ii) Let H_S be the set of reachable histories over $stable\text{-}info_{p_c}(I)$. For each $h \in H_S$, p_c maintains an independent copy of Safra's construction.

The non-signalling processes need not run Safra's construction; it is sufficient for them to maintain the first component of the state.

On reading a letter a, each process p in $\theta(a)$ updates its local states as follows (where we have left out some important details due to a lack of space):

(i) First p updates the local state components corresponding to the time-stamping automaton and the subset automaton.

(ii) If p is a signalling process and if the time-stamps assigned to $stable\text{-}info_p(I)$ have not changed, then p updates each copy of Safra's construction using the new information provided by the subset automaton.

On the other hand, if the time-stamp corresponding to any event in $stable\text{-}info_p(I)$ changes, p erases all the existing copies of Safra's construction and begins a fresh copy for each history in the new set H_S.

The single entry τ in $\mathcal{T}_B$ generates a table $\mathcal{T}_{R_\tau}$ in $\mathbf{R}\mathfrak{B}_\tau$ with multiple entries. Each possible history h over $\bigcup_{C \in \mathcal{C}} stable\text{-}info_{p_c}(\mathcal{E})$ generates a distinct entry τ_h

of the form $(\mathcal{C}, T, \{(p_{\mathcal{C}}, pairs_{\mathcal{C}})\}_{\mathcal{C} \in \mathcal{C}})$ in $\mathcal{J}_{R_r}$. In τ_h, the entries $\mathcal{C}$, T and the set of signalling processes $\{p_{\mathcal{C}}\}_{\mathcal{C} \in \mathcal{C}}$ are as in the original entry $\tau \in \mathcal{J}_B$.

The sequential version of Safra's construction uses labels $\{\ell_1, \ell_2, \ldots, \ell_{2M}\}$, where M is the number of states of the original automaton. The acceptance condition of the determinized automaton consists of pairs $\{(G^j, R^j)\}_{j \in [1..2m]}$, where each pair (G^j, R^j) expresses some constraints on the label ℓ_j.

In the distributed construction, each signalling process uses labels in the same manner as in the sequential case and generates the same number of acceptance pairs. Let $M_C = |V_{p_C}|$ be the number of possible local states for each signalling process p_C in $\mathfrak{A}$. Then $pairs_C = \{(G_C^j, R_C^j)\}_{j \in [1..2M_C]}$, where, for $j \in [1..2M_C]$, the sets G_C^j and R_C^j consist of all possible states of p_C in which label ℓ_j meets the criteria required by Safra's construction and, in addition, the set of histories stored in the state includes the projection h_{p_C} of h onto $stable\text{-}info_{p_C}(\mathcal{E})$.

It can then be verified that $\mathbf{R}\mathfrak{B}_\tau$ accepts an input $\alpha : \mathbf{N} \to \Sigma$ iff there is a run of $\mathbf{B}\mathfrak{A}$ on α satisfying τ.

Removing the progress assumption So far we have assumed that $T = \emptyset$ in the Büchi table entry τ. Suppose $T \neq \emptyset$ and there is a run of $\mathfrak{A}$ on α which satisfies τ. Then each process $p \in T$ moves only finitely often while $\mathfrak{A}$ reads α. So, we just run the subset construction for p and verify that it terminates in one of the states in G_p.

Combining the individual automata $\{\mathbf{R}\mathfrak{B}_\tau\}_{\tau \in \mathcal{J}_B}$ We can combine the individual automata $\{\mathbf{R}\mathfrak{B}_\tau\}_{\tau \in \mathcal{J}_B}$ using a standard product construction which preserves determinacy. The construction is essentially the same as in the sequential case and we omit the details.

A complementation construction Following the technique proposed by Vardi (described in [Saf]), we can use a non-deterministic Büchi asynchronous to efficiently simulate the complement of a deterministic Rabin asynchronous automaton. So, from $\mathbf{R}\mathfrak{B}$ we can construct an Büchi automaton $\mathbf{B}\overline{\mathfrak{A}}$ such that $\mathbf{B}\overline{\mathfrak{A}}$ accepts an infinite string α iff $\mathbf{R}\mathfrak{B}$ does not accept α. Since $\mathbf{R}\mathfrak{B}$ accepts the same inputs that $\mathbf{B}\mathfrak{A}$ does, $\mathbf{B}\overline{\mathfrak{A}}$ is a complement automaton for $\mathbf{B}\mathfrak{A}$.

Complexity analysis In the input automaton $\mathbf{B}\mathfrak{A} = (\mathfrak{A}, \mathcal{J}_B)$, let N be the number of processes in $\mathfrak{A}$, M the size of the largest set in the collection $\{V_p\}_{p \in \mathcal{P}}$ and K the number of entries in $\mathcal{J}_B$.

Then, in the deterministic Rabin automaton $\mathbf{R}\mathfrak{B}$ which we construct, the number of local states of each process p is bounded by $2^{KM^{O(N^3)}}$, while in the complement automaton $\mathbf{B}\overline{\mathfrak{A}}$, the number of local states of each process p is bounded by $2^{K^2 M^{O(N^4)}}$ (see [KMS2] for details).

In [KMS1], it is shown that in the subset automaton for $\mathfrak{A}$, the number of states of each process p is bounded by $2^{M^{O(N^3)}}$. So, the blow-up involved in the construction of $\mathbf{R}\mathfrak{B}$ and $\mathbf{B}\overline{\mathfrak{A}}$ is essentially that of the subset construction.

Consolidating the results of this section, we have our main result.

Theorem 8. *Let* $\mathbf{B}\mathfrak{A} = (\mathfrak{A}, \mathcal{J}_B)$ *be a non-deterministic Büchi asynchronous automaton over* (Σ, θ). *Then, we can construct a deterministic Rabin asynchronous* automaton $\mathbf{R}\mathfrak{B} = (\mathfrak{B}, \mathfrak{I}_R)$ over (Σ, θ) such that $\mathbf{R}\mathfrak{B}$ accepts the same set of infinite strings that $\mathbf{B}\mathfrak{A}$ does. From $\mathbf{R}\mathfrak{B}$, we can construct a complementary non-deterministic Büchi automaton $\mathbf{B}\overline{\mathfrak{A}}$ over (Σ, θ) which accepts an infinite string α iff the original automaton $\mathbf{B}\overline{\mathfrak{A}}$ does not accept α.

The number of local states of each process in $\mathbf{R}\mathfrak{B}$ and $\mathbf{B}\overline{\mathfrak{A}}$ is essentially exponential in the number of global states of the original automaton $\mathbf{B}\mathfrak{A}$.

References

[DM]   V. DIEKERT, A. MUSCHOLL: Deterministic asynchronous automata for infinite traces, *Acta Inf.*, **31** (1994) 379–397.

[EM]   W. EBINGER, A. MUSCHOLL: Logical definability on infinite traces, *Proc. ICALP '93, LNCS* **700** (1993) 335–346.

[GP]   P. GASTIN, A. PETIT: Asynchronous cellular automata for infinite traces, *Proc. ICALP '92, LNCS* **623** (1992) 583–594.

[HJJ]   J.G. HENRIKSEN, J. JENSEN, M. JØRGENSEN, N. KLARLUND, B. PAIGE, T. RAUHE, A. SANDHOLM: Mona: Monadic Second-order logic in practice, *Report RS-95-21*, BRICS, Department of Computer Science, Aarhus University, Aarhus, Denmark (1995).

[Kla]   N. KLARLUND: Progress measures for complementation of ω-automata with applications to temporal logic, *Proc. 32nd IEEE FOCS*, (1991) 358–367.

[KMS1] N. KLARLUND, M. MUKUND, M. SOHONI: Determinizing asynchronous automata, *Proc. ICALP '94, LNCS* **820** (1994) 130–141.

[KMS2] N. KLARLUND, M. MUKUND, M. SOHONI: Determinizing asynchronous automata on infinite traces, *Report TCS-95-6*, School of Mathematics, SPIC Science Foundation, Madras (1995).

[Maz]   A. MAZURKIEWICZ: Basic notions of trace theory, in: J.W. de Bakker, W.-P. de Roever, G. Rozenberg (eds.), *Linear time, branching time and partial order in logics and models for concurrency, LNCS* **354**, (1989) 285–363.

[MS]   M. MUKUND, M. SOHONI: Keeping track of the latest gossip: Bounded time-stamps suffice, *Proc. FST&TCS '93, LNCS* **761** (1993) 388–399.

[Mus]   A. MUSCHOLL: On the complementation of Büchi asynchronous cellular automata, *Proc. ICALP '94, LNCS* **820** (1994) 142–153.

[Rab]   M.O. RABIN: Decidability of second order theories and automata on infinite trees, *Trans. AMS*, **141**(1969) 1–37.

[Saf]   S. SAFRA: On the complexity of ω-automata, *Proc. 29th IEEE FOCS*, (1988) 319–327.

[Thi]   P.S. THIAGARAJAN: TrPTL: A trace based extension of linear time temporal logic, *Proc. 9th IEEE LICS*, (1994) 438–447.

[Tho]   W. THOMAS: Automata on infinite objects, in J. van Leeuwen (ed.), *Handbook of Theoretical Computer Science, Volume B*, North-Holland, Amsterdam (1990) 133–191.

[VW]   M. VARDI, P. WOLPER: An automata theoretic approach to automatic program verification, *Proc. 1st IEEE LICS*, (1986) 332–345.

[Zie1] W. ZIELONKA: Notes on finite asynchronous automata, *R.A.I.R.O.—Inf. Théor. et Appl.*, **21** (1987) 99–135.

[Zie2] W. ZIELONKA: Safe executions of recognizable trace languages, in *Logic at Botik, LNCS* **363** (1989) 278–289.

Achilles and the Tortoise Climbing Up the Arithmetical Hierarchy

Eugene Asarin[1] Oded Maler[2]

[1] Institute for Information Transmission Problems, 19 Ermolovoy st., Moscow, Russia, asarin@ippi.msk.su

[2] SPECTRE – VERIMAG. Miniparc-ZIRST, 38330 Montbonnot, France, Oded.Maler@imag.fr

Abstract. In this paper we show how to construct for every set R of integers in the arithmetical hierarchy a dynamical system $\mathcal{H}$ with piecewise-constant derivatives (PCD) such that deciding membership in R can be reduced to solving the reachability problem between two rational points for $\mathcal{H}$. The ability of such simple dynamical systems to "simulate" highly undecidable problems is closely related to Zeno's paradox dealing with the ability to pack infinitely many discrete steps in a bounded interval of time.

1 Introduction

There has been recently an increasing interest in models of *hybrid systems*, i.e., systems that combine intercommunicating discrete and continuous components (see. for example. [GNRR93]. [ACH+95], [AMP95]). In [AMP95] we have introduced PCD systems, a sub-class of the so-called linear hybrid automata of [ACH+95]. Such systems consist of partitioning the Euclidean space into distinct convex polyhedra ("regions") such that the derivative within any region is constant. We have shown in [AMP95] that for two-dimensional PCD systems the reachability problem is decidable while the same problem becomes undecidable when we move to three dimensions or more.

In this paper we worsen our negative result by showing that by adding each time few dimensions we can climb up indefinitely in the arithmetical hierarchy. This means that such systems are, at least in the worst-case sense, "very very hard" to analyze. Our construction shows some interesting connections between dynamics and computation, and offers an alternative powerful geometrical model of computation.

The rest of the paper is organized as follows: in section 2 we introduce the arithmetical hierarchy. PCD systems and define recognition (acceptance) by such

---

* This research was supported in part by the European Community projects BRA-REACT(6021). HYBRID EC-US-043 and INTAS-94-697 as well as by Research Grant #93-012-884 of Russian Foundation of Fundamental Research. VERIMAG is a joint laboratory of CNRS. INPG. UJF and VERILOG SA. SPECTRE is a project of INRIA.

systems. In section 3 we present our previous results concerning the recognition of r.e. sets by 3-dimensional PCDs and draw the plan for the proof of the main result. In section 4 we prove a key lemma allowing to pass from a PCD $\mathcal{H}$ that semi-recognizes a set S to a PCD $\mathcal{H}'$ that fully recognizes S. Two additional lemmas and the final result are presented in section 5 followed by a discussion in section 6.

2 Preliminaries

PCD Systems

Definition 1 (PCD System). *A piecewise-constant derivative (PCD) system is a dynamical system $\mathcal{H} = (X, f)$ where $X = \mathbb{R}^d$ for some d is the state-space and f is a (possibly partial) function from X to X such that the range of f is a finite set of vectors $C \subset X$, and for every $\mathbf{c} \in C$, $f^{-1}(\mathbf{c})$ is a finite union of convex polyhedral sets and $\frac{d^+\mathbf{x}}{dt} = f(\mathbf{x})$ is the differential equation governing the evolution of $\mathbf{x}$. A trajectory of $\mathcal{H}$ starting at some $\mathbf{x}_0 \in X$ is $\xi : \mathbb{R}^+ \to X$ such that $\xi()$ is a solution of the equation with initial condition $\mathbf{x} = \mathbf{x}_0$, i.e., $\xi(0) = \mathbf{x}_0$ and for every t, $f(\xi(t))$ is defined and is equal to the right derivative of $\xi(t)$.*

In other words, a PCD system consists of partitioning the space into convex polyhedral sets ("regions"), and assigning a constant derivative $\mathbf{c}$ ("slope") to all the points sharing the same region. The trajectories of such systems are broken lines, with the breakpoints occurring on the boundaries of the regions.

A description of a PCD system is simply a list of the regions (expressed as intersections of linear inequalities) and their corresponding slope vectors. From now on we assume that all the constants in the system's definition are rational.

Given a description of a PCD system $\mathcal{H}$, the reachability problem for $\mathcal{H}$, denoted by **Reach**$(\mathcal{H}, \mathbf{x}, \mathbf{x}')$ is the following: *Given $\mathbf{x}, \mathbf{x}' \in X$, are there a trajectory ξ and $t \geq 0$ such that $\xi(0) = \mathbf{x}$ and $\xi(t) = \mathbf{x}'$?*

The Arithmetical Hierarchy We review here some classical definitions from recursion theory (see [Rog67]). The arithmetical hierarchy consists of the classes $\Sigma_1, \Sigma_2, \ldots$ and $\Pi_1, \Pi_2, \ldots$ of sets of integers defined inductively as follows: Σ_1 consists of all the sets $R \subseteq \mathbb{N}$ such that there exists a Turing machine $\mathcal{T}$ satisfying: $R = \{n : \mathcal{T}(n) = 1\}$ and $\overline{R} = \{n : \mathcal{T}(n) = \bot\}$ where $\mathcal{T}(n) = 1$ means that $\mathcal{T}$ halts on input n and $\mathcal{T}(n) = \bot$ means that $\mathcal{T}$ does not halt on n. The class Π_i consists of all the sets R such that $\overline{R} \in \Sigma_i$ and Σ_{i+1} is the class of all sets R defined as $R = \{n : \exists m \langle m, n \rangle \in R'\}$ for some $R' \in \Pi_i$, where $\langle \rangle$ is some computable pairing function.[3]

The arithmetical hierarchy is infinite and it satisfies the strict inclusions $\Pi_i \subset \Sigma_{i+1}$ and $\Sigma_i \subset \Pi_{i+1}$. The class Σ_1 is sometimes called the class of recursively enumerable (r.e.) sets. Due to the undecidablity of the halting problem for Turing

---

[3] A pairing function is an bijection from $\mathbb{N} \times \mathbb{N}$ to $\mathbb{N}$, a standard way to encode two-dimensions in one.

machines, membership of some $n \in I\!N$ is some $R \in \Sigma_1$ is only semi-decidable, i.e., there is an algorithm which is guaranteed to terminate if $n \in R$ but not otherwise.

Recognition by PCDs We will use PCD systems as recognizers of sets of integers in the following way:

Definition 2 (Recognition by PCDs). *Let $\mathcal{H} = (I\!R^d, f)$ be a PCD, let $I = [0, 1] \times \{0\}^{d-1}$ be a one-dimensional subset of $I\!R^d$ (the "input port"), let $r : I\!N \to [0, 1] \cap Q\!\!\!\!Q$ be an injective coding function, let $\mathbf{x}^1, \mathbf{x}^0 \in I\!R^d - I$ be two distinct points. The system $\widehat{\mathcal{H}} = (I\!R^d, f, r, I, \mathbf{x}^1, \mathbf{x}^0)$ semi-recognizes the set $R \subseteq I\!N$ iff for every n. the trajectory starting at $(r(n), 0, \ldots, 0)$ can continue forever[4] and that it eventually reaches $\mathbf{x}^1$ iff $n \in R$. We say that $\mathcal{H}$ (fully) recognizes R when, in addition, this trajectory reaches $\mathbf{x}^0$ iff $n \notin R$. We assume that the derivatives at $\mathbf{x}^1$ and $\mathbf{x}^0$ are zero.*

In other words, every integer n is encoded into a distinct rational point on the input port, and the membership of n in R is indicated by whether the trajectory starting at this point is settles in an accepting (rejecting) point after a finite amount of time.

Remark: This notion of recognition is not much different from recognition (acceptance) of sets by Turing machines. A TM is nothing but a discrete dynamical system whose state-space is the set of all its configurations (state, tape, head). This system accepts an input x if the trajectory starting at a configuration where x is encoded on the tape eventually reaches the halting state.

3 PCDs Realize TMs

In this section we present our previous result ([AM94], [AMP95]) on realizing Turing machines by PCD systems. A PCD $\mathcal{H} = (X, f)$ *realizes* a discrete transition systems $\mathcal{A} = (Q, \delta)$, where Q is a countable set of states and $\delta : Q \to Q$ is a transition function, if there exists an injective and surjective partial function $\pi : X \to Q$ such that $\delta(q) = q'$ iff there is a trajectory between $\pi^{-1}(q)$ and $\pi^{-1}(q')$ that does not intersect the domain of π between these two points.

The simulation result is based on the equivalence of TMs and two-stack machines. A stack is an element of Γ^* where $\Gamma = \{0, 1\}$. We define the following two functions: PUSH: $\Gamma \times \Gamma^* \to \Gamma^*$ and POP: $\Gamma^* \to \Gamma \times \Gamma^*$ as PUSH$(v, S) = v \cdot S$ and POP$(v \cdot S) = (v, S)$.

Definition 3 (2PDA). *A deterministic two-stack pushdown-automaton (2PDA) is a transition system $\mathcal{A} = (Q \times \Gamma^* \times \Gamma^*, \delta)$ for some $Q = \{q_1, \ldots, q_n\}$ such that δ is defined using a finite collection of statements of one of the following two forms:*

---

[4] Which means that it always stays the domain of definition of f, and that it can continue from every point.

$$q_i\colon S_\alpha :=\text{PUSH}(v, S_\alpha);$$
$$\text{GOTO } q_j$$

$$q_i\colon (v, S_\alpha) :=\text{POP}(S_\alpha);$$
$$\text{IF } v = 0 \text{ GOTO } q_{i_0};$$
$$\text{IF } v = 1 \text{ GOTO } q_{i_1};$$

where $\alpha \in \{1, 2\}$.

The contents of a stack is denoted by $S = s_1 s_2 \ldots$ where s_1 is the top of the stack. We define an encoding function $\rho : \Gamma^* \to [0, 1]$ as $\rho(S) = \sum_{i=1}^{|S|} s_i 2^{-i}$. It is easily verified that the stack operations have arithmetic counterparts that operate on the representation:

$$S' = \text{PUSH}(v, S) \text{ iff } \rho(S') = (\rho(S) + v)/2$$
$$(S', v) = \text{POP}(S) \text{ iff } \rho(S') = 2\rho(S) - v$$

Remark: We present here a simplified version of the construction in [AMP95], omitting some tedious details concerning the encoding of rational numbers using bottom-less stacks.

Theorem 1 (Realization of 2PDAs). *Every 2PDA can be realized by a 3-dimensional PCD system.*

Sketch of Proof: We show first how a one-stack PDA are realized. Consider the three planar sub-systems depicted in figure 1 and a trajectory segment starting at $\mathbf{x} = (x, 0)$, $x \in [0, 1]$ and ending at $\mathbf{x}' = (x', 1)$. It can be verified that either:

$$x' = (x + 1)/2 \quad \text{PUSH 1}$$
$$x' = x/2 \quad \text{PUSH 0}$$
$$x' = 2x - 1/2 \quad \text{POP}$$

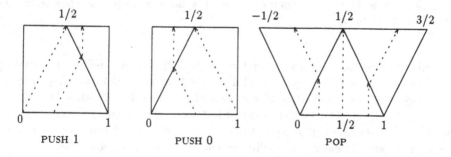

Fig. 1. The basic elements.

If $x = \rho(S)$ at the "input port" $(y = 0)$ of a PUSH element, then $x' = \rho(S')$ at the "output port" $(y = 1)$ of that element where S' is the resulting stack. For the POP element we have two output ports $-1/2 \leq x < 1/2$ and $1/2 \leq x < 3/2$. If the top of the stack was 0 the trajectory reaches the left port with $x' = \rho(S') - 1/2$, otherwise it goes to the right port with $x' = \rho(S') + 1/2$. In both cases the value of x' (relative to the "origin" of the port) encodes the new content of the stack. Thus, in order to simulate a PDA we pick for every q_i an element corresponding to its stack operation, place it with the origin in position, say, $(2i, 0, 0)$ and use the third dimension in order to connect the output ports back to the input ports according to the GOTO's (see figure 2). Finally the state-mapping is defined as $\pi(x, y, z) = (q_i, S)$ iff $y = z = 0$, $2i \le x < 2i + 1$ and $\rho(S) = x - 2i$.

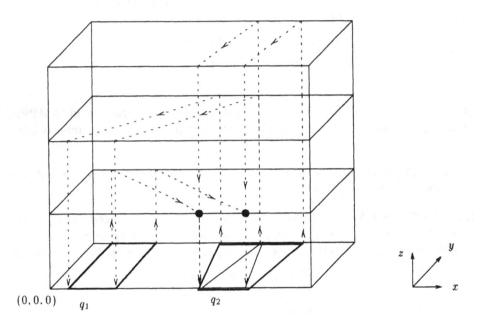

$(0, 0, 0)$ q_1 q_2

Fig. 2. Simulating a PDA with 2 states, defined by: q_1 : S :=PUSH$(1, S)$; GOTO q_2; q_2 : (v, S) :=POP(S); If $v = 1$ THEN GOTO q_2 ELSE GOTO q_1. Note the place where the two GOTOs to q_1 merge.

This construction generalizes naturally to 2PDAs. We define an encoding function $\bar{\rho} : \Gamma^* \times \Gamma^* \to [0, 1] \times [0, 1]$ by letting $\bar{\rho}(S_1, S_2) = (\rho(S_1), \rho(S_2))$. This way every configuration of the two stacks can be encoded by a point $\mathbf{x} = (x_1, x_2, 0)$ in a two-dimensional input port. The elements that simulate the stack operations PUSH(v, S_1), PUSH(v, S_2), POP(S_1) and POP(S_2) operate on the appropriate dimension, according to the stack involved, and leave the other dimension intact. As an example, an element corresponding to PUSH$(0, S_1)$ appears in figure 3. From this we can immediately conclude that a 2PDA can be realized by a 4-dimensional PCD. In [AMP95] we introduce additional tricks to avoid the fourth dimension, but this paper deals with the infinite and small constants do not matter. □

Corollary 2 (PCD and Σ_1). *Every Σ_1 set R is semi-recognized by some PCD.*

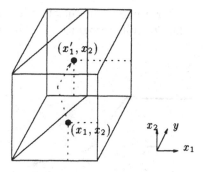

Fig. 3. An element simulating the operation PUSH$(0, S_1)$

Proof: We take the 2PDA associated with R and assume w.l.o.g. that it always halts (if it halts) at a given configuration (q, S_1, S_2). By constructing $\mathcal{H}$ as in the proof of theorem 1, encoding $I\!N$ into some input port and letting $F = \{\pi(q, S_1, S_2)\}$ we obtain a semi-recognizing PCD. □

Corollary 3 (PCD and Recursive Functions). *Every recursive function φ : $I\!N \to I\!N$ can be computed[5] by a 3-dimensional PCD.*

Proof: We take the 2PDA that computes φ where the inputs are written each on one stack and the output is written on S_1 and use theorem 1. □

Without loss of generality we can assume that our PCDs always work in a bounded subset of $(I\!R^+)^d$ sufficiently far from the origin. Corollary 2 gives us the basis for climbing up the hierarchy. In order to continue we need the the following lemmata ordered according to their decreasing difficulty:

1. From a PCD that semi-recognizes R one can construct a PCD that recognizes R.
2. From a PCD that recognizes R we can construct
   (a) a PCD that semi-recognizes $\{x : \exists y \langle x, y \rangle \in R\}$, and
   (b) a PCD that recognizes $\overline{R}$.

4 From Semi-recognition to Recognition

The intuitive idea is the following. Suppose $\mathcal{H}$ semi-recognizes R. The trajectory corresponding to some $n \notin R$ is wandering forever in $I\!R^d$ without reaching F. We will create a higher dimensional PCD $\mathcal{H}'$ such that for a unit interval $\mathcal{H}'$ "mimics" $\mathcal{H}$ (in the projection), then it goes to some other regions, and comes back after having divided all the variables by 2. Then it mimics $\mathcal{H}$ again on a smaller scale for a temporal interval of length $1/2$, divides the variables by 2 and so on. Clearly every diverging trajectory that does not reach F in $\mathcal{H}$ will reach

---

[5] Computing a function by a PCD is a natural extension of deciding membership in a set. You just introduce an output port and use r^{-1} to decode the result.

the origin $(0, \ldots, 0)$ in $\mathcal{H}'$ after a finite amount of time (and an infinite number of region switchings) and thus $\mathcal{H}'$ fully recognizes R. This is sketched in figure 4.

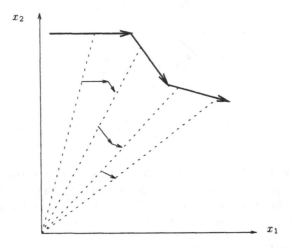

Fig. 4. A 2-dimensional trajectory in $\mathcal{H}$ (bold line) and the projection of its "Zenofied" corresponding trajectory in $\mathcal{H}'$.

In order to gain some intuition for the style of "PCD programming" let us demonstrate how a PCD divides a variable x by 2. For this we need another variable y initialized to 0. The system in figure 5, whenever started at some point $(x, 0)$ completes one cycle and returns to $(x/2, 0)$. The time to complete such a cycle is $2.5x$. This system consists of 4 regions. In region A we have $(\dot{x}, \dot{y}) = (-1, 1/2)$ until the point $(0, x/2)$. Then in B, C and D we have respectively $(\dot{x}, \dot{y}) = (-1, -1)$, $(\dot{x}, \dot{y}) = (1, -1)$ and $(\dot{x}, \dot{y}) = (1, 1)$. This can be generalized to d variables using $4d$ regions and $d + 1$ dimensions, as we will see later.

Note that if we make k cycles we arrive at $(2^{-k}x, 0)$ within $2.5x \sum_{i=0}^{k-1} 2^{-i}$ time. If we let $f(0, 0) = (0, 0)$ the trajectory starting at $(x, 0)$, spiralling infinitely many times in A, B, C, D during the temporal interval $[0, 5x)$ and staying in $(0, 0)$ in the interval $[5x, \infty)$ is indeed a valid trajectory of the system according to definition 1, i.e., a solution to the initial value problem of the differential equation.

The last construction we need before we prove the main lemma is the *homogenization* of a PCD (and a dynamical system in general). Let $\mathcal{H} = (\mathbb{R}^d, f)$ be a PCD. Its homogenization is a system $\mathcal{H}^0 = (\mathbb{R}^{d+1}, f^0)$ such that for every $\mathbf{x}, \mathbf{x}' \in \mathbb{R}^d$, if $\mathbf{x}'$ is $\mathcal{H}$-reachable from $\mathbf{x}$ in time t then for every $x_{d+1}, 0 \leq x_{d+1} \leq 1$ $(x_{d+1}\mathbf{x}', x_{d+1})$ is[6] $\mathcal{H}^0$-reachable from $(x_{d+1}\mathbf{x}, x_{d+1})$ in time $x_{d+1}t$. Geometrically $\mathcal{H}^0$ is obtained from $\mathcal{H}$ by choosing a point at the origin and drawing replacing

---

[6] We slightly abuse notations: for $\mathbf{x} = (x_1, \ldots, x_d)$ we use $(\mathbf{x}, x_{d+1})$ to denote $(x_1, \ldots, x_d, x_{d+1})$.

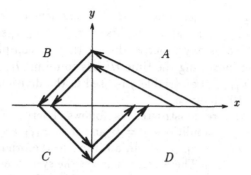

Fig. 5. A 2-dimensional PCD for dividing a 1-dimensional quantity.

every region by a "pyramid" (see figure 6). Syntactically all you do is replace every inequality of the form $\mathbf{a} \cdot \mathbf{x} \leq b$ in $\mathcal{H}$ by $(\mathbf{a}, 0) \cdot (\mathbf{x}, x_{d+1}) \leq b x_{d+1}$ and add the conjunct $0 \leq x \leq 1$ to every definition of a region. Finally every slope $\mathbf{c}$ is replaced by $(\mathbf{c}, 0)$. The importance of this procedure (which can be applied to any PCD) is that each time after division by 2 the homogenized system keeps on behaving like the original one.

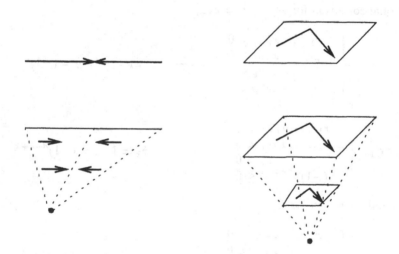

Fig. 6. A homogenization of a 1-dimensional system (left) and a 2-dimensional system (right).

Lemma 4 (Semi-Recognition $\Rightarrow$ Recognition). *From a PCD* $\mathcal{H} = (\mathbb{R}^d, f, I, r, \mathbf{x}^1, \mathbf{x}^0)$ *semi-recognizing R one can construct a PCD* $\mathcal{H}' = (\mathbb{R}^{d+3}, f', I', r', \mathbf{x}'^1, \mathbf{x}'^0)$ *that recognizes R.*

Sketch of Proof: We assume that the original variables $\{x_1, \ldots, x_d\}$ of $\mathcal{H}$ remain positive along the trajectories of $\mathcal{H}$. We augment the system with three additional variables x_{d+1}, h and y. The first variable, x_{d+1}, serves for homogenization and is treated as any other x_i during the division phase. The second, h, is used as a timer indicating the time $\mathcal{H}'$ has to mimic $\mathcal{H}$. During this phase it grows from 0 to x_{d+1}. The variable y is used in the division sub-system as in the system described in figure 5.

The regions of $\mathcal{H}'$ are constructed as follows. Every original region of $\mathcal{H}$ is homogenized and the conditions $y = 0$ and $h < x_{d+1}$ are added. All these regions also satisfy $x_1, \ldots, x_{d+1} > 0$. In addition to the original derivatives we have $(\dot{x}_{d+1}, \dot{h}, \dot{y}) = (0, 0, 1)$. Therefore, whenever the system enters such a region with $x_{d+1} = c$ it emulates the original system for c time (see the first phase, denoted by $\mathcal{H}$, in the signal diagram of figure 7). Then we define the following families of regions for every k, $1 \leq k \leq d+1$ (only non-zero derivatives are written down):

$$A_k : \begin{cases} x_1, \ldots, x_{k-1} < 0 \\ x_{k+1}, \ldots, x_{d+1} > 0 \\ x_k > 0 \\ (-1)^k y \leq 0 \end{cases} \quad \longrightarrow \quad \dot{x}_k = -1, \ \dot{y} = (-1)^{k+1}/2$$

Additional condition for A_1 : $\quad h = x_{d+1}$

$$B_k : \begin{cases} x_1, \ldots, x_{k-1} < 0 \\ x_{k+1}, \ldots, x_{d+1} > 0 \\ x_k \leq 0 \\ (-1)^k y < 0 \end{cases} \quad \longrightarrow \quad \dot{x}_k = -1, \ \dot{y} = (-1)^k$$

$$C_k : \begin{cases} x_1, \ldots, x_{k-1} > 0 \\ x_{k+1}, \ldots, x_{d+1} < 0 \\ x_k < 0 \\ (-1)^{d+k+1} y \leq 0 \end{cases} \quad \longrightarrow \quad \dot{x}_k = 1, \ \dot{y} = (-1)^{d+k+1}$$

Additional condition for C_1 : $\quad h = 0$

$$D_k : \begin{cases} x_1, \ldots, x_{k-1} > 0 \\ x_{k+1}, \ldots, x_{d+1} < 0 \\ x_k \geq 0 \\ (-1)^{d+k+1} y > 0 \end{cases} \quad \longrightarrow \quad \dot{x}_k = 1, \ \dot{y} = (-1)^{d+k}$$

In addition we add a special region Z (for resetting h to 0):

$$Z : \begin{cases} x_1, \ldots, x_{d+1} < 0 \\ y = 0 \\ h > 0 \end{cases} \quad \longrightarrow \quad \dot{h} = -1$$

For every k, the passage through the sequence of regions A_k, B_k, C_k, D_k will result in dividing x_k by 2. However we do first $A_1, B_1, A_2, B_2, \ldots, A_{d+1}, B_{d+1}$

making all variables negative, thus we enter Z, reset h to zero and then complete $C_1, D_1, C_2, D_2, \ldots, C_{d+1}, D_{d+1}$. The reader should verify the example in figure 7 for $d = 1$, where (x_1, x_2) start the division phase at $(1,1)$ and terminate it at $(1/2, 1/2)$, ready to simulate the (scaled-down) behavior of $\mathcal{H}$, now for $1/2$ time interval. Clearly the trajectory of $\mathcal{H}'$ converges in finite time to $(0, \ldots, 0)$ which we can consider as the rejecting point $\mathbf{x}'^0$. We should take care of not treating accepting trajectories (those that reach $\mathbf{x}^1$ in $\mathcal{H}$) this way. This is done by adding the condition $\mathbf{x} \neq \mathbf{x}^1 x_{d+1}$ to each of the regions defined above. Then when $\mathbf{x} = \mathbf{x}^1 x_{d+1}$ we have two additional regions: if $h > 0$ we just lower h to zero. When $h = 0$ we let $\dot{\mathbf{x}} = \mathbf{x}^1$ and $\dot{x}_{d+1} = 1$ until $x_{d+1} = 1$. This way we reach the point $(\mathbf{x}^1, 1, 0, 0)$ which is the new accepting point $\mathbf{x}'^1$. □

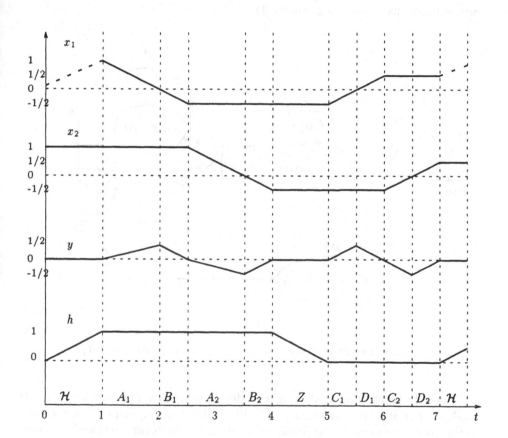

Fig. 7. An initial part of the behavior of $\mathcal{H}'$ when $d = 1$. The regions through which the system passes are written below ($\mathcal{H}$ denotes any region in which $\mathcal{H}'$ emulates $\mathcal{H}$).

5 Quantifier Elimination

Lemma 5 (Quantifier Elimination). *Let $\mathcal{H}$ be a PCD in $\mathbb{R}^d$ that recognizes a set R. Then one can construct a PCD $\widetilde{\mathcal{H}}$ in $\mathbb{R}^{d+2}$ that semi-recognizes the set $\widetilde{R} = \{n : \exists m \langle n, m \rangle \in R\}$.*

Sketch of Proof: The idea of the proof is standard: given an n we just test one after the other all the possible values of m and use the PCD $\mathcal{H}$ with inputs $\langle n, m \rangle$ to verify whether these inputs belong to R. For any input $n \in \widetilde{R}$ we will eventually find a good m while for $n \notin \widetilde{R}$ the process will continue forever.

In order to avoid an unreadable collection of linear inequalities we will describe the PCD $\widetilde{\mathcal{H}}$ more schematically – see figure 8. Bold line segments and squares in the figure stand for one- and two-dimensional ports respectively; arrows denote connections. Ellipses stand for PCDs that compute various recursive integer functions (based on corollary 3).

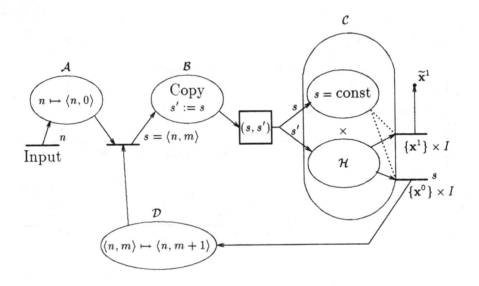

Fig. 8. The PCD $\widetilde{\mathcal{H}}$ semi-recognizing $\widetilde{R} = \{n : \exists m \langle n, m \rangle \in R\}$

The only block that needs a special description is $\mathcal{C}$. This block is a direct product of $\mathcal{H}$ and the line segment $I = [0, 1]$. When an input (s, s') is provided to this block the variable s is preserved unchanged (for future use) and s' is used as an input for a copy of $\mathcal{H}$. So the trajectory exits $\mathcal{C}$ either at $\{\mathbf{x}^1\} \times I$ (if $\mathcal{H}$ accepts s') or at $\{\mathbf{x}^0\} \times I$ otherwise.

Let us describe the trajectory entering $\widetilde{\mathcal{H}}$ through the input port at a point $r(n)$. First $s = \langle n, 0 \rangle$ is calculated by $\mathcal{A}$. The block $\mathcal{B}$ creates another copy of s denoted by s' (that is, the trajectory exits $\mathcal{B}$ through its two-dimensional output port at the point $(r(s), r(s)))$. This copy is used as input for the original PCD $\mathcal{H}$ in block $\mathcal{C}$. Meanwhile s is preserved for further use. If $\langle n, 0 \rangle \in R$, the trajectory exits $\mathcal{C}$ at $\{\mathbf{x}^1\} \times I$ and then goes to the new accepting point $\tilde{\mathbf{x}}^1$. In the case when $\langle n, 0 \rangle \notin R$ further search is necessary and the trajectory goes from $\{\mathbf{x}^0\} \times I$ to the block $\mathcal{D}$ which transforms $\langle n, 0 \rangle$ to $\langle n, 1 \rangle$. This last value is used at the next iteration of the loop ad infinitum. Recall that all these blocks can be simply realized according to corollary 3.

If $n \in \tilde{R}$ then an m satisfying $\langle n, m \rangle \in R$ exists and the mth iteration of the main loop the PCD $\tilde{\mathcal{H}}$ will stop in the accepting point $\tilde{\mathbf{x}}^1$. Otherwise $\tilde{\mathcal{H}}$ will check all the natural m's in turn and will never halt. Hence $\tilde{\mathcal{H}}$ does semi-recognize the set $\tilde{R}$. The system $\tilde{\mathcal{H}}$ fits in $d + 2$ dimensions. In fact its largest block is $\mathcal{C}$ which uses d dimension for $\mathcal{H}$, one dimension for s and one more dimension for merging in- and outgoing connections. $\qquad\Box$

Lemma 6 (Complementation). *From a PCD that recognizes R one can construct a PCD that recognized $\mathbb{N} - R$.*

Proof: Exchange $\mathbf{x}^1$ and $\mathbf{x}^0$. $\qquad\Box$
From this we conclude:

Theorem 7 (Main Result). *Every set R in the arithmetical hierarchy can be recognized by a PCD system of finite dimension and a finite number of regions.*

The number of dimensions required to recognize $R \in \Sigma_i \cup \Pi_i$ is $5i + 1$.

6 Discussion

What is the significance of this result? On the one hand we have a rather simple class of dynamical systems which are "locally effective" in the following sense: Given a description of the system, and a rational initial point $\mathbf{x}$, there exists some positive $\epsilon > 0$ such that, for every Δt, $0 < \Delta t < \epsilon$, one can calculate *precisely* the point $\mathbf{x}'$ which a trajectory starting at $\mathbf{x}$ will reach after time Δt. This is unlike more general dynamical systems where one can only approximate trajectories. On the other hand these systems give rise to highly undecidable reachability problems. The reasons for this "expressiveness excess" of the model should worry researchers in hybrid systems and urge them to find ways to tackle these problems, either by restricting the models or by changing the questions. One common solution is to exclude Zeno trajectories from the semantics of the system which brings back the reachability problem into the level Σ_1 of simple undecidability: you just need to simulate the system forward (as described in [AMP95] for the two-dimensional case) and see whether the target point is reached within a finite number of discontinuities. Other approaches, such as "stable" (in the sense of insensitivity to small perturbations) realizations of transition systems by PCDs are currently investigated.

Beside the negative results, PCDs suggest an interesting model of computation which could theoretically[7] decide every statement in first-order arithmetics, i.e., solve every open problem in Number theory. This model which is more geometrical and topological in nature, may bring new insights on computability and synchronization and promote a new style of parallel programming. The art of PCD programming is to ensure that regions do not overlap except for the boundaries and that the derivatives takes you were you want, usually using other variables as timers (or loop delimiters). We have shown, for example, how to implement parallel sorting of n numbers by PCDs using $3n/2 + 1$ dimensions in linear time.

Acknowledgement The question answered in this paper was posed to us by Philippe Darondeau.

References

[ACH+95] R. Alur, C. Courcoubetis, N. Halbwachs, T.A. Henzinger, P.-H. Ho, X. Nicollin, A. Olivero, J. Sifakis and S. Yovine, The Algorithmic Analysis of Hybrid Systems, *Theoretical Computer Science* 138, 3–34, 1995.

[AM94] A. Asarin and O. Maler, On some Relations between Dynamical Systems and Transition Systems, in S. Abiteboul and E. Shamir, editors, *Proc. of ICALP'94*, Lect. Notes in Comp. Sci. 820, pages 59–72, Springer-Verlag, 1994.

[AMP95] A. Asarin, O. Maler and A. Pnueli, Reachability Analysis of Dynamical Systems having Piecewise-Constant Derivatives, *Theoretical Computer Science* 138, 35-66, 1995.

[GNRR93] R.L Grossman, A. Nerode, A. Ravn and H. Rischel (editors), *Hybrid Systems*, Lect. Notes in Comp. Sci. 736, Springer-Verlag, 1993.

[Rog67] H. Rogers, *Theory of Recursive Functions and Effective Computability*, McGraw-Hill, 1967.

---

[7] If we ignore physical limitations concrening the granularity of measurements.

Generalized Temporal Verification Diagrams*

I.A. Browne, Z. Manna, H.B. Sipma

Computer Science Department
Stanford University, Stanford, CA 94305
e-mail: {anca,manna,sipma}@cs.stanford.edu

Abstract. Verification diagrams are a succinct and intuitive way of representing proofs that reactive systems satisfy a given temporal property. We present a generalized verification diagram that allows representation of a proof of any property expressible by a temporal formula. We show that representation of a proof by generalized verification diagram is sound and complete.

1 Introduction

Verification diagrams are a succinct and intuitive way of representing proofs that reactive systems satisfy a given temporal property. Verification diagrams were first introduced for this purpose in [MP94] (see also [MP95]). In that paper, diagrams dedicated to particular classes of properties were presented, e.g., WAIT-FOR diagrams for precedence properties, CHAIN and RANKED diagrams for response properties. In this paper we present generalized verification diagrams, which allow the representation of a proof of any property expressible by a temporal formula.

The method of proof by verification diagram, called here *diagram verification*, is based on the representation of reactive systems by fair transition systems [MP91], summarized in Section 2, and on the representation of the specification by a temporal formula or a formula automaton (ω-automaton), explained in Section 3. A verification diagram, described in Section 4, is a formula automaton with additional components. The languages accepted by both formula automata and verification diagrams are sets of infinite sequences of states.

For a given reactive system P and temporal formula φ, diagram verification is accomplished by constructing a verification diagram (VD) Ψ and showing that Ψ faithfully represents all computations of the corresponding fair transition system (FTS) Φ and that it "satisfies" φ. To show that a VD Ψ satisfies a temporal formula φ we must show that the set of sequences accepted by Ψ, denoted by $\mathcal{L}(\Psi)$, is a subset of the set of sequences satisfying φ, denoted by $\mathcal{L}(\varphi)$, i.e., we must show

$$\mathcal{L}(\Psi) \subseteq \mathcal{L}(\varphi).$$

---

* This research was supported in part by the National Science Foundation under grant CCR-92-23226, by the Defense Advanced Research Projects Agency under NASA grant NAG2-892, by the United States Air Force Office of Scientific Research under grant F49620-93-1-0139, and by Department of the Army under grant DAAH04-95-1-0317.

A set of first-order verification conditions associated with the VD Ψ establishes that the set of computations of Φ, denoted by $\mathcal{L}(\Phi)$, is a subset of $\mathcal{L}(\Psi)$, i.e.,

$$\mathcal{L}(\Phi) \subseteq \mathcal{L}(\Psi).$$

Diagram verification consists of establishing the validity of a set of assertions (first-order formulas), and checking decidable graph-theoretical problems.

In Section 5 we show that if an FTS Φ satisfies a formula φ then there always exists a VD Ψ such that

$$\mathcal{L}(\Phi) \subseteq \mathcal{L}(\Psi) \subseteq \mathcal{L}(\varphi),$$

i.e., diagram verification is shown to be complete.

Throughout the paper the concepts are illustrated with a simple example.

In this paper, we present the results and proofs using temporal formulas as our specification language. The same results hold also for the larger class of properties that are specifiable by formula automata. However there is a difference in the complexity of the verification process.

2 Computational Model: Fair Transition Systems

The computational model used for reactive systems is that of a *fair transition system* (FTS), $\Phi = \langle V, \Theta, \mathcal{T}, \mathcal{J}, \mathcal{C} \rangle$, where V is a finite set of variables, Θ is an initial assertion, $\mathcal{T}$ is a finite set of transitions, $\mathcal{J} \subseteq \mathcal{T}$ contains the just (weakly fair) transitions and $\mathcal{C} \subseteq \mathcal{T}$ contains the compassionate (strongly fair) transitions. A *state* s is an interpretation of V, and Σ denotes the set of all states. A transition $\tau \in \mathcal{T}$ is a function $\tau : \Sigma \mapsto 2^{\Sigma}$, and each state in $\tau(s)$ is called a τ-successor of s. Each transition τ is represented by a transition relation $\rho_\tau(s, s')$, an assertion that expresses the relation between the values of V in s and the values of V (referred to by V') in any of its τ-successors s'. The enabledness of a transition τ is expressed by $En(\tau) : \exists s' . \rho_\tau(s, s')$.

A run of Φ is an infinite sequence of states such that the first state satisfies Θ and any two consecutive states satisfy a ρ_τ for some $\tau \in \mathcal{T}$. A computation of Φ is a run of Φ with the additional property that for each $\tau \in \mathcal{J}$ ($\tau \in \mathcal{C}$), it is not the case that τ is continually enabled beyond some point (infinitely many times enabled) but taken only finitely many times. The set of all runs of Φ is denoted by $\mathcal{L}_R(\Phi)$ and the set of all computations of Φ is denoted by $\mathcal{L}(\Phi)$.

Example Consider the program MTX (MAY-TERMINATE-X) shown in Figure 1. The **idle** statement does not modify any data variables so its only observable effect is when it terminates. It is not required to terminate however, which is represented in the transition system by excluding the transition associated with the **idle** statement from the justice and compassion set. The **request** and **release** statements are usually associated with a semaphore variable. Execution of **request** y decrements y by 1; it can only be executed if $y > 0$. Execution of **release** y increments y by 1.

The FTS Φ associated with this program is shown in Figure 2. The boolean variables ℓ_i and m_i indicate if control currently resides at the corresponding program location. The formula $pres(U)$, with $U \subseteq V$, states that the variables in U are not modified by the transition, i.e., $pres(U) : \bigwedge_{u \in U}(u' = u)$. ∎

$$\boxed{\begin{array}{c} \textbf{local} \quad x : \text{integer where } x \geq 0 \\ \textbf{local} \quad y : \text{integer where } y = 1 \end{array}}$$

$$P_1 :: \begin{bmatrix} \ell_0: \textbf{loop forever do} \\ \begin{bmatrix} \ell_1: \textbf{idle} \\ \ell_2: \textbf{request } y \\ \ell_3: \textbf{idle} \\ \ell_4: \textbf{release } y \end{bmatrix} \end{bmatrix} \quad \| \quad P_2 :: \begin{bmatrix} m_0: \textbf{while } x > 0 \textbf{ do} \\ \begin{bmatrix} m_1: \textbf{request } y \\ m_2: \textbf{release } y \\ m_3: x := x - 1 \end{bmatrix} \\ m_4: \end{bmatrix}$$

Fig. 1. Program MTX

$$
\begin{aligned}
V &= \{\ell_0, \ell_1, \ell_2, \ell_3, \ell_4\} \cup \{m_0, m_1, m_2, m_3, m_4\} \cup \{x, y\} \\
\Theta &= \ell_0 \wedge m_0 \wedge x \geq 0 \wedge y = 1 \\
\mathcal{T} &= \{\tau_{\ell_i} \mid i \in [0..4]\} \cup \{\tau_{m_i} \mid i \in [0..4]\} \cup \{\tau_I\} \text{ where} \\
\rho_{\tau_{\ell_0}} &= \ell_0 \wedge \neg \ell_0' \wedge \ell_1' \wedge pres(V - \{\ell_0, \ell_1\}) \\
\rho_{\tau_{\ell_1}} &= \ell_1 \wedge \neg \ell_1' \wedge \ell_2' \wedge pres(V - \{\ell_1, \ell_2\}) \\
\rho_{\tau_{\ell_2}} &= \ell_2 \wedge \neg \ell_2' \wedge \ell_3' \wedge y > 0 \wedge y' = y - 1 \wedge pres(V - \{\ell_2, \ell_3, y\}) \\
\rho_{\tau_{\ell_3}} &= \ell_3 \wedge \neg \ell_3' \wedge \ell_4' \wedge pres(V - \{\ell_3, \ell_4\}) \\
\rho_{\tau_{\ell_4}} &= \ell_4 \wedge \neg \ell_4' \wedge \ell_0' \wedge y' = y + 1 \wedge pres(V - \{\ell_0, \ell_4, y\}) \\
\rho_{\tau_{m_0}} &= m_0 \wedge \neg m_0' \wedge ((m_1' \wedge x > 0 \wedge m_4' = m_4) \vee \\
&\qquad\qquad\qquad\qquad (m_4' \wedge x \leq 0 \wedge m_1' = m_1)) \\
&\qquad \wedge pres(V - \{m_1, m_2, m_4\}) \\
\rho_{\tau_{m_1}} &= m_1 \wedge \neg m_1' \wedge m_2' \wedge y > 0 \wedge y' = y - 1 \\
&\qquad \wedge pres(V - \{m_1, m_2, y\}) \\
\rho_{\tau_{m_2}} &= m_2 \wedge \neg m_2' \wedge m_3' \wedge y' = y + 1 \wedge pres(V - \{m_2, m_3, y\}) \\
\rho_{\tau_{m_3}} &= m_3 \wedge \neg m_3' \wedge m_0' \wedge x' = x - 1 \wedge pres(V - \{m_0, m_3, x\}) \\
\rho_{\tau_I} &= pres(V) \\
\mathcal{J} &= \{\tau_{\ell_0}, \tau_{\ell_4}, \tau_{m_0}, \tau_{m_2}, \tau_{m_3}\} \\
\mathcal{C} &= \{\tau_{\ell_2}, \tau_{m_1}\}
\end{aligned}
$$

Fig. 2. FTS Φ for program MAY-TERMINATE-X(MTX)

3 Specification Language

3.1 Temporal Logic

We use *linear-time temporal logic* as our specification language for reactive systems. A *temporal formula* is constructed out of assertions and the usual boolean and temporal operators.

An FTS Φ *satisfies* a temporal formula φ if all its computations satisfy φ.

Example Consider again the program MTX. The property that process P_2 will eventually terminate can be expressed by the temporal formula $\Diamond m_4$, i.e., eventually the boolean variable m_4 will be true. Unfortunately, the program does not satisfy this property as the process P_2 might get stuck at location m_1 while process P_1 is idle at location ℓ_3. A weaker formula that excludes this case is

$$\varphi : \Box \Diamond (\neg \ell_3) \rightarrow \Diamond m_4,$$

which expresses the property that if P_1 will not stay idle forever at location ℓ_3 then P_2 will eventually terminate. ◢

3.2 Formula Automata

A formula automaton (FA) is an ω-automaton (for a survey see [Tho90]) with Streett-like acceptance conditions on edges. The set of models of a property expressed by an FA is the set of state sequences that are accepted by the ω-automaton. It has been proven that any property expressible by a temporal formula can be expressed as an FA, however formula automata are strictly more expressive than quantifier-free temporal logic ([MW84]).

An FA $\mathcal{A} = \langle N, N_0, E, \mu, \mathcal{F} \rangle$ over a set of variables V has the components

- N: A finite set of nodes.
- $N_0 \subseteq N$: A set of initial nodes.
- $E \subseteq N \times N$: A set of edges connecting nodes.
- $\mu : N \mapsto F(V)$: A node-labeling function, where $F(V)$ denotes the set of all assertions over V.
- $\mathcal{F} \subseteq 2^{E \times E}$: An edge acceptance condition given by an acceptance list $\{(P_1, R_1), \ldots, (P_m, R_m)\}$, where $P_j \subseteq E$ are called the *persistent* edges, and $R_j \subseteq E$ are called the *recurrent* edges.

We say an infinite sequence of nodes $\pi : n_0, n_1, \ldots$ is a *path* of $\mathcal{A}$ if $n_0 \in N_0$, and for each $i \geq 0$, $(n_i, n_{i+1}) \in E$. We say that $e = (n_i, n_{i+1})$ occurs in π. For a path π,

- $inf_n(\pi)$ stands for $\{n \in N \mid n \text{ occurs infinitely often in } \pi\}$
- $inf_e(\pi)$ stands for $\{e \in E \mid e \text{ occurs infinitely often in } \pi\}$

We say a path π is *accepting* if it satisfies

$$\forall j \in [1..m] \, . \, inf_e(\pi) \subseteq P_j \lor (inf_e(\pi) \cap R_j \neq \emptyset),$$

i.e., for each pair (P_j, R_j), some edge in R_j occurs infinitely often in π or all edges that occur infinitely often in π are in P_j.

For any infinite sequence of states $\sigma : s_0, s_1, \ldots$ we say that a path of $\mathcal{A}$, $\pi : n_0, n_1, \ldots$, is a *trail* of σ in $\mathcal{A}$ if for all $i \geq 0$, $s_i \models \mu(n_i)$, i.e., every state s_i in the sequence satisfies the assertion associated with n_i. The trail of a finite sequence of states is defined similarly.

An infinite sequence of states σ is a *computation* of an FA $\mathcal{A}$ if it has an accepting trail in $\mathcal{A}$. The set of computations of $\mathcal{A}$ is denoted by $\mathcal{L}(\mathcal{A})$. We say that $\mathcal{A}$ is an FA of a temporal formula φ if

$$\mathcal{L}(\mathcal{A}) = \{\sigma \mid \sigma \models \varphi\}.$$

If $n \in N$ then $next(n)$ stands for $\{n_1 \in N \mid (n, n_1) \in E\}$, that is, $next(n)$ is the set of all successor nodes of n. For $N_1 \subseteq N$, $\mu(N_1)$ stands for $\bigvee_{n \in N_1} \mu(n)$.

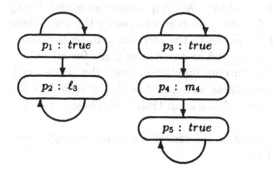

initial nodes:

$$N_0 = \{p_1, p_2, p_3, p_4\}$$

acceptance list:

$$\mathcal{F} = \{(P_1, R_1)\} \text{ where}$$
$$P_1 = \{(p_2, p_2), (p_5, p_5)\}$$
$$R_1 = \emptyset$$

Fig. 3. Formula Automaton for $\varphi : \Diamond \Box \ell_3 \vee \Diamond m_4$

Example　An FA for $\varphi : \Box \Diamond (\neg \ell_3) \rightarrow \Diamond m_4$, which can be rewritten as

$$\varphi : \Diamond \Box \ell_3 \vee \Diamond m_4$$

is shown in Figure 3. It is straightforward to see that the subgraphs defined by p_1 and p_2 represent all sequences that satisfy the first disjunct, and the subgraph for p_3, p_4 and p_5 represent all sequences that satisfy the second disjunct.　◢

We define $SF(\varphi)$ to be the set of all the atomic subformulas of φ. For example,

$$SF(\Diamond \Box (\neg \ell_3) \rightarrow \Diamond m_4) = \{\ell_3, m_4\}.$$

A φ-*propositional state*, denoted by s_φ, assigns a truth-value $s_\varphi[q]$ to each $q \in SF(\varphi)$. We denote by Σ_φ the set of all such states. Clearly Σ_φ is finite with size $2^{|SF(\varphi)|}$. A φ-*propositional model* is an infinite sequence of φ-propositional states. *Satisfaction* of a formula φ over a φ-propositional model $\sigma_\varphi : s_{\varphi,0}, s_{\varphi,1}, \ldots$ is defined inductively, as before, where for an atomic subformula q, $s_\varphi \models q$ iff $s_\varphi[q] = \mathrm{T}$.

We define the propositional language of φ, denoted by $\mathcal{L}^p(\varphi)$, to be

$$\mathcal{L}^p(\varphi) = \{\sigma_\varphi \mid \sigma_\varphi \models \varphi\}.$$

The *propositional projection* of a state $s \in \Sigma$ is a φ-propositional state, denoted by s_φ^p, such that for every $q \in SF(\varphi)$, $s_\varphi^p \models q$ iff $s \models q$. Similarly, we say a φ-propositional model $\sigma_\varphi^p : s_{\varphi,0}^p, s_{\varphi,1}^p, \ldots$, is the *propositional projection* of a model $\sigma : s_0, s_1, \ldots$, if for every $i \geq 0$, $s_{\varphi,i}^p$ is the propositional projection of s_i.

If σ_φ^p is a propositional projection of σ then

$$\sigma_\varphi^p \in \mathcal{L}^p(\varphi) \quad \text{iff} \quad \sigma \in \mathcal{L}(\varphi).$$

Example　A model of $\varphi : \Box (x > 5) \vee \Diamond (y \neq 3)$ with atomic subformulas $p : x > 5$ and $q : y \neq 3$ is $\langle x : 7, y : 3\rangle, \langle x : 6, y : 3\rangle, \langle x : 7, y : 0\rangle, \ldots \in \mathcal{L}(\varphi)$ with propositional projection $\langle p : \mathrm{T}, q : \mathrm{F}\rangle, \langle p : \mathrm{T}, q : \mathrm{F}\rangle, \langle p : \mathrm{T}, q : \mathrm{T}\rangle, \ldots \in \mathcal{L}^p(\varphi)$.

4 Verification Diagrams

A verification diagram (VD) Ψ approximates the set of computations of an FTS Φ. It can be viewed as an FA with three additional components: an edge-labeling function to represent fairness, a set of ranking functions to justify the acceptance conditions, and a mapping relating the nodes to the formula φ to be proven.

For an FTS Φ, a (Φ, φ)-valid VD Ψ, viewed as an FA, has trails for all runs of Φ. The conditions associated with the ranking functions ensure that these trails are accepting trails. The fairness conditions associated with the edge-labeling function ensure that every computation of Φ has a fair trail in Ψ. These conditions together ensure that $\mathcal{L}(\Phi) \subseteq \mathcal{L}(\Psi)$.

The conditions associated with the mapping function provide the other necessary inclusion, namely $\mathcal{L}(\Psi) \subseteq \mathcal{L}(\varphi)$.

4.1 Definition

We define a *(generalized) verification diagram* $\Psi = \langle N, N_0, E, \mu, \eta, \mathcal{F}, \Delta, f \rangle$ over an FTS Φ and formula φ to be a labeled directed graph with N, N_0, E, μ and $\mathcal{F}$ defined as in the FA, and η, Δ and f defined as follows:

- $\eta : E \mapsto 2^T$: An edge-labeling function. Each edge is labeled by a set of transitions of Φ.
- $\Delta \subseteq \{\delta \mid \delta : \Sigma \mapsto \mathcal{D}\}$: A set of ranking functions, mapping states to elements of a well-founded domain $\mathcal{D}$. For each pair (P_j, R_j) of the acceptance list, and for each node $n \in N$, Δ contains a ranking function $\delta_{j,n}$.
- $f : N \mapsto PF(SF(\varphi))$. A mapping from nodes to propositional formulas over the atomic subformulas of φ.

The definitions and notation introduced in Section 3 for the FA apply to the VD. Here we define some additional notions related to the three extra components of the VD.

If $n \in N$, $\tau \in T$ then $\tau(n)$ stands for $\{n_1 \in next(n) \mid \tau \in \eta(n, n_1)\}$, i.e., $\tau(n)$ is the set of all successor nodes of n that are reachable via an edge labeled by τ.

We say an infinite path π is *fair* if it satisfies

1. $\forall \tau \in \mathcal{J} : \quad (\forall n \in inf_n(\pi))\,(\tau(n) \neq \emptyset) \quad \rightarrow \quad (\exists e \in inf_e(\pi).\tau \in \eta(e))$ and
2. $\forall \tau \in \mathcal{C} : \quad (\exists n \in inf_n(\pi))\,(\tau(n) \neq \emptyset) \quad \rightarrow \quad (\exists e \in inf_e(\pi).\tau \in \eta(e))$.

The first condition says that if a node on which a compassionate transition τ is enabled is visited infinitely many times, then there must be at least one edge labeled by τ that appears infinitely often as well, i.e., τ must be taken infinitely often. The second condition says that if a just transition τ is enabled on all nodes that appear infinitely often, then τ must be taken infinitely often.

An infinite sequence of states σ is a *run* of Ψ if there exists an accepting trail of σ in Ψ. An infinite sequence of states σ is a *computation* of Ψ if there exists a fair and accepting trail of σ in Ψ. The *languages* defined by Ψ are the following:

$$\mathcal{L}_R(\Psi) = \{\sigma \mid \sigma \text{ is a run of } \Psi\}$$

$$\mathcal{L}(\Psi) = \{\sigma \mid \sigma \text{ is a computation of } \Psi\}.$$

A computation $\sigma : s_{\varphi,0}, s_{\varphi,1}, \ldots$ is a φ-propositional model of Ψ if there exists a fair and accepting path $\pi : n_0, n_1, \ldots$ in Ψ such that for all $i \geq 0$

$$s_{\varphi,i} \models f(n_i).$$

The set of all φ-propositional models of a VD Ψ is denoted by $\mathcal{L}^p(\Psi)$.

4.2 Verification Conditions

To show that a VD Ψ faithfully represents all computations of an FTS Φ, and that all its computations satisfy the property φ, the following verification conditions associated with Ψ must be established:

- *Initiation*: At least one initial node satisfies the initial condition of Φ:

$$\Theta \to \mu(N_0).$$

- *Consecution*: Any τ-successor of a state satisfying $\mu(n)$ satisfies the label of some successor node of n: For every node $n \in N$, and every transition $\tau \in \mathcal{T}$,

$$\mu(n)(s) \wedge \rho_\tau(s, s') \to \mu(next(n))(s')$$

- *Acceptance*: For each of the pairs (P_j, R_j) of the acceptance list and for any $e = (n_1, n_2) \in E$ and $\tau \in \mathcal{T}$, when taking τ from an arbitrary state s, if $e \notin R_j$ then δ_j does not increase, and if $e \notin (P_j \cup R_j)$ then δ_j decreases:
  (A1) If $e \in P_j - R_j$ then

$$\rho_\tau(s, s') \wedge \mu(n_1)(s) \wedge \mu(n_2)(s') \to \delta_{j,n_1}(s) \succeq \delta_{j,n_2}(s').$$

  (A2) If $e \in \overline{R_j \cup P_j}$ then

$$\rho_\tau(s, s') \wedge \mu(n_1)(s) \wedge \mu(n_2)(s') \to \delta_{j,n_1}(s) \succ \delta_{j,n_2}(s').$$

Note that, due to the presence of the idling transition, this requirement implies that self-loops must be accepting, i.e.,

$$\text{if } (n, n) \in E \text{ then } (n, n) \in P_j \cup R_j.$$

This is in accordance with our earlier statement that every run has an accepting trail: a finite computation padded with idling transitions is a run.

- *Fairness*: For each $e = (n_1, n_2) \in E$ and $\tau \in \eta(e)$
  (F1) τ is guaranteed to be enabled:

$$\mu(n_1)(s) \to En(\tau).$$

  (F2) Any τ-successor of a state satisfying $\mu(n_1)$ satisfies the label of some node in $\tau(n_1)$:

$$\mu(n_1)(s) \wedge \rho_\tau(s, s') \to \mu(\tau(n_1))(s').$$

(F3) If τ is compassionate, then, for any other node in the same connected component with n_1, either τ is guaranteed to be disabled or τ labels some outgoing edge from n_1: if $\tau \in \mathcal{C}$, $n_3 \in N$ is such that $\tau(n_3) = \emptyset$ and n_3 is in the maximal strongly connected component (MSCS) that contains n_1, then $\mu(n_3)(s) \to \neg En(\tau)$.

- *Satisfaction:*
  (S1) for all $n \in N$, if $s \models \mu(n)$ then $s_\varphi^p \models f(n)$, and
  (S2) $\mathcal{L}^p(\Psi) \subseteq \mathcal{L}^p(\varphi)$

Definition 1 A verification diagram Ψ over an FTS Φ is (Φ, φ)-*valid* if all the verification conditions associated with Ψ are Φ-valid.

Lemma 1. If Ψ is a (Φ, φ)-valid verification diagram then (i) $\mathcal{L}_R(\Phi) \subseteq \mathcal{L}_R(\Psi)$, (ii) $\mathcal{L}(\Phi) \subseteq \mathcal{L}(\Psi)$, and (iii) $\mathcal{L}(\Psi) \subseteq \mathcal{L}(\varphi)$.

Proof:
(i) Consider an arbitrary run $\sigma : s_0, s_1, \ldots$ of the FTS Φ. A straightforward induction proof shows that the *Initiation* and the *Consecution* conditions imply that for every finite prefix $s_0, \ldots, s_k$ of σ, there exists a finite path $n_0, \ldots, n_k$ of the VD such that for all $0 \le i \le k$, $s_i \models \mu(n_i)$. Therefore σ has a trail π in Ψ.

To show that π is accepting, suppose to the contrary that there exists $j \in [1 \ldots m]$ such that $inf_e(\pi) \cap R_j = \emptyset$ and $inf_e(\pi) \not\subseteq P_j$. Then there exists $\ell \ge 0$ such that for all $i \ge \ell$, $(n_i, n_{i+1}) \in \overline{R_j}$, and there exist infinitely many i's such that $(n_i, n_{i+1}) \in \overline{P_j \cup R_j}$. But then there exists, by the *Acceptance* conditions, an infinite sequence $\delta_{j,n_\ell} \succeq \delta_{j,n_{\ell+1}} \succeq \ldots$ that is strictly decreasing infinitely many times, contradicting the well-foundedness of $\mathcal{D}$.

(ii) We have to show that if σ is a computation then σ has a fair accepting trail in Ψ. By (i), σ has an accepting trail in Ψ, and the proof of (i) shows that any trail of a prefix of σ can be completed to a trail of σ.

Let π be a trail of σ such that for any other trail π' of σ, if $inf_n(\pi')$ can be reached from $inf_n(\pi)$ then $inf_n(\pi')$ is in the same connected component as $inf_n(\pi)$. We show how to construct another trail of σ that is fair (under (i) we have already shown that any trail of σ is accepting). Let $k \ge 0$ be such that for any $i \ge k$, $n_i \in inf_n(\pi)$. We want to show that for any $l \ge k$, there exists $m \ge l$ and a finite trail $n_0, \ldots, n_k, n'_{k+1}, \ldots, n'_l, \ldots, n'_m$ of σ, such that for any $\tau \in \mathcal{J}$ ($\tau \in \mathcal{C}$), $n_l, \ldots, n'_m$ gratifies τ; i.e., the segment is fair: it contains an edge labeled by τ, or $\tau(n) = \emptyset$ holds for some (for every) node n in the segment.

We first gratify the transitions in $\mathcal{J}$. Notice that once a transition is gratified on a segment, it will stay gratified on any segment that includes that segment. Suppose that we have already gratified some transitions in the segment $n'_l, \ldots, n'_i$ and τ has not been considered yet. If $\tau(n'_i) = \emptyset$ then τ is gratified; if $\tau(n'_i) \ne \emptyset$ then, by F2, it must be the case, that there exist a node n'_{i+i} such that $\tau \in \eta(n'_i, n'_{i+1})$ and $s_{i+1} \models n'_{i+1}$.

Next, we gratify the transitions in $\mathcal{C}$ such that they remain gratified on any longer segment. We have two cases:

- In the original trail π, for every $n \in inf_n(\pi)$ we have $\tau(n) = \emptyset$. Then any other trail of σ that starts with $n_0, \ldots, n_k$ has the same property. Indeed, let π' be a trail of σ of the form $\pi' : n_0, \ldots, n_k, n'_{k+1}, \ldots$. Then $inf_n(\pi')$ is reachable from $inf_n(\pi)$ and therefore it must be in the same connected component. This implies, by F3, that there is no node n'_i with $i \geq k$ such that $\tau(n') \neq \emptyset$. Therefore any segment of any trail of σ that starts with $n_0, \ldots, n_k$ gratifies τ.
- In the original trail π, there exists a node $n \in inf_n(\pi)$ such that $\tau(n) \neq \emptyset$. Thus, by F1, τ is infinitely often enabled on σ. As any trail π' of the form $\pi' : n_0, \ldots, n_k, n'_{k+1}, \ldots$ stays in the same connected component as $inf_n(\pi)$ it follows, by F3, that any time we take a τ transition on σ after k, it must be the case that the respective node has an outgoing edge labeled τ. Then, by F2, we must be able to take that edge and thus gratify τ.

It follows, by induction, that we can construct a trail
$$\pi' : \quad n_0, \ldots, n_{k_1}, \ldots, n_{k_i}, \ldots, n_{k_{i+1}-1}, n_{k_{i+1}}, \ldots \text{ of } \sigma \text{ such that each segment}$$
$n_{k_i}, \ldots, n_{k_{i+1}-1}$ gratifies all the transitions in $\mathcal{C} \cup \mathcal{J}$. Therefore π' is fair.

(iii) $\sigma \in \mathcal{L}(\Psi)$ implies, by S1, $\sigma_\varphi^p \in \mathcal{L}^p$ which implies, by S2, $\sigma_\varphi^p \in \mathcal{L}^p(\varphi)$ and therefore $\sigma \in \mathcal{L}(\varphi)$. □

Proposition 2. (Soundness) Let Φ be an FTS and φ a temporal formula. If there exists a (Φ, φ)-valid VD then Φ satisfies φ.

Proof: It follows directly from lemma 1 that $\mathcal{L}(\Phi) \subseteq \mathcal{L}(\varphi)$. □

Example Consider the FTS Φ (Figure 2) of program MTX. A VD Ψ that summarizes the proof of the property

$$\varphi : \quad \square \Diamond (\neg \ell_3) \rightarrow \Diamond m_4$$

over Φ is shown in Figure 4. We sketch a proof that Ψ is (Φ, φ)-valid.
We need to check the five conditions mentioned before.

- *Initiation:*

$$\underbrace{\ell_0 \wedge m_0 \wedge x \geq 0 \wedge y = 1}_{\Theta} \rightarrow \underbrace{m_0 \wedge \ell_{0..2} = y \wedge x \geq 0}_{\mu(n_6)}$$

Thus the initiation condition holds.
- *Consecution:* We have to check the consecution condition for all nodes and all transitions. As an example we show the verification condition for node n_2. The verification conditions for the other nodes are similar.
The only transitions enabled in n_2 are τ_I and τ_{ℓ_4}, since m_1 is disabled due to $y = 0$. Obviously ρ_{τ_I} preserves $\mu(n_2)$ and for τ_{ℓ_4} we have

$$\underbrace{m_1 \wedge \ell_4 \wedge y = 0 \wedge x > 0}_{\mu(n_2)(s)} \wedge \underbrace{\ell_4 \wedge \neg \ell'_4 \wedge \ell'_0 \wedge y' = y + 1 \wedge pres(\overline{\{\ell_0, \ell_4, y\}})}_{\rho_{\tau_{\ell_4}}}$$

$$\rightarrow \underbrace{\cdots}_{\mu(n_2)(s')} \vee \underbrace{m'_1 \wedge \ell'_{0...2} \wedge y' = 1 \wedge x' > 0}_{\mu(n_3)(s')}$$

which is obviously valid.

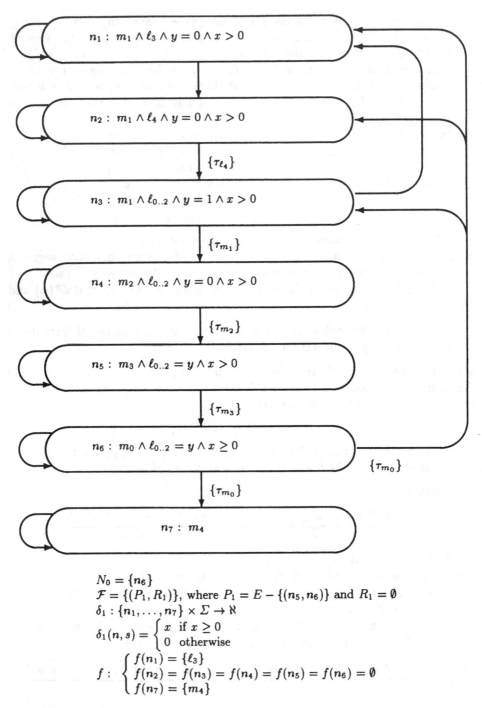

$$N_0 = \{n_6\}$$
$$\mathcal{F} = \{(P_1, R_1)\}, \text{ where } P_1 = E - \{(n_5, n_6)\} \text{ and } R_1 = \emptyset$$
$$\delta_1 : \{n_1, \ldots, n_7\} \times \Sigma \to \aleph$$
$$\delta_1(n, s) = \begin{cases} x & \text{if } x \geq 0 \\ 0 & \text{otherwise} \end{cases}$$
$$f : \begin{cases} f(n_1) = \{\ell_3\} \\ f(n_2) = f(n_3) = f(n_4) = f(n_5) = f(n_6) = \emptyset \\ f(n_7) = \{m_4\} \end{cases}$$

Fig. 4. VD for program MTX and property $\square \lozenge (\neg \ell_3) \to \lozenge m_4$

– *Fairness:* The fairness condition requires us to check for node n_2 that

(F1) $\underbrace{m_1 \wedge \ell_4 \wedge y = 0 \wedge x > 0}_{\mu(n_2)} \rightarrow \underbrace{\ell_4}_{En(\ell_4)}$ and

(F2) $\underbrace{m_1 \wedge \ell_4 \wedge y = 0 \wedge x > 0}_{\mu(n_2)(s)} \wedge \underbrace{\ell_4 \wedge \neg \ell_4' \wedge \ell_0' \wedge y' = y + 1 \wedge pres(\overline{\{\ell_0, \ell_4, y\}})}_{\rho_{\tau_{\ell_4}}}$

$\rightarrow \underbrace{m_1' \wedge \ell_{0...2}' \wedge y' = 1 \wedge x' > 0}_{\mu(n_3)(s')}$

which are easily seen to be valid.

The fairness verification conditions associated with the nodes n_4, n_5, and n_6 can be checked similarly.

For the fairness verification condition associated with node n_3 and transition m_1 we also have to check condition F3, since $m_1 \in C$. The nodes in the same MSCS as n_3 are $\{n_1, n_2, n_4, n_5, n_6\}$. It is straightforward to check that on none of these nodes m_1 is enabled.

– *Acceptance:* Notice that m_3 is the only transition that modifies x. Therefore A1 holds. For A2 we have $\overline{P_1 \cup R_1} = \{(n_5, n_6)\}$, and the only transition that leads from n_5 to n_6 is m_3, which decrements x, so A2 also holds.

– *Satisfaction:* We have to show that conditions S1 and S2 hold. Obviously condition S1 holds for f.

It is clear from Figure 3 that

$$\mathcal{L}^p(\varphi) = \langle -, - \rangle^* \langle \text{T}, - \rangle^\omega \cup \langle -, - \rangle^* \langle -, \text{T} \rangle \langle -, - \rangle^\omega$$

where $\langle v_1, v_2 \rangle$ interprets $\langle \ell_3, m_4 \rangle$ and $-$ denotes a don't-care value. The propositional language accepted by Ψ is

$$\mathcal{L}^p(\Psi) = \underbrace{\langle -, - \rangle^*}_{n_1...n_6} \underbrace{\langle \text{T}, - \rangle^\omega}_{n_1} \cup \underbrace{\langle -, - \rangle^*}_{n_1...n_6} \underbrace{\langle -, \text{T} \rangle^\omega}_{n_7}$$

which is a subset of $\mathcal{L}^p(\varphi)$.

Note that the strongly connected components $\{n_1, n_2, n_3\}$ and $\{n_2\}$, $\{n_3\}$, $\{n_4\}$, $\{n_5\}$ and $\{n_6\}$ are excluded from the infinity set by the fairness requirement: e.g., in $\{n_1, n_2, n_3\}$ the transition $m_1 \in C$ is infinitely often enabled but never taken. The strongly connected components $\{n_1, n_2, n_3, n_4, n_5, n_6\}$, $\{n_2, n_3, n_4, n_5, n_6\}$ and $\{n_3, n_4, n_5, n_6\}$ are excluded by the acceptance condition, which states that edge (n_5, n_6) cannot occur infinitely often.

5 Completeness

A verification diagram is a complete representation of a proof that a reactive system satisfies a temporal formula. The following definitions and results are needed for the proof.

Definitions. An FA $\mathcal{A}$ is *deterministic* if any sequence of states σ has at most one trail of σ in $\mathcal{A}$, i.e., for any $n_1, n_2 \in N$, if $n_1, n_2 \in N_0$ or for some $n_3 \in N$, $n_1, n_2 \in next(n_3)$ then $\neg(\mu(n_1) \wedge \mu(n_2))$. An FA $\mathcal{A}$ is *complete* if any sequence of states σ has at least one trail of σ in $\mathcal{A}$, i.e., $\mu(N_0) \wedge \bigwedge_{n \in N} \mu(next(n))$ is valid.

Lemma 3. Let Φ be an FTS and $\mathcal{A}$ an FA. Then there exists a mapping from nodes to assertions $acc : N \mapsto F(V)$ such that $s \models acc(n)$ iff there exists a computation segment $s_0(\models\Theta), \ldots, s_k = s$ of Φ and a path segment $n_0(\in N_0), \ldots, n_k = n$ of $\mathcal{A}$ such that for any $i \in [0..k]$, $s_i \models \mu(n_i)$.

Justification. Such an assertion can always be constructed in an assertion language that is sufficiently expressive to encode finite sequences.

Theorem 4. [LPS81] Let $\sqsupset$ be a well-founded ordering over a set S. Then there exists a function into the ordinals, $\delta : S \to \mathcal{O}$ such that

(W1) $s \sqsupset s'$ implies $\delta(s) > \delta(s')$.

(W2) If $s \sqsupset s''$ implies $s' \sqsupset s''$ for every $s'' \in S$ then $\delta(s) \leq \delta(s')$.

Lemma 5. Let Φ be an FTS and $\mathcal{A}$ a deterministic FA with acceptance list $\{(P_1, R_1), \ldots, (P_m, R_m)\}$ such that $\mathcal{L}_R(\Phi) \subseteq \mathcal{L}(\mathcal{A})$ and for every node n in $\mathcal{A}$ and $s \models \mu(n)$, there exists a computation segment $s_0(\models\Theta), \ldots, s_k = s$ of Φ and a path $n_0(\in N_0), \ldots, n_k = n$ of $\mathcal{A}$ such that for any $i \in [0..k]$, $s_i \models \mu(n_i)$

Then there exist well-founded domains $(\mathcal{D}_j)_{j \in [1..m]}$ and functions $(\delta_j : N \times \Sigma \to \mathcal{D}_j)_{j \in [1..m]}$ such that for any edge $e = (n, n')$ in $\mathcal{A}$ and for any $\tau \in \mathcal{T}$

(a) if $e \in P_j - R_j$ then $\rho_\tau(s, s') \wedge \mu(n)(s) \wedge \mu(n')(s') \to \delta_j(n, s) \geq \delta_j(n', s')$

(b) if $e \in \overline{R_j \cup P_j}$ then $\rho_\tau(s, s') \wedge \mu(n)(s) \wedge \mu(n')(s') \to \delta_j(n, s) > \delta_j(n', s')$.

Proof. Let $\sqsupset_j$ be the order on $N \times \Sigma(V)$ given by $(n, s) \sqsupset_j (n', s')$ iff there is an "unfair" computation leading from (n, s) to (n', s'), that is there exists a computation segment $s_0 = s, \ldots, s_k = s'$ of Φ and a path segment $n_0 = n, \ldots, n_k = n'$ in $\mathcal{A}$ such that $s_i \models \mu(n_i)$, $\{(n_0, n_1), (n_1, n_2) \ldots, (n_{k-1}, n_k)\} \not\subseteq P_j$ and $\{(n_0, n_1), (n_1, n_2), \ldots, (n_{k-1}, n_k)\} \cap R_j = \emptyset$.

$\sqsupset_j$ **is a partial order.** It is clearly reflexive and transitive. In order to show that it is also antisymmetric, suppose that $(n, s) \sqsupseteq_j (n', s')$, $(n', s') \sqsupseteq_j (n, s)$ and $(n, s) \neq (n', s')$. There exists a path segment $n_0(\in N_0), \ldots, n_k = n$ and a computation segment $s_0(\models\Theta), \ldots s_k = s$ such that for all $i \in [0..k]$, $s_i \models \mu(n_i)$. As $(n, s) \sqsupset_j (n', s')$, there exists a path segment $n_k, \ldots, n_l = n'$ of $\mathcal{A}$ and a computation segment $s_k, \ldots, s_l = s'$ of Φ such that for all $i \in [k..l]$, $s_i \models \mu(n_i)$. Similarly there exists a path segment $n_l, \ldots, n_m = n$ of $\mathcal{A}$ and a computation segment $s_l, \ldots, s_m = s$ of Φ such that for all $i \in [l..m]$, $s_i \models \mu(n_i)$. But then we have a run $s_0, \ldots, (s_k, \ldots, s_l, \ldots, s_{m-1})^\omega$ of Φ, and a path $n_0, \ldots, (n_k, \ldots, n_l, \ldots, n_{m-1})^\omega$ in $\mathcal{A}$ that is not accepting, such that for any $i \in [0..m-1]$, $s_i \models n_i$. This contradicts the assumption that every run of Φ is accepted by $\mathcal{A}$. Thus it must be the case that $(n, s) = (n', s')$.

$\sqsupset_j$ **is well-founded.** Indeed, suppose that $(n_1, s_1) \sqsupset_i (n_2, s_2) \sqsupset_j \ldots$ Then, as before, we could construct a run of Φ (that goes through $s_1, s_2, \ldots$) that is not a computation of $\mathcal{A}$.

By Theorem 4, there exists a function $\delta_j : S \to \mathcal{O}$ that satisfies W1 and W2. We have to show that δ_j satisfies (a) and (b). Let $n, n' \in N$ and $\tau \in \mathcal{T}$.

(a) Suppose $(n, n') \in P_j - R_j$ and $\rho_\tau(s, s') \wedge \mu(n)(s) \wedge \mu(n)(s')$. For any (n'', s''), if $(n', s') \sqsupset_j (n'', s'')$ then also $(n, s) \sqsupset_j (n'', s'')$. Indeed, suppose that

$(n', s') \sqsupseteq_j (n'', s'')$. Then there exists a path segment $n_1 = n', \ldots, n_k = n''$ of $\mathcal{A}$ and a computation segment $s_1 = s', \ldots, s_k = s''$ of Φ such that for any $i \in [1..k]$, $s_i \models \mu(n_i)$. It follows that $n, n_1, \ldots, n_k$ is a path segment of $\mathcal{A}$, $s, s_1, \ldots, s_k$ is a computation segment of Φ, $s \models \mu(n)$, and for any $i \in [1..k]$, $s_i \models \mu(n_i)$. Clearly, $\{(n, n_1), (n_1, n_2), \ldots, (n_{k-1}, n_k)\}$ is not accepted by (P_j, R_j) and thus we have shown that $(n, s) \sqsupseteq_j (n', s')$. This implies, by W2, that $\delta(n, s) \geq \delta(n', s')$.

(b) Suppose $(n, n') \in \overline{R_j \cup P_j}$ and $\rho_\tau(s, s') \wedge \mu(n)(s) \wedge \mu(n)(s')$. Then $(n, s) \sqsupseteq_j (n', s')$ which implies, by W1, that $\delta(n, s) > \delta(n's')$. □

Proposition 6. (Completeness) Let Φ be an FTS that satisfies a temporal formula φ. Then there exists a (Φ, φ)-valid VD Ψ. Moreover all the verification conditions of Ψ are valid.

Proof. (1) *Earmarked transitions.* Let $(\tau_1, \ldots, \tau_m)$ be an ordering of all the transitions in $\mathcal{T}$ an let $t \notin V$ be a variable that ranges over $\{1, \ldots, m\}$. Then we can construct an FTS $\Phi^t = \langle V \cup \{t\}, \Theta, \mathcal{T}^t, \mathcal{C}^t, \mathcal{J}^t \rangle$ obtained from Φ by "earmarking" each transition with a different value for t. More precisely, $\mathcal{T}^t = \{\tau_i^t \mid \rho_{\tau_i^t} = \rho_{\tau_i} \wedge t' = i\}$, $\mathcal{C}^t = \{\tau_i^t \mid \tau_i \in \mathcal{C}\}$ and $\mathcal{J}^t = \{\tau_i^t \mid \tau_i \in \mathcal{J}\}$. Φ^t is used to construct an FA $\mathcal{A}_p$ that accepts all the runs in Φ^t and an edge-labeling function that exactly translates the fairness of Φ^t. The variable t is then eliminated to obtain an FA $\mathcal{A}_f$ that accepts all the runs of Φ and an edge-labeling function with all the desired properties. The ranking functions are defined on a different FA $\mathcal{A}_r$ that accepts all the runs of Φ and also admits ranking functions satisfying the *Acceptance* conditions. Finally, the two automata are combined into a VD.

(2) *Characterization of fairness in Ψ^t.* Let $\mathcal{A}_{\mathcal{T}^t}$ be the FA with nodes $N^t = N_0^t = 2^{\mathcal{T}^t} \times \mathcal{T}^t$, edges $E^t = \{((T, \tau), (T', \tau')) \mid \tau' \in T\}$, node-labeling function defined by $\mu^t(T, \tau_i^t) = \bigwedge_{\tau \in T} En(\tau) \wedge \bigwedge_{\tau \notin T} \neg En(\tau) \wedge t = i$ and the fairness condition $\mathcal{F}^t = \{(R_\tau, P_\tau) \mid \tau \in \mathcal{C}^t \cup \mathcal{J}^t\}$ where

if $\tau \in \mathcal{J}^t$ then $\begin{cases} P_\tau = \emptyset \\ R_\tau = \{((T_1, \tau_1), (T_2, \tau_2)) \mid \tau \notin T_1\} \cup \{((T_1, \tau_1), (T_2, \tau))\} \end{cases}$

if $\tau \in \mathcal{C}^t$ then $\begin{cases} P_\tau = \{((T_1, \tau_1), (T_2, \tau_2)) \mid \tau \notin T_1\} \\ R_\tau = \{((T_1, \tau_1), (T_2, \tau))\} \end{cases}$

Let η^t be an edge-labeling defined by $\eta^t((T, \tau), (T', \tau')) = \begin{cases} \tau' & \text{if } \tau' \in \mathcal{C}^t \cup \mathcal{J}^t \\ \emptyset & \text{otherwise} \end{cases}$

Then $\mathcal{A}_{\mathcal{T}^t}$ has the following properties: it is deterministic; any run of Φ^t has a trail in $\mathcal{A}_{\mathcal{T}^t}$; a run of Φ^t is a computation of Φ^t iff its trail in $\mathcal{A}_{\mathcal{T}^t}$ is η^t-fair; a path in $\mathcal{A}_{\mathcal{T}^t}$ is η^t-fair iff it is accepting; the *Fairness* conditions hold.

(3) *Underlying FA for a candidate VD.* Any computation of such an FA, has to be a model of φ and any run of Φ is either a computation of this FA or it is non-fair (in which case it is eliminated by an edge-labeling function). A first step towards achieving these conditions is to consider a deterministic FA $\mathcal{A}_u$, such that $\mathcal{L}(\mathcal{A}_u) = \mathcal{L}(\varphi) \cup \overline{\mathcal{L}(\mathcal{A}_{\mathcal{T}^t})}$. ($\mathcal{A}_u$ can be constructed from $\mathcal{A}_{\mathcal{T}^t}$ and an FA for φ by standard ω-automata constructions.) Let $\mathcal{A}_s$ be the FA obtained from $\mathcal{A}_u$ by strengthening its labeling to the labeling given by Lemma 3.

Then $\mathcal{A}_s$ has the following properties: $\mathcal{L}_R(\Phi^t) = \mathcal{L}(\mathcal{A}_s)$; $\mu^s(n)(s)$ iff there exists a computation segment $s_0(\models \Theta^t), s_1, \ldots, s_k = s$ of Φ^t and a path segment

$n_0(\in N_0), n_1, \ldots, n_k = n$ of $\mathcal{A}_s$ such that $\forall i \in [0..k]$. $s_i \models \mu^s(n_i)$; the *Consecution* condition holds. We can always assume that for any node n there exists a constant t_n such that $\mu(n) \rightarrow t = t_n$;

(4) *Edge-labeling.* In order to be able to use the edge-labeling on $\mathcal{A}_{T^t}$ we take $\mathcal{A}_p$ to be the cross product of $\mathcal{A}_s$ and $\mathcal{A}_{T^t}$ with the accepting condition induced by $\mathcal{A}_s$. The edge-labeling function on $\mathcal{A}_p$ is defined to be the one induced by η^t.

Then $\mathcal{A}_p$ has the following properties: it is deterministic; the *Fairness* conditions hold; $\mathcal{L}_R(\Phi^t) \subseteq \mathcal{L}(\mathcal{A}_p)$; for any node n there exists a constant t_n such that $\mu(n) \rightarrow t = t_n$; $\mathcal{L}(\Phi^t) \subseteq \{\sigma \in \mathcal{L}(\mathcal{A}_p) \mid \sigma \text{ has a fair and accepting trail in } \mathcal{A}_p\}$; $\{\sigma \in \mathcal{L}(\mathcal{A}_p) \mid \sigma \text{ has a fair and accepting trail in } \mathcal{A}_p\} \subseteq \mathcal{L}(\varphi)$.

(5) *Removal of t.* Let $\mathcal{A}_f$ be the FA over V with the same structure as $\mathcal{A}_p$ but with the node-labeling function defined by $\mu^f(n) = \mu^p(n)[t/t_n]$ where t_n is such that $\mu^p(n) \rightarrow t = t_n$. Let η^f be the function induced by η^p. Then $\mathcal{A}_f$ has the following properties: $\mathcal{L}_R(\Phi) \subseteq \mathcal{L}(\mathcal{A}_f)$; the *Consecution* and *Fairness* conditions hold; $\mathcal{L}(\Phi) \subseteq \{\sigma \in \mathcal{L}(\mathcal{A}_f) \mid \sigma \text{ has a fair and accepting trail in } \mathcal{A}_f\}$; $\{\sigma \in \mathcal{L}(\mathcal{A}_f) \mid \sigma \text{ has a fair and accepting trail in } \mathcal{A}_f\} \subseteq \mathcal{L}(\varphi)$.

(6) *Ranking functions.* Let $\mathcal{A}_{df}$ be a deterministic FA that accepts the same language as $\mathcal{A}_f$ and let $\mathcal{A}_r$ be an FA that has the same structure as $\mathcal{A}_{df}$ except for the labeling function, which is given by the formula acc from Lemma 3. Then $\mathcal{A}_r$ has the following properties: $\mathcal{L}_R(\Phi) = \mathcal{L}(\mathcal{A}_r)$; $\mu^r(n)(s)$ iff there exists a computation segment $s_0(\models \Theta), s_1, \ldots, s_k = s$ of Φ and a path segment $n_0(\in N_0), n_1, \ldots, n_k = n$ of $\mathcal{A}_r$ such that $\forall i \in [0..k]$. $s_i \models \mu^r(n_i)$; the *Consecution* and *Acceptance* conditions hold.

(7) *The verification diagram.* Let Ψ be the VD obtained by taking the cross product of $\mathcal{A}_f$ and $\mathcal{A}_r$ with the edge-labeling function induced by η_f, the ranking functions induced by $\delta_{n,i}$ defined under (6) and the mapping function defined by $f(n) = \bigwedge_{q \in SF(\varphi) \,:\, \mu(n) \rightarrow q} q \wedge \bigwedge_{q \in SF(\varphi) \,:\, \mu(n) \rightarrow \neg q} \neg q$. We can always assume that the labels of the nodes of Ψ are satisfiable (otherwise we can remove them).

Ψ is (Φ, φ)-valid. *Initiation:* Both $\mathcal{A}_f$ and $\mathcal{A}_r$ accept any run of Φ and therefore the product has a trail for any run in Φ, which implies that any initial node has to satisfy the label of some initial node of Ψ. *Consecution* results from the fact that both $\mathcal{A}_f$ and $\mathcal{A}_r$ have this property. *Acceptance* follows from the *Acceptance* condition for $\mathcal{A}_r$. *Fairness* holds because both automata satisfy the *Consecution* condition and $\mathcal{A}_f$ satisfies the *Fairness* conditions. The (S1) *Satisfaction* condition is true by the definition of f. To prove (S2), let $\sigma_\varphi \in \mathcal{L}^p(\Psi)$ and let π be its trail in Ψ. As the label of each node on the trail is satisfiable, there exists a sequence $\sigma \in \mathcal{L}(\Psi)$ with trail π. Then $\sigma \in \mathcal{L}_R(\Phi)$. Let σ^t be a sequence of states over $V \cup \{t\}$ defined by the trail π and σ. Then σ^t is a run of Φ^t with a fair trail in $\mathcal{A}_{T^t}$. Therefore σ^t is a computation of Φ^t, which implies $\sigma^t \models \varphi$ and as a consequence $\sigma \models \varphi$. This implies, by the definition of f, $\sigma_\varphi \models \varphi$.

6 Discussion

The verification diagram as presented can be used as a proof rule. It reduces the proof of a property over a fair transition system to a set of proofs of first-order validities and regular ω-automata operations, such as language inclusion.

As regards complexity, a proof by verification diagram consists of checking the validity of the *Initiation* condition, $|N| \cdot |\mathcal{T}|$ *Consecution* conditions, $\mathcal{O}(|\mathcal{F}| \cdot |\mathcal{T}| \cdot |E|)$ *Acceptance* conditions, $\mathcal{O}(|\mathcal{T}| \cdot (|E| + |N|))$ *Fairness* conditions, and $\mathcal{O}(|N|)$ *Satisfaction* conditions. In addition, it requires finding the maximal strongly connected components of the verification diagram (F3), and the verification of the inclusion of two propositional languages (S2).

The generalized verification diagrams presented here, generalize the verification diagrams presented in [MP94]. A verification diagram is an instance of a generalized verification diagram, specialized to the property to be proven, that is, in the verification diagram some of the verification conditions of the corresponding generalized verification diagram have been translated into requirements on structure and node labeling, specific to the property.

The advantage of the verification diagram is that, in general, structural requirements are easier to check than the corresponding verification conditions, in particular language inclusion; in addition the predetermined structure provides guidance in constructing the proof. The advantage of the generalized verification diagram is that it is universally applicable to any property specifiable by a formula automaton, whereas the verification diagram is limited to the properties for which the structural requirements have been determined. When automating checking of the diagram, a single algorithm suffices for generalized verification diagrams, whereas verification diagrams require multiple algorithms, one for each property. Generalized verification diagrams can also be used to systematically derive verification diagrams for specific properties.

Acknowledgements

We thank Anuchit Anuchitanukul, Luca de Alfaro and Tomás Uribe for their careful reading of the paper and their helpful suggestions.

References

[LPS81] D. Lehmann, A. Pnueli, and J. Stavi. Impartiality, justice and fairness: The ethics of concurrent termination. In *Proc. 8th Int. Colloq. Aut. Lang. Prog.*, volume 115 of *LNCS*, pages 264–277. Springer-Verlag, 1981.

[MP91] Z. Manna and A. Pnueli. *The Temporal Logic of Reactive and Concurrent Systems: Specification*. Springer-Verlag, New York, 1991.

[MP94] Z. Manna and A. Pnueli. Temporal verification diagrams. In *Proc. Int. Symp. on Theoretical Aspects of Computer Software*, volume 789 of *LNCS*, pages 726–765. Springer-Verlag, 1994.

[MP95] Z. Manna and A. Pnueli. *Temporal Verification of Reactive Systems: Safety*. Springer-Verlag, New York, 1995.

[MW84] Z. Manna and P. Wolper. Synthesis of communicating processes from temporal logic specifications. *ACM Trans. Prog. Lang. Sys.*, 6(1):68–93, Jan. 1984.

[Tho90] Wolfgang Thomas. Automata on infinite objects. In J. van Leeuwen, editor, *Handbook of Theoretical Computer Science*, volume B, pages 133–191. Elsevier Science Publishers (North-Holland), 1990.

Model Checking of Probabilistic and Nondeterministic Systems*

Andrea Bianco[1] and Luca de Alfaro[2]

[1] Politecnico di Torino, Italy. bianco@polito.it
[2] Stanford University, USA. luca@cs.stanford.edu

Abstract. The temporal logics pCTL and pCTL* have been proposed
as tools for the formal specification and verification of probabilistic systems: as they can express quantitative bounds on the probability of system evolutions, they can be used to specify system properties such as
reliability and performance. In this paper, we present model-checking
algorithms for extensions of pCTL and pCTL* to systems in which
the probabilistic behavior coexists with nondeterminism, and show that
these algorithms have polynomial-time complexity in the size of the system. This provides a practical tool for reasoning on the reliability and
performance of parallel systems.

1 Introduction

Temporal logic has been successfully used to specify the behavior of concurrent and reactive systems. These systems are usually modeled as nondeterministic processes: at any moment in time, more than one future evolution may be
possible, but a probabilistic characterization of their likelihood is normally not
attempted. While many important system properties can be studied in this setting, others, such as reliability and performance, require instead a probabilistic
characterization of the system.

The first applications of temporal logic to probabilistic systems consisted
in studying which temporal logic properties are satisfied with probability 1 by
systems modeled either as finite Markov chains [14, 18, 12, 1, 20] or as augmented
Markov models exhibiting both nondeterministic and probabilistic behavior [22,
19, 5, 20].

Subsequently, [10, 2] considered systems modeled by discrete Markov chains,
and introduced the logics pCTL and pCTL*, that can express quantitative
bounds on the probability of system evolutions. These logics can thus be used
to reason on the reliability and performance of systems. They are obtained by
adding to the branching time logics CTL and CTL* a probabilistic operator ℙ,

---

* This research was supported in part by the National Science Foundation under
grant CCR-92-23226, by the Advanced Research Projects Agency under NASA grant
NAG2-892, by the United States Air Force Office of Scientific Research under grant
F49620-93-1-0139, by Department of the Army under grant DAAH04-95-1-0317, and
by the Italian National Research Council.

such that the formula $\mathbb{P}_{\geq a}\phi$ is true at a given point of the system evolution if, starting from that point, the probability that a future evolution satisfies ϕ is at least a.

The model-checking algorithms presented in [10, 2] can be used to determine the validity of pCTL and pCTL* formulas on systems modeled by finite Markov chains. Moreover, [2] considers generalized Markov processes, representing families of Markov chains, and shows that the decision problem for pCTL* formulas on generalized Markov processes is decidable using results from the theory of real closed fields. However, no efficient computational method is given for this latter problem.

In this paper, we extend the logics pCTL and pCTL* to systems in which nondeterministic and probabilistic behavior coexist. We model these systems by *probabilistic-nondeterministic systems*, similar to the augmented Markov models of [19, 20]. Due to the presence of nondeterminism it is not possible, in general, to talk about the probability with which a formula is satisfied, but only about the lower and upper bounds of such probability. Therefore, according to our definition, the formula $\mathbb{P}_{\geq a}\phi$ (resp. $\mathbb{P}_{\leq a}\phi$) is true at a given point of the system evolution if a system evolution starting from that point satisfies ϕ with a probability bounded from below (resp. above) by a. We then present model-checking algorithms that verify whether a system satisfies a specification written in pCTL or pCTL* in polynomial time in the size of the description of the system.

The logics pCTL and pCTL*, together with these model-checking algorithms, provide a practical tool for the formal specification and verification of the performance and reliability of parallel systems. Nondeterminism, as already recognized by [22, 19, 5, 20], is in fact the key to the natural modeling of parallel probabilistic systems by interleaving, as it allows us to model the choice of which system in the parallel composition takes a transition. Nondeterminism also gives the flexibility of leaving some transition probabilities unspecified. This leads to simpler system models, and it is necessary when some transition probabilities are unknown. Leaving some transition probabilities unspecified can also be useful when it is not desirable that a correctness proof of the system with respect to some specification depends on the value of those probabilities.

2 Probabilistic-Nondeterministic Systems

Following an approach similar to [19, 20], we use *Probabilistic-Nondeterministic Systems* (PNS) to model systems in which probabilistic and nondeterministic components of the behavior coexist. To give a formal definition of PNS, we first introduce *next-state probability distributions*.

Definition 1 (next-state probability distribution). If S is the state space of a system, a *next-state probability distribution* is a function $p : S \mapsto [0, 1]$ such that $\sum_{s \in S} p(s) = 1$. For $s \in S$, $p(s)$ represents the probability of making a direct transition to s from the current state. $\quad\square$

A PNS can then be defined as follows.

Definition 2 (PNS). A PNS is a quadruple $\Pi = (S, s_{\text{in}}, V, \tau)$, where:

1. S is the denumerable or finite state space of the system;
2. $s_{\text{in}} \in S$ is the initial state;
3. V is a labeling function that associates with each $s \in S$ the set $V(s) \subseteq \mathcal{P}$ of propositional variables that are true in s;
4. τ is a function that associates with each $s \in S$ the set $\tau(s) = \{p_1^s, \ldots, p_{k_s}^s\}$ of next-state probability distributions from s. We denote $|\tau(s)|$ by k_s. □

The successor of a state $s \in S$ is chosen according to a two-phase process: first, a next-state probability distribution $p_i^s \in \tau(s)$ is selected nondeterministically among $p_1^s, \ldots, p_{k_s}^s$; second, a successor state $t \in S$ is chosen according to the probability distribution p_i^s on S.

This model, based on the one proposed in [19], generalizes the approach of [22] by allowing a simpler encoding of the parallel composition of systems. To see how parallelism can be modeled by a PNS, consider as an example the parallel composition of m Markov chains $A_1, \ldots, A_m$. In a PNS Π representing $A_1 \parallel A_2 \parallel \cdots \parallel A_m$, we can associate with each state $s \in S$ the next-state distributions $\tau(s) = \{p_1^s, \ldots, p_m^s\}$, where the distribution p_i^s arises from a move taken by the chain A_i, $1 \le i \le m$. In this way, the probabilistic information on the behavior of each chain is preserved in Π, and the choice of the Markov chain that takes a transition is nondeterministic.

We define a *reachability* relation $\rho \subseteq S \times S$ by

$$\rho = \{(s, t) \mid \exists p^s \in \tau(s) . p^s(t) > 0\} .$$

Then, we associate with each state $s \in S$ the set

$$\Omega_s = \{s_0 s_1 s_2 \ldots \mid s = s_0 \wedge \forall n \in \mathbb{N} . \rho(s_n, s_{n+1})\}$$

of *legal* infinite sequences of states beginning at s. The set of computations of a system Π is thus $\Omega_{s_{\text{in}}}$. For $\omega \in \Omega_s$, we denote with $\omega|_n$ the n-th state of ω, with $\omega|_0 = s$.

Moreover, we let $\mathcal{B}_s \subseteq 2^{\Omega_s}$ be the smallest algebra of subsets of Ω_s that contains all the *basic cylinder sets* $\{\omega \in \Omega_s \mid \omega|_0 = s_0 \wedge \ldots \wedge \omega|_n = s_n\}$ for all $n \ge 0$, $s_0, \ldots, s_n \in S$, and that is closed under complement and countable unions and intersections. This algebra is called the *Borel σ-algebra* of basic cylinder sets, and its elements are the *measurable* sets of sequences, to which it will be possible to assign a probability [13].

Minimal and Maximal Probabilities

Due to the presence of nondeterminism, we cannot define a probability measure on the Borel σ-algebra $\mathcal{B}_s$. However, for each set of sequences $\Delta \in \mathcal{B}_s$, we can define its *maximal probability* $\mu_s^+(\Delta)$ and its *minimal probability* $\mu_s^-(\Delta)$. Intuitively, $\mu_s^+(\Delta)$ (resp. $\mu_s^-(\Delta)$) represents the probability that the system follows a sequence in Δ provided that the nondeterministic choices are as favorable (resp. unfavorable) as possible. To formalize the idea of favorable and unfavorable choices, we introduce the concept of *strategies* (similar to the *schedules* of [22, 19, 5, 20]), that determine which next-state probability distribution is chosen for each state.

If the system reaches the root s of Ω_s following the sequence $s_{in}s_1 \ldots s_n s$, we can assume that a strategy does not depend on the "past" sequence $\omega_p = s_{in}s_1 \ldots s_n$. In fact, we are interested in a strategy that maximizes or minimizes the probability that the system, starting from s, follows a sequence in Δ: as neither Δ nor the next-state distributions depend on ω_p, such strategy also need not depend on ω_p. Formally, a strategy is defined as follows.

Definition 3 (strategy). A *strategy* η is a set of conditional probabilities $Q_\eta(i \mid s_0 s_1 \ldots s_n)$ such that $\sum_{i=1}^{k_{s_n}} Q_\eta(i \mid s_0 s_1 \ldots s_n) = 1$, for all $n \in \mathbb{N}$, $s_0, s_1, \ldots, s_n \in S$, and $1 \leq i \leq k_{s_n}$. $\qquad\square$

When a system behaves according to a strategy η in the evolution from $s_0 \in S$, and has reached s_n following the sequence $s_0 \ldots s_n$, it will choose the next-state distribution $p_i^{s_n}$ with probability $Q_\eta(i \mid s_0 s_1 \ldots s_n)$. The probability $\Pr_\eta(t \mid s_0 \ldots s_n)$ that a direct transition to t is taken next is thus equal to $\sum_{i=1}^{k_{s_n}} Q_\eta(i \mid s_0 s_1 \ldots s_n) p_i^{s_n}(t)$.

Therefore, we can associate with each finite sequence $s_0 \ldots s_n$ starting at the root $s = s_0$ of Ω_s the probability $\prod_{i=0}^{n-1} \Pr_\eta(s_{i+1} \mid s_0 \ldots s_i)$. These probabilities for the finite sequences give rise to a unique probability measure $\mu_{s,\eta}$ on $\mathcal{B}_s$ that associates with each $\Delta \in \mathcal{B}_s$ its probability $\mu_{s,\eta}(\Delta)$ [13]. We can then define minimal and maximal probabilities as follows.

Definition 4 (minimal and maximal probability). The *minimal and maximal probabilities* $\mu_s^-(\Delta)$, $\mu_s^+(\Delta)$ of a set of sequences $\Delta \in \mathcal{B}_s$ are defined by

$$\mu_s^-(\Delta) = \inf_\eta \mu_{s,\eta}(\Delta) \qquad \mu_s^+(\Delta) = \sup_\eta \mu_{s,\eta}(\Delta) \qquad\square$$

Thus, $\mu_s^-(\Delta)$ and $\mu_s^+(\Delta)$ represent the probability with which the system follows an evolution $s s_1 s_2 \ldots \in \Delta$ when the nondeterministic choices are as unfavorable or as favorable as possible, respectively. In general, μ^+ and μ^- are not additive on $\mathcal{B}_s$, as the following lemma states.

Lemma 5. *If* $\Delta_1, \Delta_2 \in \mathcal{B}_s$, *with* $\Delta_1 \cap \Delta_2 = \emptyset$, *then*

$$\mu_s^-(\Delta_1 \cup \Delta_2) \geq \mu_s^-(\Delta_1) + \mu_s^-(\Delta_2) \qquad \mu_s^+(\Delta_1 \cup \Delta_2) \leq \mu_s^+(\Delta_1) + \mu_s^+(\Delta_2)$$

and equality does not hold in general.

The minimal and maximal probability are related by the following lemma.

Lemma 6. *For* $\Delta \in \mathcal{B}_s$, *it is* $\mu_s^-(\Delta) = 1 - \mu_s^+(\Omega_s - \Delta)$.

Proof. From $\mu_{s,\eta}(\Delta) = 1 - \mu_{s,\eta}(\Omega_s - \Delta)$, we have $\mu_s^-(\Delta) = \inf_\eta \mu_{s,\eta}(\Delta) = \inf_\eta \big(1 - \mu_{s,\eta}(\Omega_s - \Delta)\big) = 1 - \sup_\eta \mu_{s,\eta}(\Omega_s - \Delta) = 1 - \mu_s^+(\Omega_s - \Delta)$. $\qquad\square$

3 Probabilistic Temporal Logic

Syntax. The logics pCTL and pCTL* are derived from the branching-time logics CTL and CTL* [6] by introducing a probabilistic operator $\mathbb{P}$, with the intuitive reading that $\mathbb{P}_{\geq a}\phi$ (resp. $\mathbb{P}_{\leq a}\phi$) means that the probability of ϕ holding in the future evolution of the system is at least (resp. at most) a [10, 11, 9, 2]. Formally, we distinguish two classes of formulas: the class *Stat* of state formulas (whose truth-value is evaluated on the states), and the class *Seq* of sequence formulas (whose truth-value is evaluated on infinite sequences of states). For pCTL*, the classes *Stat* and *Seq* are defined as follows:

$$\mathcal{P} \subseteq Stat \tag{1}$$

$$\phi,\ \psi \in Stat \implies \phi \wedge \psi,\ \neg\phi \in Stat \tag{2}$$

$$\phi \in Seq \implies \mathrm{A}\phi,\ \mathrm{E}\phi,\ \mathbb{P}_{\bowtie a}\phi \in Stat \tag{3}$$

$$\phi \in Stat \implies \phi \in Seq \tag{4}$$

$$\phi,\ \psi \in Seq \implies \phi \wedge \psi,\ \neg\phi,\ \Box\phi,\ \Diamond\phi,\ \phi\,\mathcal{U}\,\psi \in Seq\,. \tag{5}$$

In the above definition, $\bowtie$ stands for one of $<, \leq, \geq, >$, and $a \in [0,1]$. The logic pCTL is a restricted version of pCTL*, and its definition can be obtained by replacing the clauses (4), (5) in the above definition with the single clause

$$\phi,\ \psi \in Stat \implies \Box\phi,\ \Diamond\phi,\ \phi\,\mathcal{U}\,\psi \in Seq\,. \tag{6}$$

Semantics. For a formula $\phi \in Stat$, we indicate with $s \models \phi$ its satisfaction on state $s \in S$, and for $\phi \in Seq$ we indicate with $\omega \models \phi$ its satisfaction on the infinite state sequence ω. The semantics of the logical connectives and of the temporal operators is defined in the usual way; the semantics of A, E and $\mathbb{P}$ are defined as follows:

$$s \models \mathrm{A}\phi \text{ iff } \forall \omega \in \Omega_s \,.\, \omega \models \phi \tag{7}$$

$$s \models \mathrm{E}\phi \text{ iff } \exists \omega \in \Omega_s \,.\, \omega \models \phi \tag{8}$$

$$s \models \mathbb{P}_{\geq a}\phi \text{ iff } \mu_s^-(\{w \in \Omega_s \mid \omega \models \phi\}) \geq a \tag{9}$$

$$s \models \mathbb{P}_{\leq a}\phi \text{ iff } \mu_s^+(\{w \in \Omega_s \mid \omega \models \phi\}) \leq a\,. \tag{10}$$

The semantics of $s \models \mathbb{P}_{>a}\phi$, $s \models \mathbb{P}_{<a}\phi$ are defined in a similar way. This definition has a very intuitive reading: if $s \models \mathbb{P}_{\geq a}\phi$, it means that regardless of the choices made in nondeterministic states, the probability that the future evolution satisfies ϕ is at least a (and similarly for $s \models \mathbb{P}_{\leq a}\phi$).

To see that the semantics is well-defined, it is possible to show by induction on the structure of ϕ that $\{\omega \in \Omega_s \mid \omega \models \phi\} \in \mathcal{B}_s$ for every $\phi \in Seq$ [22]. We say that a formula $\phi \in Stat$ is satisfied by a PNS Π, written $\Pi \models \phi$, if $s_{\mathrm{in}} \models \phi$.

4 Model Checking

We now present algorithms to decide whether a PNS Π with finite state space S satisfies a specification ϕ written in pCTL or pCTL*. We will prove that these algorithms have polynomial time complexity in the size of the description of Π. We first give the algorithm for pCTL, and then we examine the one for pCTL*.

The algorithms share the same basic structure of those proposed in [8, 7] for CTL and CTL*. Given a formula $\phi \in Stat$, they recursively evaluate the truth-values of the state subformulas $\psi \in Stat$ of ϕ at all states $s \in S$, starting from the propositional formulas of ϕ and following the recursive definitions (1)–(3) of state formulas, until the truth-value of ϕ itself can be computed at all $s \in S$.

In fact, since pCTL and pCTL* differ from CTL and CTL* only for the presence of the $\mathbb{P}$ operator, we can use the same techniques proposed for CTL and CTL* to deal with the operators $\wedge$, $\neg$, A, E. In the algorithms below, therefore, we need to examine only the case corresponding to $\mathbb{P}$.

4.1 pCTL Formulas

Let $\mathrm{Pr}_s^+ \phi \overset{\text{def}}{=} \mu_s^+(\{w \in \Omega_s \mid \omega \models \phi\})$, $\mathrm{Pr}_s^- \phi \overset{\text{def}}{=} \mu_s^-(\{w \in \Omega_s \mid \omega \models \phi\})$. From (9), (10) we see that in order to check whether $s \models \mathbb{P}_{\bowtie a}\phi$ it suffices to compute $\mathrm{Pr}_s^+ \phi$, $\mathrm{Pr}_s^- \phi$. Using $\Box\psi \leftrightarrow \neg\Diamond\neg\psi$, $\Diamond\psi \leftrightarrow \mathbf{true}\, \mathcal{U}\, \psi$, and the relations

$$\mathrm{Pr}_s^+ \neg\phi = 1 - \mathrm{Pr}_s^- \phi \qquad \mathrm{Pr}_s^- \neg\phi = 1 - \mathrm{Pr}_s^+ \phi \, ,$$

derived from Lemma 6, we need only to consider the case of $\phi = \gamma\, \mathcal{U}\, \psi$. Let $S_d = \{s \in S \mid s \models \psi\}$ be the set of "destination" states, and let $S_p = \{s \in S \mid s \models \gamma\}$ be the set of "intermediate" states.

Computation of $\mathrm{Pr}_s^- \phi$. It is useful to determine, first of all, for which states $s \in S$ is $\mathrm{Pr}_s^- \phi > 0$. To this end, let the monotone set function $\Lambda : 2^S \mapsto 2^S$ be such that, for $A \subseteq S$,

$$\Lambda(A) = A \cup \{s \in S_p \mid \forall i \in \{1, \dots, k_s\}\,.\,\exists t\,.\,(t \in A \wedge p_i^s(t) > 0)\} \, .$$

As S is finite, the fixpoint $\Lambda^\infty(A) = \bigcup_{i=0}^\infty \Lambda^i(A)$ is computable in at most $|S|$ iterations. Let $S_{>0} = \Lambda^\infty(S_d)$. The following lemma states that this is exactly the set of states from which ϕ can be true with probability greater than 0.

Lemma 7. $s \in S - S_{>0}$ implies $\mathrm{Pr}_s^- \phi = 0$, $s \in S_{>0}$ implies $\mathrm{Pr}_s^- \phi > 0$, $s \in S_d$ implies $\mathrm{Pr}_s^- \phi = 1$.

We still have to determine the value of $\mathrm{Pr}_s^- \phi$ for the states in $S_p' \overset{\text{def}}{=} S_{>0} - S_d$. Each $s \in S_p'$ will choose the next-state distribution $p_i^s : 1 \leq i \leq k_s$ that minimizes the probability of getting to S_d. Thus, for all $s \in S_p'$ we have:

$$\mathrm{Pr}_s^- \phi = \min_{1 \leq i \leq k_s} \left[\sum_{t \in S_p'} p_i^s(t)\mathrm{Pr}_t^- \phi + \sum_{t \in S_d} p_i^s(t) \right] . \tag{11}$$

We can find a solution for the above set of equations by solving a linear programming problem, as the following lemma states.

Lemma 8. *To determine* $\Pr_s^- \phi$ *for all* $s \in S'_p$, *it suffices to find the set of values* $\{x_s : s \in S'_p\}$ *that maximizes* $\sum_{s \in S'_p} x_s$ *subject to the set of constraints*

$$x_s \leq \sum_{t \in S'_p} p_i^s(t) x_t + \sum_{t \in S_d} p_i^s(t)$$

for all $s \in S'_p$ *and* $1 \leq i \leq k_s$. *Then, it is simply* $\Pr_s^- \phi = x_s$, *for all* $s \in S'_p$. *These values are well-defined, as the above problem admits a unique optimal solution.*

To solve the above linear programming problem, it is possible to use well-known algorithms, such as the simplex method. To state the results about the complexity of pCTL model checking, assume that Π is described by listing all the next-state distributions for all states as vectors of rational numbers, each represented as the ratio of two integers. The *size of* Π, denoted by $|\Pi|$, will be simply the length of this description, considered as a string. Using algorithms based on the ellipsoid method, the above linear programming problem can be solved in polynomial time in $|\Pi|$ [21]. Therefore, we have the following theorem.

Theorem 9. *If the truth-values of* γ, ψ *are known at all* $s \in S$, *the truth-value of* $\mathbb{P}_{\leq a}(\gamma \, \mathcal{U} \, \psi)$ *at all* $s \in S$ *can be computed in polynomial time in* $|\Pi|$.

Computation of $\Pr_s^+ \phi$. In the case of $\Pr_s^+ \phi$, the set $S_{>0} = \{s \in S \mid \Pr_s^+ \phi > 0\}$ is simply the set of states of the directed graph $(S_d \cup S_p, \rho)$ from which it is possible to reach S_d following a path belonging to the graph itself. Again, $\Pr_s^+ \phi = 0$ for $s \in S - S_{>0}$, and $\Pr_s^+ \phi = 1$ for $S \in S_d$. Letting $S'_p \stackrel{\text{def}}{=} S_{>0} - S_d$, for all $s \in S'_p$ we can write, in analogy to (11),

$$\Pr_s^+ \phi = \max_{1 \leq i \leq k_s} \left[\sum_{t \in S'_p} p_i^s(t) \Pr_t^+ \phi + \sum_{t \in S_d} p_i^s(t) \right].$$

Again, we can compute $\Pr_s^+ \phi$ for all $s \in S'_p$ by solving a linear programming problem, and the analogous of Theorem 9 holds for $\mathbb{P}_{\geq a} \phi$.

Complexity of pCTL model checking. Combining the results about the complexity of CTL model checking [4] with Theorem 9, we get the following theorem about the complexity of pCTL model checking on PNS.

Theorem 10. *Model checking of pCTL formulas over a PNS* Π *can be done in time polynomial in* $|\Pi|$ *and linear in the size of the formula.*

4.2 pCTL* Formulas

We now turn to the problem of computing $\mathrm{Pr}_s^- \, \phi$ and $\mathrm{Pr}_s^+ \, \phi$ for a general pCTL* path formula $\phi \in \mathit{Seq}$. As $\mathrm{Pr}_s^- \, \phi = 1 - \mathrm{Pr}_s^+ \, \neg\phi$ by Lemma 6, we need to consider only the case of $\mathrm{Pr}_s^+ \, \phi$. As usual, we assume that the truth-values of all state subformulas of ϕ have already been evaluated at all states of the system.

The algorithm we propose consists of three steps. First, we put the formula ϕ in a canonical form ϕ''. Second, we construct from Π a new system Π', such that the states of Π' keep track of the truth-values of the subformulas of ϕ'', and the probability of sets of sequences in Π is equal to the probability of the corresponding sets of sequences in Π'. Third, we show that computing $\mathrm{Pr}_s^+ \, \phi$ in Π corresponds to computing the probability of reaching certain sets of states of Π', and this can be done using the method previously outlined for pCTL.[3]

Canonical form for ϕ. Let $\Gamma = \{\gamma_1, \ldots, \gamma_n\}$ be the set of *maximal state subformulas* of ϕ, that is, the set of state subformulas of ϕ that are not proper subformulas of any other state subformula of ϕ. For each γ_i, we introduce a new propositional variable r_i, and let $\phi' = \phi[r_1/\gamma_1] \cdots [r_n/\gamma_n]$ be the result of replacing each occurrence of γ_i in ϕ with r_i, for all $1 \le i \le n$. As for each state $s \in S$ we have already computed whether $s \models \gamma_i$, we can extend the labeling V by letting $\tilde{V}(s) = V(s) \cup \{r_i \mid s \models \gamma_i, 1 \le i \le n\}$.

The resulting formula ϕ' is a *linear-time* temporal formula constructed with the propositional connectives and the temporal operators $\Box$, $\Diamond$, $\mathcal{U}$ on the propositional variables $r_1, \ldots, r_n$ [17]. By the results of [16, 3], $\neg\phi'$ can be put into the canonical form $\bigwedge_{i=1}^l (\Box\Diamond\chi_i \vee \Diamond\Box\lambda_i)$ for some *past* temporal formulas $\chi_1, \ldots, \chi_l$, $\lambda_1, \ldots, \lambda_l$ built with propositional connectives and the *past* temporal operators $\mathcal{S}$ (*since*) and $\ominus$ (*previous*) [15, 17]. Thus, ϕ' can be put into the form

$$\phi'' : \bigvee_{i=1}^l \Diamond\Box(\delta_i \wedge \Diamond\psi_i) \, ,$$

where again $\delta_1, \ldots, \delta_l, \psi_1, \ldots, \psi_l$ are past temporal formulas. Moreover, the size of ϕ'' is at most doubly exponential in the size of ϕ.

Construction of Π'. The truth-value of a past formula at point s_k of a sequence $s_0, s_1, s_2, \ldots$ depends only on the finite "past" sequence $s_0, s_1, \ldots, s_k$. Therefore, it is possible to construct from $\Pi = (S, s_{\mathrm{in}}, V, \tau)$ a system $\Pi' = (S', s'_{\mathrm{in}}, V', \tau')$ whose states keep track of the truth-values of the past formulas in ϕ'' as Π follows a sequence of states.

To do so, let $\theta_1, \ldots, \theta_m$ be the set of past subformulas of ϕ'' having $\mathcal{S}$ or $\ominus$ as the main connective, ordered in such a way that no θ_i is a subformula of θ_j for $i > j$. The space state of Π' is $S' = S \times \{\mathbf{true}, \mathbf{false}\}^m$, so that a state $s' = \langle s, u_1, \ldots, u_m \rangle \in S'$ consists of a state s of Π and a sequence $u_1, \ldots, u_m$

---

[3] An alternative approach, not pursued in this paper, would have been to construct Π' from Π and from a deterministic Street automaton for $\neg\phi$.

of truth-values of $\theta_1, \ldots, \theta_m$. Any state in S' can be taken as the initial state s'_{in} of Π'. We define the projection function $\pi : S' \mapsto S$ by $\pi(\langle s, u_1, \ldots, u_m \rangle) = s$. Let $q_1, \ldots, q_m$ be new propositional variables, that will be used to replace the formulas $\theta_1, \ldots, \theta_m$. The labeling function V' is defined by

$$V'(\langle s, u_1, \ldots, u_m \rangle) = \tilde{V}(s) \cup \{q_i \mid u_i = \text{true}, \ 1 \leq i \leq m\} .$$

For $1 \leq i \leq m$, define $\hat{\theta}_i = (\ldots (\theta_i[q_{i-1}/\theta_{i-1}]) \ldots [q_1/\theta_1])$ to be the formulas resulting from successively substituting in θ_i $q_{i-1}, \ldots, q_1$ for $\theta_{i-1}, \ldots, \theta_1$. For $1 \leq k \leq l$, define

$$\hat{\delta}_k = (\ldots (\delta_k[q_m/\theta_m]) \ldots [q_1/\theta_1]) \qquad \hat{\psi}_k = (\ldots (\psi_k[q_m/\theta_m]) \ldots [q_1/\theta_1])$$

to be the propositional formulas resulting from successively substituting $q_m, \ldots, q_1$ for $\theta_m, \ldots, \theta_1$ in δ_k, ψ_k. Note that $\hat{\theta}_i$ does not contain any q_j for $1 \leq i \leq j \leq m$. Define also $\hat{\phi}$ to be the formula obtained by orderly replacing each δ_i, ψ_i in ϕ'' with $\hat{\delta}_i, \hat{\psi}_i, 1 \leq i \leq l$, respectively.

The definition of the reachability relation ρ' in Π' encodes the semantics of the past operators. Recall that a formula $\alpha \, S \, \beta$ holds at a given state of a sequence if β holds at that state, or if α holds at that state and $\alpha \, S \, \beta$ holds at the previous one; a formula $\ominus \alpha$ holds at a given state if α holds at the previous one. Consider any two states $s' = \langle s, u_1, \ldots, u_m \rangle$, $t' = \langle t, v_1, \ldots, v_m \rangle$ of Π'. As u_i, v_i represent the truth-values of $\hat{\theta}_i$ at s', t' respectively, we let t' be reachable from s', written $\rho'(s', t')$, if $\rho(s, t)$ and, for all $1 \leq i \leq m$:

1. if $\hat{\theta}_i$ has the form $\ominus \alpha$, $v_i = \text{true}$ iff $s' \models \alpha$;
2. if $\hat{\theta}_i$ has the form $\alpha \, S \, \beta$, $v_i = \text{true}$ iff $[t' \models \beta$ or $(u_i = \text{true}$ and $t' \models \alpha)]$.

The next-state probability distributions for Π' are then defined, for $s' \in S'$, $k_{s'} = k_{\pi(s')}$, and for $1 \leq i \leq k_{s'}$, by:

$$p_i^{s'}(t') = \begin{cases} p_i^{\pi(s')}(\pi(t')) & \text{if } \rho'(s', t'); \\ 0 & \text{otherwise.} \end{cases}$$

The fact that the above equation defines next-state probability distributions is a consequence of the following lemma.

Lemma 11. *Given $s \in S$ and $s' \in S'$ such that $s = \pi(s')$, for every $t \in S$ such that $\rho(s, t)$ there is exactly one $t' \in S'$ such that $t = \pi(t')$ and $\rho'(s', t')$.*

Proof. Let $t' = \langle r, v_1, \ldots, v_m \rangle$ be a state in S' such that $t = \pi(t')$ and $\rho'(s', t')$. The value of r is uniquely determined by $r = t$. For $1 \leq i \leq m$, the truth value of v_i is determined by s', t and by the truth values of $v_1, \ldots, v_{i-1}$. Hence, t' is uniquely determined. $\square$

Relationship between Π and Π'. A formula $\ominus\alpha$ is always false on the first state of a sequence. A formula $\alpha\,\mathcal{S}\,\beta$ holds on the first state of a sequence if that state satisfies β. Thus, in order for $u_1,\ldots,u_m$ to represent the truth-value of $\hat{\theta}_1,\ldots,\hat{\theta}_m$, a sequence in Π that starts at the state $s \in S$ should start in Π' at a state $\xi(s) = \langle s, u_1, \ldots, u_m \rangle$ such that, for all $1 \le i \le m$, u_i is true iff $\hat{\theta}_i$ has the form $\alpha\,\mathcal{S}\,\beta$ and $\xi(s) \models \beta$. As the above requirement uniquely determines $\xi(s)$, it defines a one-to-one function $\xi : S \mapsto S'$.

Moreover, for all $s \in S$, there is a bijective correspondence between the legal sequences of Π that start at $s \in S$ and those of Π' that start at $\xi(s)$. This correspondence relates each legal sequence $\omega : s_0, s_1, s_2, \ldots$ of Π with the unique legal sequence $\zeta(\omega) : s_0', s_1', s_2', \ldots$ of Π' such that $\xi(s_0) = s_0'$, and $\pi(s_i') = s_i$ for all $i > 0$. If $\Delta \in \Omega_s$ is a set of sequences of Π, denote with $\zeta(\Delta)$ the set of ζ-related sequences in Π'. The following lemma follows from the construction of Π' and $\hat{\phi}$.

Lemma 12. $\omega \models \phi$ iff $\zeta(\omega) \models \hat{\phi}$, so that

$$\zeta(\{\omega \in \Omega_s \mid \omega \models \phi\}) = \{\omega' \in \Omega'_{\xi(s)} \mid \omega' \models \hat{\phi}\} \, .$$

Proof. Given two corresponding sequences $\omega : s_0, s_1, s_2, \ldots, \zeta(\omega) : t_0, t_1, t_2, \ldots$ with labelings $\tilde{V}$, V' respectively, ϕ holds at s_0 iff ϕ'' holds at s_0. By induction on i it can be proved that θ_i holds at s_k iff $\hat{\theta}_i$ holds at t_k iff $u_i = \textbf{true}$ at t_k, for $1 \le i \le m$, $k \ge 0$. Hence ϕ'' holds at s_0 iff $\hat{\phi}$ holds at t_0, and this concludes the proof. $\qquad\square$

Furthermore, there is a correspondence between the strategies of Π and Π'. To η for Π corresponds η' for Π' such that

$$Q_{\eta'}(i \mid s_0' \ldots s_n') = Q_\eta(i \mid \pi(s_0') \ldots \pi(s_n')) \, , \tag{12}$$

for all $n \ge 0$, all sequences $s_0' \ldots s_n'$ of states of Π', and $1 \le i \le k_{s'}$. Related sets of sequences starting from related states of Π, Π' have thus the same probability, as the following lemma states.

Lemma 13. If $\Delta \in B_s$ is a measurable set and η, η' are related as in (12), $\mu_{s,\eta}(\Delta) = \mu_{\xi(s),\eta'}(\zeta(\Delta))$. Therefore, by definition of maximal measure,

$$\mu_s^+(\Delta) = \mu_{\xi(s)}^+(\zeta(\Delta)) \, .$$

Proof. The result follows easily from the definition of next step probabilities in Π' and from the fact that ξ is one-to-one and ζ is bijective. $\qquad\square$

Computing in Π'. From the above relations, in order to compute $\text{Pr}_s^+ \phi$ in Π it suffices to compute $\text{Pr}_{\xi(s)}^+ \hat{\phi}$ in Π'; and to compute this we can take advantage of the special form of $\hat{\phi} : \bigvee_{i=1}^l \Diamond\square(\hat{\delta}_i \wedge \Diamond\hat{\psi}_i)$.

For $1 \le i \le l$, define $C_i = \{s \in S' \mid s \models \hat{\delta}_i\}$, set $B_i := C_i$, and iterate the following three-step procedure until no more states can be removed from B_i.

1. Define, for each $s \in B_i$, the set of indices

$$M_s = \left\{ j \in \{1, \ldots, k_s\} \ \middle| \ \{t \in S' \mid p_j^s(t) > 0\} \subseteq B_i \right\}$$

of next-state distributions that do not lead any computation outside B_i.

2. Consider the directed graph $G = (B_i, E)$, where

$$E = \{(s, t) \mid \exists j \in M_s \cdot p_j^s(t) > 0\} \ .$$

3. Remove from B_i all states s that cannot reach a state in $\{s \in B_i \mid s \models \hat{\psi}_i\}$ by a path in G of length at least 1.

Note that the above procedure is iterated $N_i \leq |S'|$ times.

For $1 \leq i \leq l$, let F_i be the subsets of B_i obtained, and let $F = \bigcup_{i=1}^{l} F_i$. For $A \subseteq S'$, $s \in S'$, define $\Gamma_s(A) = \{\omega \in \Omega'_s \mid \exists k . \omega|_k \in A\}$ to be the set of sequences that reach A from s. The following theorem allows us to compute $\Pr_s^+ \phi$.

Theorem 14. *For $s \in S$, $\Pr_s^+ \phi = \mu_{\xi(s)}^+ (\Gamma_{\xi(s)}(F))$.*

The quantity $\mu_{\xi(s)}^+ (\Gamma_{\xi(s)}(F))$ can then be computed with the algorithm given in the previous section for pCTL, taking F as S_d and S' as S_p. The proof of the theorem uses the two following lemmas.

Lemma 15. *For all $s \in S'$, there is a strategy η such that a sequence $\omega \in \Gamma_s(F)$ satisfies $\hat{\phi}$ with probability 1, i.e.*

$$\mu_s^+ (\Gamma_s(F)) = \mu_s^+ (\{\omega \in \Gamma_s(F) \mid \omega \models \hat{\phi}\}) \ .$$

Moreover, this strategy does not need to depend on the portion of $\omega \in \Gamma_s(F)$ outside F.

Proof. Assume that $t_0 \in F$ is the first state at which $\omega \in \Gamma_s(F)$ enters F. Let $i = \min\{m \mid 1 \leq m \leq l \wedge t_0 \in F_m\}$. For $t \in F_i$, let

$$M_t = \left\{ j \in \{1, \ldots, k_t\} \ \middle| \ \{t' \in S' \mid p_j^t(t') > 0\} \subseteq F_i \right\}$$

be the set of indices of next-state distributions that do not leave F_i. The strategy η, at $t \in F_i$, will choose one of the $j \in M_t$ with equal probabilities. Note that while the strategy depends on the state t_0 of first entry in F, it does not depend on the portion of ω outside F. After the entry in F, the sequence is confined to F_i; from each $t \in F_i$ there is a path to a state of F_i where $\hat{\psi}_i$ holds; and F_i has finite size. Therefore, the sequence ω will satisfy $\Diamond(\Box\hat{\delta}_i \wedge \Box\Diamond\hat{\psi}_i)$, and $\hat{\phi}$, with probability 1. $\qquad \Box$

Lemma 16. *For $1 \leq i \leq l$, $s \in S'$, and for any strategy η, the measure of the set of sequences from s that satisfy $\Diamond(\Box\hat{\delta}_i \wedge \Box\Diamond\hat{\psi}_i)$ without ever entering F_i is 0, i.e.*

$$\mu_{s,\eta}(\{\omega \in \Omega'_s \mid \omega \models \Diamond(\Box\hat{\delta}_i \wedge \Box\Diamond\hat{\psi}_i)\} - \Gamma_s(F_i)) = 0 \ .$$

Proof. For $1 \leq j \leq N_i$, let D_j be the set of states that have been removed from B_i at the j-th iteration of the procedure; let also $D_0 = S' - C_i$ be the set of states that does not satisfy $\hat{\delta}_i$. Let $D_{<j} = \bigcup_{k=0}^{j-1} D_j$, $D_{>j} = \bigcup_{k=j+1}^{N_i} D_j$. Moreover, call a $\hat{\psi}_i$-state any state $t \in S'$ such that $t \models \hat{\psi}_i$. Define also

$$b = \inf \{ p_m^s(t) \mid s, t \in S' \wedge 1 \leq m \leq k_s \wedge p_m^s(t) > 0 \}$$

and note that $b > 0$, as S' has finite size. We will prove the following assertion by complete induction on j, from N_i down to 1:

For $1 \leq j \leq N_i$, a sequence passing from $s \in D_j$, never entering F_i and satisfying $\square \lozenge \hat{\psi}_i$ will contain a state in $D_{<j}$ with probability 1.

Clearly, this assertion implies the result stated by the lemma.

Consider the case of j, $1 \leq j \leq N_i$, and assume that the assertion has been proved for all j', $j < j' \leq N_i$.

Let a_1 be the fraction of sequences passing through $s \in D_j$ and reaching a $\hat{\psi}_i$-state without leaving $D_j \cup D_{>j}$. Since s has been removed from B_i, each of these sequences, before reaching the $\hat{\psi}_i$-state, must pass through a *critical point*, i.e. a point where the strategy η has chosen a next-state probability distribution p such that $\{ t \in S' \mid p(t) > 0 \} \not\subseteq D_j \cup D_{>j}$. Therefore, $a_1 \leq 1 - b$, as at most $1 - b$ sequences that pass through a critical point remain in $D_j \cup D_{>j}$.

The $\hat{\psi}_i$-state reached by the a_1 sequences belongs to either D_j or $D_{>j}$. If it belongs to D_j, we say that the first cycle is concluded. Otherwise, by the induction hypothesis we know that the sequences that pass through $D_{>j}$ and satisfy $\square \lozenge \hat{\psi}_i$ without entering F_i eventually go to a state in $D_j \cup D_{<j}$ with probability 1. For these sequences, the first cycle is concluded when they reach $D_j \cup D_{<j}$. In either case, at most a_1 sequences complete the first cycle without leaving D_j or $D_{>j}$.

A fraction a_2 of the sequences that complete the first cycle without leaving $D_j \cup D_{>j}$ will reach another $\hat{\psi}_i$-state without leaving $D_j \cup D_{>j}$. As they must pass again through a critical point, $a_2 \leq 1-b$. In general, the fraction of sequences that goes through k cycles without leaving $D_j \cup D_{>j}$ is at most $\prod_{m=1}^{k} a_m \leq (1-b)^k$. Therefore, the set of sequences passing through $s \in D_j$ that satisfy $\square \lozenge \hat{\psi}_i$ without leaving $D_j \cup D_{>j}$ has measure 0. $\square$

Corollary 17. *For any $s \in S'$ and η, $\mu_{s,\eta}(\{ \omega \in \Omega_s' \mid \omega \models \hat{\phi} \} - \Gamma_s(F)) = 0$.*

Proof. From Lemma 16 we have

$$\mu_{s,\eta}(\{ \omega \in \Omega_s' \mid \omega \models \hat{\phi} \} - \Gamma_s(F))$$
$$\leq \sum_{i=1}^{l} \mu_{s,\eta}(\{ \omega \in \Omega_s' \mid \omega \models \lozenge(\square \hat{\delta}_i \wedge \square \lozenge \hat{\psi}_i) \} - \Gamma_s(F_i)) = 0. \qquad \square$$

Proof of Theorem 14. For $s' \in S'$, by the definition of maximal probability we have $\mathrm{Pr}^+_{s'}\hat{\phi} = \sup_\eta \mu_{s',\eta}(\{\omega \in \Omega'_{s'} \mid \omega \models \hat{\phi}\})$. By Corollary 17 we have, for any strategy η,

$$\mu_{s',\eta}(\{\omega \in \Omega'_{s'} \mid \omega \models \hat{\phi}\})$$
$$= \mu_{s',\eta}(\{\omega \in \Gamma_{s'}(F) \mid \omega \models \hat{\phi}\}) + \mu_{s',\eta}(\{\omega \in \Omega'_{s'} \mid \omega \models \hat{\phi}\} - \Gamma_{s'}(F))$$
$$= \mu_{s',\eta}(\{\omega \in \Gamma_{s'}(F) \mid \omega \models \hat{\phi}\}) .$$

Hence, by Lemma 15, $\mathrm{Pr}^+_{s'}\hat{\phi} = \sup_\eta \mu_{s',\eta}(\{\omega \in \Gamma_{s'}(F) \mid \omega \models \hat{\phi}\}) = \mu^+_{s'}(\Gamma_{s'}(F))$. From Lemmas 12 and 13 we finally have $\mathrm{Pr}^+_s\phi = \mathrm{Pr}^+_{\xi(s)}\hat{\phi} = \mu^+_{\xi(s)}(\Gamma_{\xi(s)}(F))$, as was to be proved. $\qquad\square$

Complexity of pCTL* model checking. By combining results about the complexity of CTL* model checking [7], pCTL model checking, and an analysis of the above algorithm, we get the following result, that summarizes the complexity of pCTL* model checking for PNS.

Theorem 18. *Model checking of pCTL* formulas over a PNS Π can be done in polynomial time in $|\Pi|$.*

On the other hand, from the results of [5] we know that determining whether a linear-time temporal formula is satisfied with probability 1 by a PNS requires at least doubly exponential time in the size of the formula. As this problem can be reduced to pCTL* model checking, we have the following result.

Theorem 19. *Model checking of pCTL* formulas over PNS has a time complexity that is at least doubly exponential in the size of the formula.*

In the algorithm we presented, we can trace the source of this complexity to the step that computes the canonical form of a temporal formula, and to the construction of Π'. In fact, $|\Pi'|$ is triply-exponential in $|\phi|$, in the worst case.

Strategies for pCTL and pCTL*. We say that a strategy η is *deterministic* if $Q_\eta(i \mid s_0 \ldots s_n)$ is either 0 or 1 for all $1 \le i \le k_{s_n}$, $n \ge 0$ and all sequences $s_0 \ldots s_n$ of states of S. We say that a strategy is *Markovian* if

$$Q_\eta(i \mid s_0 \ldots s_n) = Q_\eta(i \mid s_n)$$

for all $n \ge 0$ and all sequences $s_0 \ldots s_n$ of states of S.

Given a system Π, and $\phi \in Seq$, $s \in S$, say that a strategy η is *most favorable* (resp. *most unfavorable*) if $\mu_{s,\eta}(\{\omega \in \Omega_s \mid \omega \models \phi\}) = \mathrm{Pr}^+_s\phi$ (resp. if $\mu_{s,\eta}(\{\omega \in \Omega_s \mid \omega \models \phi\}) = \mathrm{Pr}^-_s\phi$). The following corollary, derived from an analysis of the model-checking algorithms, gives us a characterization of the most favorable and unfavorable strategies corresponding to pCTL and pCTL* formulas.

Corollary 20. *The following results hold.*

1. *For all PNS Π and all pCTL formulas $\phi \in Seq$, there are Markovian and deterministic strategies that are most favorable and most unfavorable for all $s \in S$.*
2. *For all PNS Π, all pCTL* formulas $\phi \in Seq$ and all $s \in S$, there are most favorable and most unfavorable strategies that are deterministic. However, there are PNS Π, $s \in S$, and pCTL* formulas $\phi \in Seq$ such that there are no most favorable nor most unfavorable strategies that are Markovian.*

The second part of this corollary shows that nondeterminism cannot be encoded by leaving some transition probabilities of a Markov chain unspecified, if pCTL* is used as the specification language.

5 Conclusions

It is known from [10, 2] that pCTL and pCTL* model checking on Markov chains can be done in polynomial time in the size of the system. It is interesting to note that adding nondeterminism still preserves the polynomial time bound, provided the size of the system takes into account not only the number of states, but also the encoding of the transition probabilities.

The situation is different for the time bounds expressed in terms of the size of the formula. Model checking of pCTL formulas can be done in linear time on the size of the formula both for Markov chains [10] and PNS. However, while pCTL* model checking on Markov chains can be done in single exponential time in the size of the formula [5, 2], pCTL* model checking on PNS requires at least doubly exponential time in the size of the formula. In our algorithm, the complexity of putting formulas in canonical form is partially mitigated by the fact that many common formulas used in system specification can be efficiently put into canonical form.

Acknowledgements. We would like to thank Anca Browne, Anil Kamath, Zohar Manna, Serge Plotkin and Amir Pnueli for helpful comments and suggestions.

References

1. R. Alur, C. Courcoubetis, and D. Dill. Verifying automata specifications of probabilistic real-time systems. In *Real Time: Theory in Practice*, Lecture Notes in Computer Science 600, pages 28–44. Springer-Verlag, 1992.
2. A. Aziz, V. Singhal, F. Balarin, R.K. Brayton, and A.L. Sangiovanni-Vincentelli. It usually works: The temporal logic of stochastic systems. In *Computer Aided Verification, 7th International Workshop*, volume 939 of *Lect. Notes in Comp. Sci.* Springer-Verlag, 1995.
3. E. Chang, Z. Manna, and A. Pnueli. The safety-progress classification. In *Logic, Algebra, and Computation*, NATO ASI Series, Subseries F: Computer and System Sciences. Springer-Verlag, Berlin, 1992.

4. E.M. Clarke, E.A. Emerson, and A.P. Sistla. Automatic verification of finite state concurrent systems using temporal logic. In *Proc. 10th ACM Symp. Princ. of Prog. Lang.*, 1983.

5. C. Courcoubetis and M. Yannakakis. Verifying temporal properties of finite-state probabilistic programs. In *Proc. 29th IEEE Symp. Found. of Comp. Sci.*, 1988.

6. E.A. Emerson. Temporal and modal logic. In J. van Leeuwen, editor, *Handbook of Theoretical Computer Science*, volume E, chapter 16, pages 995–1072. Elsevier Science Publishers (North-Holland), Amsterdam, 1990.

7. E.A. Emerson and C.L. Lei. Modalities for model checking: Branching time strikes back. In *Proc. 12th ACM Symp. Princ. of Prog. Lang.*, pages 84–96, 1985.

8. E.A. Emerson and A.P. Sistla. Deciding branching time logic. In *Proc. 16th ACM Symp. Theory of Comp.*, pages 14–24, 1984.

9. H. Hansson. *Time and Probabilities in Formal Design of Distributed Systems*. Real-Time Safety Critical Systems. Elsevier, 1994.

10. H. Hansson and B. Jonsson. A framework for reasoning about time and reliability. In *Proc. of Real Time Systems Symposium*, pages 102–111. IEEE, 1989.

11. H. Hansson and B. Jonsson. A logic for reasoning about time and probability. *Formal Aspects of Computing*, 6(5):512–535, 1994.

12. S. Hart and M. Sharir. Probabilistic temporal logic for finite and bounded models. In *Proc. 16th ACM Symp. Theory of Comp.*, pages 1–13, 1984.

13. J.G. Kemeny, J.L. Snell, and A.W. Knapp. *Denumerable Markov Chains*. D. Van Nostrand Company, 1966.

14. D. Lehman and S. Shelah. Reasoning with time and chance. *Information and Control*, 53(3):165–198, 1982.

15. O. Lichtenstein, A. Pnueli, and L. Zuck. The glory of the past. In *Proc. Conf. Logics of Programs*, volume 193 of *Lect. Notes in Comp. Sci.*, pages 196–218. Springer-Verlag, 1985.

16. O. Maler and A. Pnueli. Tight bounds on the complexity of cascaded decomposition of automata. In *Proc. 31th IEEE Symp. Found. of Comp. Sci.*, pages 672–682, 1990.

17. Z. Manna and A. Pnueli. *The Temporal Logic of Reactive and Concurrent Systems: Specification*. Springer-Verlag, New York, 1991.

18. A. Pnueli. On the extremely fair treatment of probabilistic algorithms. In *Proc. 15th ACM Symp. Theory of Comp.*, pages 278–290, 1983.

19. A. Pnueli and L. Zuck. Probabilistic verification by tableaux. In *Proc. First IEEE Symp. Logic in Comp. Sci.*, pages 322–331, 1986.

20. A. Pnueli and L.D. Zuck. Probabilistic verification. *Information and Computation*, 103:1–29, 1993.

21. A. Schrijver. *Theory of Linear and Integer Programming*. J. Wiley & Sons, 1987.

22. M.Y. Vardi. Automatic verification of probabilistic concurrent finite-state systems. In *Proc. 26th IEEE Symp. Found. of Comp. Sci.*, pages 327–338, 1985.

Authors Index

Springer-Verlag
and the Environment

We at Springer-Verlag firmly believe that an international science publisher has a special obligation to the environment, and our corporate policies consistently reflect this conviction.

We also expect our business partners – paper mills, printers, packaging manufacturers, etc. – to commit themselves to using environmentally friendly materials and production processes.

The paper in this book is made from low- or no-chlorine pulp and is acid free, in conformance with international standards for paper permanency.

Lecture Notes in Computer Science

For information about Vols. 1–949

please contact your bookseller or Springer-Verlag

Vol. 985: T. Sellis (Ed.), Rules in Database Systems. Proceedings, 1995. VIII, 373 pages. 1995.

Vol. 986: Henry G. Baker (Ed.), Memory Management. Proceedings, 1995. XII, 417 pages. 1995.

Vol. 987: P.E. Camurati, H. Eveking (Eds.), Correct Hardware Design and Verification Methods. Proceedings, 1995. VIII, 342 pages. 1995.

Vol. 988: A.U. Frank, W. Kuhn (Eds.), Spatial Information Theory. Proceedings, 1995. XIII, 571 pages. 1995.

Vol. 989: W. Schäfer, P. Botella (Eds.), Software Engineering — ESEC '95. Proceedings, 1995. XII, 519 pages. 1995.

Vol. 990: C. Pinto-Ferreira, N.J. Mamede (Eds.), Progress in Artificial Intelligence. Proceedings, 1995. XIV, 487 pages. 1995. (Subseries LNAI).

Vol. 991: J. Wainer, A. Carvalho (Eds.), Advances in Artificial Intelligence. Proceedings, 1995. XII, 342 pages. 1995. (Subseries LNAI).

Vol. 992: M. Gori, G. Soda (Eds.), Topics in Artificial Intelligence. Proceedings, 1995. XII, 451 pages. 1995. (Subseries LNAI).

Vol. 993: T.C. Fogarty (Ed.), Evolutionary Computing. Proceedings, 1995. VIII, 264 pages. 1995.

Vol. 994: M. Hebert, J. Ponce, T. Boult, A. Gross (Eds.), Object Representation in Computer Vision. Proceedings, 1994. VIII, 359 pages. 1995.

Vol. 995: S.M. Müller, W.J. Paul, The Complexity of Simple Computer Architectures. XII, 270 pages. 1995.

Vol. 996: P. Dybjer, B. Nordström, J. Smith (Eds.), Types for Proofs and Programs. Proceedings, 1994. X, 202 pages. 1995.

Vol. 997: K.P. Jantke, T. Shinohara, T. Zeugmann (Eds.), Algorithmic Learning Theory. Proceedings, 1995. XV, 319 pages. 1995.

Vol. 998: A. Clarke, M. Campolargo, N. Karatzas (Eds.), Bringing Telecommunication Services to the People – IS&N '95. Proceedings, 1995. XII, 510 pages. 1995.

Vol. 999: P. Antsaklis, W. Kohn, A. Nerode, S. Sastry (Eds.), Hybrid Systems II. VIII, 569 pages. 1995.

Vol. 1000: J. van Leeuwen (Ed.), Computer Science Today. XIV, 643 pages. 1995.

Vol. 1002: J.J. Kistler, Disconnected Operation in a Distributed File System. XIX, 249 pages. 1995.

VOL. 1003: P. Pandurang Nayak, Automated Modeling of Physical Systems. XXI, 232 pages. 1995. (Subseries LNAI).

Vol. 1004: J. Staples, P. Eades, N. Katoh, A. Moffat (Eds.), Algorithms and Computation. Proceedings, 1995. XV, 440 pages. 1995.

Vol. 1005: J. Estublier (Ed.), Software Configuration Management. Proceedings, 1995. IX, 311 pages. 1995.

Vol. 1006: S. Bhalla (Ed.), Information Systems and Data Management. Proceedings, 1995. IX, 321 pages. 1995.

Vol. 1007: A. Bosselaers, B. Preneel (Eds.), Integrity Primitives for Secure Information Systems. VII, 239 pages. 1995.

Vol. 1008: B. Preneel (Ed.), Fast Software Encryption. Proceedings, 1994. VIII, 367 pages. 1995.

Vol. 1009: M. Broy, S. Jähnichen (Eds.), KORSO: Methods, Languages, and Tools for the Construction of Correct Software. X, 449 pages. 1995. Vol.

Vol. 1010: M. Veloso, A. Aamodt (Eds.), Case-Based Reasoning Research and Development. Proceedings, 1995. X, 576 pages. 1995. (Subseries LNAI).

Vol. 1011: T. Furuhashi (Ed.), Advances in Fuzzy Logic, Neural Networks and Genetic Algorithms. Proceedings, 1994. (Subseries LNAI).

Vol. 1012: M. Bartosˇek, J. Staudek, J. Wiedermann (Eds.), SOFSEM '95: Theory and Practice of Informatics. Proceedings, 1995. XI, 499 pages. 1995.

Vol. 1013: T.W. Ling, A.O. Mendelzon, L. Vieille (Eds.), Deductive and Object-Oriented Databases. Proceedings, 1995. XIV, 557 pages. 1995.

Vol. 1014: A.P. del Pobil, M.A. Serna, Spatial Representation and Motion Planning. XII, 242 pages. 1995.

Vol. 1015: B. Blumenthal, J. Gornostaev, C. Unger (Eds.), Human-Computer Interaction. Proceedings, 1995. VIII, 203 pages. 1995.

VOL. 1016: R. Cipolla, Active Visual Inference of Surface Shape. XII, 194 pages. 1995.

Vol. 1017: M. Nagl (Ed.), Graph-Theoretic Concepts in Computer Science. Proceedings, 1995. XI, 406 pages. 1995.

Vol. 1018: T.D.C. Little, R. Gusella (Eds.), Network and Operating Systems Support for Digital Audio and Video. Proceedings, 1995. XI, 357 pages. 1995.

Vol. 1019: E. Brinksma, W.R. Cleaveland, K.G. Larsen, T. Margaria, B. Steffen (Eds.), Tools and Algorithms for the Construction and Analysis of Systems. Selected Papers, 1995. VII, 291 pages. 1995.

Vol. 1020: I.D. Watson (Ed.), Progress in Case-Based Reasoning. Proceedings, 1995. VIII, 209 pages. 1995. (Subseries LNAI).

Vol. 1021: M.P. Papazoglou (Ed.), OOER '95: Object-Oriented and Entity-Relationship Modeling. Proceedings, 1995. XVII, 451 pages. 1995.

Vol. 1022: P.H. Hartel, R. Plasmeijer (Eds.), Functional Programming Languages in Education. Proceedings, 1995. X, 309 pages. 1995.

Vol. 1023: K. Kanchanasut, J.-J. Lévy (Eds.), Algorithms, Concurrency and Knowlwdge. Proceedings, 1995. X, 410 pages. 1995.

Vol. 1024: R.T. Chin, H.H.S. Ip, A.C. Naiman, T.-C. Pong (Eds.), Image Analysis Applications and Computer Graphics. Proceedings, 1995. XVI, 533 pages. 1995.

Vol. 1025: C. Boyd (Ed.), Cryptography and Coding. Proceedings, 1995. IX, 291 pages. 1995.

Vol. 1026: P.S. Thiagarajan (Ed.), Foundations of Software Technology and Theoretical Computer Science. Proceedings, 1995. XII, 515 pages. 1995.

Vol. 1027: F.J. Brandenburg (Ed.), Graph Drawing. Proceedings, 1995. XII, 526 pages. 1996.